Spatial and Landscape Ecology

Spatial and Landscape Ecology

Edited by Alex Vedder

SYRAWOOD
PUBLISHING HOUSE
New York

Published by Syrawood Publishing House,
750 Third Avenue, 9th Floor,
New York, NY 10017, USA
www.syrawoodpublishinghouse.com

Spatial and Landscape Ecology
Edited by Alex Vedder

International Standard Book Number: 978-1-68286-428-9 (Hardback)

Cataloging-in-publication Data

Spatial and landscape ecology / edited by Alex Vedder.
p. cm.
Includes bibliographical references and index.
ISBN 978-1-68286-428-9
1. Spatial ecology. 2. Landscape ecology. 3. Ecology. 4. Biodiversity. 5. Biodiversity conservation. I. Vedder, Alex.
QH541.15.S62 S73 2017
577--dc23

Printed in China.

TABLE OF CONTENTS

PREFACE

This book discusses the fundamentals as well as modern approaches of spatial and landscape ecology. It strives to provide a fair idea about this discipline and to help develop a better understanding of the latest advances within this field. Spatial ecology refers to that branch of ecology, which deals with the study of spatial patterns and their relation to ecological occurrences. On the other hand, landscape ecology refers to the study of interactions between ecological processes and ecosystems. It is an essential part of systems science. This book aims to develop the relationship between spatial and landscape ecology and its utmost significance in modern times. The various studies that are constantly contributing towards advancing technologies and evolution of this field are examined in detail in the text. As this field is emerging at a fast pace, this book will help the readers to better understand the concepts of spatial and landscape ecology. Scientists and students actively engaged in this area will find the book full of crucial and unexplored concepts.

All of the data presented henceforth, was collaborated in the wake of recent advancements in the field. The aim of this book is to present the diversified developments from across the globe in a comprehensible manner. The opinions expressed in each chapter belong solely to the contributing authors. Their interpretations of the topics are the integral part of this book, which I have carefully compiled for a better understanding of the readers.

At the end, I would like to thank all those who dedicated their time and efforts for the successful completion of this book. I also wish to convey my gratitude towards my friends and family who supported me at every step.

Editor

Spatial Scaling of Non-Native Fish Richness across the United States

Qinfeng Guo[1]*, Julian D. Olden[2]

1 USDA FS, Eastern Forest Environmental Threat Assessment Center, Asheville, North Carolina, United States of America, 2 School of Aquatic and Fishery Sciences, University of Washington, Seattle, Washington, United States of America

Abstract

A major goal and challenge of invasion ecology is to describe and interpret spatial and temporal patterns of species invasions. Here, we examined fish invasion patterns at four spatially structured and hierarchically nested scales across the contiguous United States (i.e., from large to small: region, basin, watershed, and sub-watershed). All spatial relationships in both richness and fraction between species groups (e.g., natives vs. exotics) were positive at large scales. However, contrary to predictions using null/neutral models, the patterns at small scales were hump-shaped (unimodal), not simply negative. The fractions of both domestic (introduced among watersheds within the USA) and foreign (introduced from abroad) exotics increased with area across scales but decreased within each scale. The foreign exotics exhibited the highest dominance (lowest evenness) and spatial variation in distribution, followed by domestic exotics and natives, although on average natives still occupy larger areas than domestic and foreign exotics. The results provide new insight into patterns and mechanisms of fish species invasions at multiple spatial scales in the United States.

Editor: Dennis M. Higgs, University of Windsor, Canada

Funding: The authors have no support or funding to report.

Competing Interests: The authors have declared that no competing interests exist.

* E-mail: qguo@fs.fed.us

Introduction

Understanding spatiotemporal patterns and drivers of species invasion remains a central challenge in ecology. Among the heavily debated issues in invasion ecology is the spatial scaling of native-exotic richness relationships and exotic fraction (a measure of "degree of invasion" or DI) within and across locations. Three main factors contribute to the highly variable nature of this relationship: (1) data quality (e.g., whether domestic species are accounted for), (2) spatial scale at which this relationship is investigated, and (3) metrics used to assess habitat invasibility or DI [1,2,3,4,5]. In addition, the intensity, extent, and accuracy of data related to species invasions are remarkably inconsistent among, and even within, regions and taxonomic groups, leading to additional uncertainty in native-exotic richness relationships.

Elucidating similarities and differences in native-exotic richness associations across scales, regions, and taxonomical groups would offer critical and timely information for both basic and applied research [6,7]. There exists a rich history of research on fish invasion and homogenization at biogeographical scales [8,9,10,11,12,13]. However, little attention has been given to investigations across multiple spatial scales. Recent decades have witnessed considerable advancements in both the quantity and quality of data pertaining to the intentional introduction of exotic fish species, thus providing new opportunities to tackle the critical scaling issue associated with native-exotic richness associations (scaling-up and -down) that have continued to challenge ecologists.

Multiple but not mutually exclusive mechanisms and null/neutral models have been proposed to explain the diverse patterns related to native vs. exotic diversity observed at different scales and for various organisms [2,3,14,15,16,17]. However, given the large differences among organisms in many aspects (size/dispersal), different taxonomic groups are likely to show different patterns at various spatial scales (extent and grain). Without field survey/observations, existing theories are challenged to differentiate the spatial scales at which positive or negative native-exotic richness relationships may manifest. This is particularly evident when species introductions are unevenly distributed across space and time. Given the relative nature of scale (large vs. small) and diversity of taxonomic group (e.g., fishes vs. plants in terms of species richness), knowledge regarding when and how native-exotic relationships change over space and time is needed.

Here, we examine patterns of richness and distribution for native and exotic (both foreign and native translocations) fishes across four nested scales of the contiguous United States. We address: (1) how degree of invasion by exotic fishes varies among and within spatial scales as defined by nested watershed levels, and (2) what are the main drivers for the observed patterns across scales. We also use native species across watersheds as a background reference for comparison to describe and interpret spatial patterns of exotic species invasion [18]. The answers to these questions would facilitate our understanding of fish invasion and homogenization and advance the scientific knowledge in freshwater conservation biogeography [19].

Methods

We collated native and non-native species occurrence for watersheds of the contiguous United States using the NatureServe [20] and USGS NAS (2012; http://nas.er.usgs.gov/) databases, respectively. Fish occurrences were based on hierarchically and spatially nested watersheds across the contiguous United States. For this study, the nested watersheds were classified based on the Watershed Boundary Dataset into four levels (spatial scales) according to the Hydrologic Unit Codes (HUCs): 18 hydrologic regions (2-digit HUC; the largest), 204 river basins (4-digit HUC), 333 watersheds (6-digit HUC), and 2073 sub-watersheds (8-digit HUC; the smallest). The database included a total of 935 fish species consisting of 774 native species (of which 295 were "*domestic exotics*" introduced to at least one watershed) and 161 "*foreign exotics*" that origin from outside the United States. Importantly, nativity status of a species depends on the spatial grain of investigation. For example, a fish species might be native at the largest scale (2-digit HUC) but could be either native or non-native (domestic exotic) at smaller scales (4-, 6-, and 8-digit HUCs).

We determined native and non-native species occurrence for sub-watersheds of the contiguous United States using the NatureServe Central Database [20] and the United States Geological Society (USGS) Non-indigenous Aquatic Species Database (2012), respectively (Fig. S1. The NatureServe Central Database contains primarily native species occurrences that were derived from state natural heritage programs, the scientific literature and from species experts. The USGS Non-indigenous Aquatic Species Database contains non-native species occurrences (defined as a species introduced from outside its native range) that were sourced from the literature, museums, databases, monitoring programs, state and federal agencies, professional communications, online reporting forms, and Aquatic Nuisance Species hotline reports. Together, these two databases represent the most complete distributional information possible for freshwater fish

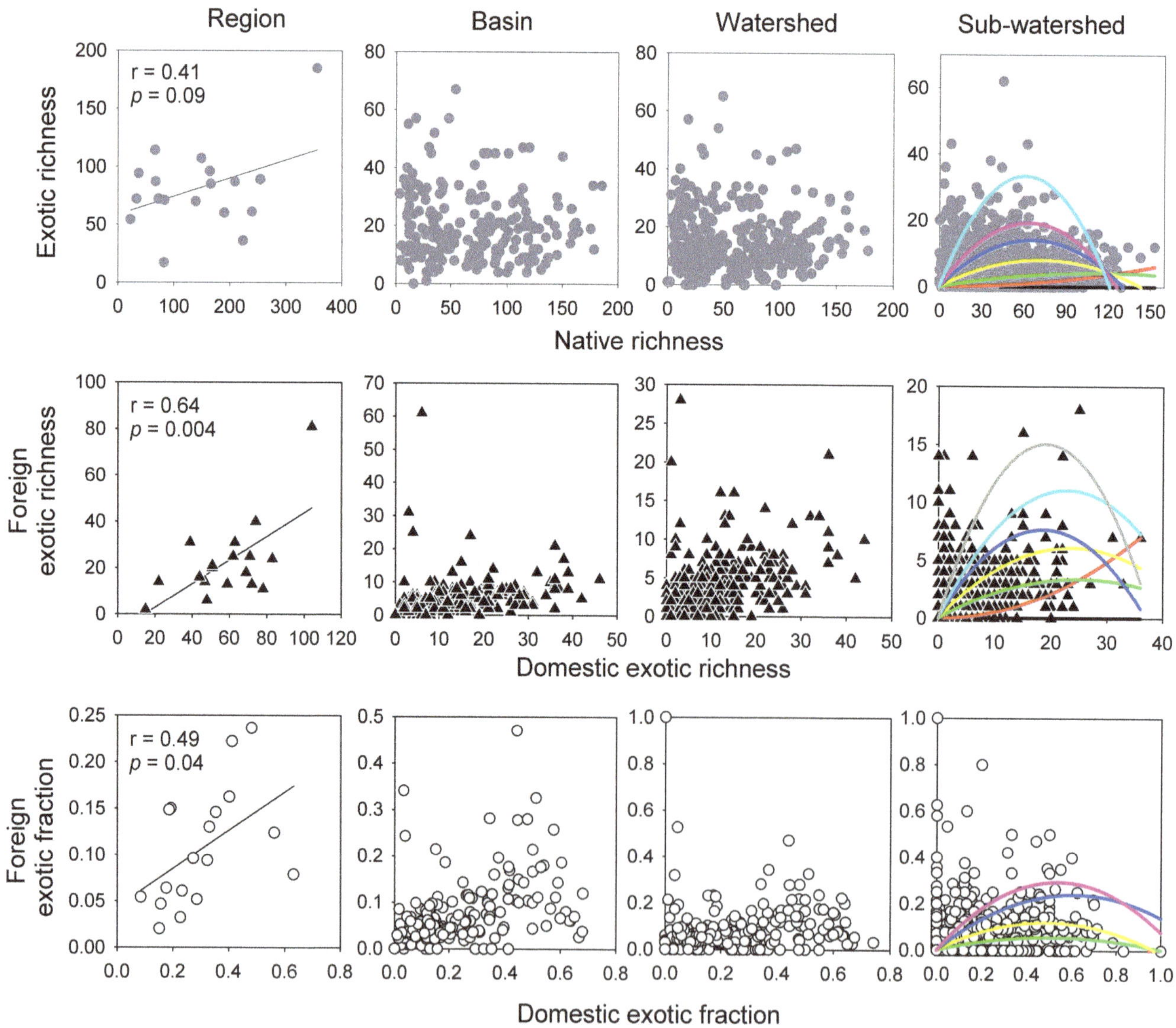

Figure 1. Relationships between native and exotic richness (top), domestic and foreign exotic richness (middle), and domestic and foreign fraction (bottom) across multiple spatial scales in the contiguous United States. In each panel for sub-watersheds, the second order regression curves were based on quantiles of 0.1, 0.25, 0.5, 0.75, 0.9, and 0.95 from bottom to top, respectively; and the corresponding equations and significance were given in Table 1. Exotic richness included both foreign and domestic exotics.

Table 1. Second-order quantile regression relationships between paired variables related to native and exotic species based on data from sub-watersheds throughout the contiguous United States ($n = 2073$) (see also Fig. 1 for graphical illustrations).

Variables	Model equations [$y = f(x)$]	Quantile	P
Native (x) – exotic richness (y)	$-x+x^2$	0.1	<0.0001
	$0.0011-0.0013x+0.0003x^2$	0.25	<0.0001
	$0.9693+0.0307x+x^2$	0.5	0.85
	$4.593+0.0267x+x^2$	0.75	0.89
	$8.19+0.0777x-0.0004x^2$	0.9	0.27
	$1.81+0.1009x-0.0007x^2$	0.95	0.21
Domestic (x) – foreign exotics (y)	$-x+x^2$	0.1	<0.0001
	$-0.0056x+0.0056x^2$	0.25	<0.0001
	$0.27x-0.0055x^2$	0.5	<0.0001
	$0.64+0.36x-0.0051x^2$	0.75	0.0008
	$2+0.34x-0.0041x^2$	0.9	0.04
	$3+0.24x+0.006x^2$	0.95	0.14
Domestic (x) - foreign exotic fraction (y)	$x-x^2$	0.1	<0.0001
	$-x+x^2$	0.25	<0.0001
	$0.22x-0.22x^2$	0.5	<0.0001
	$0.01+0.39x-0.38x^2$	0.75	<0.0001
	$0.04+0.5x-0.31x^2$	0.9	<0.0001
	$0.07+0.63x-0.46x^2$	0.95	0.004

species across the United States and provided the basis for regional freshwater conservation assessments [21].

We recognize two general drawbacks associated with of the use of biodiversity databases: (1) lack of data on survey effort, and (2) incomplete coverage of the geographical and environmental diversity that affects the distribution of the organisms [22]). Both of these concerns are well recognized to limit the inferences drawn from large-scale analyses [23]. Although data on survey effort is not available (and thus represents a data limitation), the diverse data sources ensure a comprehensive spatial coverage of freshwater ecosystems across the United States (described above). For example, non-species occurrences were sourced extensively from national biomonitoring efforts that involved systematic, probability-based surveys or were synoptic in scope and sampled locations from a broad range of stream types and sizes across the United States.

The exotic fraction – area relationships were examined using regression analysis and comparisons among spatial scales were conducted using analyses of variance (ANOVA). The comparisons between domestic and foreign exotic fractions were made for each of the four individual watershed scales using paired Student t-tests. We comparatively examined the relationships among variables (or measures) such as species richness, exotic fraction (a surrogate for measuring degree of invasion or DI), domestic and foreign exotics, and area, across the four spatial scales and within each scale when appropriate. We first used simple regression analyses to test whether a relationship existed between the paired variables. When such analysis failed to detect any relationship, we deployed quantile regressions (both first and second order) to test possible relationships [24,25]. Quantile regression estimate rates of change in all parts of the distribution of a response variable, which is well suited to examine relationships between native and exotic species richness in cases where other influencing factors are unmeasured and unaccounted for, as is often the case [9]. Rather than focusing exclusively on changes in the mean response that can lead to underestimation, overestimation, or a failure to detect changes in heterogeneous distributions [25], quantile regression estimates multiple rates of change (slopes) in responses with unequal variation. The quantile regression analyses were conducted for six quantiles (0.1, 0.25, 0.5, 0.75, 0.9, and 0.95) using PROC QUANTREG in SAS [26] for the sub-watersheds where data were most appropriate for such analysis [25].

Results

At each of the four spatial scales, native fishes had the highest richness, followed by domestic exotics and foreign exotics (Fig. S2). All the spatial relationships examined here in both richness and fraction between species groups switched from positive (linear regression) at the larger hydrologic region scale to unimodal (quantile regression) at smaller scales where the patterns became more complicated (Fig. 1). For example, for higher quantiles (i.e., 0.9, 0.95) at the sub-watershed level, the native-exotic, domestic-foreign exotic relationships were unimodal (or hump-shaped) better fitted with second-order regressions. In other words, when native richness was low, its relationship with exotic richness was positive; but as native richness continued to increase, their relationships became negative. By contrast, for lower quantiles (i.e., 0.1, 0.25) the relationships were mostly linear and positive. The same was true for the domestic-foreign exotic relationships at the sub-watershed level regardless if exotic richness or fraction was used (Figs. 1, S3; Table 1). Quantile regression on watersheds at basin- and watershed-levels yielded very similar results but with less statistical support (not shown).

Across the hierarchy of spatial scales represented by nested watersheds, the fractions of both domestic and foreign exotics, as a measure of degree of invasion (DI), increased with area (ANOVA, domestic fraction: $F = 74.08$, $df = 3$, $P < 0.0001$; Foreign fraction:

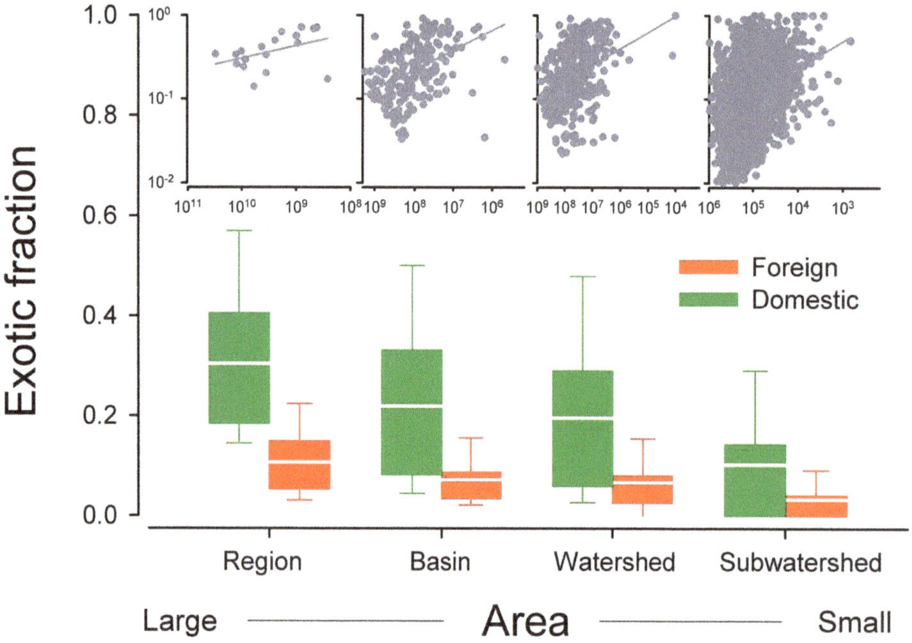

Figure 2. Relationship between the fractions of both foreign and domestic exotic fish and surface area across multiple spatial scales. Foreign exotic fish accounted for a much lower overall exotic proportion than domestic exotic fish (The horizontal bars represent means across the four spatial scales). Note one sub-watershed from Florida with extremely high exotic fish richness is regarded as an outlier and then removed from the analysis.

$F = 43.23$, $df = 3$, $P<0.0001$). Also, domestic exotic fraction was significantly higher than foreign exotic fractions across the four spatial scales (i.e., regions, basins, watersheds, and sub-watersheds; paired t - test, $T = 5.16$, $P = 0.014$) and within each scale (in all paired t – tests for each of the four scales, $P<0.0001$; Fig. 2). However, within each of the four scales, the reverse pattern was observed; i.e., the exotic fraction decreased with watershed area (Fig. 2 insert). This was true when domestic and foreign exotics were analyzed separately or in combination (results not shown).

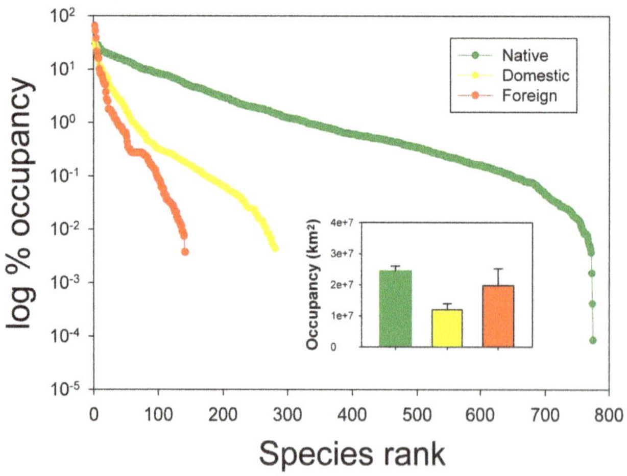

Figure 3. Species rank curves for the three groups (natives, domestic exotics, and foreign exotics) using the distribution data at sub-watersheds level in the contiguous United States. Although foreign exotics still had smaller distribution than natives (but higher than domestic exotics; see inserted panel [mean ± SE]), the variation among species was much larger which was in agreement with the occupancy curves of the three species groups.

Discussion

Native-exotic Relationships Vary Across Scales

Consistent with field observations and theoretical predictions, all spatial relationships in both richness and fraction between species groups (e.g., natives vs. exotics, foreign vs. domestic exotics) are positive at large spatial scales. However, in direct contrast with the simple negative relationships predicted by null/neutral models, the small-scale patterns may be better described by either unimodal relationships or constraint envelopes. Positive relationships at larger scales are expected if large spatial extents are associated with more space, habitat types, and resources that are available for both natives and exotic to coexist [2,15]. It is also possible that, over large scales, null/neutral models may simply explain the positive relationships without considering species interactions or by potentially averaging over sampling biases. However, at such scales, even though competition among species or species groups (natives and exotics) occurs, exotic species may only replace individuals of some native species, not their entire populations [27]. At small scales, the opposite would be mostly true thus competition and/or predation from natives or all resident species (natives plus exotics) becomes more intense thus resisting further invasions [28,29]. However, at the smallest scale (sub-watershed) in this study, we observed unimodal patterns rather than clear negative relationships, indicating the limitations of null/neutral models in explaining this observation [30]. However, at smaller scales, it is also plausible that humans have introduced species more frequently to highly disturbed areas with high human densities where native species diversity has been substantially reduced [6,31], leading to a negative relationship (the second half of the unimodal relationship) that may not have much to do with heterogeneity or resource use. The first half of the unimodal curve (positive) could be due to the fact that, in habitats far from saturation by species (i.e., low native richness) yet space and resources are not limited (similar to the situation at large-scales), both native and exotic species can continue to increase (either through speciation/colonization or human introduction). At present, the foreign exotics exhibited the highest dominance (lowest evenness) and spatial variation in distribution, followed by domestic exotics and natives, although on average natives still occupy larger areas than domestic and foreign exotics (Fig. 3). We urge that additional research focuses on how the relative distribution area of natives vs. exotics (both domestic and foreign; Fig. 3) influences native-exotic richness relationships across scales.

It is interesting to note the positive relationship between domestic and foreign exotic fishes and their similar relationships with native species. In agreement with previously proposed mechanisms for plant invasions in the United States, the positive relationships between domestic and foreign exotic richness especially at large scales (i.e., regions) may indicate that either both groups of species have been introduced in the similar ways or they may respond to the same factors that influence colonization success, of which habitat is one of several factors [32]. By contrast, unlike observation in many field research and simulations, we did not observe strong negative native-exotic relationships at the smallest scale (sub-watersheds). There are two possible reasons. First, in contrast to individual streams [30], the sub-watershed level may be too large (e.g., not fully invaded) for the negative relationships to manifest. Second, the strong negative native-exotic relationship may not actually exist for fishes in the United States, even at smaller scales [9]. Nevertheless, the emergence of increasingly negative relationships at the sub-watersheds in which native richness is higher might indicate the presence of stronger competition or even species saturation [2](Fig. 1).

The issue of native-exotic relationships being scale dependent has most frequently discussed for plants [33,34]. Our study of freshwater fishes suggests that such scale-dependency may also apply to aquatic ecosystems. However, contrary to most claims of simple negative relationships on small-scales in literature, our results demonstrate that the small-scale patterns observed in our study are much more complex than previously thought (but see [30] for an even smaller scale). Furthermore, as mentioned earlier, even null/neutral theory may explain the changes in the direction of native-exotic richness relationships. Information regarding the particular scales at which the direction of this relationship switches is relevant for invasive species management. For instance, both native and exotic species richness might be supported at spatial scales that exceed the switch point (i.e., species coexist), but at smaller scales exotic species may out-compete native species for resources/space leading to population losses [4,9,28,33].

Opposite Exotic Fraction-area Associations within and among Scales

The strong scale-dependent patterns in the degree of invasion (DI) by exotic fishes across the United States are somewhat expected, but the exactly opposite patterns observed between cross- and within-scales are surprising. The exotic fraction increases with area across the spatially nested scales from region to sub-watersheds but decreases with area within each individual scale (Fig. 2). Interestingly, these observations are very similar to the patterns for the exotic plants in the United States; i.e., when compared between the state and county levels, exotic fraction increases with area (state>county) but when compared either among states or among counties, the exotic fraction decreases with area; Q. Guo unpublished results).

Declines in exotic fraction with area across all four spatial scales is likely given that the exotic fraction inevitably decreases with increasing area as the pool of potential exotic species decreases (Fig. 2 inserted panels on top). However, the interpretation of the opposing cross-scale relationships is more difficult (i.e., exotic fraction increases with area; Fig. 2). The seemingly contradictory patterns may be caused by the relative dominance level or spread (distribution areas) of each of the three species groups (natives vs. domestic exotics vs. foreign exotics) at different spatial scales. It is also possible that, at smaller scales, exotic richness may be similar across HUCs but composition may differ, leading to an increase in exotic fraction with increasing scale. For example, the exotic-area curve is steeper than the native-area curve, which could lead to greater exotic fractions with area [35]; supporting the notion that the large-scale determinants of non-native fish richness are context-dependent [36] and may differ depending on whether species richness or composition are examined [37].

In conclusion, our multi-scale investigation supports the notion that native-exotic and exotic-area relationships vary across and within spatial scales. However, the results also reveal a very different observation particularly from that in theoretical studies [2,3,15,38]; that is, at small-scale, the native-exotic relationships are not simply or consistently negative, but unimodal. Although, at present, native fishes still have larger distribution areas than exotics, this is likely to change as a result of continued human activities leading to habitat modification and climate change [39]. Future investigations on the effects of climate change on the biogeography of native and non-native species remain an urgent research need.

Supporting Information

Figure S1 Map of the study region depicting sub-watersheds of the contiguous United States. Shading represents total (native and non-native) fish species richness.

Figure S2 Comparison of species richness of natives, domestic exotics, and foreign exotics across the four scales (region, basin, watershed, and sub-watershed) in the contiguous United States.

Figure S3 Sample estimates, i.e., the intercepts and slopes of first- and second-order quantile regressions between paired variables for sub-watersheds. The blue area represents 95% confidence intervals. Top: Native richness – Exotic richness; Middle: Domestic exotic richness - Foreign exotic richness; Bottom: Domestic exotic fraction – Foreign exotic fraction.

Acknowledgments

We thank S. Fei and B. Iannone for helpful discussions on related topics, two anonymous reviewers for comments on the manuscript, J. Wen for assistance with data analyses, and W. Christie for assistance with GIS. Support to JDO was provided in part by the University of Washington H. Mason Keeler Endowed Professorship.

Author Contributions

Conceived and designed the experiments: QG JDO. Performed the experiments: QG JDO. Analyzed the data: QG JDO. Contributed reagents/materials/analysis tools: QG JDO. Contributed to the writing of the manuscript: QG JDO.

References

1. Lonsdale WM (1999) Global patterns of plant invasions and the concept of invasibility. Ecology 80: 1522–1536.
2. Fridley JD, Brown RL, Bruno JF (2004) Null models of exotic invasion and scale-dependent patterns of native and exotic species richness. Ecology 85: 3215–3222.
3. Davies KE, Chesson P, Harrison S, Inouye BD, Melbourne BA, Rice KJ (2005) Spatial heterogeneity explains the scale dependence of the native-exotic diversity relationship. Ecology 86: 1602–1610.
4. Sax DF, Brown JH, White E, Gaines SD (2005) The dynamics of species invasions: Insights into the mechanisms that limit species diversity. In: Sax DF, Stachowicz JJ, Gaines SD, editors. Species Invasions: Insights into Ecology, Evolution and Biogeography. Sunderland: Sinauer Associates. 447–465.
5. Davies KF, Cavender-Bares J, Deacon N (2011) Native communities determine the identity of exotic invaders even at scales at which communities are unsaturated. Diversity and Distributions 17: 35–42.
6. Lee DC, Sedell JR, Rieman BE, Thurow RF, Williams JE, et al. (1997) An assessment of ecosystem components in the interior Columbia River Basin and portions of the Klamath and Great Basins. In: Quigley TM, Arbelbide SJ, editors. General technical report PNW-GTR-405. Portland: U.S. Forest Service. 1057–1496.
7. McKinney ML (2002) Influence of settlement time, human population, park shape and age, visitation and roads on the number of alien plant species in protected areas in the USA. Diversity and Distributions 8: 311–318.
8. Fuller PL (1999) Nonindigenous species: methods of introduction and impacts. Transactions of the 63rd North American Wildlife and Natural Resources Conference. Wildlife Management Institute, Washington, D.C.
9. Gido KB, Brown JH (1999) Invasion of North American drainages by alien fish species. Freshwater Biology 42: 387–399.
10. Rahel FJ (2000) Homogenization of fish faunas across the United States. Science 288: 854–856.
11. Olden JD, Poff NL (2004) Ecological mechanisms driving biotic homogenization: Testing of a mechanistic model using fish faunas. Ecology 85: 1867–1875.
12. Olden JD, Kennard MJ, Pusey BJ (2008) Species invasions and the changing biogeography of Australian freshwater fishes. Global Ecology and Biogeography 17: 25–37.
13. Villéger S, Blanchet S, Beauchard O, Oberdorff T, Brosse S (2011) Homogenization patterns of the world's freshwater fish faunas. Proceedings of the National Academy of Sciences USA 108: 18003–18008.
14. Levine JM, D'Antonio CM (1999) Elton revisited: a review of evidence linking diversity and invasibility. Oikos 87: 15–26.
15. Shea K, Chesson P (2002) Community ecology theory as a framework for biological invasions. Trends in Ecology and Evolution 17: 170–176.
16. Byers JE, Noonburg EG (2003) Scale dependent effects of biotic resistance to biological invasion. Ecology 84: 1428–1433.
17. Herben T, Mandák B, Bímova K, Münzbergova Z (2004) Invasibility and species richness of a community: a neutral model and a survey of published data. Ecology 85: 3223–3233.
18. Williamson M (1996) Biological invasions. London: Chapman & Hall. 244 p.
19. Olden JD, Kennard MK, Leprieur F, Tedesco PA, Winemiller KO, et al. (2010) Conservation biogeography of freshwater fishes: past progress and future directions. Diversity and Distributions 16: 496–513.
20. NatureServe (2010) Downloadable animal datasets. NatureServe Central Databases. Available: www.natureserve.org/getData/dataSets/watershedHucs/index.jsp ([accessed November 2010]).
21. Abell R, Theime ML, Revenga C, Bryer M, Kottelat M, et al. (2008) Freshwater Ecoregions of the world: a new map of biogeographic units for freshwater biodiversity conservation. BioScience 58: 403–414.
22. Hortal J, Lobo JM, Jimenez-Valverde A (2007) Limitations of biodiversity databases: case study on seed-plant diversity in Tenerife, Canary Islands. Conservation Biology 21: 853–863.
23. Sánchez-Fernández D, Lobo JM, Abellán P, Ribera I, Millán A (2008) Bias in freshwater biodiversity sampling: the case of Iberian water beetles. Diversity and Distributions 14: 754–762.
24. Koenker R, Bassett G (1978) Regression quantiles. Econometrica 46: 33–50.
25. Cade BS, Noon BR (2003) A gentle introduction to quantile regression for ecologists. Frontiers in Ecology and the Environment 1: 412–420.
26. SAS Institute (2011) SAS/STAT 9.3 User's Guide: Mixed Modeling (Book Excerpt). Cary: SAS Publishing.
27. Guo QF, Symstad A (2008) A two-part measure of degree of invasion for cross-community comparisons. Conservation Biology 22: 666–672.
28. Elton CS (1958) The ecology of invasions by plants and animals. London: Methuen. 222 p.
29. Tomasetto F, Duncan RP, Hulme PE (2013) Environmental gradients shift the direction of the relationship between native and alien plant species richness. Diversity and Distributions 19: 49–59.
30. Mitchell AL, Knouft JH (2009) Non-native fishes and native species diversity in freshwater fish assemblages across the United States. Biological Invasions 11: 1441–1450.
31. Luck GW (2007) A review of the relationships between human population density and biodiversity. Biological Review 82: 607–45.
32. Guo QF, Ricklefs RE (2010) Domestic exotics and the perception of invasibility. Diversity and Distributions 16: 1034–1039.
33. Brown RL, Peet RK (2003) Diversity and invasibility of southern Appalachian plant communities. Ecology 84: 32–39.
34. Von Holle B (2013) Environmental stress alters native-nonnative relationships at the community scale. Biological Invasions 15: 417–427.
35. Oberdorff T, Tedesco P, Hugueny B, Leprieur F, Beauchard O, et al. (2011) Global and regional patterns of riverine fish species richness: a review. International Journal of Ecology 2011: Article ID 967631.
36. Blanchet S, Leprieur F, Beauchard O, Staes J, Oberdorff T, et al. (2009) Broad-scale determinants of non-native fish species richness are context-dependent. Proceedings of the Royal Society B 276: 2385–2394.
37. Leprieur F, Olden JD, Lek S, Brosse S (2009) Contrasting patterns and mechanisms of spatial turnover for native and exotic freshwater fishes in Europe. Journal of Biogeography 36: 1899–1912.
38. Clark GF, Johnston EL, Leung B (2013) Intrinsic time dependence in the diversity-invasibility relationship. Ecology 94: 25–31.
39. Rahel FJ, Olden JD (2008) Assessing the effects of climate change on aquatic invasive species. Conservation Biology 22: 521–533.

2

The Efficacy of Landscape-Level Conservation in Changbai Mountain Biosphere Reserve, China

Jianliang Zhang, Fangzheng Liu, Guofa Cui*

College of Nature Conservation, Beijing Forestry University, Beijing, China

Abstract

Anthropogenic landscape alteration is rather common in many protected areas (PAs), jeopardizing the efficacy of PAs conservation. However, the general consensus is that PAs still remain effective in habitat conservation. To assess the efficacy of landscape-level conservation, we examined landscape alterations in the Changbai Mountain Biosphere Reserve (CMBR), which was established in 1960 as a "flagship" protected area in China. Based on analyses of high-resolution satellite images and data of forest inventory, field survey and interview, we developed two new indexes to assess the efficacy of landscape conservation, i.e. the quality index of protected landscape and the interference index of anthropogenic landscape. From 1993 to 2012, the quality index increased from 74.48 to 75.50, and the interference index decreased from 0.49 to 0.06, suggesting that the overall quality of protected landscape improved and the degree of anthropogenic interference decreased in CMBR. The increase in landscape quality was mainly due to the progressive vegetation recovery of previous cutover land in the windthrow area, the cease of the use of the cultivated land, and the amelioration of spatial pattern of protected landscape. We conclude that the current landscape conservation methods used in CMBR are effective, and the method we developed has the potential to be used to assess the efficacy of landscape-level conservation in nature reserves worldwide.

Editor: Edward Webb, National University of Singapore, Singapore

Funding: This study was supported by the Special Funds of Forestry Research for Public Welfare,China (Grant 201104029),the funder's website is as follows: http://www.forestry.gov.cn. The funders had no role in study design, data collection and analysis, decision to publish, or preparation of the manuscript.

Competing Interests: The authors have declared that no competing interests exist.

* E-mail: fa6716@163.com

Introduction

Habitat reduction and fragmentation are the leading causes of biodiversity loss worldwide [1,2], and establishing protected areas (PAs) or natural reserves is the principal defense [3–5]. Prior to 2012, there had been 130,709 PAs worldwide, covering more than 12.9% of total land surface [6]. In PAs, two types of landscape are identified, i.e. protected landscape and anthropogenically interfered landscape (thereafter referred to as "interfered landscape") [7,8]. The former refers mainly to natural landscape and also includes some artificial landscape that is beneficial to wildlife [8]. The latter is frequently interfered by human activities that often disturb and damage natural ecosystems and wildlife habitats [9]. Currently interfered landscape is rather common in PAs [10–12], and expected to increase because human population directly adjacent to PAs continues to expand [13,14]. However, the general consensus is that PAs still remain effective in habitat conservation [5,15]. Therefore, it is highly necessary to examine changes in the quality of protected landscape and the degree of anthropogenic interference to assess the efficacy of landscape-level conservation in nature reserves [16–18].

In China, "Rescue" efforts had resulted in 2640 nature reserves (excluding those in Hong Kong, Macao, and Taiwan) before 2012 [19]. Nevertheless, the absence of systematic conservation planning [20] and the continued adoption of unsustainable policies [21,22] have resulted in uncontrolled utilization of resources in nature reserves. For example, rapid development of ecotourism industries has led to dramatic changes in landscape type [23–25]. Quantitative information about efficacy of landscape conservation for China's nature reserves is scant, and there are no studies examining landscape changes in PAs based on high-resolution satellite images.

The combination of remote sensing and geographic information system has created a powerful analytical method for exploring landscape dynamics and testing the efficacy of PAs [17,18,26–28]. Nevertheless, due to the limitation of spatial resolution in satellite images (≥30 m, generally), classifications in PAs were simple and usually consisted of only forest and non-forest categories [10,12,29,30], and only large area of deforestation could be detected. Landscape changes driven by human activities at smaller scales (<30 m) in PAs such as residential houses, small hydropowers, ditches, industrial and commercial facilities were often incognizable. In recent years, the occurrence of satellite images with higher spatial resolution has allowed for a more accurate recognization of interfered landscape at smaller scales. However, there is still lack of a comprehensive evaluation method combining high-resolution image data and other sources of data (forest inventory, field survey and interview) to quantify the overall quality of protected landscape and the degree of anthropogenic interference.

To examine the efficacy of landscape conservation, we develop two new indexes: the quality index of protected landscape and the interference index of anthropogenic landscape. Based on data of high-resolution satellite images, forest inventory, field survey, and interview, landscape was first classified into different types and each type was assigned to a value according to its contribution to habitat protection. Then the two indexes were constructed to quantify the overall quality of protected landscape and the degree of anthropogenic interference. We adopted this method to assess the efficacy of landscape conservation within the Changbai Mountain Biosphere Reserve (CMBR). CMBR was selected as it is one of the earliest nature reserves established in China and classified as a demonstration [31]. The evaluation method applied in CMBR has the potential to be used in other nature reserves in China as well as those in other countries. The objectives were to (1) provide a quantitative method for assessing efficacy of landscape-level conservation, (2) assess the efficacy of landscape conservation in CMBR using this method, and (3) recommend protection and management strategies.

Materials and Methods

Ethics statement

This study obtained relevant permissions from the administrative bureau of CMBR, and was conducted under the Nature Reserve Regulation and the Wildlife Protection Law of the People's Republic of China.

General situation of the research area

CMBR is located in the central and eastern part of Jilin Province, China (127°42′55″–128°16′48″E, 41°41′49″–42°25′18″N; Fig. 1). It borders the Democratic People's Republic of Korea to the southeast. According to the PAs model launched by UNESCO's Man and Biosphere Program [32], the reserve is divided into three functional zones, i.e. a core area, a buffer zone, and a transition area, accounting for 65.3, 10.2 and 24.5% area of CMBR, respectively (Fig. 1). The core area where harvesting and poaching are absolutely prohibited is composed of old-growth forests. The buffer zone, surrounding the core area, is established to prevent the core area from human disturbances. The transition area is outside the buffer zone, and used as an area for re-establishment of endemic species, ecotourism and bases for in situ reproduction of natural resources.

Under the influence of monsoon climate, the Changbai Mountain area has a temperate continental climate with dry and windy springs, warm and rainy summers, cool and foggy autumns, and cold and long winters. The mountain exhibits high variation in climate conditions due to an expansive elevation gradient, ranging from 700 to 2691 m. The base of the mountain has the typical warm temperate climate, whereas higher elevations have complex, changeable near polar climates [33]. Thus, an obvious vertical zonation of vegetation is present, forming five vegetation zones: broad-leaved forest (<720 m), coniferous and broad-leaved mixed forest (720–1,100 m), coniferous forest (1,100–1,700 m), *Betula ermanii* forest (1,700–2,000 m) and alpine tundra (>2,000 m) [34]. CMBR is rich in biodiversity due to its diverse topography, climate, and ecosystems. For example, there are 9 species of amphibians, 12 reptiles, 24 fish, 56 mammals, 230 birds, and 1255 insects [35]. CMBR is also home to 430 fungi, 200 lichens, 311 mosses and liverworts, 78 ferns, 11 gymnosperms, and at least 1325 angiosperms [36].

Sources of data

Direct access from CMBR management organization. We obtained the reserve boundary maps, functional zone maps, road maps, data of forest inventory in 1993, and records of nature reserve administration. In the data of forest inventory, different landscape types were delineated using aerial photographs (1:34,000), LandSat TM image and topographic data (1:25,000), in combination with fixed sample plots. The different landscape types included natural forest, artificial forest, sparse forest, shrub forest, young afforested land, grassland, cutover land, forestry facilities (e.g. construction land and road), and other land types (e.g. cultivated land and bare land). Inspection and acceptance protocols for the data were in accordance with the "Quality Management Method of the Survey of the Forest Resources of Jilin Province" [37]. The system of inspection and acceptance was divided into three levels: survey team, office and bureau check. The proportions of field check by survey team, office and bureau were 10, 8, and 5%, respectively. Before finishing the field survey, sampling check was used to examine the field survey quality by office and bureau. All survey quality of every survey team was qualified to reach to accuracy requirement. The proportions of indoor check by survey team and office were 100 and 65%, respectively. The overall accuracy of the data was 96.2% according to bureau check [37].

Remote sensing imagery. We obtained the ZY-1 02C HR orthorectified panchromatic image (spatial resolution 2.36 m), ZY-1 02C P/MS multi-spectral image (spatial resolution 10 m) and LandSat TM image in 2010, with the spatial resolution of 30 m. Both ZY-1 02C HR and P/MS images were acquired in 2012, and provided by China Center for Resource Satellite Data and Application. The TM image was obtained from International Scientific Data Service Platform. The P/MS and TM images were georeferenced to HR orthorectified panchromatic image with submeter accuracy. The ZY-1 02C fusion image (spatial resolution 2.36 m) was generated from HR and P/MS images mentioned above, by using Gram-Schmidt Spectral Sharpening Method.

Field survey data. Field data of different types of landscape in CMBR were collected in August 2010, using a portable GPS to record geographic locations (including longitude, latitude and elevation) in 71 surveyed sites (Fig. 2A). These data were used to establish interpretation symbol for image classification. To classify the interfered landscape such as residential houses, small hydropowers, ditches, industrial and commercial facilities, the special purposes of such landscape were surveyed in the field in May 2013 and used as supplementary information. A total of 23 sites was survived (Fig. 2A). Actual width of the interfered landscape with linear feature (such as road and ditches) was measured in the field, when the width of the landscape was less than 10 m. In addition, vegetation recovery in the windthrow area above altitude of 1,500 m was investigated in 10 plots (10 m×10 m), in which canopy closure, plant species composition, and seedlings regeneration were recorded.

Interviews with regional stakeholders (e.g. CMBR staff and local residents). The interviews focused on changes in the interfered landscape, especially changes in type, area, and spatial distribution from 1993 to 2012.

Landscape classification

With reference to Current Land Use Classification [38], the landscape in CMBR was initially divided into natural, and artificial landscape. The former included six types of landscape: natural forest, sparse forest, shrub forest, grassland, lake area, and bare land. The latter included artificial forest, young afforested land, cutover land, cultivated land, orchard, commerce and service land, industry and storage land, residential land, scenic facilities, public facilities, special purposes land (e.g. military), highways, forest roads, ditches, hydraulic architectures, and facility agricultural land.

Classification of natural landscape. Based on the data of forest inventory in 1993, the classification layer of natural landscape of CMBR in 1993 was established using ArcMap 10.0 software

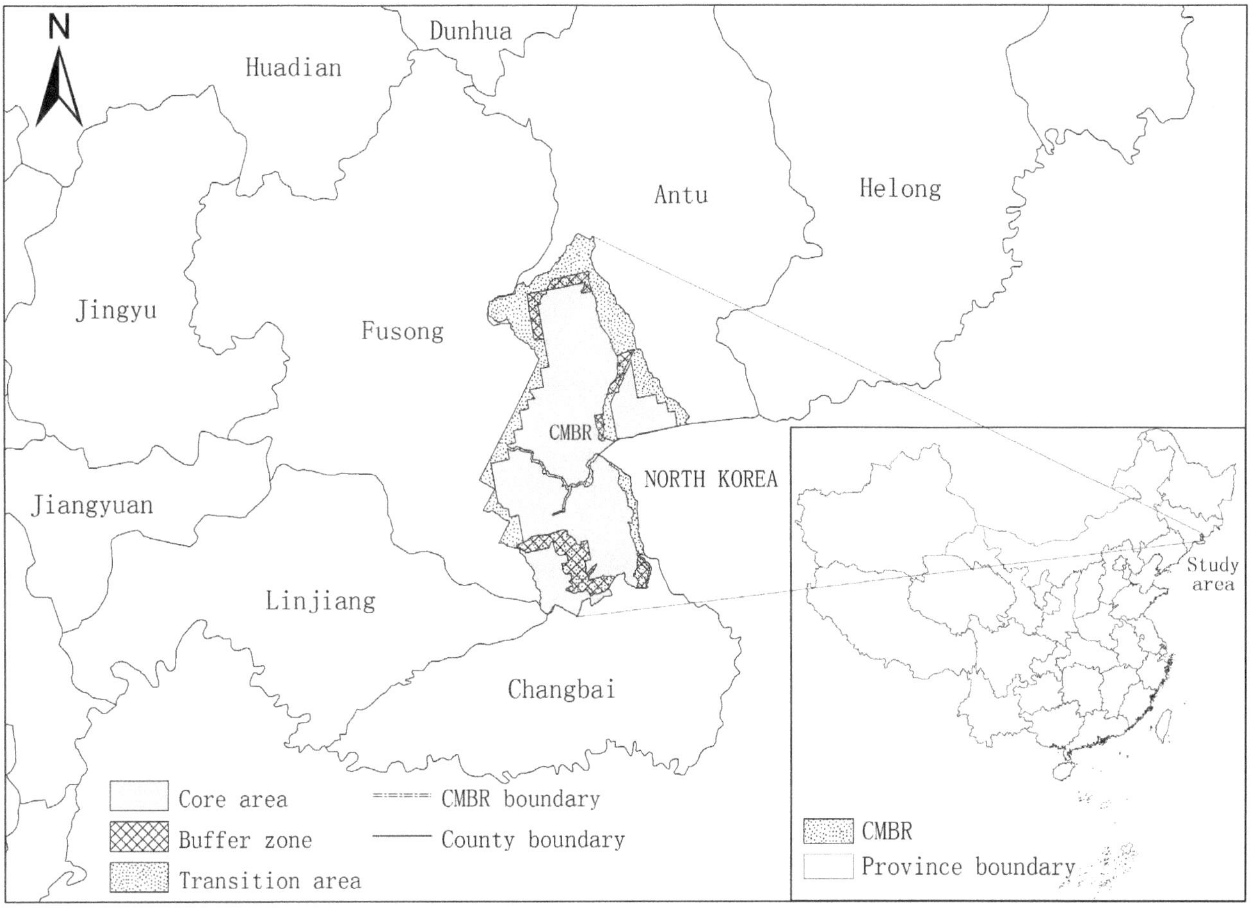

Figure 1. Location of Changbai Mountain Biosphere Reserve (CMBR). The CMBR is located in the northeast of China. It borders the Democratic People's Republic of Korea (DPRK) to the southeast.

according to the six types of landscape mentioned above. In contrast to the ZY-1 02C image, the LandSat TM image of CMBR was of high quality (no influence of clouds and snow), and could provide more band combination to identify different types of vegetation. The 30 m spatial resolution satisfied accuracy requirements for the classification of a nature reserve covering a total area of nearly 200,000 ha. Therefore, the TM image was adopted for the classification of natural landscape in 2012. Using the ENVI 5.0 software, an Example-Based Object-oriented Classification method was employed to classify the natural landscape (See Text S1 for the detailed processes of classification). The confusion matrix was computed using the validation samples selected from the ZY-1 02C fusion image and HR panchromatic image, and the classification accuracy was tested. The overall accuracy was 92.3%, and the Kappa Coefficient was 0.9. The classification accuracy for each type of natural landscape was shown in Table S1.

Extraction of artificial landscape. The ZY-1 02C fusion image and HR panchromatic image supplemented with the field survey data were used to classify the artificial landscape. Visual interpretation (i.e. direct delineation and interpretation of the images) was applied. An artificial landscape presented in 2012 was extracted by the ArcMap10.0 software to establish the layer of artificial landscape for that year. The central line was drawn on images of the linear landscape with width less than 10 m. Their patches were generated using bilateral buffering, according to the measured width.

Data from the artificial forest, young afforested land, cutover land, land for the forestry facilities, and other lands extracted from the data of forest inventory in 1993 were used, and supplemented with the artificial landscape data interpreted in 2012. In addition, the variation of artificial landscape for 1993-2012 was obtained via interview. The distribution of artificial landscape in 1993 was reconstructed, and the layer of artificial landscapes of 1993 was generated.

Establishment of composite landscape map. The classification layers of the natural, and artificial landscape were merged by ArcMap10.0. A topological approach was used to remove gaps, and to process overlapping surfaces. Composite landscape layers for 1993 and 2012 were then established. The landscape classification composite map for CMBR in 2012 is presented in Fig. 2.

The ecological level of landscape

Based on the ecological function and role in habitat conservation, the natural and artificial landscape was further divided into three functional types: protected landscape, interfered landscape, and neutral landscape. The golden section method was used to classify the ecological level of landscape into seven grades: 1.00, 0.62, 0.38, 0, −0.38, −0.62 and −1.00 (Table 1). The golden section method was a typical algorithm in selecting optimization

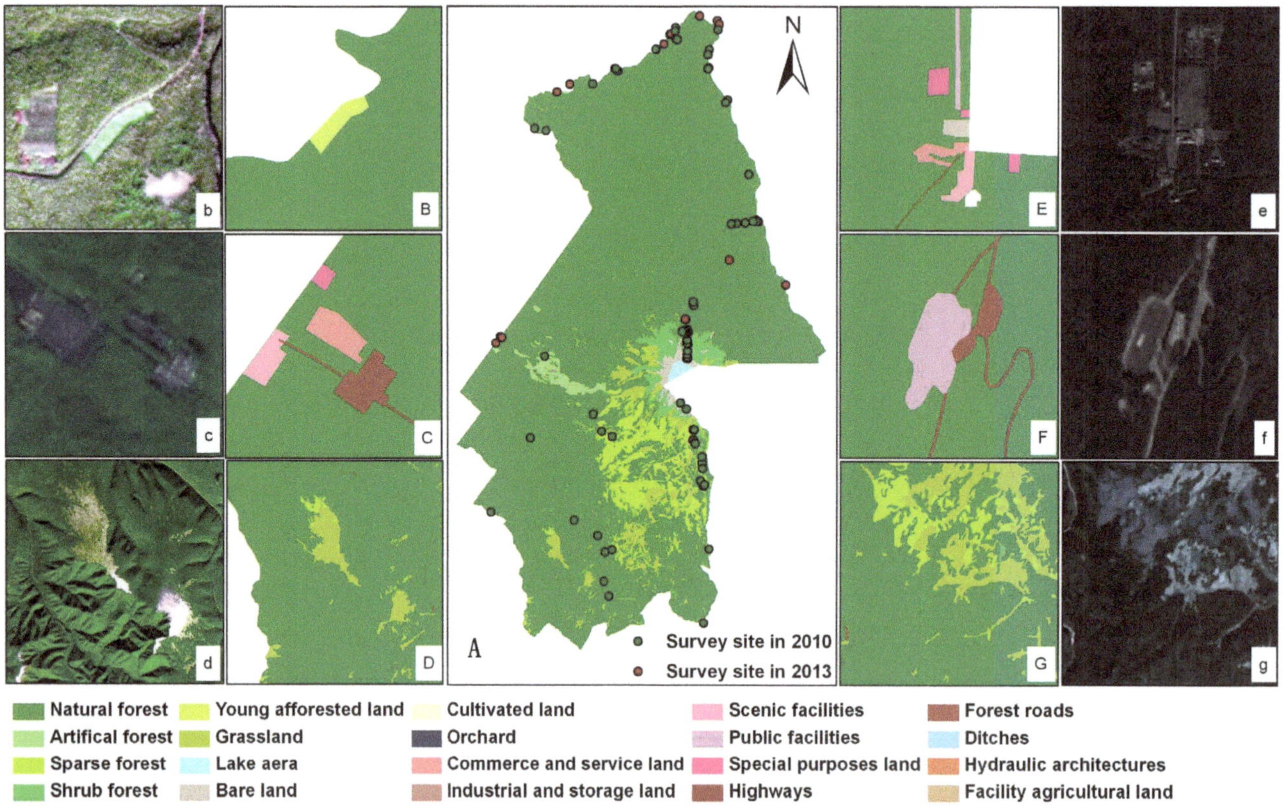

Figure 2. Composite map of landscape classification for CMBR in 2012. (A) Full map of landscape classification, (B) Young afforested land on northern slope, (b) The ZY-1 02C fusion image of B, (C) Scenic facilities, commerce and service land, highways, and special purposes land near the western gate, (c) ZY-1 02C fusion image of C, (D) Natural forest and grassland on southern slope, (d) ZY-1 02C fusion image of D, (E) Scenic facilities, commerce and service land, special purposes land near the northern gate, (e) The HR panchromatic image of E, (F) Highways and public facilities in the U-shaped valley, (f) The HR panchromatic image of F, (G) Natural forest, sparse forest and grassland on the southern slope, (g) The TM image of G.

based on Golden Section Theory [39], and it used the critical value of 0.62 and 0.38 to divide the given interval. A positive value was assigned to the protected landscape; the greater the value, the higher the ecological level. A negative value was assigned to the interfered landscape; the greater the absolute value, the smaller the ecological level (i.e. the higher the interference degree). The value 0 was assigned to the neutral landscape.

Quality index of protected landscape

We developed the quality index of protected landscape (Q) as follows:

$$Q = \frac{\sum_{i=1}^{l}(1.00S_{ci} + 0.62S_{bi} + 0.38S_{ti})D_i}{TS} \times 100$$

where S_{ci}, S_{bi} and S_{ti} are the area of protected landscape of type i in the core area, in the buffer zone and in the transition area, respectively, D_i the value assigned to the ecological level of protected landscape of type i, TS the total area of the nature reserve, and l the number of protected landscape types. The constants 1.00, 0.62 and 0.38 represent the weights of the core area, buffer zone and transition area, respectively.

The value of Q ranges from 0 to100. If there is no protected landscape in the nature reserve, then Q equals 0; if landscape in the nature reserve is all protected landscape, all distributed in the core area and with the highest ecological level (1.00), then Q equals 100. By comparing the Q values in 1993 with those in 2012, changes in the quality of protected landscape in the nature reserve could be quantified.

Interference index of anthropogenic landscape

We also developed the interference index of anthropogenic landscape (I) as follows:

$$I = -\frac{\sum_{i=1}^{k}(1.00S'_{ci} + 0.62S'_{bi} + 0.38S'_{ti})D_i}{TS} \times 100$$

where S'_{ci}, S'_{bi} and S'_{ti} are the area of interfered landscape of type i in the core area, in the buffer zone and in the transition area, respectively, D_i the value assigned to the ecological level of interfered landscape of type i, and k the number of interfered landscape types.

The value of I ranges from 0 to 100. If there is no interfered landscape in the nature reserve, then $I=0$. If landscape in the nature reserve is all interfered landscape, all distributed in the core area and with the lowest ecological level (-1.00), $I=100$. The greater the value is, the higher is the interference degree. By comparing the I values in 1993 and in 2012, the change in the degree of anthropogenic interference in the nature reserve could be quantified.

Alterations of landscape area of the same types

We calculated the alteration rate of landscape area in the natural reserve as:

Table 1. Value of ecological level of different landscape in CMBR.

Landscape type	Attribute	Ecology level
Natural forest, Lake area	Protected landscape	1.00
Sparse forest, Shrub forest	Protected landscape	0.62
Grassland, Artificial forest, Young afforested land	Protected landscape	0.38
Bare land	Neutral landscape	0
Cutover land, Cultivated land, Orchard	Interfered landscape	−0.38
Residential land, Scenic facilities, Public facilities, Forest roads, Facility agricultural land	Interfered landscape	−0.62
Commerce and service land, Industry and storage land, Special purposes land, Highways, Ditches, Hydraulic architectures	Interfered landscape	−1.00

$$R_i = \frac{S_{i2012} - S_{i1993}}{S_{i1993}} \times 100\%$$

where R_i is the alteration rate of the area of landscape type i, S_{i2012} the area of landscape type i in 2012, and S_{i1993} the area of landscape type i in 1993.

Alterations of landscape area of different types

Based on the composite landscape maps in 1993 and 2012, we calculated the area transfer matrix between different landscape types using the ENVI 5.0 software. The proportion of area transformed to ($A_{p\rightarrow q}$) and from ($B_{p\leftarrow q}$) other types of landscape proportion were calculated as:

$$A_{p\rightarrow q} = \frac{S_{p\rightarrow q}}{S_{p1993}} \times 100\%$$

$$B_{p\leftarrow q} = \frac{S_{p\leftarrow q}}{S_{p2012}} \times 100\%$$

where $A_{p\rightarrow q}$ is the proportion of area transformed to other type of landscape (i.e. the proportion of the area of landscape type p in 1993 transformed into landscape type q in 2012), $S_{p\rightarrow q}$ the area of landscape type p in 1993 transformed into landscape type q in 2012, S_{p1993} the area of landscape type p in 1993, $B_{p\leftarrow q}$ the proportion of the area newly added (i.e. the proportion of the area of landscape type p in 2012 transformed from landscape type q in 1993), $S_{p\leftarrow q}$ the area of landscape type q in 1993 transformed into landscape type p in 2012, and S_{p2012} the area of landscape type p in 2012.

Alterations of spatial pattern of protected landscape mosaic

Using the ArcMap10.0 software, protected landscape with ecological levels of: (1) 0.38 and above (including seven types: natural forest, lake area, sparse forest, shrub forest, grassland, artificial forest, young afforested forest), (2) 0.62 and above (including natural forests, lake area, sparse forest and shrub forest), and (3) 1.00 (natural forest and lake area) were merged into one landscape mosaic, respectively. Integrating different habitat resources into a complete pattern made it possible to analyze the overall influence brought to the survival species due to variations in spatial pattern of habitats.

The pattern indexes, including the number of patches (*NP*), edge density (*ED*), perimeter-area fractal dimension (*FR*), division index (*DI*), and connectivity index (*CON*), were computed by the FragStats 4.0 software, the equations of the indexes were detailed in (Text S2). *NP* is fundamentally important to a number of ecological processes. If the target landscape mosaic area increases, the rise of *NP* may reflect the early stage in progressive succession of vegetation. As vegetation recovery proceeds further, *NP* may hold constant or decrease because of the merger of the neighbor landscape. *ED* and *FR* quantify the edge size and the degree of shape complexity of the target landscape mosaic, and are suggestive of anthropogenic interference affecting the target landscape mosaic across their edges at a wide range of scales [40]. *DI* refers to the tendency of the target landscape mosaic to be spatially aggregated and reflects the degree of habitat fragmentation [41]. *CON* refers to the degree to which a landscape facilitates or impedes ecological flows (e.g., the movement of organisms among habitat patches) [42–45].

Results

Changes in the quality of protected landscape

The quality index of protected landscape increased from 74.48 in 1993 to 75.50 in 2012, indicating that the overall quality of protected landscape in CMBR improved. From 1993 to 2012, the total area of protected landscapes increased from 191,862.5 ha (97.8% of the total area of the reserve) to 194,567.3 ha (99.2%; Table 2). The natural forest area increased from 168,693.0 ha to 170,898.4 ha, and the increased area was mainly transformed from sparse forest (1,555.1 ha) and cutover land (1,300.9 ha; Table S4). The area of sparse forest increased from 7,505.2 ha to 7,968.0 ha by 6.2%. The 1,140.4 ha of cutover forest and 1,181.1 ha of grassland were transformed into sparse forest (Table S4). The total area of protected landscape in the core area, buffer zone, and transition area all increased to some extent (Table 2). The total area of protected landscape in the core area increased from 127,893.9 ha to 129,743.7 ha, and increased by 846.5 ha in the transition area. The area of sparse forest in the buffer zone, and the transition area increased by 68.6% and 64.9%, respectively; the area of the natural grassland in the transition area increased by 10.1% (Table S3).

Changes in the interference degree of interfered landscape

The interference index of anthropogenic landscape decreased from 0.49 in1993 to 0.06 in 2012, suggesting that the degree of anthropogenic interference decreased. From 1993 to 2012, the total area of interfered landscape in the nature reserve was reduced from 3,033.5 ha to 320.8 ha, with a variation rate of −89.4% (Table 2). Cutover land, cultivated land, industry and storage land, residential land, forest roads and hydraulic architectures decreased greatly. The area of cutover land transformed to natural forest was 1,300.9 ha, accounting for 49.0%, and about 43.0% of cutover land was transformed into the sparse land (Table S4). In addition to those transformations, there were still 206.6 ha of cutover land transformed to grassland (Table S4). The area of cultivated land transformed to artificial forest was 48.1 ha (56.6%), and there were 23.7 ha (27.9%) cultivated land transformed to grassland, shrub forest, sparse forest and other types of protected landscape (Table S4).

The total area of interfered landscape in the core area, buffer zone and transition area significantly decreased. The area of interfered landscape in the core area was 1,894.9 ha in 1993, and reduced to only 57.8 ha in 2012 (Table 2). The area of interfered landscape in the buffer zone was the lowest both in 1993 and in 2012 (Table 2), and primarily used for transportation. The construction land (i.e. commerce and service land, industry and storage land, residential land, scenic facilities, public facilities, special purpose land, highways, forest roads, hydraulic architectures and facility agricultural land), cultivated land, and orchard in the transition area were significantly larger than those in the core area and buffer zone. The total area of the three types mentioned above was 258.2 ha in 1993, accounting for 23.8% of the total area of interfered landscape. Among these, the total area of construction land increased by 26.9%. Especially, the area of commerce and service land increased significantly from 2.6 ha to 13.4 ha (Table S3).

Changes in spatial pattern of protected landscape mosaic

From 1993 to 2012, *NP* of protected landscape mosaic (with an ecological level of ≥0.38) increased from 281 to 345, *ED* decreased from 6.70 to 6.08, *FR* decreased from 1.74 to 1.53, and *DI* did not change significantly (Table S2). The spatial connectivity of protected landscape at scales of 1, 3, 5, 10 and 15 km all significantly increased (Fig. 3A). *NP* of protected landscape mosaic (with an ecological level of ≥0.62) increased from 271 to 288, whereas *ED* slightly increased from 14.57 to 14.61. The overall shape became simpler, and the degree of fragmentation did not change significantly (Table S2). The spatial connectivity decreased at smaller scales (1, 3, and 5 km), but increased at larger scales (10 and 15 km; Fig. 3B).

The natural forest and lake area both had an ecological level of 1. As the lake area experienced little or no change, variation in landscape pattern was mainly driven by the change in the area of natural forest. From 1993 to 2012, *NP* of natural forest increased from 333 to 456, and its *ED* increased from 16.24 to 16.39. The degree of fragmentation increased, and *DI* increased from 0.56 to 0.61. The spatial connectivity of natural forest decreased at smaller scales (1, and 3 km), but increased at scales of ≥5 km (Fig. 3C).

Discussion

Efficacy of landscape conservation

Changes in the quality index of protected landscape and the interference index of anthropogenic landscape indicate that from 1993 to 2012 the quality of protected landscape increased and the degree of anthropogenic interference decreased. These results suggest that the protection and management measures used in CMBR are generally effective.

Table 2. Total area and percentage of protected landscape and interfered landscape in each functional zone.

	Protected landscape					Interfered landscape				
	Area in 1993/ha	Area in 2012/ha	Percentage in 1993/%	Percentage in 2012/%	Variation ratio/%	Area in 1993/ha	Area in 2012/ha	Percentage in 1993/%	Percentage in 2012/%	Variation ratio/%
Biosphere reserve	191862.5	194567.3	97.8	99.2	1.4	3033.5	320.8	1.5	0.2	−89.4
Core area	127893.9	129743.7	98.0	99.4	1.4	1894.9	57.8	1.5	0.0	−96.9
Buffer zone	18036.3	18045.3	99.2	99.2	0.0	53.1	39.7	0.3	0.2	−25.2
transition area	45929.4	46775.9	96.7	98.5	1.8	1084.8	222.2	2.3	0.5	−79.5

Four reasons account for the effectiveness of landscape conservation in CMBR. First, from 1993 to 2012, the cutover land in the windthrow area gradually resembled the natural landscape following the 20 years of recovery. The progressive succession of large areas of this cutover land resulted in the formation of natural forest and sparse forest. This is also the primary reason why the degree of anthropogenic interference decreased and the area of protected landscape increased.

Second, following the recovery, canopy closure in some parts of sparse forest in the windthrow area increased gradually to over 0.2 (i.e. the threshold value for forming forest in China [46]), suggesting the establishment of forest. In addition, the pioneer tree species also colonized within 1181.1 ha of natural grassland at lower elevations (Table S4), thus forming sparse forest through progressive succession during the 20 years.

Third, the area of cultivated land in the transition area markedly decreased. The cultivated land area in 1993 was primarily distributed along the northern boundary of the biosphere reserve, and contracted to local residents to reclaim. By 2012, the administrative authority of CMBR no longer allowed such use, and most of such cultivated land was abandoned and transformed to artificial forest, grassland, shrub forest, sparse forest and other types of protected landscape.

Forth, the spatial pattern of the protected landscape mosaic (with an ecological level of ≥0.38), which was important habitat for animals and plants, was optimized during the 20 years. The edge effect of protected landscape mosaic was weakened and its shape became simpler, as indicated by the decreased *ED* and *FR*. This further implies that the possibility of anthropogenic interference attempts across borders decreased. The increased spatial connectivity may suggest that it became easier for animals to migrate, feed or survive in the landscape mosaic in 2012 compared to that in 1993 [44–45]. For natural forest, the edge density and degree of fragmentation increased in the course of the 20 years, indicating the occurrence of natural succession of cutover land, grassland and sparse forest, and subsequent formation of many small patches of natural forest. These suggest that an increase in edge density and fragmentation must occur at the earlier stage of vegetation recovery. But as the succession proceeds further, the edge density and degree of fragmentation will decrease [47].

Challenges for landscape conservation

Although most of the cutover land in the windthrow area was transformed into sparse forest and natural forest through progressive succession, there were still 206.6 ha of cutover land transformed into grassland (Table S4), distributed mainly above 1500 m asl. on the south slope in the biosphere reserve. This type of grassland may not represent the original high-quality vegetation, but may well represent the end point in vegetation succession. Field investigation indicates that the plant species in this grassland were members of the *Gramineae*, *Compositae* and *Cyperaceae* families. *Deyeuxia angustifolia* and *Synurus deltoides* were dominant species in this plant community. There were few pioneer tree species colonizing this area, although there was no lack of tree provenance. Luxuriant growth of herbaceous plants can reduce the chance of tree seed to in contact with the soil, thereby reducing germination rates. Intensive coiling of grass roots (usually 5–10 cm thick) also prevents the rooting of tree seedlings. Areas with thin, barren soil and arid climate are also unfavorable for the regeneration of trees [48]. Therefore, it is difficult for progressive vegetation succession to occur in grassland that was transformed from cutover land.

The form of human interference changed fundamentally over the past 20 years, from timber collection in windthrow area to facility and transportation corridor construction for supporting a thriving tourism industry. This change was evident from the observed increase in construction land, especially the increase of commerce and service land, scenic facilities, public facilities and highways (Table S3). In 1990s, tourist activity in CMBR was concentrated on the northern slope of Changbai Mountain. In about 2000, an additional tourist route opened on the western slope, and after the Administration Committee of Changbai Mountain was established in 2005, new tourist routes on the southern slope were also opened. An airport was even constructed outside CMBR.

Promoting the tourism industry in CMBR also resulted in an increased demand for support services, and thus, construction of numerous hotels, roads, parking lots, and scenic promenade occurred [24]. Such type of construction permanently destroyed the original vegetation, and the habitats, thereby lowering the ecosystem value of the landscape. Improper disposal of white pollutants, solid waste, and sewage also resulted in ecological destruction [49]. In addition, the number of tourists increased from 0.20 million in 2000 to 1.42 million per year in 2011 [25],

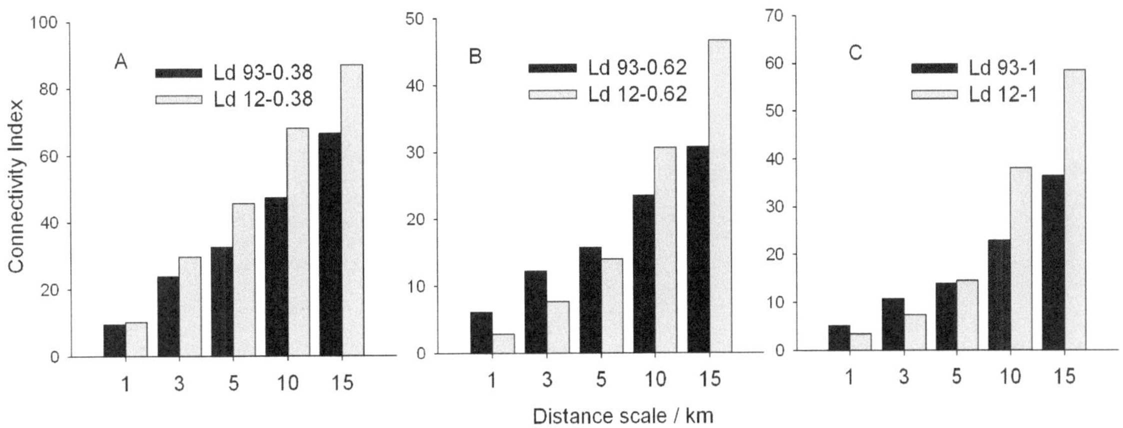

Figure 3. Connectivity index of protected landscape mosaic. Ld 93-0.38, Ld 93-0.62, Ld 93-1 represents the landscape mosaic composed by protected landscape with ecological levels ≥0.38, ≥0.62 and ≥1 in 1993, respectively. Similarly, Ld12-0.38, Ld12-0.62, Ld12-1 represents the landscape mosaic composed by protected landscape with ecological levels ≥0.38, ≥0.62 and ≥1 in 2012, respectively.

and the demand on resources such as firewood and vegetables, increased rapidly [30]. Such activity also greatly increased the occurrence of fire [31]. Therefore, the development of tourism has negatively affected habitat conservation effort in CMBR.

Land for hydraulic architectures in CMBR includes hydropower stations, a dam, and other water conservancy facilities. From 1993 to 2012, the total area of the land for this use was reduced (Table S3), primarily due to the demolition of several hydropower stations, however, in 2012, there were still 6.8 ha of land dedicated to hydraulic architecture use, including 6 large hydropower stations and 3 ditches, which had a total length of 4.4 km and cover an area of 3.3 ha. These dams, hydropower stations, and ditches were all constructed along natural rivers. Thus the development and utilization of water resources affects the migration and spawning of fish [50], and severely interferes with waterfowl, and amphibian habitat [51–53].

Suggestions for landscape protection

Artificial stimulation (e.g. weed removal and soil surface exposure) may help tree seed germination, and subsequent natural regeneration, because the grassland (above 1500m asl.) transformed from cutover land is not as robust as original natural forest.

The continued development of tourism in this region should follow the eco-tourism model and all visitor services should be provided outside of CMBR. Since the hydraulic architecture uses seriously affect protection of biodiversity, it may be beneficial to implement a program to gradually demolish these facilities or relocate those to less-sensitive ecosystems.

The administrative departments of the CMBR should formulate specific regulations, and strengthen the enforcement of those regulations (as with national regulations for the protection of nature reserve). All tourism, production and business activities within the core area and buffer zone should be prohibited. The exploration, planning and resource development in the transition area (especially the development of the construction projects) must: (1) be endorsed by the administrative bureau; (2) follow a strict environmental impact assessment system, and (3) not commence until approved by higher level department of forestry.

Conclusions

We conclude that landscape conservation in CMBR is effective, although there still exists some challenges. The methods developed in this study have the potential to be used to assess the efficacy of landscape conservation in PAs worldwide. Ideally it would be much more robust to compare also landscape changes in the surrounding, non-protected area from 1993 to 2012. But, unfortunately, data of forest inventory in 1993 in surrounding area of CMBR is not available.

Supporting Information

File S1 Table S1 Classification accuracy of each type of natural landscape in 2012. Table S2 Spatial pattern characteristics of protected landscape mosaic. Table S3 Variation of area of landscape in CMBR and in each functional zone from 1993 to 2012. Table S4 Area transfer matrix of landscape type in CMBR from 1993 to 2012. Text S1 Specific processes of classification for natural landscape. Text S2 Equations of pattern indexes.

Acknowledgments

We are grateful to Prof. Fei-Hai Yu for correcting the English and giving suggestions that greatly improved this manuscript. We thank Zhi-Hong Li and Xiao-Ming Yang from the administrative bureau of CMBR for giving permission to work in the reserve, Hai-Cheng Zhou for assistance in field surveys, and Dr. Edward Webb and two anonymous reviewers for their helpful comments. We also thank Hui-Feng Lin for giving help in revising the paper.

Author Contributions

Conceived and designed the experiments: GFC JLZ. Performed the experiments: JLZ. Analyzed the data: JLZ FZL. Wrote the paper: JLZ GFC.

References

1. Sala OE, Chapin FS, Armesto JJ, Berlow E, Bloomfield J, et al. (2000) Global biodiversity scenarios for the year 2100. Science 287: 1770–1774.
2. Fischer J, Lindenmayer DB (2007) Landscape modification and habitat fragmentation: a synthesis. Global Ecol Biogeogr 16:265–280.
3. Howard PC, Davenport TRB, Kigenyi FW, Viskanic P, Baltzer MC, et al. (2000) Protected area planning in the tropics: Uganda's national system of forest nature reserves. Conserv Biol 14: 858–875.
4. Pimm SL, Pimm TL, Ayres M, Balmford A, Branch G, et al. (2001) Can we defy nature's end? Science 233: 2207–2208.
5. Bruner AG, Gullison RE, Rice RE, da Fonseca GAB (2001) Effectiveness of parks in protecting tropical biodiversity. Science 291: 125–128.
6. IUCN and UNEP-WCMC (2012) The world database on protected areas. Cambridge, UK: UNEP-WCMC. Available from: http://www.wdpa.org/Statistics.aspx. Accessed 10 August 2013.
7. Lindenmayer DB, Fischer J (2006) Habitat fragmentation and landscape change. Island press, Washington, D.C.
8. State Forestry Bureau (2008) Classification for land cover of nature reserve(LY/T 1725–2008). Standards press of China, Beijing.
9. McPherson M, Schill S, Raber G, John K, Zenny N (2008) GIS-based modeling of environmental risk surfaces(ERS) for conservation planning in Jamaica. Journal of conservation planning, 4:60–89.
10. Liu JG, Linderman M, Ouyang ZY, An L, Yang J, et al. (2001) Ecological degradation in protected areas: the case of Wolong Nature Reserve for giant pandas. Science 292: 98–101.
11. Southworth J, Nagendra H, Carlson LA, Tucker C (2004) Assessing the impact of Celaque National Park on forest fragmentation in western Honduras. Appl Geogr 24:303–322.
12. Curran LM, Trigg SN, McDonald AK, Astiani D, Hardiono YM, et al. (2004) Lowland forest loss in protected areas of Indonesian Borneo. Science 303: 1000–1003.
13. Jha S, Bawa KS (2006) Population growth, human development, and deforestation in biodiversity hotspots. Conserv Biol 20: 906–912.
14. Wittemyer G, Elsen P, Bean WT, Burton ACO, Brashares JS (2008) Accelerated human population growth at protected area edges. Science 321:123–126.
15. Pimm SL, Lawton JH (1998) Planning for biodiversity. Science 279: 2068–2069.
16. Terborgh J, van Schaik CP (2002) Why the world need parks. In: Terborgh J, van Schaik CP, Davenport L, Rao M, editors. Making parks work, strategies for preserving tropical nature. pp. 3–14.
17. Joppa LN, Loarie SR, Pimm SL (2008) On the protection of "protected areas". PNAS 105: 6673–6678.
18. Zheng YM, Zhang HY, Niu ZG, Gong P (2012) Protection efficacy of national wetland reserves in China. Chin Sci Bull 57: 1116–1134.
19. Ministry of Environmental Protection of the People's Republic of China [MEPPRC] (2012) Nature reserves lists in China. Available from: http://datacenter.mep.gov.cn. Accessed 25 August 2013.
20. Wu RD, Zhang S, Yu DW, Zhao P, Li XH, et al. (2011) Effectiveness of China's nature reserves in representing ecological diversity. Front Ecol Environ 9: 383–389.
21. Wang Z, Jiang MK, Zhu GQ, Tao SM, Zhou HL, et al. (2004) Comparison of Chinese nature reserve classifications with IUCN protected areas categories. Rural Eco-environ 20: 72–76.
22. Jiang MK, Wang Z, Zhu GQ, Tao SM, Zhou HL, et al. (2004) Chinese nature reserve classification standard based on IUCN protected area categories. Rural Eco-environ 20: 1–6, 11.
23. Han NY, Zhuge R (2001) Ecotourism in China's nature reserve: opportunities and challenges. J Sustain Tour 9: 228–242.
24. Zhao JZ, Yuan L, Wang DY, Xu D (2011) Tourism-induced deforestation outside Changbai Mountain Biosphere Reserve, northeast China. Ann For Sci 68: 935–941.

25. Yuan JQ, Dai LM, Wang QL (2008) State-led ecotourism development and nature conservation: a case study of the Changbai Mountain Biosphere Reserve, China. Ecol Soc, 13: 55.
26. Mehring M, Stoll-Kleemann S (2011) How effective is the buffer zone? Linking institutional processes with satellite images from a case study in the Lore Lindu Forest Biosphere Reserve, Indonesia. Ecol Soc 16: 3–18.
27. Sajeev R, Subramanian V. (2003) Land use/land cover changes in Ashtamudi wetland region of Kerala: A study using remote sensing and GIS. J Geol Soc India, 2003, 61: 573–580.
28. Vunohelainen AJ, Coad L, Marthews TR, Malhi Y, Killeen TJ (2011) The effectiveness of contrasting protected areas in preventing deforestation in Madre de Dios, Peru. Environ Manag 50: 645–663.
29. Rayn D, Sutherland WJ (2011) Impact of nature reserve establishment on deforestation: a test. Biodivers Conserv 20:1625–1633.
30. Tang LN, Shao GF, Piao ZJ, Dai LM, Jenkins MA, et al. (2010). Forest degradation deepens around and within protected areas in East Asia. Biol Conserv 143: 1295–1298.
31. Yang X, Xu M. (2003) Biodiversity conservation in Changbai Mountain Biosphere Reserve, northeastern China: status, problem, and strategy. Biodivers Conserv 12: 883–903.
32. WRI IUCN, UNEP (1992) Global biodiversity strategy.
33. Zhao G, Shao GF, Hao ZQ, Wu WQ (2001) Remote sensing of forest landscape fragmentation on Changbai Mountain. Acta Ecol Sin 21: 1393–1402.
34. Shao GF, Zhao G, Zhao SD, Shugart HH, Wang SX, et al. (1996) Forest cover types derived from Landsat Thematic Mapper imagery for Changbai Mountain area of China. Can J For Res 26:206–216.
35. Tang LN, Li AX, Shao GF (2011) Landscape-level Forest Ecosystem Conservation on Changbai Mountain, China and North Korea (DPRK). Mt Res Dev, 31:169–175.
36. Wang SX (2007) Biological resources in Changbai Mountain Nature Reserve. Shengyang: Liaoning Sci Technol Press.
37. Foretry survey and planning institute of Jilin province (1995) Comprehensive survey for forestry resources in Changbai Mountain National Nature Reserve. Changchun.
38. Ministry of Land and Resources (2007) Current land use classification (GB/T21010-2007) China Standards Press, Beijing.
39. Jochen P (1978) The golden section: the "ture" ratio? Perceptual and Motor skill 46: 831–834.
40. Krummel, JR, Gardner RH, Sugihara G, O'Neill RV, Coleman PR (1987) Landscape patterns in a disturbed environment. Oikos 48:321–324.
41. Saunders D, Hobbs RJ, Margules CR (1991). Biological consequences of ecosystem fragmentation: a review. Conserv. Biol. 5:18–32.
42. Keitt TH, Urban DL, Milne BT (1997) Detecting critical scales in fragmented landscapes [online]. Conserv. Ecol. 1:4. Available from: http://www.consecol.org/vol1/iss1/art4/ Accessed 20 February 2014.
43. Fahring L, Merriam HG (1986) Habitat patch connectivity and population survival. Ecology 67: 61–67.
44. Wu JG (1993) Effects of patch connectivity and arrangement on animal meta-population dynamics: a simulation study. Ecol model 65: 221–254.
45. Hobbs RJ (1993) Effects of landscape fragmentation on ecosystem processes in the Western Australian Wheatbelt. Biol Conserv 64: 193–201.
46. State forestry bureau survey planning and design institute (2011). Technical regulations for inventory for forest management planning and design. (GB/T 26424-2010) China Standards Press, Beijing.
47. Li ZF, Tao JP, Wang W, Li XG, He YJ (2005) Spatial Patterns of Main Species of the Degenerate Community in the Recovering Succession of Upper Reaches of Minjiang River. Journal of Southwest China Normal University (Natural Science) 30: 329–332.
48. Xue JG (2009) Investigation about vegetation restoration at typhoon ruined area in Jilin Changbai Mountain National Nature Reserve. Territ Nat Resour 1: 95–96.
49. Ma HY, Wen QH (2009) Changbai Mountain environment analysis and protection measures. Public adm law, 12: 61–63.
50. Zhao ZJ (1984) The resources, reproduction and feeding of *Brachymystax lenok tsinlingensis* in Changbai Mounain. In: Institute of Changbai Mountain National Nature Reserve. Research papers of Changbai Mountain Nature Reserve. Yanji: Yanbian People's Publ House. pp.30–32.
51. Zhao ZJ (1984) Mammal resources and its situation in Changbai Mountain. In: Institute of Changbai Mountain National Nature Reserve. Research papers of Changbai Mountain Nature Reserve. pp.215–222.
52. He JJ (1984) The ecological distribution and protection of rare animals in Changbai Mountain. In: Institute of Changbai Mountain National Nature Reserve. Research papers of Changbai Mountain Nature Reserve. pp.207–214.
53. Piao ZJ, Sui YC, Wang Q, Li Z, Niu LJ, et al. (2011) Population fluctuation and resource protection of Otter (*Lutralutra*) in Changbai Mountain Nature Reserve. J Hydro Ecol 32: 51–55.

Forest Trees in Human Modified Landscapes: Ecological and Genetic Drivers of Recruitment Failure in *Dysoxylum malabaricum* (Meliaceae)

Sascha A. Ismail[1]*, Jaboury Ghazoul[1], Gudasalamani Ravikanth[2], Cheppudira G. Kushalappa[3], Ramanan Uma Shaanker[2,4], Chris J. Kettle[1]

1 ETH Zürich, Institute of Terrestrial Ecosystems, Ecosystem Management, Zürich, Switzerland, **2** Ashoka Trust for Research in Ecology and the Environment, Royal Enclave, Srirampura, Jakkur Post, Bangalore, India, **3** College of Forestry, University of Agricultural Sciences (Bangalore), Ponnampet, Kodagu district, Karnataka, India, **4** Department of Crop Physiology and School of Ecology and Conservation, University of Agricultural Sciences, GKVK Campus, Bangalore, India

Abstract

Tropical agro-forest landscapes are global priority areas for biodiversity conservation. Little is known about the ability of these landscapes to sustain large late successional forest trees upon which much forest biodiversity depends. These landscapes are subject to fragmentation and additional habitat degradation which may limit tree recruitment and thus compromise numerous ecosystem services including carbon storage and timber production. *Dysoxylum malabaricum* is a large canopy tree species in the Meliaceae, a family including many important tropical timber trees. This species is found in highly fragmented forest patches within a complex agro-forest landscape of the Western Ghats biodiversity hot spot, South India. In this paper we combined a molecular assessment of inbreeding with ecological and demographic data to explore the multiple threats to recruitment of this tree species. An evaluation of inbreeding, using eleven microsatellite loci in 297 nursery-reared seedlings collected form low and high density forest patches embedded in an agro-forest matrix, shows that mating between related individuals in low density patches leads to reduced seedling performance. By quantifying habitat degradation and tree recruitment within these forest patches we show that increasing canopy openness and the increased abundance of pioneer tree species lead to a general decline in the suitability of forest patches for the recruitment of *D. malabaricum*. We conclude that elevated inbreeding due to reduced adult tree density coupled with increased degradation of forest patches, limit the recruitment of this rare late successional tree species. Management strategies which maintain canopy cover and enhance local densities of adult trees in agro-forest mosaics will be required to ensure *D. malabaricum* persists in these landscapes. Our study highlights the need for a holistic understanding of the incipient processes that threaten populations of many important and rare tropical tree species in human dominated agro-forest landscapes.

Editor: Edward Webb, National University of Singapore, Singapore

Funding: This research was funded by ETH Zurich under the grant number ETH-22 08-2 (http://www.vpf.ethz.ch/researchfunding/index_EN). The funders had no role in study design, data collection and analysis, decision to publish, or preparation of the manuscript.

Competing Interests: The authors have declared that no competing interests exist.

* E-mail: sascha.ismail@env.ethz.ch

Introduction

Tropical forests are globally important centers of biological diversity and provide crucial ecosystem services including important terrestrial carbon stores [1]. The vast majority of tropical forest does not reside within protected areas [2] but rather as fragmented patches of forest within a more complex landscape mosaic. These forests are typically logged, degraded, highly patchy [3] and subject to a wide array of additional anthropogenic pressures such as grazing, hunting and invasive exotic species [4]. The persistence of late successional tree species within such human dominated tropical landscapes is important for maintaining tropical forest biodiversity. How well late successional tropical forest tree species are able to persist within these landscapes depends upon their resilience to these multiple and potentially synergistic genetic and ecological stressors [5,6]. Few empirical studies have endeavored to evaluate these intrinsic genetic and extrinsic ecological stressors in unison despite these rapidly increasing pressures.

Studies investigating genetic consequences of fragmentation in woody species reveal that fragmentation can lead to elevated inbreeding [7,8] and reduced fitness of the progeny [9,10] through inbreeding depression. One possible underlying process is that pollinators predominantly forage among near neighbors, which may increase bi-parental inbreeding due to increased mating among few related individuals [11]. Habitat fragmentation has a number of ecological consequences for the reproductive ecology of plant species, for example by reducing pollinator visitation which directly reduces fruit and seed set [12]. Fragmentation can also have significant impacts on the community of seed dispersers influencing dispersal [13] and even on the selection pressures upon seed size [14]. When forests become fragmented, reproduction of many tree species can be altered due to unfavorable environmental conditions for seedling establishment. For example, tree seedling survival can be reduced due to drier conditions [15], increased litter fall, which smothers seedlings [16], and alien species invasion [17] especially at the forest edges. Recruitment of late successional shade tolerant trees is thought to be especially sensitive to fragmentation effects because of competition with faster growing light-demanding species that perform better under more open canopies [18]. Forest fragmentation may thus cause shifts in species compositions where slow growing late successional tree species are replaced by early successional pioneer species [19,20].

Although genetic and non-genetic consequences of fragmentation have been shown to affect tree species reproductive ecology, these stressors are rarely considered in unison within a single study. Our objective was to empirically investigate if genetic and ecological processes affected recruitment of a late successional tree species within a fragmented landscape mosaic. We combined a molecular assessment of inbreeding in progeny of the canopy tree species *Dysoxylum malabaricum* with ecological and demographic data to address the question: can a late successional tropical tree species persist in a complex agro-forest landscape within the Western Ghats biodiversity hot spot, South India?

In a previous study we demonstrated that pollen dispersal of *D. malabaricum* in this fragmented landscape can be extensive. However, in the highly fragmented patches - when densities of adult trees were low (5 or fewer conspecific trees within 500 m) - the frequency of short distance pollen dispersal and consequently mating among related individuals was greater than in patches which were less fragmented (when local population densities were high, 6 or more conspecific trees within 500 m) [21]. The implications of this elevated inbreeding for growth and performance of progeny have until now not been examined. To evaluate the genetic and ecological factors which undermine the ability of *D. malabaricum* to persist in this landscape we present two data sets. The genetic data consists of a nursery trial of progeny collected from low and high density *D. malabaricum* forest patches. We used multilocus genotypes of these seedlings together with adult tree genotypes to determine the level of inbreeding and its implications for seedling vigour. The ecological data consists of detailed demographic data on the adult trees and densities of seedlings coupled with seven indicators of degradation in 17 forest patches, to investigate whether degradation correlated with seedling density. We used these data sets to test the following hypotheses: A) Mating between related individuals due to fragmentation reduces seedling performance in *D. malabaricum.* B) Ecological degradation leads to reduced *D. malabaricum* seedling densities in forest patches. By combining the two approaches above we hoped to gain a more holistic view of the factors influencing the demographic structure of *D. malabaricum* in this landscape. *D. malabaricum* is representative of numerous other threatened canopy tree species of late successional forest whose habitat is increasingly limited to forest patches within complex landscape mosaics. This study thus advances our understanding of how the genetic and ecological consequences of habitat fragmentation and degradation in unison undermine canopy tree persistence in human modified landscape mosaics.

Methods

Ethics statement

The forest patches we worked in are conserved due to cultural beliefs, owned by the Forest Department of Karnataka and managed by local temple committees, which granted access for sample collection. This was the only formal access permission required to conduct our research.

The study species *D. malabaricum* is not assessed under the IUCN Red List and is not listed as a protected species, but is classified as endangered under the Indian national threat assessment [22].

Study area

This study focusses on an agro-forest landscape encompassing 216 km^2 including coffee, rice and forests patches (Figure 1). Our study site is situated in Kodagu district, within the Western Ghats biodiversity hotspot, South India. This district is a major coffee-growing region where coffee is grown predominantly under native shade trees [23]. Kodagu is renowned for the high density of small native forest patches conserved for cultural use [23]. These forest patches are set within an agricultural matrix consisting mainly of native shade coffee plantations and paddy fields [24]. Although these forest patches contribute only marginally to the overall forest area of the region [25], they are recognized as important repositories of biodiversity [26]. The forest patches within our study area are subject to anthropogenic disturbances through intensified resource extraction by the local community, such as fuel wood, small poles and non-timber forest products [27]. Illegal timber extraction and the encroachment of forest patches by coffee plantations are also common [28]. Within the landscape of the study area we have located and mapped all adult *D. malabaricum* trees [21].

Study species

Dysoxylum malabaricum is a highly prized timber species within the tropical tree family Meliaceae. The species produces hermaphrodite flowers with a nectar reward suggesting insect pollination, but the main pollinators are unknown. Within the study area *D. malabaricum* maintains high genetic connectivity with around 9% of the pollen dispersal beyond 5 km distance [21]. The ripe fruit split along the four septa displaying the shiny brown seed (c. 30×20 mm) consumed by large-gape birds [29]. Our own field observations support the idea that the Malabar Grey Hornbill *Ocyceros griseus* is an important dispersal agent. The birds ingest the entire seed and remove the brown lipid rich seed coat before regurgitation. If the seed fall passively to the forest floor they rot quickly suggesting obligate zoochory (Ismail *pers. obs.* 2010). The Malabar Grey Hornbill is known to be relatively robust to habitat disturbances and is known to cross over open habitat [30] indicating the potential to disperse seed over long distances. The potential for long distance gene dispersal by pollen and seed is expected to buffer tree species against negative effects of fragmentation [31]. *D. malabaricum* has been extensively logged [32], and demand for *D. malabaricum* timber remains high with round wood logs fetching up to US$ 620 per cubic metre at local timber auctions in Kodagu (Ismail *pers. obs.* 2008).

Within our study area *D. malabaricum* is predominantly found in forest patches: of the 235 adult trees recorded, we found 223 trees in 35 forest patches. The remaining twelve trees were found in coffee plantations (of which nine were within 300 meters of a forest patch). Our previous study on pollen dispersal suggests that our sample of adult trees is exhaustive [21]. We have mapped, genotyped and recorded the diameter at breast height (DBH) of all adult *D. malabaricum* trees in our study area.

Evaluation of seedling performance and inbreeding

Nursery trials. 617 seeds were collected near to 37 fruiting trees (97% of seed were found within 10 m of a fruiting tree) across 16 of the 35 forest patches within our study area (see Figure 1 and Table S1 in Supporting Information for details). Each seed was planted in a 1 liter polyethylene bag into a mixture of 1 part red soil, 2 parts cow dung and 2 parts river sand during July and August 2008, within 1 day after collection. A total of 363 seeds (61%) germinated. Of those that failed to germinate, at least 119 seeds (20%) were predated by larvae of a tephritid fly (*Dacus sp.*). Because of this initial exogenous loss we did not continue

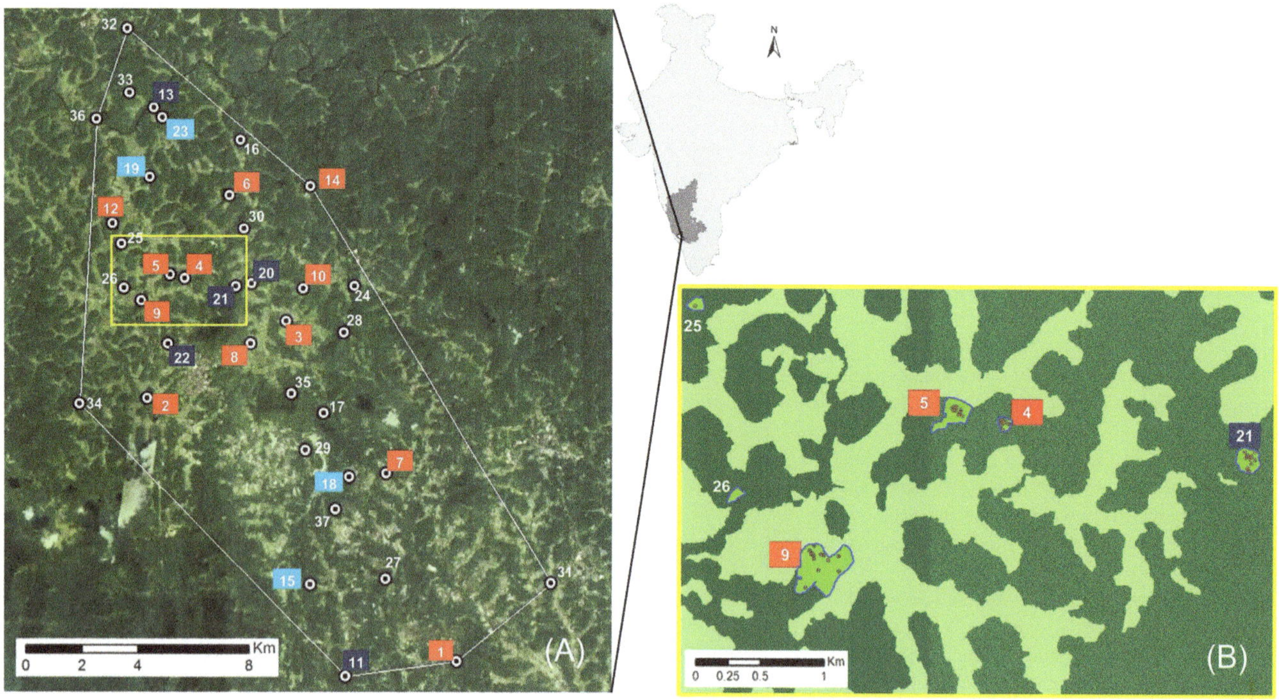

Figure 1. Location of the study area within India (light grey) and Karnataka (dark grey). Image (A): Study area marked with a white minimum convex polygon (216 km^2) around the numbered forest patches where adult trees were found. The color labels of the numbers indicate if only *Dysoxylum malabaricum* seed for the nursery trial were collected (light blue), if only survey plots were established (dark blue) or if both records were taken (red). Image (B): Zoom of the yellow rectangle on image (A) with coffee plantations marked dark green and open areas (mainly paddy) marked light green. Investigated forest patches are the bright green polygons with a blue border and dots displaying adult *D. malabaricum* trees.

assessment of germination rates as a performance variable attributable to genetic factors. We grew the subsequent seedlings (n = 363) in a shade house (50% shade cloth) and watered daily. We rotated the seedlings monthly within the shade house to prevent any local effects, such as daylight orientation and neighbor competition. To reduce mortality due to pests we also applied insecticide and fungicide uniformly across the plants. Plants were repotted after 1 year into 5 liter polyethylene bags. Growth of seedlings in this controlled nursery experiment was monitored over 21 months by monthly measurements of total stem length from soil to top of the apical meristem, to an accuracy of 0.5 cm. After the 21 month period leaf material was sampled from each seedling for subsequent genotyping (see Appendix S1 in Supporting Information for details on genotyping and genetic analysis).

Evaluation of seedling inbreeding. To investigate if inbreeding and mating between related individuals reduced growth of the seedlings, we first applied a parentage analysis with the delta maximum-likelihood approach implemented in CERVUS [33] to determine the two most likely parents at the 90% confidence level (see Appendix S1 for details of the parameter settings). Candidate parents were all adult trees in our study area, genotypes of which had already been published in our earlier study [21]. The microsatellite data of the adult samples and the nursery seedlings were deposited in the DRYAD Digital Repository (doi:10.5061/dryad.3ck30 and doi:10.5061/dryad.mq8fn respectively).

The kinship coefficients [34] among the two assigned parent pairs provide one measure of inbreeding. In addition, individual inbreeding coefficients [35] were calculated for each seedling using the program SPAGEDI 1.3a [36] based on the seedling genotypes alone. The individual inbreeding coefficient $\hat{\rho}$ is calculated as:

$$\hat{\rho} = \sum_{i,l} \frac{S_{il} - P_{il}^2}{P_{il}} \Big/ \sum_{l} (n_l - 1)$$

where P_{il} is the estimated frequency of the i_{th} allele at the l_{th} locus, S is an indicator variable which is one if the individual is homozygous for the i_{th} allele at the l_{th} locus and zero otherwise, and n_l is the number of alleles at the l_{th} locus.

Characterizing local density of D. malabaricum adults in forest patches where seed was collected. The 16 forest patches where we collected seed to investigate inbreeding effects were characterized as either low density (LD) or high density (HD) patches of *D. malabaricum* based upon the number of conspecifics within 500 m of any given adult tree within a patch following Ismail *et al.* [21]. This threshold of 500 m ensures that all conspecifics within the same forest patch are included and accounts for the rare cases where high local densities are the result of conspecific trees located in adjacent coffee plantations or in neighbouring forest patches [21]. This parameter reflects the local abundance of adult individuals of *D. malabaricum* and the distance to the neighboring forest patches and abundance of adult individuals within those neighboring patches. Only five of the 16 forest patches had another forest patch within 500 m and the distance between any two trees within a patch never exceeded 450 m. Our previous work using a detailed genetic analysis of pollen dispersal distances demonstrated increased genetic isolation and increased bi-parental inbreeding when local densities were below six conspecific trees within 500 meters [21] providing the justification for this categorization of genetic fragmentation.

Statistical analysis of inbreeding and growth performance under nursery conditions. Because the effect of local tree density in a forest patch has already been demonstrated to influence patterns of inbreeding in *D. malabaricum* [21], we classified the nursery seedlings based upon whether they were collected from low density (LD) or high density (HD) *D. malabaricum* forest patches. We tested for significant differences of median seedling height, median individual inbreeding coefficient and median kinship coefficient between seedlings of LD and HD origin with non-parametric Wilcoxon rank sum test implemented in R 2.13.1 [37]. To test for significant correlation of seedling height with individual inbreeding coefficients as well as with the kinship of parent pairs Pearson's product-moment correlation (PPMC) test was computed in R [37]. Linear regression was not performed on this data due to variance inhomogeneity.

Evaluation of recruitment and habitat degradation

Sample plots. To quantify habitat degradation and recruitment (indicated by seedling, sapling and pole stage tree densities) of *D. malabaricum* under natural conditions, we established five random plots in each of 17 selected forest patches (85 random plots) in 2010 (12 patches are common among both seed collection for inbreeding assessments and survey plots). For each patch we selected plots randomly in north orientated maps with an overlaid grid. Each grid cell was assigned a number, line by line from the top left to bottom right. We then used a random number generator in the program Excel (Microsoft) to select five random plots for each patch. The lower left-hand corner of the plot was defined by these coordinates. In case of overlap with other plots or when more than 10% of the plot was outside the forest patch the plot was turned 90° clockwise until the entire plot lay within the patch. Plots were 14 m×14 m except in the smallest forest patch where we had to reduce plot size to 10 m×10 m. We standardized all plot based measures to 100 m^2. We quantified recruitment of *D. malabaricum* by recording the number of seedlings (< 50 cm height), number of saplings (> 50 cm, <150 cm) and poles (> 150 cm and DBH <5 cm) within each plot.

Ecological parameters to quantify habitat degradation. We recorded seven parameters which reflect the forest structure and species composition as indicators for habitat degradation. Within each plot this included three variables indicative of low habitat quality: 1) Canopy openness measured with a densiometer (model A) [38] (by averaging four measurements in the centre of the 7 m by 7 m quarters of each plot. In each quarter we measured in all four cardinal directions). 2) The number of coffee seedlings (< 150 cm height), because coffee *Coffea canephora* (a common and important exotic crop species) is naturalizing within forest patches of our study area (Ismail *pers. obs.* 2009). 3) Juvenile (> 0.25 m; <2 m) abundance of the pioneer tree species *Clerodendrum viscosum*, a light demanding species which dominates degraded forest. Four variables were selected as indicators of good habitat quality: 4) the number of adult *D. malabaricum* trees within a patch; 5) the total area of each forest patch (by mapping the forest patch with a handheld GPS (60CSx, Garmin, USA) to an accuracy of five meters). 6) The proportion of the forest patch border by shade coffee plantations. We chose this variable as shade coffee plantations of the region are predominately grown under dense native shade providing a more 'forest like' matrix than other land uses in the study area. This might buffer edge effects more effectively. 7) In addition we quantified the number of arboreal termite nests attached to branches and trees within each plot. Termites have been shown to be effective indicators of forest disturbances in other regions, because they are sensitive to forest disturbance, especially to canopy loss [39–41]. Using all these variables as indicators of habitat quality and *D. malabaricum* seedling density as an indicator of recruitment we tested the hypothesis that recruitment is reduced in degraded forest patches.

Statistical analysis of recruitment and ecological degradation. All statistical analysis was performed with R 2.13.1 [37]. To account for the nestedness of the study design we averaged the measurements of the five plots per forest patch before analysis. This resulted in 17 observations (for an overview of this data see Table S2). This approach was chosen as an alternative to a generalized linear mixed model based on the observations from all 85 plots, because non-normality and inhomogeneity of the residuals negated the use of these models. To investigate reproductive success of *D. malabaricum* in the wild, we performed multiple linear regression (MLR) models with seedling densities as the response variable and the seven variables on habitat degradation described above as explanatory variables. Model assumptions were checked for normality and homogeneity by visual inspections of plots of residuals against fitted values. To meet model assumptions seedling densities were square root transformed prior to analysis. On this initial model we applied model selection with the forward as well as the backward step function implemented in R to exclude irrelevant variables and to improve the model. Pearson's product-moment correlation (PPMC) test was applied to check for co-linearity among pairs of explanatory variables.

Results

Effects of inbreeding and seedling performance

Of the 363 seeds which germinated 297 seedlings survived for 21 months. Using a parentage analysis we were able to assign 99% of seedlings at the 90 % confidence level to the two most likely parents (293 individuals). For subsequent analysis of growth performance 14 seedlings were excluded because the height measurements were ambiguous. This resulted in the exclusion of one of the forest patches from subsequent analysis of inbreeding. The classification of the remaining 273 seedlings into HD forest and LD forest provenance resulted in 197 seedlings from eight HD forest patches and 76 seedlings from seven LD forest patches.

Comparing the assigned seedlings originating from HD forest patches vs LD forest patches, the median ranks of the seedling height after 21 months were significantly higher (Median HD = 38.0 cm, Median LD = 25.5 cm, p-value = 0.00003), individual inbreeding coefficients and kinships of parent pairs were significantly lower (p-value = 0.001 and p-value = 0.0001 respectively) in seedlings from HD forest patches (Figure 2). Individual inbreeding coefficient was significantly negatively correlated with seedling height (PPMC coefficient = −0.139, t = −2.36, df = 281, p-value = 0.02) (the scatterplot of this data is given in Figure S1). In contrast, we observed no significant correlation between kinship of parent pairs and height of the seedlings (t = −1.32, df = 271, p-value = 0.19) (the scatterplot of this data is given in Figure S1).

State of recruitment in *D. malabaricum*

Plotting size class frequency distributions of all recorded reproductive (> 5 cm DBH) *D. malabaricum* trees across the entire 216 km^2 area showed an inverted 'U' shape distribution (Figure 3). There was a noticeable absence of the smallest size classes (< 20 cm DBH) with only one tree with a DBH of 8.5 cm and no

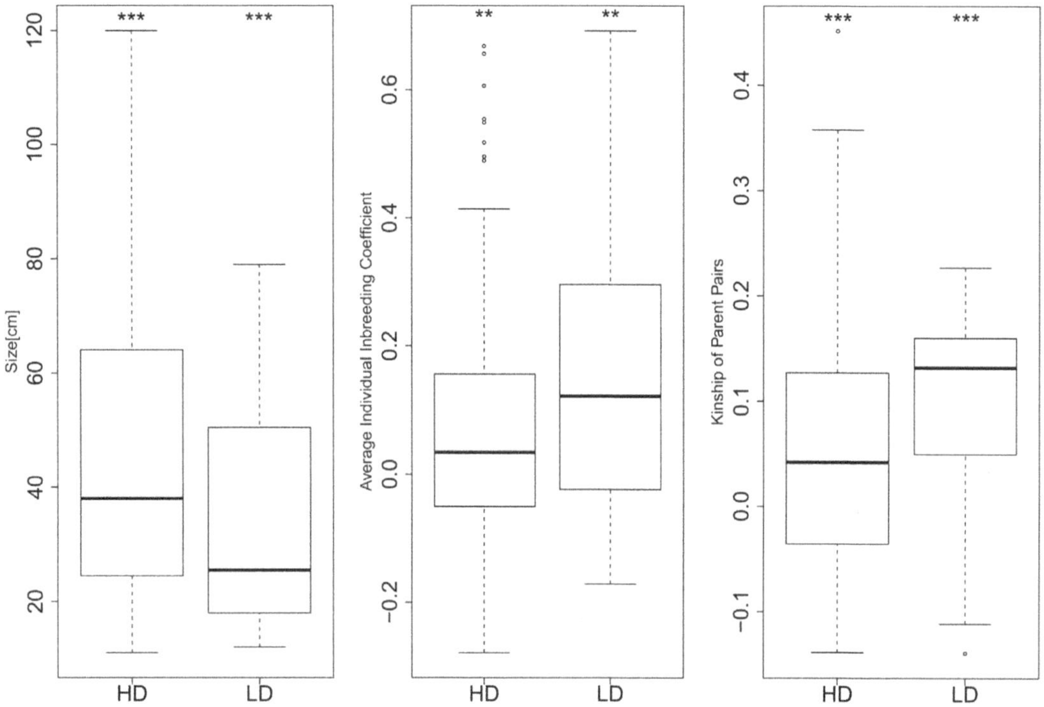

Figure 2. Effect of high density (HD) and low density (LD) of adult Dysoxylum malabaricum trees on A) seedling height after 21 months of growth (Median HD = 38.0 cm, Median LD = 25.5 cm), B) individual inbreeding coefficient (Median HD = 0.034, Median LD = 0.122), and C) pairwise parental kinship coefficients [32] (Median HD = 0.042, Median LD = 0.132), of nursery-reared D. malabaricum seedlings. Boxplots show the median and the upper and the lower quartile, the whiskers are 1.5 times the interquartile range from the box, dots outside of the whiskers are considered outliers. Significant differences are based on Wilcoxon rank sum test; * $p<0.05$, ** $p<0.01$, *** $p<0.001$.

trees in the 10 – 20 cm DBH class. There was a constant decline in the frequency of individuals from 60 cm DBH downward.

Although seedlings were present in all forest patches and saplings were present in nearly all surveyed forest patches, pole stage trees (> 1.5 m height and <5 cm DBH) were extremely rare. We detected only 11 poles in total in six of 17 surveyed forest patches. Seedling and sapling densities (averaged per site) varied greatly and ranged from 0.1 to 17.6 seedlings per 100 m^2 and

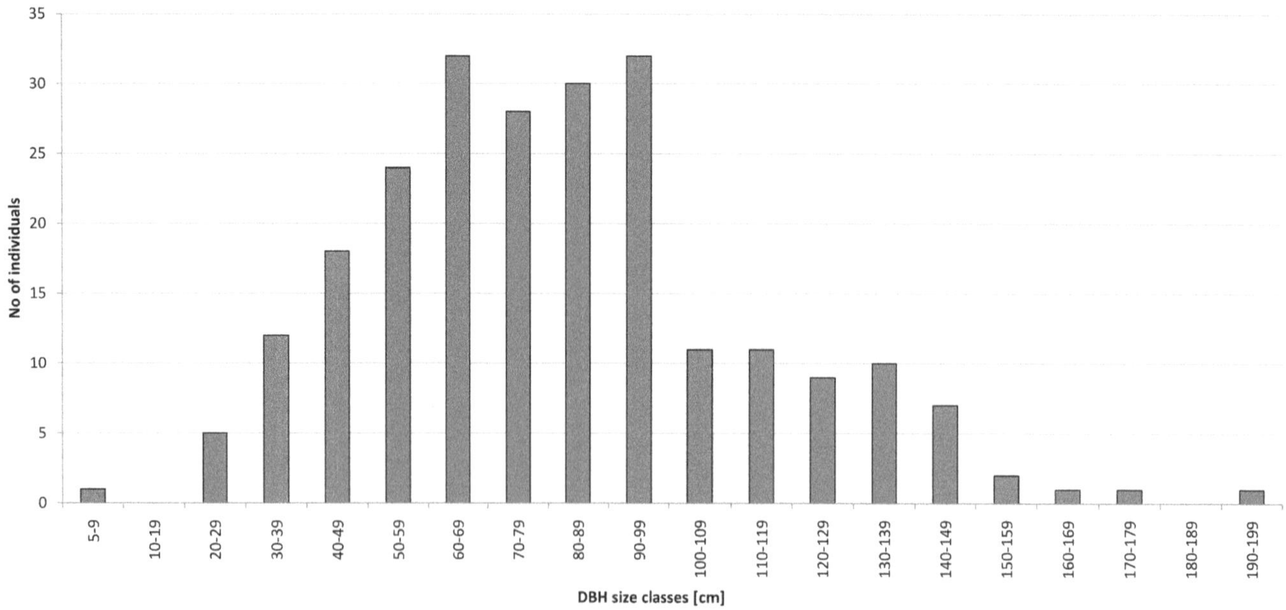

Figure 3. Histogram of diameter at breast height (DBH) of all the 235 enumerated adult Dysoxylum malabaricum trees within the 216 km² study area.

from 0 to 1.7 saplings per 100 m^2. We only found pole stage trees at very low densities from 0 to 0.4 individuals per 100 m^2.

Effects of habitat degradation on recruitment of *D. malabaricum*

We used seven indicators of habitat degradation as explanatory variables of our MLR model to investigate their effect on seedling densities. In the model selection the number of adult *D. malabaricum* trees, the area of the forest patch and the density of coffee seedlings, did not improve the regression model and were excluded from the subsequent model. Within the final reduced model (Table 1) the percentage of closed canopy ($t = 3.67$, $p = 0.0032$) and the density of termite nests ($t = 7.21$, $p = 0.00001$) were positively associated with seedling density. The density of *C. viscosum* juveniles ($t = -5.05$, $p = 0.0003$) was negatively associated with seedling density. The proportion of the forest patch bordered by coffee plantations improved the model but was not significantly positively correlated with seedling density ($t = 1.54$, $p = 0.149$). PPMC among the pairs of explanatory variables was non-significant. The adjusted r squared of the final model was 0.81. Restricting the MLR only to the three significant variables resulted in an adjusted r squared of 0.79.

Discussion

The ability of late successional tropical tree species to persist in complex agro-forest landscapes will have significant implications for biodiversity and ecosystem services. Here we provide an example of one rare but important tropical tree *Dysoxylum malabaricum*. The nursery trial of genotyped seedlings indicates that fragmentation can cause reduced seedling vigour through inbreeding effects. Further, we demonstrate that habitat degradation reduces seedling densities, which appears to be associated with increased canopy disturbance. Thus fragmentation and degradation both appear to compromise the ability of this late successional tree to reproduce and persist in human modified landscapes. Below we examine the evidence for both genetic and ecological stressors influencing the recruitment of *D. malabaricum*. We discuss the implications of these findings for understanding of the long term persistence of late successional forest trees in tropical agro-forest landscapes.

Importance of genetic factors for *D. malabaricum* recruitment

The present study, using nursery-grown seedlings and the kinship coefficients of their most likely parent pairs, confirms that increased inbreeding due to fragmentation reduces seedling growth. We interpret the reduced seedling average growth rates as a sign of inbreeding depression. Inbreeding depression has been shown to be a key factor reducing plant population viability [42] and is a plausible constraint to recruitment of *D. malabaricum*. Although alternative mechanisms for reduced performance exist, such as low pollen diversity [11] or maternal origin [9], the implications for reproduction success remain the same.

Table 1. Summary of the multiple linear regression after model selection used to fit *Dysoxylum malabricum* seedling densities under natural conditions.

	Estimate	Std. Error	t value	Pr(>|t|)	
(Intercept)	−35.007	9.796	−3.573	0.0038	**
% of closed canopy	0.375	0.102	3.671	0.0032	**
C. viscosum juveniles	−0.286	0.057	−5.053	0.0003	***
Termite nests	3.918	0.543	7.213	0.0000	***
% of border with coffee	0.008	0.005	1.542	0.1491	

* p<0.05, ** p<0.01, *** p<0.001.

Impact of habitat degradation on *D. malabaricum* recruitment

Our evaluation of habitat suitability suggests that degradation of forest patches as indicated by increased openness of the forest canopy is resulting in sites less favorable for *D. malabaricum* recruitment. This idea is supported by the negative association of *D. malabaricum* seedlings with the density of *C. viscosum* seedlings and the percentage of open canopy as well as the positive association with the density of arboreal termite nests, independent of adult tree abundance.

Clerodendrum viscosum is an early successional light demanding species establishing after gap formation [44]. We refrain from concluding a causal link between increased abundance of *C. vicosum* and lower *D. malabaricum* seedling densities. However, these associations are consistent with the idea that *D. malabaricum* seedlings in forest patches with more open canopy experienced increased competition from light demanding species, and a less favorable microclimate.

Interestingly, *D. malabaricum* seedling densities are greater where arboreal termite nests are more abundant, which could indicate more favorable and less degraded habitat. Termite communities were shown to be sensitive to canopy disturbance in other studies [39–41] which may well represent a response of termite communities to the loss of big trees. However, the processes underlying this correlation are likely to be more complicated, termites are also important agents of soil formation and nutrient cycling in tropical forests [45] and therefore may influence edaphic conditions important for *D. malabaricum* seedling establishment.

The additional factors we recorded (density of coffee seedlings, the number of adult *D. malabaricum* trees within a patch and the total area of each forest patch) were excluded by the model selection. However, it is noteworthy that though there was no statistically significant relationship between the number of coffee seedlings and *D. malabaricum* seedling densities, in plots where we observed high densities of coffee seedlings (> 200 per 100 m^2) we consistently detected very low densities of *D. malabaricum* seedlings. These plots had on average more than six times lower *D. malabaricum* seedling densities than plots with lower coffee seedling densities (where we recorded very low to high densities of *D. malabaricum* seedlings). The potential competitive interactions between *D. malabaricum* and coffee seedlings are worthy of future empirical validation. The observed variation of seedling densities across patches does not correspond with the density of reproductive adult trees and the area of the forest patches (no significant fits to the MLR) suggesting that seed production and suitable area for establishment *per se* are not driving the limited recruitment at the patch level.

Evidence for recruitment limitation in *D. malabaricum*

Based upon estimates of seedling and sapling densities within our 85 random plots, seed production and germination appears to be substantial in *D. malabaricum*. However, evaluation of the population structure based upon tree diameter classes indicates an almost complete absence of the smallest size class (> 5 cm<20 cm DBH) trees across the entire 216 km^2 study area (Figure 3). Based upon preliminary dendrological estimates in this species (see Appendix S2 in Supporting Information for details), we infer that trees greater than 20 cm DBH are at least 30 years old. We interpret this as a sign of poor recruitment from the sapling to the adult stage over the last 30 years. The survey plot data demonstrate a clear decline in abundance of adult trees from about 60 cm DBH to 20 cm DBH. This demographic structure is concurrent with a decline in recruitment since the intensification of commercial logging and plantation crops in the Western Ghats some 80– 100 years ago [43] and a lack of recruitment since the last extensive expansion of coffee plantations in the region during the last 30 years [23].

Conclusions

Tropical agro-forest landscape mosaics offer great promise as habitats to support a wide array of tropical biodiversity [46] provided they retain key forest features such as diverse tree communities. Our analysis provides a general view of intrinsic genetic and extrinsic ecological factors which affect the recruitment of a rare late successional tropical tree species in forest fragments. *D. malabaricum* shows relatively long-distance pollen dispersal [21]. In addition, the seed disperser (the Malabar Grey Hornbill) moves in open habitat and can persist in fragmented forests [30]; thus one might predict that recruitment of *D. malabaricum* would be little affected by fragmentation and degradation [31]. However, this expectation was not supported by our study. Fragmentation and associated habitat degradation are likely to have direct but subtle genetic and ecological consequences for recruitment. Fragmentation can lead to elevated inbreeding which reduces seedling vigour. Degradation of remnant forest patches reduces the suitability for late successional tree species leading to a double-blow to recruitment. Because low levels of inbreeding might cause more pronounced fitness consequences under more stressful habitat conditions [47] the common co-occurrence of both genetic and ecological stressors might be highly relevant for forest dynamics in fragmented landscapes. Such multiple co-occurring ecological, genetic and anthropogenic stressors were shown to become disproportionally effective when local population densities become low (so called allee effects) [48]. These represent insidious processes, which in unison may undermine recruitment of late successional tree species. Considering the global pattern of late successional trees species being replaced by few pioneer tree species in tropical forest fragments [20], our study highlights the need for future research to recognise potentially interacting stressors which will advance our understanding of forest dynamics in mosaic landscapes in general. Conservation management therefore needs to address both the genetic and ecological stressors and consider that these stressors are likely to be amplified by low population densities.

A failure to conserve this important biodiversity within forest patches is likely to have implications for ecosystem processes and services. Indeed, it was shown that rare tree species contribute disproportionately to the maintenance of ecosystem processes [49]. Therefore a more holistic understanding of the threats to tree species in human modified landscapes, is urgently required if these landscapes are to sustain biodiversity and ecosystem services in the future.

Supporting Information

Figure S1 Scatterplots of *Dysoxylum malabricum* seedling height after 21 months of growth under nursery conditions against (A) individual inbreeding coefficient (Pearson product moment correlation coefficient $= -0.139$, $t = -2.36$, $df = 281$, p-value $= 0.02$) and (B) kinship of parent pairs (Pearson product moment correlation coefficient $= -0.079$, $t = -1.32$, $df = 271$, p-value $= 0.19$).

Appendix S1 Methods of genotyping and genetic analysis.

Appendix S2 Dendrological estimates.

Table S1 Details of *Dysoxylum malabaricum* seed sampling.

Table S2 Data on degradation and reproductive success of *Dysoxylum malabaricum* averaged per forest patch.

Acknowledgments

We thank the hard-working field assistants Navin H. K., Monappa C.S., Chengappa S.K. and particularly Range G., Harsha C. and the late Umesh who took care for the nursery experiment in the absence of S.A.I. The painstaking single DNA extractions conducted by Sandeep S. and Shruthi J. are esteemed a lot. The constructive comments of three anonymous referees helped to improve the manuscript and are appreciated a lot. Fragment analysis was conducted at the Genetic Diversity Centre (GDC) of ETH Zürich. We thank David Burslem for useful comments on this manuscript.

Author Contributions

Conceived and designed the experiments: SAI JG GR RUS CJK. Performed the experiments: SAI. Analyzed the data: SAI CJK. Contributed reagents/materials/analysis tools: JG GR RUS CGK. Wrote the paper: SAI JG GR RUS CJK.

References

1. Naidoo R, Balmford A, Costanza R, Fisher B, Green RE, et al. (2008) Global mapping of ecosystem services and conservation priorities. Proc Natl Acad Sci U S A 105: 9495–9500.
2. Brooks TM, Bakarr MI, Boucher T, Da Fonseca GAB, Hilton-Taylor C, et al. (2004) Coverage provided by the global protected-area system: Is it enough? Bioscience 54: 1081–1091.
3. Wright SJ, Muller-Landau HC (2006) The Future of Tropical Forest Species. Biotropica 38: 287–301.
4. Tabarelli M, Da Silva MJC, Gascon C (2004) Forest fragmentation, synergisms and the impoverishment of neotropical forests. Biodivers Conserv 13: 1419–1425.
5. Brook BW, Sodhi NS, Bradshaw CJA (2008) Synergies among extinction drivers under global change. Trends in Ecology & Evolution 23: 453–460.
6. Laurance WF, Useche DC (2009) Environmental synergisms and extinctions of tropical species. Conserv Biol 23: 1427–1437.
7. Kettle CJ, Hollingsworth PM, Jaffré T, Moran B, Ennos RA (2007) Identifying the early genetic consequences of habitat degradation in a highly threatened tropical conifer, *Araucaria nemorosa* Laubenfels. Mol Ecol 16: 3581–3591.
8. Dick CW, Hardy OJ, Jones FA, Petit RJ (2008) Spatial Scales of Pollen and Seed-Mediated Gene Flow in Tropical Rain Forest Trees. Tropical Plant Biology 1: 20–33.
9. Cascante A, Quesada M, Lobo JJ, Fuchs EA (2002) Effects of dry tropical forest fragmentation on the reproductive success and genetic structure of the tree *Samanea saman*. Conserv Biol 16: 137–147.
10. Breed MF, Gardner MG, Ottewell KM, Navarro CM, Lowe AJ (2012) Shifts in reproductive assurance strategies and inbreeding costs associated with habitat fragmentation in Central American mahogany. Ecol Lett 15: 444–452.
11. Breed MF, Marklund MH, Ottewell KM, Gardner MG, Harris JB, et al. (2012) Pollen diversity matters: revealing the neglected effect of pollen diversity on fitness in fragmented landscapes. Mol Ecol 21: 5955–5968.

12. Ghazoul J (2005) Pollen and seed dispersal among dispersed plants. Biological Reviews 80: 413.
13. Markl JS, Schleuning M, Forget PM, Jordano P, Lambert JE, et al. (2012) Meta-Analysis of the Effects of Human Disturbance on Seed Dispersal by Animals. Conserv Biol 26: 1072–1081.
14. Galetti M, Guevara R, Côrtes MC, Fadini R, Von Matter S, et al. (2013) Functional Extinction of Birds Drives Rapid Evolutionary Changes in Seed Size. Science 340: 1086–1090.
15. Uriarte M, Bruna EM, Rubim P, Anciães M, Jonckheere I (2010) Effects of forest fragmentation on the seedling recruitment of a tropical herb: assessing seed vs. safe-site limitation. Ecology 91: 1317–1328.
16. Scariot A (2000) Seedling mortality by litterfall in Amazonian forest fragments. Biotropica 32: 662–669.
17. Hobbs RJ (2001) Synergisms among habitat fragmentation, livestock grazing, and biotic invasions in southwestern Australia. Conserv Biol 15: 1522–1528.
18. Benitez-Malvido J, Martinez-Ramos M (2003) Influence of edge exposure on tree seedling species recruitment in tropical rain forest fragments. Biotropica 35: 530–541.
19. Laurance WF, Nascimento HEM, Laurance SG, Andrade AC, Fearnside PM, et al. (2006) Rain forest fragmentation and the proliferation of successional trees. Ecology 87: 469–482.
20. Tabarelli M, Peres CA, Melo FP (2012) The 'few winners and many losers' paradigm revisited: Emerging prospects for tropical forest biodiversity. Biol Conserv 155: 136–140.
21. Ismail SA, Ghazoul J, Ravikanth G, Uma Shaanker R, Kushalappa CG, et al. (2012) Does long-distance pollen dispersal preclude inbreeding in tropical trees? Fragmentation genetics of *Dysoxylum malabaricum* in an agro-forest landscape. Mol Ecol 21: 5484–5496.
22. Kumar R, Ved DK, editors (2000) 100 Red listed Medicinal Plants of Coservation Concern in Southern India. Bangalore: Foundation for Revitalization of Local Health Traditions.
23. Garcia CA, Bhagwat SA, Ghazoul J, Nath CD, Nanaya KM, et al. (2010) Biodiversity Conservation in Agricultural Landscapes: Challenges and Opportunities of Coffee Agroforests in the Western Ghats, India. Conserv Biol 24: 479–488.
24. Kushalappa CG, Bhagwat SA (2001) Sacred groves: biodiversity, threats and conservation. In: Shaanker RU, Ganeshaiah KN, Bawa KS, editors. Forest genetic resources: status, threats, and conservation strategies. New Delhi: Oxford and IBH Publishing. pp. 21–29.
25. Bhagwat SA, Kushalappa CG, Williams PH, Brown ND (2005) Landscape approach to biodiversity conservation of sacred groves in the Western Ghats of India. Conserv Biol 19: 1853–1862.
26. Bhagwat SA, Kushalappa CG, Williams PH, Brown ND (2005) The role of informal protected areas in maintaining biodiversity in the Western Ghats of India. Ecol Soc 10.
27. Garcia CA, Pascal JP (2006) Sacred forests of Kodagy : Ecological value and social role. In: Cederlöf G., K S, editors. Ecological nationalisms : nature, livelihoods, and identities in South Asia. Seattle: University of Washington Press. pp. 199–229.
28. Chandrakanth MG, Bhat MG, Accavva MS (2004) Socio-economic changes and sacred groves in South India: Protecting a community-based resource management institution. Natural Resources Forum 28: 102–111.
29. Ganesh T, Davidar P (2001) Dispersal modes of tree species in the wet forests of southern Western Ghats. Curr Sci 80: 394–399.
30. Raman TRS, Mudappa D (2003) Correlates of hornbill distribution and abundance in rainforest fragments in the southern Western Ghats, India. Bird Conserv Int 13: 199–212.
31. Kramer AT, Ison JL, Ashley MV, Howe HF (2008) The Paradox of Forest Fragmentation Genetics. Conserv Biol 22: 878–885.
32. (1952) The Wealth of India : Raw Materials. New Dehli: C.S.I.R. 236 p.
33. Marshall TC, Slate J, Kruuk LEB, Pemberton JM (1998) Statistical confidence for likelihood-based paternity inference in natural populations. Mol Ecol 7: 639–655.
34. Loiselle BA, Sork VL, Nason J, Graham C (1995) Spatial genetic structure of a tropical understory shrub, *Psychotria officinalis* (rubiaceae). Am J Bot 82: 1420–1425.
35. Ritland K (1996) Estimators for pairwise relatedness and individual inbreeding coefficients. Genet Res 67: 175–185.
36. Hardy OJ, Vekemans X (2002) SPAGEDi: a versatile computer program to analyse spatial genetic structure at the individual or population levels. Mol Ecol Notes 2: 618–620.
37. R Development Core Team (2011) R: A Language and Environment for Statistical Computing. Vienna, Austria: R Foundation for Statistical Computing.
38. Lemmon PE (1956) A spherical densiometer for estimating forest overstory density. For Sci 2: 314–320.
39. Eggleton P, Bignell DE, Sands WA, Mawdsley NA, Lawton, et al. (1996) The diversity, abundance and biomass of termites under differing levels of disturbance in the Mbalmayo Forest Reserve, southern Cameroon. Philosophical Transactions of the Royal Society of London Series B-Biological Sciences 351: 51–68.
40. Jones D, Susilo F, Bignell D, Hardiwinoto S, Gillison A, et al. (2003) Termite assemblage collapse along a land-use intensification gradient in lowland central Sumatra, Indonesia. J Appl Ecol 40: 380–391.
41. Alves WF, Mota A, Lima RAA, Bellezoni R, Vasconcellos A (2011) Termites as Bioindicators of habitat Quality in the caatinga, Brazil: Is There agreement Between Structural habitat variables and the Sampled assemblages? Neotrop Entomol 40: 39–46.
42. Angeloni F, Ouborg NJ, Leimu R (2011) Meta-analysis on the association of population size and life history with inbreeding depression in plants. Biol Conserv 144: 35–43.
43. Chandran MDS (1997) On the ecological history of the Western Ghats. Curr Sci 73: 146–155.
44. Chandrashekara UM, Ramakrishnan PS (1994) Vegetation and Gap Dynamics of a Tropical Wet Evergreen Forest in the Western Ghats of Kerala, India. J Trop Ecol 10: 337–354.
45. Lee KE, Wood TG (1971) Termites and soils. London, UK, Academic Press.
46. Didham RK, Ghazoul J, Stork NE, Davis AJ (1996) Insects in fragmented forests: A functional approach. Trends Ecol Evol 11: 255–260.
47. Fox CW, Reed DH (2011) Inbreeding depression increases with environmental stress: An experimental study and meta-analysis. Evolution 65: 246–258.
48. Berec L, Angulo E, Courchamp F (2006) Multiple Allee effects and population management. Trends Ecol Evol 22: 185–191.
49. Mouillot D, Bellwood DR, Baraloto C, Chave J, Galzin R, et al. (2013) Rare species support vulnerable functions in high-diversity ecosystems. PLoS Biol 11: e1001569.

4

Hierarchical Dynamics of Ecological Communities: Do Scales of Space and Time Match?

David G. Angeler*, Emma Göthe, Richard K. Johnson

Department of Aquatic Sciences and Assessment, Swedish University of Agricultural Sciences, Uppsala, Sweden

Abstract

Theory posits that community dynamics organize at distinct hierarchical scales of space and time, and that the spatial and temporal patterns at each scale are commensurate. Here we use time series modeling to investigate fluctuation frequencies of species groups within invertebrate metacommunities in 26 boreal lakes over a 20-year period, and variance partitioning analysis to study whether species groups with different fluctuation patterns show spatial signals that are commensurate with the scale-specific fluctuation patterns identified. We identified two groups of invertebrates representing hierarchically organized temporal dynamics: one species group showed temporal variability at decadal scales (slow patterns of change), whilst another group showed fluctuations at 3 to 5-year intervals (faster change). This pattern was consistently found across all lakes studied. A spatial signal was evident in the slow but not faster-changing species groups. As expected, the spatial signal for the slow-changing group coincided with broad-scale spatial patterns that could be explained with historical biogeography (ecoregion delineation, and dispersal limitation assessed through a dispersal trait analysis). In addition to spatial factors, the slow-changing groups correlated with environmental variables, supporting the conjecture that boreal lakes are undergoing environmental change. Taken together our results suggest that regionally distinct sets of taxa, separated by biogeographical boundaries, responded similarly to broad-scale environmental change. Not only does our approach allow testing theory about hierarchically structured space-time patterns; more generally, it allows assessing the relative role of the ability of communities to track environmental change and dispersal constraints limiting community structure and biodiversity at macroecological scales.

Editor: Adam Siepielski, University of San Diego, United States of America

Funding: This study was supported by the Swedish Environmental Protection Agency. The funders had no role in study design, data collection and analysis, decision to publish, or preparation of the manuscript.

Competing Interests: The authors have declared that no competing interests exist.

* E-mail: david.angeler@slu.se

Introduction

Theory posits that ecological communities consist of species groups that operate in different scaling regimes, wherein the sets of abiotic and biotic organizing variables differ across hierarchical scales [1,2]. These sets of variables often change abruptly from one hierarchical scale to the next, creating discontinuous or cross-scale structure and non-linear patterns in the communities [3,4]. Critical to this hierarchical organization, and thus our understanding of ecological and other complex systems, is the duality of processes that operate both in space and time ("space-time duality"). This duality reflects the imprints of processes that act at spatial scales from local to regional to biome and temporal scales ranging from seconds to years to millennia [5]. For example, biological interactions entrain community assembly relatively rapidly at the local scale of habitats, biogeographical processes act over regional spatial and paleoecological temporal scales, and phylogenetic factors are mainly evident over spatially broad domains with slow dynamics [4]. These processes can also self-organize, in the sense that they can produce patterns that reinforce the processes that produced the patterns [6].

Ecologist have developed tools that allow for an assessment of the hierarchical, multiscale structure of ecological systems from either a spatial [7,8] or temporal perspective [9,10]. Most spatial studies have the drawback that they provide only single snapshots of community structure across spatial scales that limits an assessment of the dynamic component of the space-time duality. Although a method has been developed to evaluate the stability of abundances through time by assessing space-time interactions [11], temporal patterns at different scales are not explicit in this method. Temporal studies, on the other hand, have supported the theory that the dynamic system structure in terms of speeds of processes varies with scale [5]. For instance, Angeler et al. [12] have used time series modeling to study invertebrate community dynamics during a 20-year period in 26 lakes across Sweden. Consistent with theoretical predictions, one group of invertebrates showed decadal-scale variability associated with climatic variability and regional acid-deposition; that is, environmental factors that operate at regional spatial scales. In contrast, a second species group showed short-term (3–5 year) fluctuation patterns that were unrelated to environmental variables. Notwithstanding, time series modeling has also fallen short of dealing with the space-time duality by not accounting for spatial signals in the scale-specific temporal patterns. Thus neither spatial nor temporal modeling has thus far succeded to analyze the space-time duality of hierarchically organized systems simulatenously and in a coherent way. Identifying relevant scales of space and time influencing scale-specific patterns and processes is a pervasive problem in the ecological sciences [13].

The aim of this paper is to study the space-time duality in the hierarchical organization of communities by assessing spatial signals in the cross-scale structure of time series. More specifically, we test the hypothesis that spatial scales of observations are commensurate with the temporal scales of community dynamics at different hierarchies of ecological organization. That is, broad-scale spatial patterns should match temporal patterns that unfold on broader (e.g. decadal) time scales, and finer-scale spatial patterns should be associated with temporal processes on shorter (e.g. yearly) time scales. We test these conjectures using macroinvertebrate communities in lakes that serve as excellent model systems. First, previous research has shown that community dynamics follow partly theoretical predictions; that is, the temporal dynamics are hierarchically organized with faster and slower dynamics clearly operating in different temporal scaling regimes [12]. Second, lakes have a clear insular metacommunity structure [14], meaning that both their position in the landscape is fixed and their habitat boundaries clearly delineated. This assures that spatial signals in the temporal patterns of community dynamics arise from the dynamics of the invertebrates without being confounded by dynamical habitat changes in the landscape over time. Third, invertebrates are structured by spatial factors at different hierarchical scales [15,16], which allows testing whether extents of spatial scales matches temporal scales across ecological hierarchies. However, the underlying cause of a spatial signal can vary between hierarchical scales highlighting the need to closer scrutinize potential causes.

Spatial processes on community structure can take a variety of forms, including dispersal limitation and source-sink dynamics or mass effects, species extinctions and evolutionary processes [17,18]. These factors are not mutually exclusive because they are revealed at different spatial scales, commensurate with the time needed for these factors to manifest [19]. For example, evolutionary processes and extinctions can unfold over centuries, leading to historically contingent, biogeographical patterns in species diversity and community structure [20,21]. In fact, biogeographical patterns are evident in the community composition and structure of invertebrates in Swedish lakes, coinciding with ecoregions that differ both in vegetation, altitude and climate [22]. Biogeographical patterns, in turn, suggest that communities are dispersal limited at broad spatial scales [23]. However, dispersal limitation is not only manifested at biogeographical scales; it can influence community structure also at much smaller spatial scales such as boreal headwater stream catchments [16]. This highlights that dispersal limitation can emerge from different processes depending on the scale of observation. Because spatial signals in macroinvertebrate metacommunity structure have often been associated with dispersal limitation [15,16], we infer these processes based on multiple lines of evidence.

We expect that if historical biogeography in the form of ecoregion delineations is the spatial analogue to temporal community change at decadal scales [12], then the spatial signal in the temporal patterns should track broad-scale (i.e. ecoregion) patterns. In contrast, if according to theory shorter-term fluctuations are related to finer scale patterns in space, we expect the spatial structure to reflect, for example, within-ecoregion structures. If the spatial signals arise because the invertebrate communities are dispersal limited across these scales, then the temporal patterns of change should be associated with species with poor dispersal abilities (i.e. overland flight). Thus, in addition to assessing spatial signals across temporal scales qualitatively, we evaluate these signals quantitatively through a dispersal trait analysis [16]. This research will allow us to test the conjecture that spatial and temporal processes are commensurate across hierarchical scales of ecological systems. Because we analyze 20-year time series, a period of documented environmental change across Swedish lakes [24], we will also be able to assess whether the imprints of environmental change manifest distinctly across the temporal scales identified over this period.

Materials and Methods

Ethics Statement

All field sampling and laboratory analyses reported in this study are part of the Swedish National Lake Monitoring Program, and are therefore regulated by the Swedish Agency for Marine and Water Management (HaV). All data are made freely available to the public and no permission for use of the data is therefore required. It is also confirmed that the field studies did not involve endangered or protected species.

Study Area

Twenty-six lakes from the Swedish National Lake Monitoring Program were selected for this study based on longest available time series. These lakes were environmentally heterogeneous, spanning gradients in water clarity (Secchi depth: 1.1 to 10.7 m; water color: <0.01 to 7.98± mg Pt L^{-1}), acidity status (pH: 4.6 to 7.3; alkalinity: <0.01 to 0.30 meq L^{-1}) and lake size (0.11 to 5.43 km^2) [12]. Samples for water chemistry and littoral invertebrate assemblages have been collected during the last 20 years (1988–2007) from these lakes. For more information see [25].

Sampling

Standard sampling and analyses protocols for abiotic and biological variables were used throughout the period of our study. These standard protocols are certified and quality controlled through the Swedish Board for Accreditation and Conformity Assessment (SWEDAC). For this study we used water quality data that were obtained from surface water samples (taken at 0.5 m depth), which were collected once in summer (usually August) and autumn (October) at a mid-lake station in each lake. Previous research has shown that the mid-lake samples are representative for other lake areas [26].

Water was collected with a Plexiglas® sampler and kept cool during transport to the laboratory. Samples were analyzed for electrical conductivity, water temperature, and variables including those indicative of acidity (pH, alkalinity, SO_4^{2-} concentration), nutrients (Total P, NH_4-N), and water clarity (Secchi depth transparency, total organic carbon). All physicochemical analyses were done at the Department of Aquatic Sciences and Assessment following international (ISO) or European (EN) standards when available [27]. Measurement intervals and uncertainties for each variable can be found at [28].

Sampling of benthic invertebrates followed Swedish standards throughout the study period (SS-EN 27828). Benthic invertebrates were collected from wind-exposed, vegetation-free littoral habitats in late autumn (October–November) each year. Five samples were taken using standardized kick sampling with a hand net (0.5 mm mesh size). Each sample was taken by disturbing the bottom substratum for 20 seconds along a 1 m long stretch of the littoral region at a depth of c. 0.5 m; thus a total area of 1.25 m^2 was sampled in each lake. Samples were preserved in 70% ethanol in the field and processed in the laboratory by sorting against a white background with 10× magnification. Invertebrates were sorted, identified to the finest taxonomic unit possible and counted using dissecting and light microscopes by the same person, a trained taxonomist, throughout the study, thereby reducing a researcher-based bias in sample evaluation.

Statistical Analyses

Because no unified statistical procedure exists that allows testing the space-time duality of hierarchically organized systems, we used an approach that involves a sequence of independent statistical tests. The steps of this approach are summarized in Figure 1 and involved the determination of variability of invertebrates at different temporal scales (Step 1), the creation of environmental (Steps 2) and spatial matrices (Step 3), and the determination of environmental and spatial factors of community change at each scale using variance partitioning analyses (Step 4).

Revealing Temporal Variability of Invertebrates at Different Scales (Step 1)

We used a multivariate time series modeling approach, which breaks down the overall temporal structure of communities into independent temporal patterns shown by different groups of species within an assemblage [10]. This approach is based on Redundancy Analysis (RDA) where time is modeled by means of Principal Coordinate of Neighbor Matrices (PCNM). First, the analysis converts a time vector comprised of 20 time steps (i.e. 20 sampling years of invertebrates between 1988 and 2007) into a series of PCNM variables akin to a Fourier transform; that is, a number of different sine waves with different frequencies is obtained through conversion of the time vector (details in [29,30]). The number of obtained sine waves (or PCNM variables) depends on the length of the time vector that comprises the study period; thus a total of 12 PCNM variables were obtained for the analyses.

Through a forward selection procedure, these PCNMs are related to the community data sets by means of RDA. The RDA retains significant PCNM variables and these are linearly combined in ways to extract temporal patterns from the species matrices; that is, the RDA identifies species with similar temporal patterns in the species × time matrix and uses their temporal pattern to calculate a modeled species group trend for these species based on linearly combined PCNMs. The significance of the temporal patterns of all modeled species groups revealed by the RDA are tested by means of permutation tests.

The RDA relates each modeled species group trend with a significant canonical axis. The R software generates linear combination (lc) score plots, which visually present the modeled temporal patterns of species groups that are associated with each canonical axis. Based on the number of significant canonical axes, the number of modeled species groups with different temporal patterns can be deduced. The ecological relevance of these temporal patterns can be quantified, using adjusted R^2 values of the canonical axes. The overall temporal structure of the whole community can then be deduced from the number of significant canonical axes in the RDA models.

Because the canonical axes are orthogonal, the PCNM-RDA approach reveals temporal patterns of species groups that are independent from each other. Because temporal trends can be assessed at different scales, this method is useful for more accurately assessing the organization of ecological communities from a complex adaptive systems perspective [12]. This study shows that specific groups of invertebrate species show patterns of decadal change, while other species groups show fluctuation cycles on shorter temporal scales. Thus, rather than assessing change of the "overall" community, we infer patterns of species-group-specific trends that could indicate community organization at different temporal scales.

All relevant steps in the analysis, from conversion of the linear time vector to PCNM variables, to calculation of modeled species group trends to visual presentation of the results in form of lc score plots were carried out with the "quickPCNM" function implemented in R 2.6.0 statistical software package [31]. The calculations, which were based on Hellinger-transformed invertebrate abundance data [32], are therefore based exclusively on an automatic statistical procedure, thereby avoiding potential researcher-induced bias in model construction.

Identifying Spatial and Environmental Signals of Temporal Variability across Scales

Variance decomposition was used to assess the relative importance of environmental and spatial factors shaping the long-term assembly of species groups. Separate analyses were carried out for each modeled species groups identified across lakes. To account for the temporal structure of each species group explicitly in the analysis, species, environmental and spatial matrices were prepared in the following ways (Steps 2 and 3; Figure 1):

Although variance partitioning analysis is usually done on site by species abundance matrices, we used a modified approach in the present study using correlation coefficients rather than species abundances. Thus the species matrices were created as follows: Spearman rank correlations were first carried out to assess which species correlate significantly with the modeled species group trends for each lake identified through the RDA-PCNM time series modeling. Those species with significant correlations were considered as taxa contributing to temporal change and were used to define the "community composition" of the modeled species groups for each lake. The absolute values of their correlation scores with significant canonical axes from the RDA-PCNM time series modeling were extracted and compiled in species matrices, which were then used in the variance partitioning analysis. Specifically, because time series modeling revealed temporal patterns associated with two canonical axes, we created two species × site matrices to match the spatial analysis with these temporal patterns identified. That is, one species matrix was based on the correlation analysis of the first temporal patterns. It consisted of species as columns and lakes (sites) as rows. The values comprised the correlation coefficients resulting from the correlation between species raw abundances and the modeled temporal scores (linear combination scores) of canonical axis one from the time series models of each lake. The second species matrix was constructed in a similar way but using correlation coefficients derived from the correlations with the second canonical axis. Using the correlation scores associated with significant temporal patterns revealed through time series modeling, rather than time-averaged abundance data of each taxon, better accounts for the explicit temporal structure of species-group change over time in the metacommunity analyses. That is, the correlation scores, but not the time-averaged species abundance, capture the taxa contributions to the modeled temporal changes of species group; because the correlation scores reflect the relative strength of individual taxa contributing to species-group change, structural changes in abundance as a function of modeled group change is accounted for in the analyses.

Regarding the environmental matrix, we followed a similar approach as with the construction of the species matrices. Spearman rank correlations between environmental variables and the modeled species group trends were carried out to identify the set of abiotic variables contributing to species group change through time. Those variables with significant correlations were retained and the absolute values of their Spearman rank correlation coefficients were used as predictor variables in the environmental matrix. To avoid overfitting in posterior variance partitioning analyses, significantly correlated environmental variables were excluded from the matrix. Also here, the correlation

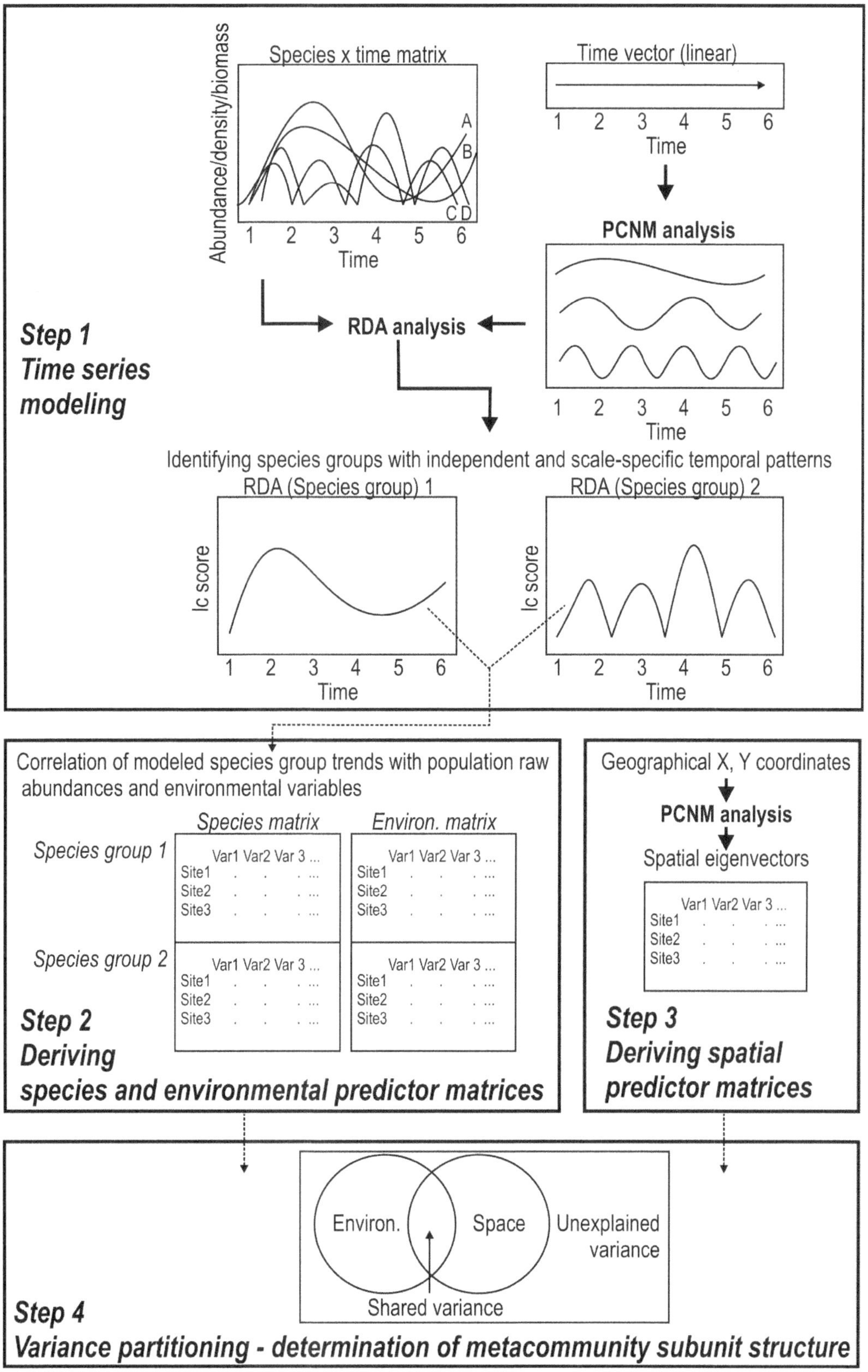

Figure 1. Outline of statistical approaches. Flow chart summarizing the sequence of statistical methods used in this study. Step 1 comprised the time series modeling, based on RDA where time was modeled with a PCNM approach, to determine the independent temporal trends of species groups of invertebrate communities in the lakes. Steps 2 and 3 describe the preparation of environmental and spatial matrices, which were then used in the variance partitioning (step 4). For further details see methods.

coefficients, rather than the time-averaged values of environmental variables, account explicitly for the relative strength of abiotic variables and their relative contribution to species-group change over time, and should therefore better reflect the temporal structure of species-environment associations compared to time-averaged units of environmental variables.

Regarding the spatial matrix, we carried out spatial analyses using the RDA-PCNM procedure outlined above. For these analyses the geographical X (north-south) and Y (east-west) coordinates of each lake composed the explanatory variable matrix, which was then converted into spatial PCNMs using the quickPCNM function. These spatial PCNMs were then related to the matrices of species groups by means of a forward selection in RDA. The spatial PCNMs that were retained in the model, and which therefore explained significant spatial structure in the data sets, were extracted and compiled in a spatial predictor matrix for the variance partitioning analyses.

To determine the relative importance of local environmental variables and spatial (dispersal) processes on the temporal dynamics of species groups, we conducted variation partitioning according to Peres-Neto et al. [33] using the varpart function, implemented in the vegan package, in R [34] (Step 4; Figure 1). Because both the environmental and species group matrices contained many zero values due to insignificant correlations of abiotic variables and individual species with modeled species group trends, the data were Hellinger-transformed prior to the analyses to avoid bias when data with double zero structure are used in ordination methods based on Euclidean distance [32]. The total variation can be partitioned into fractions which comprised of: 1) purely spatial; 2) purely environmental; 3) space uncorrected for environment; 4) environment uncorrected for space; 5) shared variance of space and environment, and 6) residual variance. From these fractions 1), 2) and 6) were the most relevant to our study questions and thus of immediate interest. Variation partitioning was conducted separately for all modeled species groups identified; that is, we associated the community composition of each modeled species group across lakes, revealed through correlation analysis (see above), with environmental and spatial factors.

We classified species according to traits related to flying strength following [35], allowing us to associate the spatial signal with either dispersal limitation or source-sink dynamics which is otherwise not possible in variance partitioning analyses [16]. According to this scheme, taxa can be classified according to the dispersal traits "adult flying strength" (AFS) and "female dispersal" (FD) and assigned to one of two dispersal ability groups: low and high dispersal ability. Low FD are genera that fly <1 km before laying eggs, high FD are genera that can fly >1 km before laying eggs, low AFS are genera that cannot fly into light breeze (weak flyers), and high AFS are genera that can fly into light breeze (strong flyers). Genera that were not included in [35] were excluded from the analysis because of limited knowledge of their dispersal ability (e.g. water mites) [36].

Results

Temporal Patterns of Individual Species Groups

The time series analyses using the RDA-PCNM approach detected significant temporal structure associated with different species groups of invertebrates in all of the twenty-six lakes between 1988 and 2007. Significant temporal structure was associated with canonical axes 1 and 2 in the RDA models, indicating the presence of two species groups with independent temporal fluctuation frequencies in the invertebrate communities across all lakes studied. The patterns associated with the first group of species explained on average >50% and the second species group on average <30% of the adjusted variance across all lakes in the models. The temporal structure of the first species group comprised change at broad temporal (decadal) scales (Fig. 2). These patterns cover temporal dynamics that were associated with environmental change over the 20-year study period (see below). We therefore refer to this first temporal frequency group of invertebrates as slowly changing ("slow") groups that track these slower changes in the environment. By contrast, the second frequency of species groups showed shorter-term periodicity at roughly 5-year intervals, presumably tracking faster ecological processes, and were more variable across lakes compared to the slow patterns, indicated by the standard deviations (Fig. 2); these groups will be referred to as faster-changing ("faster") groups.

Environmental Correlates of Group-specific Time Patterns

The importance and strength of environmental correlates of temporal patterns varied with species groups and were evident in the slow rather than the faster species groups across lakes (Table 1). Temporal patterns of the slow group correlated strongly and negatively with sulfate concentrations and electrical conductivity. Correlations of the other environmental variables with the temporal patterns of the slow group were less consistent across lakes. For example, positive correlations between the slow species group and pH and alkalinity were found in a subset of lakes. Subsets of lakes correlated negatively with changes in Secchi disc transparency and positively with total organic carbon. Significant negative correlations between the slow species group patterns and total phosphorus and NH_4-N were found for only selected lakes.

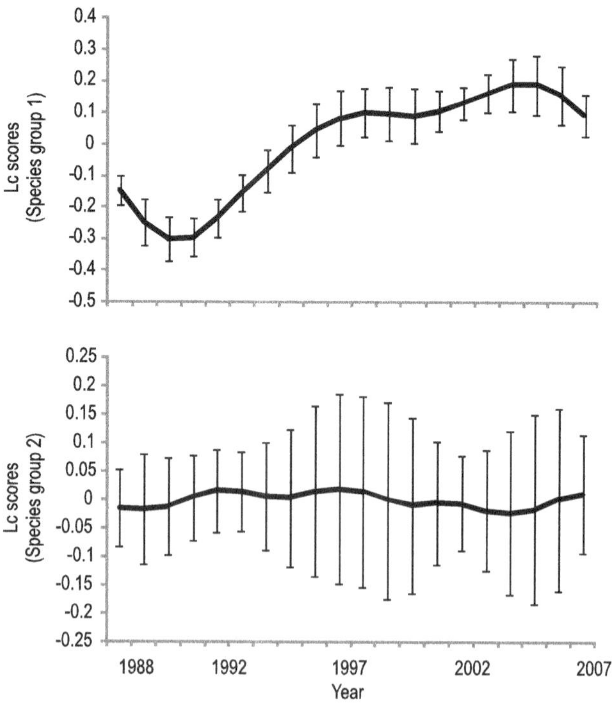

Figure 2. Modeled temporal species group patterns. Temporal patterns of species group 1 associated with RDA axis 1, and species group 2 associated with RDA axis 2 (lc, linear combination, scores) for benthic invertebrates between 1988 and 2007. Shown are the overall patterns (means ± SE) from 26 lakes with significant temporal structure. Both species groups have been identified and separated by the time series modeling analysis based on different temporal structure of the taxa that comprise a community.

Table 1. Spearman rank correlations between water quality variables and species group patterns, revealed by time series modeling, for invertebrate communities between 1988 and 2007.

Lake	SO_4 (meq L^{-1})	Electr. cond. (mS cm^{-1})	pH	TOC (mg L^{-1})	Secchi depth (m)	TP (μg L^{-1})	Alkalinity (meq L^{-1})	NH_4-N (μg L^{-1})	Water temp. (°C)
Abiskojaure	ns/0.87***	ns/0.80***	0.58*/ns	ns/ns	ns/ns	−0.69**/ns	ns/0.82***	ns/ns	ns/ns
Älgarydssjön	−0.80***/ns	−0.82***/ns	ns/0.52*	0.55*/ns	ns/ns	ns/−0.6**	ns/ns	0.77***/ns	ns/ns
Allgjuttern	−0.88***/ns	−0.90***/ns	ns/−0.49*	ns/ns	ns/ns 0.70***/0.47*	ns/ns	ns/ns	ns/0.59**	ns/ns
Brännträsket	−0.76***/ns	−0.47*/ns	0.65**/ns	ns/0.56*	ns/ns	ns/ns	ns/ns	ns/ns	0.51*/ns
Brunnsjön	−0.80***/ns	−0.79***/ns	ns/ns	0.57*/ns	−0.68**/ns	ns/ns	ns/ns	ns/0.48*	ns/ns
Bysjön	−0.85***/ns	−0.79***/ns	ns/ns	ns/ns	−0.65**/ns	ns/ns	0.55*/0.46*	ns/ns	ns/ns
Fiolen	−0.82***/ns	−0.81***/ns	0.66**/ns	ns/ns	ns/ns	ns/ns	0.68**/ns	0.55*/−0.46*	ns/ns
Fräcksjön	−0.87***/ns	−0.71***/−0.52*	0.62**/0.47*	0.62**/ns	ns/ns	−0.58**/ns	0.59**/ns	−0.57**/ns	0.58**/ns
Grissjön	−0.80***/ns	−0.74***/ns	ns/ns	0.46*/ns	−0.78***/ns	−0.61**/−0.49*	0.54*/−0.55*	ns/ns	ns/ns
Hagasjön	−0.76***/ns	−0.63**/ns	0.67**/ns	ns/ns	ns/ns	−0.61**/ns	0.56*/ns	ns/−0.57**	ns/ns
Harasjön	−0.88***/ns	−0.83***/ns	ns/ns	ns/ns	−0.72***/ns	ns/ns	ns/ns	0.76***/ns	0.54*/ns
Härsvatten	0.70***/0.59*	ns/ns 0.63**/0.74***	0.70***/−0.46*	ns/ns	ns/ns	−0.52*/ns	−0.77***/ns	ns/0.58**	ns/ns
Humsjön	ns/−0.67**	ns/ns	ns/ns	ns/ns	ns/−0.48*	ns/ns	ns/ns	ns/ns	ns/ns
Jutsajaure	ns/ns	ns/ns	0.51*/ns	ns/ns	ns/ns	ns/ns	ns/ns	ns/ns	ns/ns
Mäsen	−0.86***/ns	−0.63**/ns	0.58*/ns	0.55*/ns	ns/ns	−0.63**/ns	ns/ns	ns/ns	ns/ns
Övre Skärsjön	−0.81***/ns	ns/ns	0.79***/ns	0.54*/ns	ns/ns	ns/ns	ns/ns	ns/ns	0.53*/ns
Remmarsjön	−0.64**/ns	ns/ns	ns/ns	ns/ns	ns/ns	ns/ns	ns/ns	ns/ns	ns/ns
Rotehogstjärnen	−0.77***/ns	−0.76***/ns	ns/ns	0.65**/ns	−0.67**/ns	ns/ns	ns/ns	0.45*/ns	ns/ns
Sännen	−0.79***/ns	−0.78***/ns	0.53*/ns	0.49*/−0.47*	−0.63**/ns	−0.53*/ns	0.65**/ns	ns/ns	ns/ns
Skärgölen	−0.72***/ns	−0.62**/ns	ns/ns	ns/ns	ns/ns	ns/ns	ns/0.68**	ns/ns	ns/ns
Stensjön	ns/ns	−0.53*/0.58**	ns/ns	ns/ns	ns/ns	−0.67**/ns	ns/ns	ns/ns	ns/ns
Storasjö	−0.86***/ns	−0.91***/ns	ns/ns	0.52*/ns	ns/ns	ns/0.57**	ns/ns	ns/ns	ns/ns
Stora Skärsjön	−0.89***/ns	−0.86***/ns	ns/ns	ns/ns	−0.53*/ns	−0.54*/ns	0.71***/ns	−0.59**/ns	0.61**/ns
Stora Envättern	−0.79***/ns	−0.65**/−0.61**	ns/ns	0.49*/ns	−0.57*/ns	ns/ns	ns/ns	0.49*/0.56**	ns/ns
StorTjulträsket	ns/ns	ns/ns	ns/ns	ns/ns	ns/ns	ns/ns	ns/ns	ns/ns	ns/ns
Tväringen	ns/ns 0.57**/0.57**	−0.59**/ns	ns/ns	ns/ns	ns/ns	ns/ns	ns/0.57**	ns/ns	ns/ns

Shown are Spearman rank correlation coefficients (rho) and significance levels (* $P<0.05$; ** $P<0.01$; *** $P<0.001$) for species group 1/species group 2. ns, no significant correlation. Absolute values of rho were used as abiotic variables in the environmental predictor matrices for the variance partitioning analyses of species groups 1 and 2, respectively. Only variables with significant correlations have been included in the table. The table has been modified from [45].

Water temperature was generally a weak correlate of the temporal patterns of the slow group, perhaps because the temporal sampling resolution did not allow us to capture its full variability.

Taxonomic Composition of Species Groups

Spearman rank correlation analyses revealed that the number of taxa explaining the modeled temporal frequency patterns ranged 3–40 (slow groups) and 2–17 (faster groups). Many invertebrate taxa (e.g., *Caenis luctuosa*, *Caenis* spp., *Cloeon dipterum*, *Cloeon* spp. (Ephemeroptera), *Athripsodes bilineatus*, *Oecetis* sp. (Trichoptera), *Capnia artra* (Plecoptera), *Cladotanytarsus* sp., *Psectrocladius* sp., *Conchapelopia* sp., *Tanytarsus* sp. (Diptera), *Oulimnius* sp. (Coleoptera), *Helobdella stagnalis* (Annelida), Hydracarina, Ceratopogonidae) were found in the slow and faster species groups, but their strength and the significance of their contribution to species group patterns varied among lakes.

Spatial and Environmental Correlates of Species Group Patterns

The relative importance of environmental and spatial factors correlating with temporal species group change varied between species groups (Table 2). For the slow groups, both pure environmental (adj. $R^2 = 0.006$; $P<0.001$) and spatial factors (adj. $R^2 = 0.016$; $P<0.001$) were significant in the variance partitioning; however, the variance explained by these fractions was generally low. The spatial fraction captured the importance of broad-scale spatial processes, whereby the sets of species contributing to species group change differed along an east-west geographical gradient (Fig. 3). The midge *Tanypodinae* (correlation score with RDA axis 1: −0.94), the stonefly *Capnia atra* (−0.52) and the mayfly *Metretopus borealis* (−0.45) correlated with the temporal change of the slow groups in lakes towards the west (lakes with white squares in Fig. 3), while those in eastern lakes where associated with the isopod *Asellus aquaticus* (0.54), the mayfly *Leptophlebia vespertina* (0.52), and the mollusc *Pisidium* sp. (0.52) (lakes with black squares in Fig. 4). The residual variation that cannot be attributed to environmental and spatial factors was high for the slow species groups (>0.9). The temporal patterns found for the faster species groups were neither explained by environmental nor spatial factors (Table 2).

Dispersal Characteristics of Taxa in Species Groups

Despite different taxa explaining the slow and faster species groups across lakes, dispersal characteristics of these taxa were similar both between species groups and across lakes within each species group (indicated by the relatively low standard deviation) (Fig. 4). Approximately 60% of taxa explaining the temporal patterns of the slow and faster species groups consisted of females that fly >1 km before depositing their eggs, while ca. 40% fly <1 km (Fig. 4A). Overall, however, adult dispersal strength was low for both species groups with >80% of the taxa being weak flyers relative to the <10% that were strong flyers (Fig. 4B). No significant differences were detected when each dispersal class was compared between the slower and faster species groups (one-way ANOVA on arcsine square-root-transformed data: p>0.05).

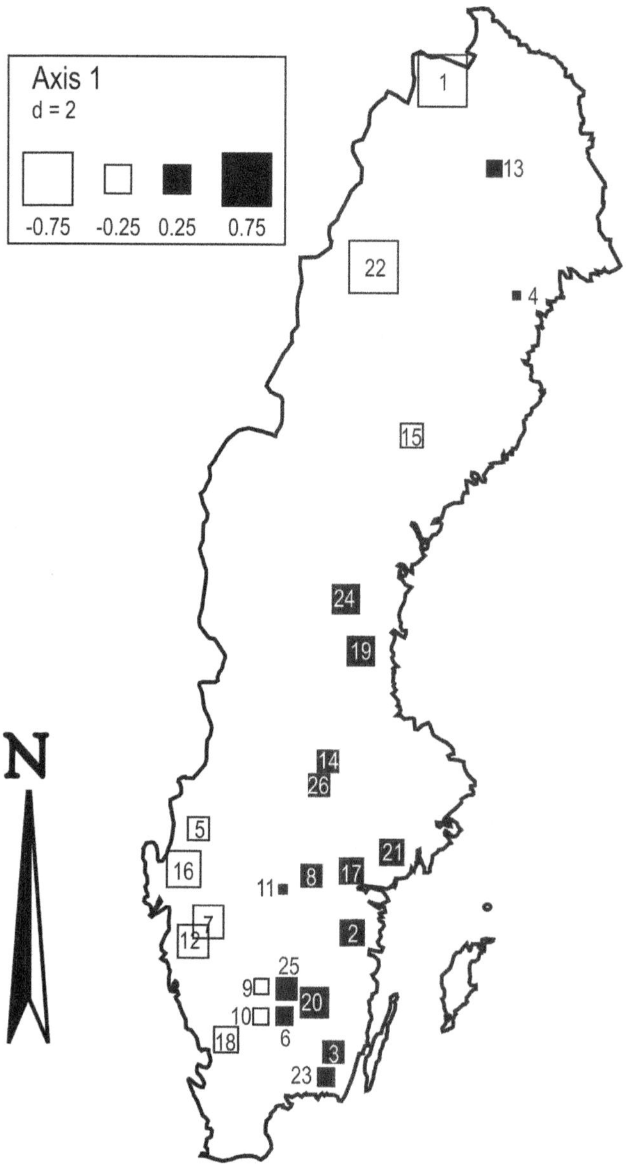

Figure 3. Spatial patterns of temporal patterns. Result from spatial analysis showing broad-scale spatial structure of invertebrates in the time-explicit metacommunity analysis. The size of the symbols are proportional to the lower-order PCNM vectors that describe this spatial structure and the color represents the sign (white = negative, black = - positive). Numbers within squares refer to lake identity: 1, Abiskojaure; 2, Allgjuttern, 3, Brunnsjön; 4, Brännträsket; 5, Bysjön; 6, Fiolen; 7, Fräcksjön; 8, Grissjön; 9, Hagasjön; 10, Harasjön; 11, Humsjön; 12, Härsvatten; 13, Jutsajaure; 14, Mäsen; 15, Remmarsjön; 16, Rotehogstjärnen; 17, Skärgölen; 18, Stora Skärsjön; 19, Stensjön; 20, Storasjö; 21, Stora Envättern; 22, Stor-Tjulträsket; 23, Sännen; 24, Tväringen; 25, Älgarydssjön; 26, Övre Skärsjön.

Discussion

Despite theory predicting that community dynamics are commensurate in space and time depending on the scales of ecological hierarchies at which these dynamics are explored [4,5], empirical testing of this space-time duality has so far been limited because of the lack of adequate time series with enough spatial and temporal resolution and statistical methods suitable for identifying cross-scale structure and hierarchically organized dynamics. Using 20 years of monitoring data of lakes situated at a broad spatial gradient in Sweden, we not only identified temporal dynamics in the form of slow and faster species groups across lakes that were consistent with theory, but also that when a spatial signal was detected in the time series the spatial pattern was commensurate with the temporal frequencies observed. That is, the patterns of change of the slow species groups of invertebrates were associated

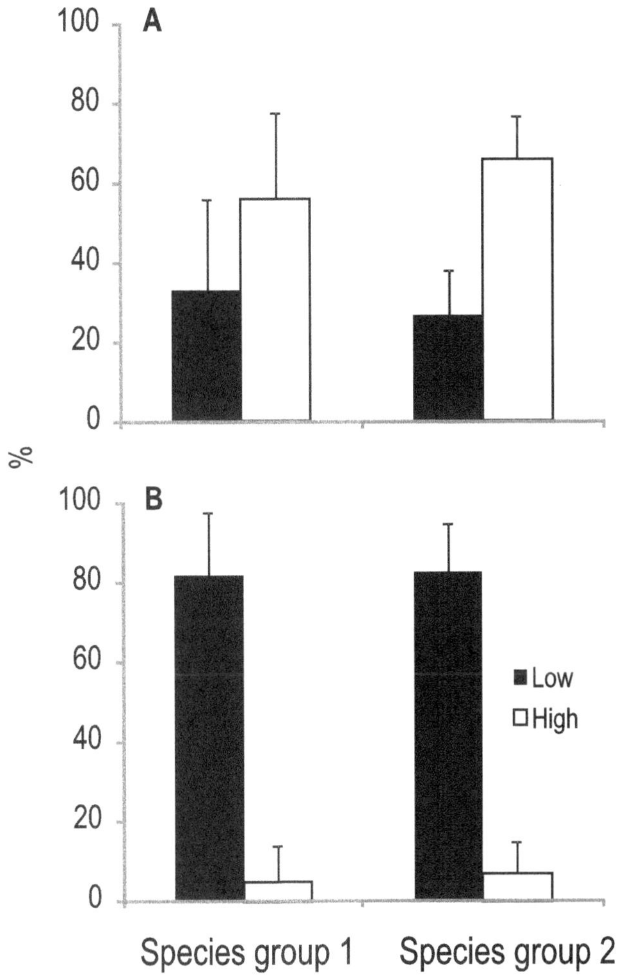

Figure 4. Dispersal trait analysis. Dispersal characteristics (A, female dispersal; B, adult dispersal) of taxa contributing to the temporal dynamics of species groups 1 and 2, respectively. Shown are the mean percentages ± standard deviations of species showing the traits across the study lakes.

with a broad-scale spatial pattern. What does this spatial signal mean?

Researchers have often invoked historically contingent effects to describe species diversity in plant and animal communities at broad spatial scales [20,21]. More specifically, evolutionary processes, extinctions and dispersal limitation have been described as biogeographical factors constraining species diversity and community structure in the absence of environmental pressures [18] or when biological adaptations to environmental change is lagging [23]. Evidence exists that the spatial signal in our study results from effects related to historical biogeography. For instance, a relatively good correspondence was found between ecoregion delineations and macroinvertebrate assemblages of the lake littoral zones in Sweden [22]. Ecoregions in Sweden can be delineated mainly based on altitude, vegetation characteristics and climatic factors. In northern latitudes a western high-elevation (arctic-alpine) ecoregion can be differentiated from an eastern low-altitudinal (boreal) ecoregion. Not only did the spatial signal in our analysis correspond well with the delineation between these ecoregion, also species that have been shown to typify these ecoregions (*Capnia atra* in the arctic-alpine ecoregion and *Asellus aquaticus* in the boreal ecoregion; [22]) contributed to the spatial

Table 2. Results from variance partitioning analyses.

	Species group 1		Species group 2	
Fractions	**adj. R^2**	**P**	**adj. R^2**	**P**
Environment uncorrected for space	0.010	0.246	0.007	0.313
Space uncorrected for environment	0.062	0.001	0.010	0.280
Pure environment	0.006	0.001	0.002	0.217
Pure space	0.016	0.001	0.001	0.234
Shared variance	0.046		0.001	
Residual	0.945		0.992	

Shown are the variance and significance levels by different fractions explaining the patterns of temporal change of two species groups of invertebrates, identified through the RDA-PCNM time series modeling, with contrasting temporal patterns. Note: no significance levels are calculated for shared and residual variance in the models.

pattern identified in our analysis. Southern Sweden comprises the boreonemoral and nemoral ecoregion with warmer climates and deciduous forests as the dominant vegetation type.

While the relatively good correspondence of the spatial signal with ecoregion patterns in this study supports the interpretation of historical biogeography influencing community dynamics, we acknowledge that the determination of underlying causes is difficult [44]. More specifically, spatial signals have been associated with diametrically opposing processes: dispersal limitation and mass effects [17], but these processes cannot be discerned in variance partitioning studies [37,38]. Mass effects explaining the spatial patterns at the biogeographical scale of our study would be counterintuitive, so we ascertained by means of a dispersal trait analysis of the invertebrates that the underlying process was dispersal limitation.

The dispersal traits analysis of the taxa that explained the slow species groups of invertebrates revealed that, despite the species groups being dominated by females that can fly>than 1 km before laying eggs, >80% of the taxa were weak flyers at the adult stage [35]. We acknowledge that our characterization of dispersal characteristics is relatively coarse because our current knowledge of invertebrate dispersal traits is still limited [36]. However, it is unlikely that the dominance of the slower groups of invertebrate species with females that disperse more than 1 km indicates mass effects at the macroecological scale of our study. Even strong flyers have been shown to be dispersal limited in small boreal catchments [16], a finding that is supported by some studies that have shown genetic differentiation within and between catchments that could reflect dispersal limitation [39–41]. Also, most taxa, especially those that were associated with the spatial signal in the variance partitioning (*Capnia artra, Leptophlebia vespertina, Metretopus borealis*) are either weak flyers [35] or they lack active overland dispersal traits (*Asellus aquatics, Pisidium* sp.). Taken together, these results suggest that dispersal between lakes across the observed west-east gradient is limited and that the spatial signal resulting from the spatial patterns of our study is most likely due to dispersal limitation. This interpretation is in agreement with an increasing body of evidence that dispersal limitation due to biogeographically defined boundaries rather than mass effects (source-sink dynamics) between macroecological spatial units describes historical community assembly of metazoans [23,42,43].

We expected to observe a spatial signal in the form of a finer-scale spatial pattern also in the faster species groups but the variance partitioning analysis failed to detect such an effect. Our hypothesis and underlying theory are therefore only partly supported. However, we acknowledge that this may be due to limitations with our approach. The temporal patterns of the faster species group showed fluctuation cycles on much shorter (roughly 5 years) time spans that explained less variance in the time series models compared to the slower species groups. We assume that while time series modeling was sensitive enough to identify distinct temporal patterns in the invertebrates, the sampling resolution for the faster species groups might have been either too coarse or the identified patterns too weak for detecting a significant spatial signal for these species groups. Also, the temporal patterns of the faster species groups were more variable across the lakes compared to the slower groups, suggesting that the faster group may be driven more by system intrinsic factors relative to an extrinsic control of the slower species group [12]. Weak patterns and high variability are, however, a common problem in studies based on survey data. These can often have a high residual variation due to the accumulation of noise related to sampling, ecosystem history and intrinsic variability [44]. We also took a conservative approach by correcting R^2-values by the number of explanatory variables [33], and the use of a sequence of statistical procedures for assessing the hierarchical space-time duality may have further introduced noise, decreasing the variance explained in the models. Consequently, the variance explained is lower compared to other variance partitioning studies [45,46]. Importantly, however, the main conclusions of our study are based on the formal analysis of patterns of R^2-values across species groups rather than on the actual values of these estimates.

We conclude by highlighting two aspects of our study related to methodology and inference. Regarding the former, our approach was based on the PCNM methodology, which has identified patterns of hierarchically structured community dynamics. Although, refinements of the original PCNM procedure in spatial or time series modeling using canonical ordination have been suggested [7,8], the PCNM method was powerful enough to identify patterns that were both credible and consistent with theoretical predictions [4,5]. We therefore regard the PCNM approach suitable for inference making in the present study.

Regarding inference, our primary interest was to test the space-time duality of ecological patterns across hierarchical scales. However, our results allowed for inferences beyond this goal. Having included both spatial and environmental factors in the analysis provided opportunities to assess the relative importance of both factors in long-term community change. We found that not only space but also environmental variables were associated with the temporal dynamics of the slow group of invertebrates. This scale-specific imprint of environmental variables has also been found in previous studies and was due to broad-scale processes related to reduced acid deposition and climatic variability [12]. These patterns were coherent across Swedish lakes suggesting that the patterns of temporal change were not contingent on the biogeographical patterns associated with ecoregions in Sweden [47]. Previous studies have shown that estimates of the relative importance of partial space or environmental components may fail to accurately represent environmental and spatial components of community variation [48,49], limiting inference about the role of niche-based or neutral processes in community assembly using variance partitioning analysis. We do not expect such a problem in our analysis because the environmental and spatial effects have been assessed at different temporal scales. That is, we accounted explicitly for time in the analysis covering a dynamic temporal component of environment and species over the 20-year study period. By contrast, space comprised a static variable. These different temporal windows of measurement reduce the risk that environmental and spatial predictors confound each other in the analysis. Based on previous studies that temporal trends were synchronized across Swedish lakes and the results from the present study, we conclude that regionally distinct sets of taxa responded similarly to broad-scale environmental change. More generally, our results show how an assessment of the hierarchically structured space-time duality can contribute to the debate of the relative role of the ability of communities to track environmental change (niche-based processes) versus dispersal constraints (neutral processes) limiting community structure and biodiversity at macro-ecological scales.

Acknowledgments

We acknowledge the constructive criticism received by the reviewers that helped improve the manuscript.

Author Contributions

Conceived and designed the experiments: DGA. Analyzed the data: DGA EG. Wrote the paper: DGA EG RKJ.

References

1. Wiens JA (1989) Spatial scaling in ecology. Functional Ecology 3: 385–397.
2. Levin SA (1992) The problem of pattern and scale in ecology. Ecology 73: 1943–1967.
3. Holling CS (1992) Cross-scale morphology, geometry, and dynamics of ecosystems. Ecological Monographs 62: 447–502.
4. Allen CR, Garmestani AS, Havlicek TD, Marquet PA, Peterson GD et al. (2006) Patterns in body mass distributions: sifting among alternative hypotheses. Ecology Letters 9: 630–643.
5. Gunderson LH, Holling CS (2002) Panarchy: understanding transformations in human and natural systems: Island Press.
6. Kauffman SA (1993) Origins of order: self-organization and selection in evolution: Oxford University Press.
7. Dray S, Legendre P, Peres-Neto PR (2006). Spatial modelling: a comprehensive framework for principal coordinate analysis of neighbor matrices (PCNM). Ecological Modelling 196: 483–493.
8. Blanchet FG, Legendre P, Borcard D (2008) Modelling directional spatial processes in ecological data. Ecological Modelling 215: 325–336.
9. Keitt TH, Fischer J (2006) Detection of scale-specific community dynamics using wavelets. Ecology 87: 2895–2904.
10. Angeler DG, Viedma O, Moreno JM (2009) Statistical performance and information content of time lag analysis and redundancy analysis in time series modeling. Ecology 90: 3245–3257.
11. Laliberté E, Paquette A, Legendre P, Bouchard A (2009) Assessing the scale-specific importance of niches and other spatial processes on beta diversity: a case study from a temperate forest. Oecologia 159: 377–388.
12. Angeler DG, Drakare S, Johnson RK (2011) Revealing the organization of complex adaptive systems using multivariate time series modeling. Ecology and Society16 (3): 5. Available: http://www.ecologyandsociety.org/vol16/iss3/art5/ Accessed 2013 January 25.
13. Adler PB, Lauenroth WK (2003) The power of time: spatiotemporal scaling of species diversity. Ecology Letters 6: 749–756.
14. Logue JB, Mouquet N, Peter H, Hillebrand H (2011) Empirical approaches to metacommunities: a review and comparison with theory. Trends in Ecology and Evolution 26: 482–491.
15. Heino J, Muotka T, Paavola R (2003) Determinants of macroinvertebrate diversity in headwater streams: regional and local influences. Journal of Animal Ecology 72: 425–434.
16. Göthe E, Angeler DG, Sandin L (2013) Metacommunity structure in a small boreal stream network. Journal of Animal Ecology 82: 449–458.
17. Leibold MA, Holyoak M, Mouquet N, Amarasekare P, Chase JM et al. (2004) The metacommunity concept: a framework for multi-scale community ecology. Ecology Letters 7: 601–613.
18. Hubbell S (2001) The Unified Neutral Theory of Biodiversity and Biogeography: Princeton University Press.

19. Ricklefs RE (2004) A comprehensive framework for global patterns in biodiversity. Ecology Letters 7: 1–15.
20. McGlone MS (1996) When history matters: scale, time, climate and tree diversity. Global Ecology and Biogeography Letters 5: 309–314.
21. Graham CH, Moritz C, Williams SE (2006) Habitat history improves prediction of biodiversity in rainforest fauna. Proceeding of the National Academy of Sciences USA 103: 632–636.
22. Johnson RK (2000) Spatial congruence between ecoregions and littoral macroinvertebrate assemblages. Journal of the North American Benthological Society 19: 475–486.
23. Svenning JC, Skov F (2007) Could the tree diversity pattern in Europe be generated by postglacial dispersal limitation? Ecology Letters 10: 453–460.
24. Angeler DG, Drakare S (2013) Tracing alpha, beta, and gamma diversity responses to environmental change in boreal lakes. Oecologia: in press doi: 10.1007/s00442-012-2554-y.
25. Website of the Department of Aquatic Sciences and Assessment. Available: http://www.slu.se/en/faculties/nl/about-the-faculty/departments/department-of-aquatic-sciences-and-assessment/data-host/Accessed 2013 January 25.
26. Göransson E, Johnson RK, Wilander A (2004) Representativity of a mid-lake surface water chemistry sample. Environmental Monitoring and Assessment 95: 221–238.
27. Wilander A, Johnson RK, Goedkoop W (2003) Riksinventering 2000. En synoptisk studie av vattenkemi och bottenfauna i svenska sjöar och vattendrag: Institutionen för Miljöanalys, Uppsala.
28. Website of the Department of Aquatic Sciences and Assessment. Available: http://www.slu.se/en/faculties/nl/about-the-faculty/departments/department-of-aquatic-sciences-and-assessment/laboratories/geochemical-laboratory/water-chemical-analyses/Accessed 2013 January 25.
29. Borcard D, Legendre P (2002) All-scale spatial analysis of ecological data by means of principal coordinates of neighbour matrices. Ecological Modelling 153: 51–68.
30. Borcard D, Legendre P, Avois-Jacquet C, Tuomisto H (2004) Dissecting the spatial structure of ecological data at multiple scales. Ecology 85: 826–1832.
31. R Development Core Team (2008) R: A language and environment for statistical computing. R Foundation for Statistical Computing, Vienna, Austria. ISBN 3-900051-07-0. Available: http://cran.r-project.org Accessed 2012 June 9.
32. Legendre P, Gallagher ED (2001) Ecologically meaningful transformations for ordination of species data. Oecologia 129: 271–280.
33. Peres-Neto PR, Legendre P, Dray S, Borcard D (2006) Variation partitioning of species data matrices: estimation and comparison of fractions. Ecology 87: 2614–2625.
34. Oksanen J, Blanchet FG, Kindt R, Legendre P, Minchin PR et al. (2011) Vegan: Community ecology package. R package version 2.0-0. Available: http://CRAN.R-project.org/package = vegan Accessed 2012 June 9.
35. Poff NL, Olden JD, Vieira NKM, Finn DS, Simmons MP et al. (2006) Functional trait niches of North American lotic insects: traits-based ecological applications in light of phylogenetic relationships. Journal of the North American Benthological Society 25: 730–755.
36. Bohonak AJ, Jenkins DG (2003) Ecological and evolutionary significance of dispersal by freshwater invertebrates. Ecology Letters 6: 783–796.
37. Cottenie K (2005) Integrating environmental and spatial processes in ecological community dynamics. Ecology Letters 8: 1175–1182.
38. Ng I, Carr C, Cottenie K (2009) Hierarchical zooplankton metacommunities: distinguishing between high and limiting dispersal mechanisms. Hydrobiologia 619: 133–143.
39. Monaghan MT, Spaak P, Robinson CT, Ward JV (2002) Population genetic structure of 3 alpine stream insects: influences of gene flow, demographics, and habitat fragmentation. Journal of the North American Benthological Society 21: 114–131.
40. Hughes JM, Mather PB, Hillyer MJ, Cleary C, Peckarsky B (2003) Genetic structure in a montane mayfly *Baetis bicaudatus* (Ephemeroptera: Baetidae), from the Rocky Mountains, Colorado. Freshwater Biology 48: 2149–2162.
41. Finn DS, Theobald DM, Black WC, Poff NL (2006) Spatial population genetic structure and limited dispersal in a Rocky Mountain alpine stream insect. Molecular Ecology 15: 3553–3566.
42. Hawkins BA, Porter EE, Diniz-Filho JAF (2003) Productivity and history as predictors of the latitudinal diversity gradient of terrestrial birds. Ecology 84: 1608–1623.
43. O'Dwyer JP, Green JL (2010) Field theory for biogeography: a spatially explicit model for predicting patterns of biodiversity. Ecology Letters 13: 87–95.
44. Leibold MA, Economo EP, Peres-Neto P (2010) Metacommunity phylogenetics: separating the roles of environmental filters and historical biogeography. Ecology Letters 13: 1290–1299.
45. Beisner BE, Peres-Neto PR, Lindström ES, Barnett A, Longhi ML (2006) The role of environmental and spatial processes in structuring lake communities from bacteria to fish. Ecology 87: 2985–2991.
46. De Bie T, De Meester L, Brendonck L, Martens K, Goddeeris B et al. (2012) Body size and dispersal mode as key traits determining metacommunity structure of aquatic organisms. Ecology Letters 15: 740–747.
47. Angeler DG, Johnson RK (2012) Temporal scales and patterns of invertebrate biodiversity dynamics in boreal lakes recovering from acidification. Ecological Applications 22: 1172–1186.
48. Gilbert B, Bennett JR (2010) Partitioning variation in ecological communities: do the numbers add up? Journal of Applied Ecology 47: 1071–1082.
49. Smith TW, Lundholm JT (2010) Variation partitioning as a tool to distinguish between niche and neutral processes. Ecography 33: 648–655.

5

The Impact of Spatial Scales and Spatial Smoothing on the Outcome of Bayesian Spatial Model

Su Yun Kang[1,2]*, James McGree[1,2], Kerrie Mengersen[1,2]

1 Mathematical Sciences School, Queensland University of Technology, Brisbane, Queensland, Australia, **2** Cooperative Research Centre for Spatial Information, Melbourne, Victoria, Australia

Abstract

Discretization of a geographical region is quite common in spatial analysis. There have been few studies into the impact of different geographical scales on the outcome of spatial models for different spatial patterns. This study aims to investigate the impact of spatial scales and spatial smoothing on the outcomes of modelling spatial point-based data. Given a spatial point-based dataset (such as occurrence of a disease), we study the geographical variation of residual disease risk using regular grid cells. The individual disease risk is modelled using a logistic model with the inclusion of spatially unstructured and/or spatially structured random effects. Three spatial smoothness priors for the spatially structured component are employed in modelling, namely an intrinsic Gaussian Markov random field, a second-order random walk on a lattice, and a Gaussian field with Matérn correlation function. We investigate how changes in grid cell size affect model outcomes under different spatial structures and different smoothness priors for the spatial component. A realistic example (the Humberside data) is analyzed and a simulation study is described. Bayesian computation is carried out using an integrated nested Laplace approximation. The results suggest that the performance and predictive capacity of the spatial models improve as the grid cell size decreases for certain spatial structures. It also appears that different spatial smoothness priors should be applied for different patterns of point data.

Editor: Gerardo Chowell, Arizona State University, United States of America

Funding: The work has been supported by the Cooperative Research Centre for Spatial Information, whose activities are funded by the Australian Commonwealth's Cooperative Research Centres Programme. The funders had no role in study design, data collection and analysis, decision to publish, or preparation of the manuscript.

Competing Interests: The authors have the following interests. The work has been supported by the Cooperative Research Centre for Spatial Information. All authors are affiliated with the CRC for Spatial Information. There are no patents, products in development or marketed products to declare. This does not alter their adherence to all the PLOS ONE policies on sharing data and materials, as detailed online in the guide for authors.

* E-mail: s7.kang@qut.edu.au

Introduction

Spatial data are available in various forms; at point level, grid level or area level. In the context of epidemiological studies, area level data are usually utilized due to its availability. This is because some phenomena are expressed naturally as area level data such as contextual variables in social epidemiology. In addition, disease incidence is often aggregated to administrative districts in order to protect patient confidentiality. For convenience, the aggregated data are further used to study small-scale geographical variation. Consequences of this practice include loss of individual information and potential ecological fallacy [1], where the latter refers to the difference between individual and group level estimates of risk measures. The aggregated data may also suffer from changes in geographical boundaries over time which calls into question the value of any analyses. Another problem concerning the aggregated data is the modifiable area unit problem, which is defined as sensitivity of statistical results to the definition of geographical units over which data are collected [2]. For instance, various datasets may exhibit different spatial patterns when viewed at one spatial scale compared to another, which is known as a 'scale' effect [3].

In contrast, point level disease data contain desirable individual information and precise domicile addresses in some instances, alleviating the issue of ecological bias. However, they are often difficult to access due to confidentiality issues. Even if they are available, patients' residential locations have to be protected and are not allowed to be published, which has restricted the types of analyses that can be carried out on point level disease datasets. Another limitation is that the study of small-scale geographical variation is not practicable if using individual level disease data. As a compromise, we utilize point level disease data in this study but employ a grid level modelling approach to study the geographical variation of residual disease risk using regular grid cells. As a result, the issue of patient confidentiality and ecological bias are both addressed in this study.

We model the individual disease risk using a logistic model with the inclusion of spatially unstructured and/or spatially structured random effects. Geographical variation of residual disease risk is modelled using a spatial component that allows for the heterogeneity of random effects and borrows strength from neighboring grid cells. The grid cells are far smaller than the typical administrative districts and therefore allow for better specification and identification of spatial random effects. Many ecological responses of interest do not recognize areas or borders defined for administrative purposes, and thus a finer geographical scale of study is often more appropriate for ecological studies [4]. The findings are more relevant and specific to the local population in a finer geographical area.

Despite being less common than studying the geographical variation using the area level data, the grid level modelling approach has rapidly increased in popularity in recent years [5–8]. Modelling of disease data at a grid level is a desirable approach as it is geographically more accurate than using area level data and yet protects patient confidentiality. Other advantages include the formation of a generalized linear model and approximation of the covariance structure by a Markov random field, which eases computation [9,10]. The grid level modelling avoids the need to deal with the problem of changes in administrative boundaries over time. This approach has the flexibility of allowing the spatial scale at which the data are modelled to be manipulated to a practically, biologically, geographically or computationally sensible scale.

One of the challenges in the grid level modelling is the specification of an appropriate spatial scale for a specific spatial dataset. At present, not much is known about the impact of different spatial scales on the outcome of spatial models at different spatial patterns. Without repeating the analyses at multiple scales, it is difficult to know whether the findings at various scales are consistent. According to [11], selection of the spatial scale of analysis should be guided by the purpose of analysis, i.e., whether to draw conclusions at the individual level or the aggregated level. It is thus important to consider analyses at various spatial resolutions in order to identify the most appropriate geographical scale that contributes to significant findings for the problem at hand.

Given this identified challenge, the study has two main aims: (i) to investigate the impact of changes in spatial scale on model outcome for a set of spatial structures; (ii) to evaluate the performance of various Bayesian spatial smoothness priors for spatial dependence, namely an intrinsic Gaussian Markov random field (IGMRF), a second-order random walk (RW2D) on a lattice, and a Gaussian field with Matérn correlation function. Bayesian inference is carried out using integrated nested Laplace approximation (INLA) throughout the study.

We designed a simulation study and utilized a case study to fulfill the aims. The simulated datasets consist of point data with various spatial structures including inhomogeneous point patterns, patterns with local repulsion, patterns with local clustering, and patterns with local clustering in the presence of a larger-scale inhomogeneity. The case study involves the analysis of the Humberside data on childhood leukaemia and lymphoma. This dataset portrays a sparse spatial pattern with potential spatial clustering.

Methods

Model

Let X be a spatial point-based dataset embedded in an observation window S, which is discretized into $I = n_1 \times n_2$ regular grid cells. Let y_{ij} denote the event outcome of the j-th individual in the i-th grid. Here y_{ij} is a binary response that follows a Bernoulli distribution with probability of disease p_{ij}. The individual risk, p_{ij}, is modelled via the logistic regression model,

$$\text{logit}(p_{ij}) = \mu_j + u_i + v_i. \tag{1}$$

Spatial variation in the individual risk is modelled using different components including μ_j, u_i, and v_i, where μ_j refers to the intercept term for individual j, u_i is an unstructured term that accounts for unexplained variability in the model, and v_i is a spatially structured term that describes the effect of the location by assuming that geographically close areas are more similar than distant areas. In this study, three Bayesian spatial smoothness priors for the spatially structured effect are considered, namely an intrinsic Gaussian Markov random field (IGMRF), a second-order random walk (RW2D) on a lattice [12], and a Gaussian field with Matérn correlation function [13] which includes a range parameter. These three priors are chosen due to their popularity in spatial modelling [14].

The IID model

The IID model considers a fixed intercept and unstructured random effects u_i.

$$\text{IIDmodel}: \text{logit}(p_{ij}) = \mu_j + u_i.$$

The IID model defines u to be a vector of independent, identical and Gaussian distributed random variable (possibly scaled) with mean zero and unknown precision (inverse variance), τ_u:

$$\pi(\mathrm{u}\,|\,\tau_u) = \prod_{i=1}^{n_1 \times n_2} \frac{1}{\sqrt{2\pi}} \sqrt{s_i \tau_u} \exp\left(\frac{1}{2}(s_i \tau_u) u_i^2\right),$$

where $s_i > 0$ is an optional fixed scale. The precision parameter τ_u is assigned a Gamma prior.

The IGMRF model

In the IGMRF model corresponding to (1), the spatially unstructured component, u_i, is assumed to be independent and identically distributed (i.i.d.) and normally distributed with mean zero and unknown precision, τ_u. The spatially structured component, v_i, is given an IGMRF prior [12] with unknown precision, τ_v. Gamma priors are assigned to the precision parameters τ_u and τ_v.

An IGMRF for v_i is defined as

$$v_i | \mathrm{v}_{-i}, \tau_v \sim \mathcal{N}\left(\frac{1}{n_i} \sum_{i \sim k} v_k, \frac{1}{n_i \tau_v}\right),$$

where n_i is the number of neighbors of grid cell i, v_{-i} denotes all elements in v except for v_i, and $i \sim k$ indicates that the two grid cells are neighbors that share a common boundary. A sum-to-zero constraint is imposed on v_i to ensure identifiability of the intercept μ. See [12], [15] and [16] for further details. This model has been widely applied in disease mapping to study spatial variation of disease risk [17–19]. The neighborhoods in these papers were defined in terms of administrative districts, we consider a finer neighborhood structure in terms of (regular) grid cells, however.

The RW2D model

The RW2D model corresponding to (1) employs a different formulation for the spatially structured effect. Here, v_i is imposed an RW2D prior on the $n_1 \times n_2$ regular lattice, which is alternatively known as a second-order polynomial intrinsic GMRF [12]. This choice is motivated by its application in a discretized log Gaussian Cox process (LGCP) [20]. The model approximates a LGCP only when the grid cells are fine enough.

The RW2D model is defined on a regular grid (see Rue and Held [12], section 3.4.2). The full conditionals of the nodes in the interior (with obvious notation) of the regular grid are as follows

$$E(v_i|v_{-i},\tau_v)=\frac{1}{20}\left(8 \quad -2 \quad -1\right),$$

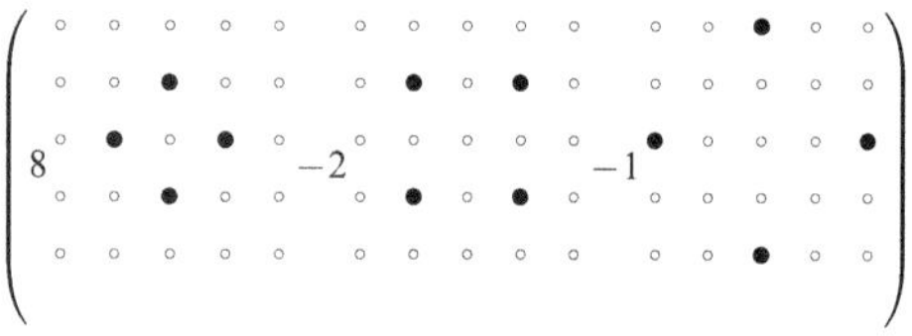

$$\mathrm{Prec}(v_i|v_{-i},\tau_v)=20\tau_v.$$

The precision τ_v is unknown. As stated by [20], the full conditionals are constructed to mimic the thin plate spline. Corrections to the boundary can be found by using the stencils in [21]. A sum-to-zero constraint is again imposed on the spatial term to ensure identifiability of μ. A Gamma prior is assigned to the precision parameter τ_v.

The MATERN2D model

The MATERN2D model corresponding to (1) is considered. The spatially structured effect v_i is imposed a prior as a Gaussian field with Matérn correlation function on the $n_1 \times n_2$ regular lattice. The Matérn isotropic correlation function on an infinite lattice is given as [13,22,23],

$$\mathrm{Corr}(d)=\frac{1}{2^{v-1}\Gamma(v)}\left(\frac{d}{r}\right)^{v}K_v\left(\frac{d}{r}\right),$$

where d is the separation distance, K_v is the modified Bessel function of the second kind of order v, $\Gamma(\cdot)$ is the Gamma-function, r is the range or distance parameter ($r>0$) which measures how quickly the correlations decay with distance, and v is the smoothness parameter ($v>0$). The latent field has marginal variance $1/\tau_v$ and range r. Gamma priors are assigned to both parameters. The Matérn model has great flexibility in modelling the spatial covariance due to the smoothness parameter v. A large value of v ($v\rightarrow\infty$) implies a smoother spatial process.

Computation

In light of the computational cost of Markov chain Monte Carlo (MCMC) methods for spatial inference, we adopt the integrated nested Laplace approximation (INLA) approach proposed by [20]. We note that MCMC might also be possible with desktop computing, but the Laplace approximation is adequate for our purposes. INLA performs approximate Bayesian inference for latent Gaussian models [24], which are defined in three stages as

$$y_{ij}|\zeta \sim \pi(y_{ij}|\zeta) \qquad \text{(Observation equation)}$$

$$\zeta|\theta \sim N(\mu(\theta),Q(\theta)^{-1}) \qquad \text{(Latent Gaussian field)}$$

$$\theta \sim \pi(\theta) \qquad \text{(Parameter model)}$$

where $Q(\theta)$ is the precision matrix of the Gaussian random vector ζ, which is sparse. The posterior can be written as

$$\pi(\zeta,\theta|y)\propto\pi(\theta)\pi(\zeta|\theta)\prod_{i\in I}\pi(y|\zeta,\theta).$$

The models considered in this study are regarded as latent Gaussian models by assigning $\zeta=\{\mu_j,u_i,v_i\}$, a Gaussian prior with precision matrix $Q(\theta)$. The precision parameters of $Q(\theta)=\{\tau_u,\tau_v\}$ are assigned a prior, Gamma distribution. The desired posterior marginals can be written as

$$\pi(\zeta_i|y)=\int\pi(\zeta_i|\theta,y)\pi(\theta|y)d\theta. \qquad (2)$$

The vector θ refers to the hyperparameters used in defining prior distributions for the precision of the Gaussian priors. The posterior marginals of θ are approximated by

$$\pi(\theta_i|y)=\int\pi(\theta|y)d\theta_{-i}. \qquad (3)$$

In order to estimate (2) and (3), nested approximations are constructed, and numerical integration is used to integrate out θ. The Laplace approximation to the posteriors of hyperparameters can be written as

$$\tilde{\pi}(\theta|y)\propto\frac{\pi(\zeta,\theta,y)}{\tilde{\pi}_G(\zeta|\theta,y)}\Big|_{\zeta=\zeta^*(\theta)}, \qquad (4)$$

where $\tilde{\pi}_G(\zeta|\theta,y)$ is the Gaussian approximation to the full conditional of ζ and $\zeta^*(\theta)$ is the mode of the Gaussian approximation for each ζ. We refer the reader to [20] for more details on INLA computation. See [25–29] for Bayesian inference using INLA in various applications. See also [10,30,31] on how the approximation of a LGCP on fine grids is carried out using INLA.

Computation in this study is performed in the R package, by calling the inla program. Two steps are taken to run the models. First, the linear predictor of a model is specified using the formula object in R. The specified model can then be run by calling the inla( ) function. We choose "strategy = laplace" to apply a Laplace approximation in (4) to estimate the marginals of the components of the latent field. The output of the inla( ) function generates various statistics such as marginal likelihood, deviance information criterion, effective number of parameters, predictive measures such as logarithmic score [32] and probability integral transform [33], useful to compare and validate models.

As an illustration, the call in R-INLA to fit the IGMRF model is

```
data = list(y, j, region.iid, region.struct)
formula = y ~ -1 + f(j, model = "iid")
+ f(region.iid, model = "iid",
hyper = list(theta = list(prior = "loggamma",param = c(1,0.01))))
+ f(region.struct, model = "besag", graph = "nb5×5.graph",
hyper = list(theta = list(prior = "loggamma",param = c(1,0.01))))
result = inla(formula, family = "binomial", Ntrials = 1, data =
data, verbose = TRUE,
control.compute = list(dic = TRUE, cpo = TRUE), control.inla
= list(strategy = "laplace"))
```

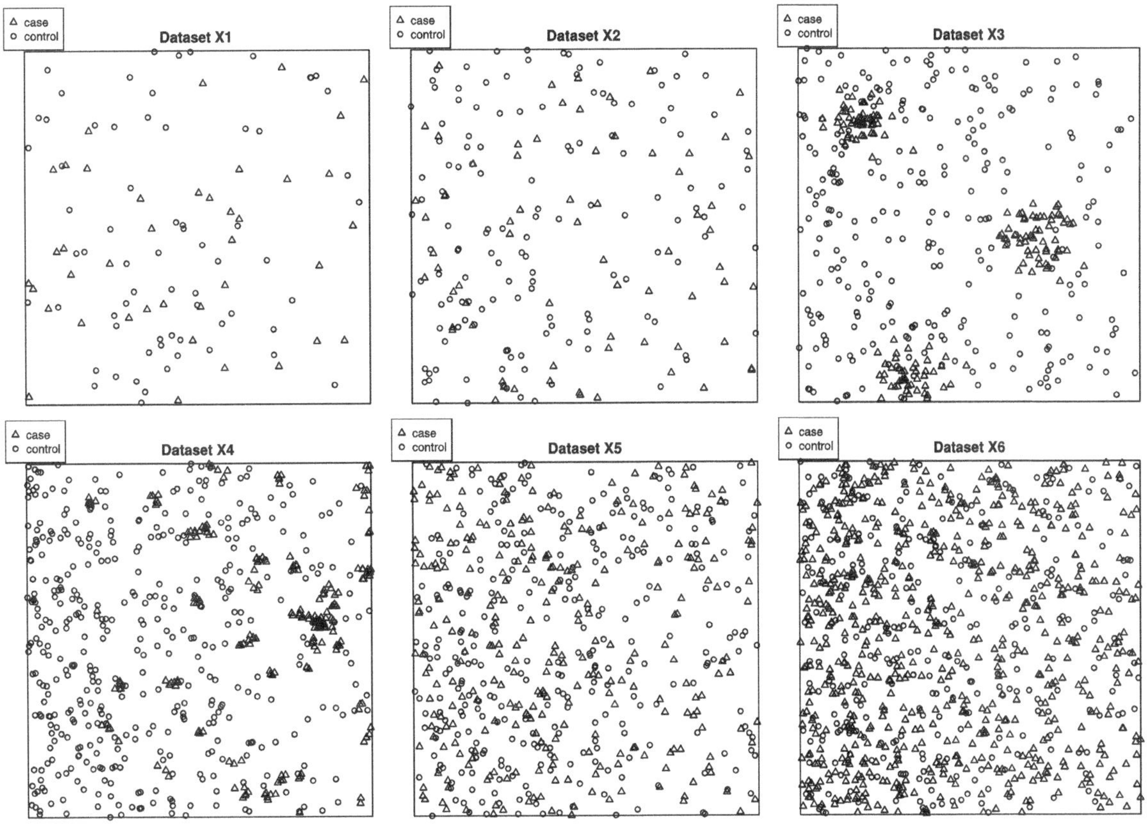

(a) **Simulated point-based data**

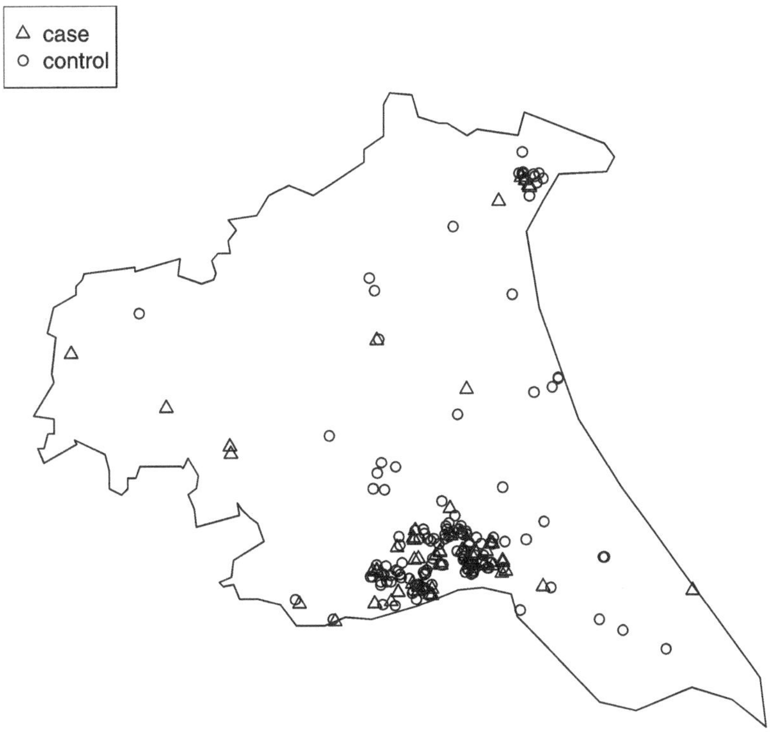

(b) **Humberside data**

Figure 1. (a) Six patterns of simulated point-based data (top). Various spatial patterns are considered, including inhomogeneous point patterns, patterns with local repulsion, patterns with local clustering, and patterns with local clustering in the presence of a larger-scale inhomogeneity. **(b) The Humberside data on childhood leukaemia and lymphoma** (bottom). The dataset portrays a sparse spatial pattern with a cluster.

Description of Data

Simulated data

We conducted a simulation study to investigate the impact of spatial scales and spatial smoothing on modelling outcomes. As discussed earlier, a spatial pattern may be present at a given aggregation level and may vanish at other scales. Therefore, using a range of spatial scales, the purpose of this simulation study was to investigate the performance of the models when dealing with different spatial structures of point-based data. We simulated spatial point-based data from various classical point-process models on the unit square. As guided by Illian et al. [31], we considered various scenarios: inhomogeneous point patterns, patterns with local repulsion, patterns with local clustering, and patterns with local clustering in the presence of a larger-scale inhomogeneity. These point-based data include cases and controls which resemble the Bernoulli outcome of an event (or a disease) in practice. We simulated cases and controls from two separate point-process models.

In dataset **X1**, the cases were generated from an inhomogeneous Poisson process with trend function $\lambda = 50\exp(-x)$ on the unit square. The controls were generated from an inhomogeneous Poisson process with trend function $\lambda = 100\exp(-x)$ which were then superimposed with the cases. This resulted in point-based data that were inhomogeneously distributed across the space with an average intensity of 107 points per unit square.

Dataset **X2** consisted of cases distributed in patterns with local repulsion, which were generated from a homogeneous Strauss process, with medium repulsion $\beta = 100$ (intensity parameter), interaction parameter $\gamma = 0.7$ and interaction radius $r = 0.05$, on the unit square. The cases were superimposed with the controls that were generated from an inhomogeneous Poisson process with trend function $\lambda = 200\exp(-x)$. The average intensity of this dataset was 210 points per unit square.

To generate the clustered cases in dataset **X3**, we simulated a homogeneous Thomas process with parameters $\kappa = 5$ (the intensity of the Poisson process of cluster centers), $\sigma = 0.05$ (the standard deviation of the distance of a point from the cluster center) and $\mu = 50$ (the expected number of points per cluster), on the unit square. Similarly, the controls were generated from an inhomogeneous Poisson process with trend function $\lambda = 400\exp(-x)$. After superimposing the cases and the controls, the average intensity of this dataset was 431 points per unit square.

The cases in dataset **X4** were generated from an inhomogeneous Thomas process with parameters $\sigma = 0.01$ and $\mu = 5$ and a simple trend function for the intensity of parent points given by $\kappa(x_1, x_2) = 100x_1$, on the unit square. The cases were then superimposed with controls generated from an inhomogeneous Poisson process with trend function $\lambda = 1000\exp(-2x)$. The average intensity of the cases and the controls was 648 points per unit square.

In dataset **X5**, both the cases and controls were generated from an inhomogeneous Poisson process with trend function $\lambda = 500\exp(-x)$ on the unit square. The resulting point-based data were inhomogeneously distributed across the space with an average intensity of 605 points per unit square.

The cases in dataset **X6** were generated from an inhomogeneous Poisson process with trend function $\lambda = 1000\exp(-x)$ on the unit square. The controls were generated from an inhomogeneous Poisson process with trend function $\lambda = 500\exp(-x)$ which were then superimposed with the cases. This resulted in point-based data that were inhomogeneously distributed across the space with an average intensity of 982 points per unit square.

Datasets **X1**, **X5** and **X6** were of similar patterns but different degree of denseness. Dataset **X3** had bigger clusters than dataset **X4**. See Figure 1(a) for illustrations of the simulated point-based data.

Humberside data on childhood leukaemia and lymphoma

We considered a realistic example (Figure 1(b)) available in the spatstat R package to illustrate the four models. The data were first presented and analyzed by [34]. [34] perform the method for detecting spatial clustering of events on this dataset. It is not the aim of this paper to pursue the detection of spatial clusters. We use this dataset as a case study that portrays natural phenomena to investigate the impact of spatial scales and spatial smoothing on modelling outcomes to complement the simulation study. The data contained 62 cases of childhood leukaemia and lymphoma diagnosed in the North Humberside region of England between 1974 and 1986, and 141 controls selected at random from the birth register for the same period. Spatial location of each individual's home address (actually, the centroid for the postal code) was given in the dataset. The dataset had a polygonal observation window; for the analysis, we created a 72.1 km×60.8 km rectangular window to enclose all events.

Model Fitting and Evaluations

To evaluate the impact of modelling the random effects at different spatial scales, we considered the partitions at the grid level by discretizing the study region using grids 5×5, 10×10, 15×15, 20×20, 25×25, 30×30, 35×35, 40×40, 45×45, and 50×50. The grid 5×5 resulted in 25 grid cells over the unit square, the grid 10×10 resulted in 100 grid cells over the unit square, and so on. So, the grid 5×5 had the largest grid cell size whereas the grid 50×50 had the smallest grid cell size. The cell2nb function in the spdep R package [35] was used to generate a list of neighbors for the grid cells, by applying a queen definition of neighborhood, where two grid cells were termed neighbors if they shared a common edge or vertex. The adjacency matrices were required in the fitting of the IGMRF model.

In terms of prior specification, the precision parameters of the unstructured random effect and spatial effect, τ_u and τ_v, were both assigned Gamma priors with parameters (1,0.01) to impose the same level of spatial smoothing on the spatial field for each model throughout the study. We carried out sensitivity analyses to assess the impact of various choices of prior distributions on the models and found that the influence of priors are negligible on the basis of minimal changes in the deviance information criterion (DIC).

For the purpose of model comparison, DIC was used to select the most parsimonious model after penalizing for model complexity. We note that DIC has been criticized [36,37] and can be problematic in models with many random effects [38]. Though it fails in some contexts, the use of DIC is appropriate in most generalized linear modelling problems and is a popular Bayesian model choice criterion for comparing complex hierarchical models [39]. A smaller DIC indicates a better fit of the model. As suggested by [39], DIC should not be used as an absolute measure of the 'best' model, but rather a method for screening alternative formulations in order to provide an indication of the relative fit of a set of candidate models. Candidate models receiving DIC within 1–2 of the 'best' deserve consideration, while 3–7 have considerably less support [39].

The logarithmic score (LS) for each model was also computed [32] to assess the predictive performance of these models. Each model was assigned a numerical score based on the predictive distribution using the cross-validated scoring rules. For discrete observations Y_{ij}, the LS is defined as

$$LS = -\log(\pi_{y_{ij}}),$$

where $\pi_{y_{ij}} = \text{Prob}(Y_{ij} = y_{ij} | y_{-ij})$ denotes the cross-validated predictive probability mass at the observed event. A smaller LS indicates a better predictive power of the model [40,41].

Results

We describe the results for model fitting on the six simulated datasets and the realistic example in this section. The DIC for fitting the four models on the six simulated datasets at various spatial scales are presented in Figure 2 and Figure 3. Figure 4 and Figure 5 present the LS for fitting the four models on the six simulated datasets at various spatial scales.

Dataset X1

The IID, IGMRF and MATERN2D models perform quite similarly at all spatial scales; whereas the RW2D model has larger DIC and LS than the other models at the first four spatial scales but its performance gradually improves as the grid cell size decreases. Across the spatial scales from the largest grid cell size to the smallest grid cell size, it is observed that the performance of the IID, IGMRF and MATERN2D models is fairly consistent. However, the RW2D prior appears to perform increasingly well as the grid cell size reduces. Therefore, for point data that are sparse and inhomogeneously distributed across the space such as dataset **X1**, the RW2D model seems to be a reasonable choice when fitted at small grid cell sizes. The results also suggest the need to repeat the analyses at multiple scales given the sensitivity of the models to changing spatial scales.

Dataset X2

Based on the results obtained for dataset **X2**, the RW2D model produces the smallest DIC and LS at all spatial scales. It appears that the scores produced by this model are quite similar at all spatial scales, suggesting that the changes in grid cell size do not affect the model performance. The performance of the other three models is inferior to the RW2D model and is rather consistent across the spatial scales. We note that the point data in this dataset are sparse and distributed with the presence of local repulsion and mild inhomogeneity. The results suggest that the changes in the spatial scales do not affect the model outcomes for this spatial pattern.

Dataset X3

In dataset **X3** which contains clustered cases and inhomogeneously distributed controls, the results clearly show the improvement in model fit and predictive performance for the RW2D model as the grid cell size decreases based on the decreasing DIC and LS. The RW2D model yields better fit than the three other models at grids 15×15 and above. A great improvement is seen for the RW2D model from the grid 10×10 to the grid 15×15. We note that at the grid 10×10, the number of individuals in the grid cells is highly varying with a maximum of 26 and median of 3.5, while less extreme at the grid 15×15 with a maximum of 17 and median of 2. There is a lot of inhomogeneity across the grid cells at the grid 10×10 which cannot be effectively smooth out by all four models. However, the disretization at grids 15×15 and above produces less inhomogeneity across the grid cells and thus the RW2D prior is able to smooth out the clusters more effectively. This is due in part to the ability of the RW2D prior in taking into account the first and second-order neighbors in spatial smoothing. For the clustered dataset presented here, there seems to be a need to discretize the study region into fine grid cells and the RW2D prior appears to be the most appropriate choice for spatial smoothing.

Dataset X4

Dataset **X4** contains relatively small clusters of cases as compared with dataset **X3**. The results suggest that the most appropriate spatial scale for fitting this dataset is at the grid 20×20 using the MATERN2D model or the grid 25×25 using the IGMRF model. Too large or too fine grid cell sizes impair the model performance as shown by the MATERN2D model. The MATERN2D and IGMRF priors both appear to be good choices for the spatial smoothness priors for this dataset. However, the spatial scales should be chosen with caution as they affect model outcomes substantially.

As suggested by the results, some models behave differently at different scales, e.g., working relatively well at certain scales but not others. This could be related to the fact that the smoothness priors perform in different mechanisms at various scales due to the impact from the neighboring grid cells. For illustration, we present the estimated precision parameters for the MATERN2D model at various spatial scales in Table 1. It is shown that the mean and standard deviation of the precision parameters for u_i and v_i at grids 5×5 and 10×10 are very different from grids 15×15 and above, hence resulting in the varying DIC seen for the MATERN2D model in dataset **X4** (Figure 3). The performance of the MATERN2D model is very similar at grids 15×15 and 20×20 due to their similarity in the precision parameters. We note that at different scales, various degree of inhomogeneity across the grid cells is observed. When there is large inhomogeneity across the cells, a higher degree of spatial smoothing is imposed while less smoothing when there is small inhomogeneity. The change in the degree of spatial smoothing results in the changes in the precision parameters.

Datasets X5 and X6

Datasets **X5** and **X6** contain point data that are inhomogeneously distributed across the space (similar to dataset **X1**). They are both denser than dataset **X1** and dataset **X6** is denser than dataset **X5**. The results for both datasets show that the IID, IGMRF and MATERN2D models produce good fit to both datasets at most spatial scales. The RW2D model is shown to yield the worst fit to the data across the spatial scales in both datasets. For both datasets, it appears that all the models perform rather consistently at all spatial scales. As a result, we note that the spatial scale is not an issue for dense point data and either one of the IID, IGMRF and MATERN2D models could be used in modelling.

Humberside data

In order to understand the discretization of the Humberside dataset better, we provide the summaries of the number of events included in the grid cells for all the non-zero cell counts at the different spatial scales (Table 2). The maximum number of cases decreases from 33 at the largest grid cell size to four at the smallest grid cell size while the maximum number of controls decreases from 83 at the largest grid cell size to nine at the smallest grid cell size. It is shown that the number of events contained in the grid cells has only slight differences for the grid 30×30 up to the grid 50×50, which suggests that the grid 30×30 might be a suitable scale and a finer grid cell is not required.

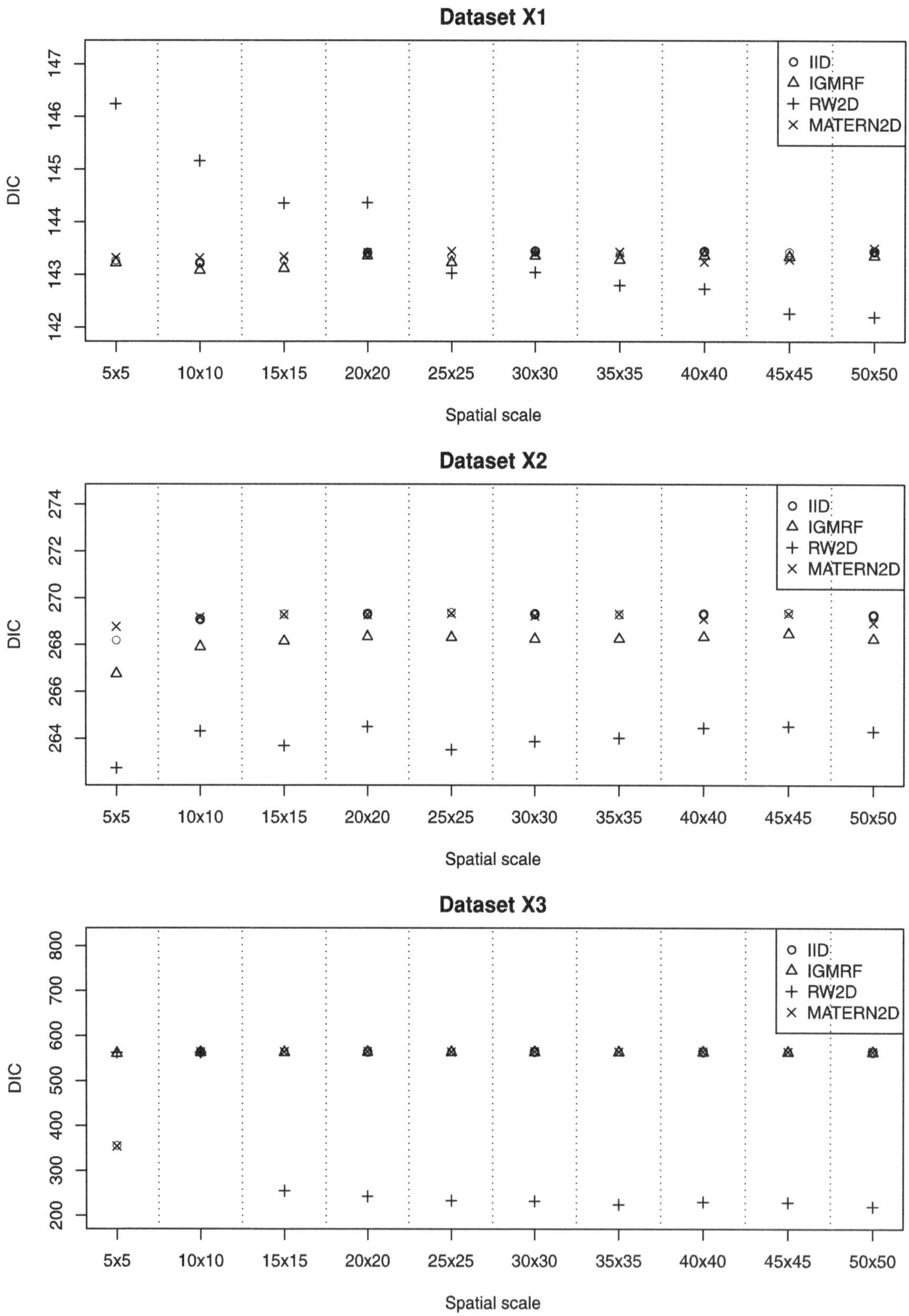

Figure 2. The estimated DIC of the four models for datasets X1, X2 and X3 at various spatial scales. The RW2D model fitted at small grid cell sizes appears to be a reasonable choice for dataset **X1**. For dataset **X2**, the RW2D model produces the smallest DIC at all spatial scales. The RW2D model also performs better than the three other models at grids 15×15 and above.

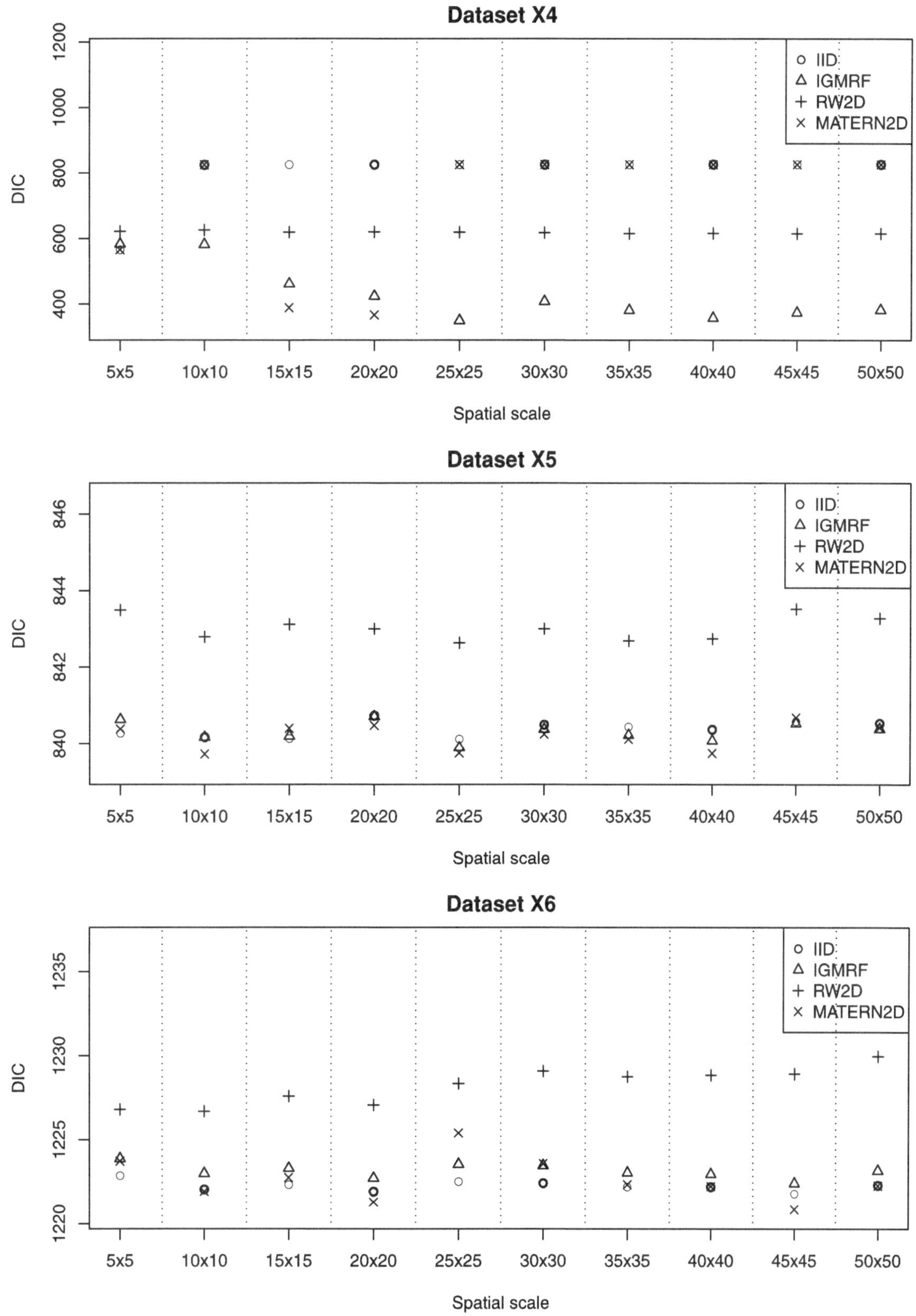

Figure 3. The estimated DIC of the four models for datasets X4, X5 and X6 at various spatial scales. The most appropriate spatial scale for fitting dataset **X4** is at the grid 20×20 using the MATERN2D model or the grid 25×25 using the IGMRF model. For datasets **X5** and **X6**, the IID, IGMRF and MATERN2D models perform well at most spatial scales.

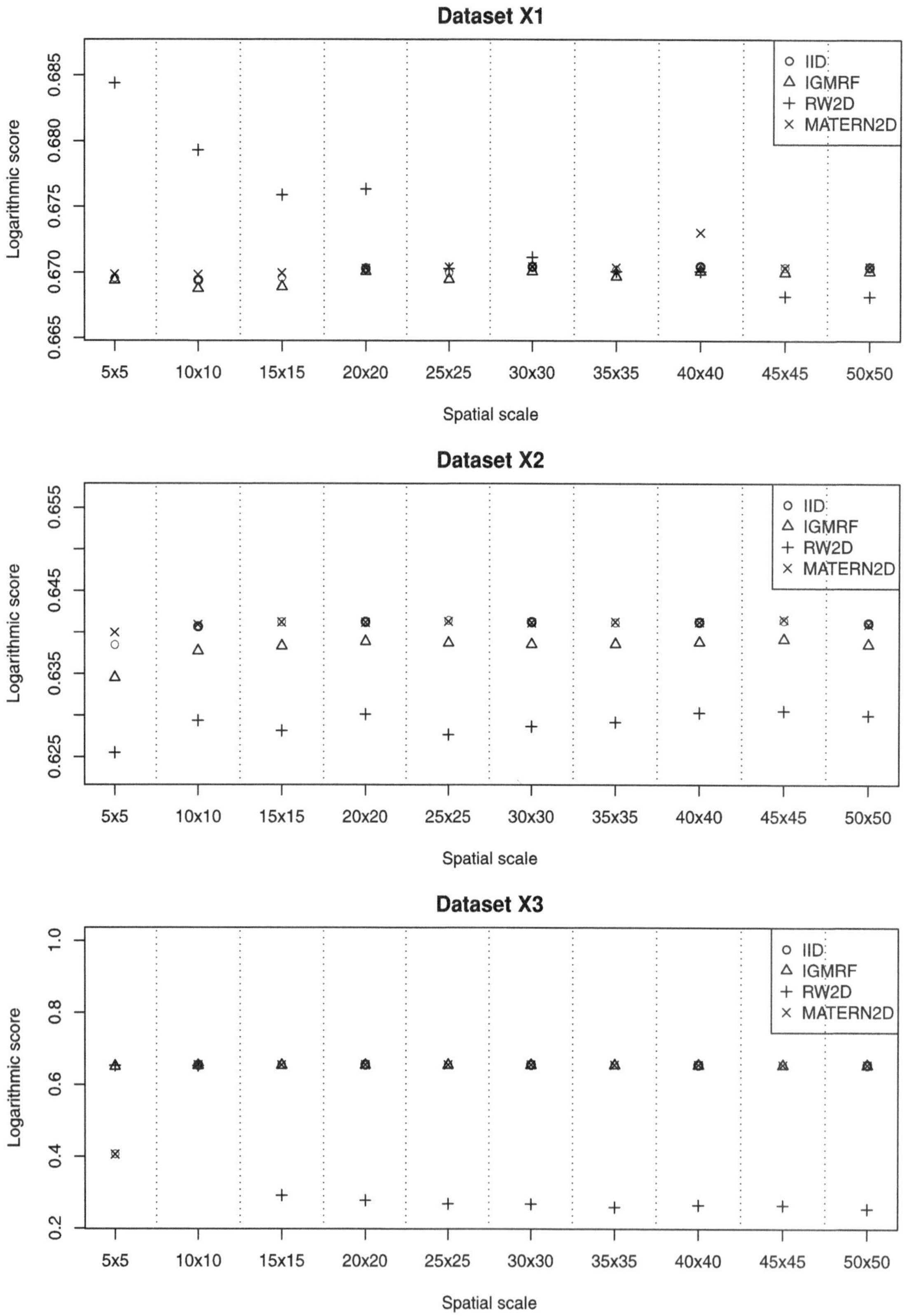

Figure 4. The estimated LS of the four models for datasets X1, X2 and X3 at various spatial scales. The RW2D model fitted at small grid cell sizes appears to be a reasonable choice for dataset **X1**. For dataset **X2**, the RW2D model produces the smallest LS at all spatial scales. The RW2D model also performs better than the three other models at grids 15×15 and above.

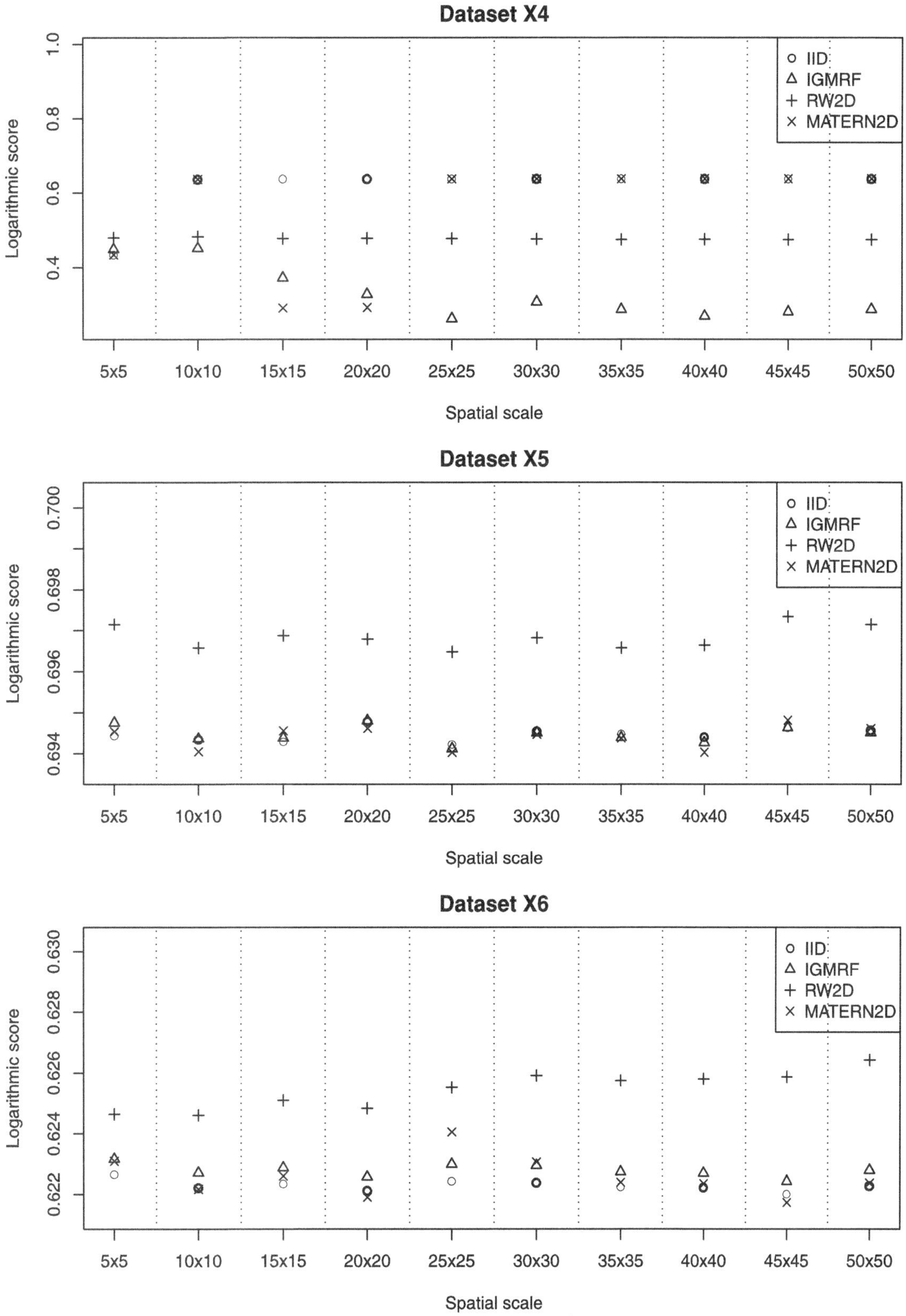

Figure 5. The estimated LS of the four models for datasets X4, X5 and X6 at various spatial scales. The most appropriate spatial scale for fitting dataset **X4** is at the grid 20×20 using the MATERN2D model or the grid 25×25 using the IGMRF model. For datasets **X5** and **X6**, the IID, IGMRF and MATERN2D models perform well at most spatial scales.

Table 1. The estimated precision parameters of the MATERN2D model at various spatial scales for dataset X4.

Spatial scale	Precision for u_i (τ_u)		Precision for v_i (τ_v)	
	Mean	Std dev	Mean	Std dev
5×5	0.455	0.092	172.474	44.838
10×10	0.217	0.043	23.700	12.438
15×15	46.034	4.236	0.084	0.007
20×20	49.972	5.169	0.065	0.006
25×25	40.582	3.022	0.041	0.003
30×30	41.473	3.413	0.051	0.004

Table 2. Summary of the number of events in the grid cells for all the non-zero cell counts at various spatial scales for the Humberside dataset.

Spatial scale	Case			Control		
	Min	Mean	Max	Min	Mean	Max
5×5	1	6.20	33	1	11.75	83
10×10	1	4.77	22	1	6.13	44
15×15	1	2.82	10	1	4.15	29
20×20	1	2.95	11	1	3.62	17
25×25	1	2.39	8	1	3.36	14
30×30	1	1.88	4	1	2.77	10
35×35	1	2.00	6	1	2.71	12
40×40	1	1.77	5	1	2.31	11
45×45	1	1.68	4	1	2.14	8
50×50	1	1.55	4	1	2.10	9

Based on the results obtained from modelling the Humberside dataset, the IID, IGMRF and MATERN2D models produce similar DIC and LS at the various spatial scales (Figure 6) but the RW2D model performs slightly worse than the other three models at all instances. At the grid 30×30, a slight improvement is observed for all the models. Also, at this scale, the MATERN2D model outperforms the other models substantially. It could be seen that the performance of the MATERN2D model is rather sensitive to the changes in the spatial scales. In short, the results suggest that the dataset should be fitted at the grid 30×30 using the MATERN2D prior for spatial smoothing. Table 3 summarizes the results for fitting the models at various spatial scales for various datasets described above. The implications of these results are considered further in the Discussion.

Discussion

We evaluated the performance of a range of spatial smoothness priors (an intrinsic Gaussian Markov random field (IGMRF), a second-order random walk on a lattice (RW2D), and a Gaussian field with Matérn correlation function (MATERN2D)) and spatial scales for various spatial structures using deviance information criterion (DIC) and logarithmic score (LS). The simulated datasets consist of points that are distributed across the space at various spatial patterns. The Humberside data are real phenomena where the data points are spatially sparse and exhibit a cluster. The results in this study suggest that different spatial smoothness priors and spatial scales may be appropriate for different patterns of spatial point-based data.

We note that for spatially sparse and inhomogeneously distributed point pattern (dataset **X1**), our study shows that it is necessary to include a spatially structured component, in addition to the unstructured component, to the model. The RW2D model at small grid cell sizes is an appropriate choice of modelling as it is the most parsimonious model (based on the DIC) and has the best predictive performance (based on the LS). Dataset **X5** and dataset **X6**, which are essentially similar structures to dataset **X1** but in a denser pattern, do not produce similar results to dataset **X1**. The results suggest that when denser point data are involved, the changes in spatial scales have little impact on the model outcomes. In addition, the spatial effect does not necessarily have to be included in the model as the unstructured component alone suffices. However, if desirable, the IGMRF and MATERN2D priors may be used as priors for the spatial effect as they do not impair the model performance.

When the point patterns with local repulsion and mild inhomogeneity (dataset **X2**) are modelled, the spatial component should be included and assigned the RW2D prior, in addition to the inclusion of the IID component in the model, as suggested by the results. Nevertheless, the spatial scales do not seem to matter in modelling this sort of spatial structure where the points are distributed across the space with local repulsion and mild inhomogeneity. For the inhomogeneous point pattern with a few large clusters (as portrayed by dataset **X3**), the RW2D model at fine grid cell sizes is shown to be a good modelling choice.

The sparse inhomogeneous point pattern with a number of small clusters across the space (dataset **X4**) appears to be quite sensitive to the changes in the spatial scales. As shown in the results, the MATERN2D model at the grid 20×20 and the IGMRF model at the grid 25×25 are the two appropriate modelling approaches for this point pattern. The model performance (based on the DIC and LS) becomes worse when larger or smaller grid cell sizes are used in modelling. This has addressed the need to select the spatial scales with caution when complicated spatial structures are of interest.

The realistic example studied here (the Humberside dataset) has further confirmed that for sparse point pattern with potential spatial clustering, the spatial scale and spatial smoothness prior have to be chosen carefully in modelling. The model fit (as guided by the DIC) and predictive performance of the models (as guided by the LS) differ at the different spatial scales. The results for this dataset show that the best modelling approach for this dataset is the MATERN2D model at the grid 30×30. This complements the results for dataset **X4**, in which both of these sparse datasets with clustering appear to be quite sensitive to the changes in the spatial scales. Furthermore, the MATERN2D model is shown to be a good modelling approach for both datasets.

The various spatial smoothness priors considered in this study have been shown to be applicable for different spatial structures. We note that it is possible to choose the appropriate prior based on the spatial structures but a range of priors should generally be considered. As suggested by our study, the RW2D prior is a reasonable choice for spatial smoothing when spatially sparse point patterns are involved, regardless of whether the points are homogenous or inhomogeneously distributed across the space. The RW2D prior imposes spatial smoothing by taking into account the first and second-order neighbors. Our study also shows that the IGMRF prior is suitable for spatial smoothing in

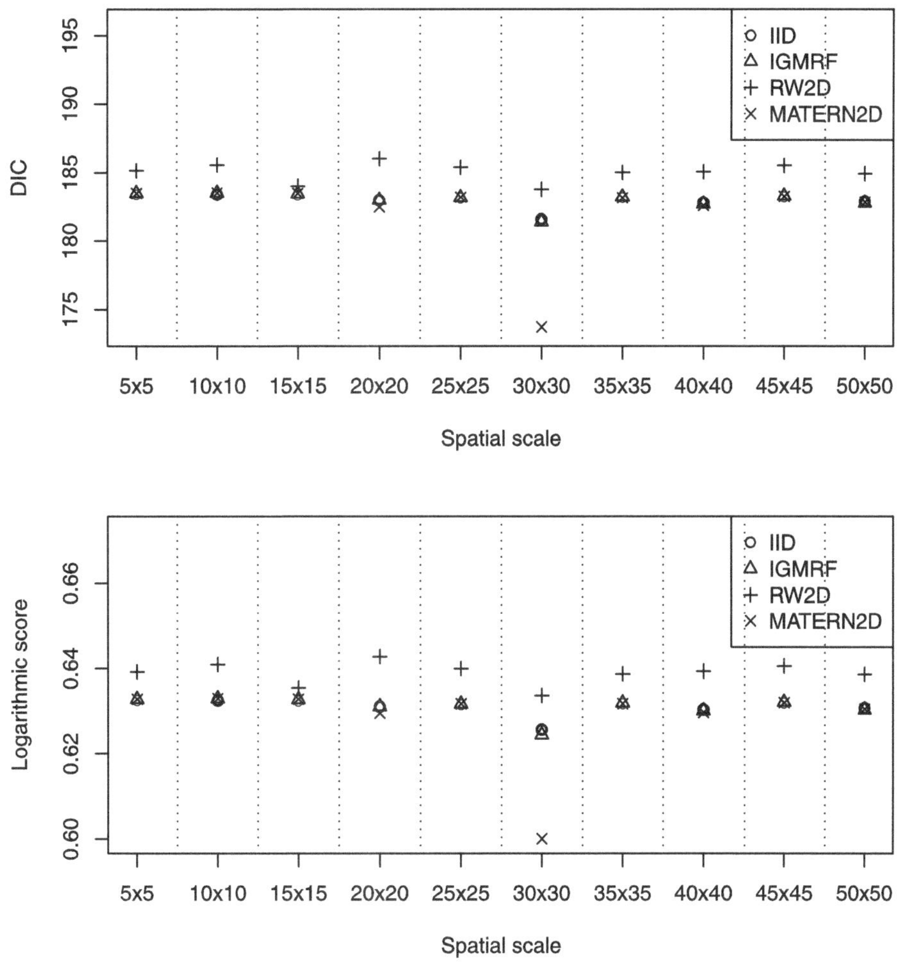

Figure 6. The estimated DIC and LS of the four models for the Humberside dataset at various spatial scales. The dataset should be fitted at the grid 30×30 using the MATERN2D prior for spatial smoothing.

Table 3. Summary of results for fitting the four models at various spatial scales for various spatial patterns.

Spatial patterns	Recommended spatial smoothness priors and spatial scales	Sensitivity of the models towards the changing spatial scales
Sparse inhomogeneous point pattern (dataset **X1**)	The RW2D model at small grid cell sizes.	The IID, IGMRF and MATERN2D models perform consistently at all spatial scales; the RW2D model is sensitive towards the changing grid cell sizes.
Sparse point pattern with local repulsion and mild inhomogeneity (dataset **X2**)	The RW2D model; spatial scales have little impact on model outcomes.	All four models perform rather consistently at all spatial scales.
Sparse inhomogeneous point pattern with large clusters (dataset **X3**)	The RW2D model at grids 15×15 and above.	The IGMRF model performs consistently at all spatial scales; the IID, RW2D and MATERN2D models are sensitive towards the changing grid cell sizes.
Sparse inhomogeneous point pattern with small clusters (dataset **X4**)	The MATERN2D model at the grid 20×20 or the IGMRF model at the grid 25×25.	The RW2D model performs consistently at all spatial scales; the IID, IGMRF and MATERN2D models are sensitive towards the changing grid cell sizes.
Dense inhomogeneous point pattern (datasets **X5** and **X6**)	The IID, IGMRF and MATERN2D models; spatial scales have little impact on model outcomes.	All four models perform rather consistently at all spatial scales.
Sparse point pattern with clusters (the Humberside dataset)	The MATERN2D model at the grid 30×30.	The IID, IGMRF and RW2D models perform rather consistently at all spatial scales; the MATERN2D model is sensitive towards the changing grid cell sizes.

spatially dense and inhomogeneous point patterns as it considers only first-order neighbors. The RW2D prior is essentially a second-order IGMRF on a lattice. It is quite flexible due to its invariance to addition of a linear trend. The RW2D prior imposes a higher level of spatial smoothing than the IGMRF prior due to the presence of the second-order neighbors. Sparse data need more spatial smoothing than dense data, therefore the RW2D prior works well in this context. If spatially dense and homogeneous point patterns are considered, the model may not include the spatially structured component but only the unstructured component assigned the IID prior. The MATERN2D prior appears to be well-suited for capturing the spatial effect in spatially clustered point patterns but it is very sensitive to the changes in spatial scales. This could be due to the representation of the smoothness parameter which gives the model great flexibility in modelling clustered point data that require a relatively high level of spatial smoothing.

In conclusion, we note that it is crucial to repeat the spatial analyses at multiple spatial scales when modelling inhomogeneously distributed point patterns as the model fit and predictive performance of the models appear to vary at different spatial scales. Methods for testing spatial heterogeneity such as Tango's Index [42] and Moran's I [43] could be used to decide if a given spatial dataset is inhomogeneously distributed. Inspection of detailed plots of the spatial data may also be a good guide to examine the presence of spatial inhomogeneity. For the inhomogeneous point patterns that do not contain clusters, the model performance improves as the grid cell size reduces. For the inhomogeneous point patterns that contain clusters, the appropriate spatial scale can be chosen by repeating the analyses at a range of spatial scales. On the other hand, the spatial scales appear to have little impact for homogeneously distributed point patterns. Also, it may not be necessary to include the spatially structured component in modelling of homogeneous point patterns unless desirable.

An acknowledged limitation of the study is that we simulated one scenario for each point process structure of interest. Therefore, we are reserved about the generality of the conclusions drawn above. For future work, more than one simulation scenario for a continuum of point process models with varying spatial structures could be studied in order to achieve more general conclusions. In this study, we consider grid cells with equal sizes as it was argued by [44] that in the specification of neighborhood structure, all regions should be of similar size and arranged in a regular pattern. For regions with different sizes, possible neighborhood structure definitions are some known function of the distance between centroid of areas [45]; an intrinsic conditional autoregressive (ICAR) prior [15] using weight definition, and alternative specifications within the CAR class [46]. Further investigation could be carried out to examine the impact of the changes in the shape of the regions including regular and irregular sizes.

Given the different results observed and different inferences made at the different spatial scales, it is crucial to repeat the analysis at different scales as the data may contain useful information at more than just one scale. It is also important to take into account the spatial scale that is of interest in a particular problem, i.e., the scale at which decisions or inferences will be made in practice. Often, disease management and policy making of subpopulation require modelling at a coarser scale than that required for understanding individual influences or associations. The choice of spatial scale is typically influenced by geo-political considerations, for instance, administrative districts are often used to describe and to understand geographical variation of a disease, with the aim being to assist public health decision making. Similarly, the identification of population-based clusters may differ from local clusters, with different interpretations and decision/action implications.

Acknowledgments

The authors would like to thank Nicole White and the two reviewers of this paper for their helpful comments.

Author Contributions

Conceived and designed the experiments: SYK JM KM. Performed the experiments: SYK. Analyzed the data: SYK. Wrote the paper: SYK.

References

1. Robinson WS (1950) Ecological correlations and the behavior of individuals. American Sociological Review 15: 351–357.
2. Openshaw S, Taylor P (1979) A million or 80 correlation coefficient: Three experiments on the modifiable areal unit problem. In: Wrigley N, editor, Statistical Applications in the Spatial Sciences, London, UK: Pion Press. pp. 127–44.
3. Perry JN, Liebhold AM, Rosenberg MS, Dungan J, Miriti M, et al. (2002) Illustrations and guidelines for selecting statistical methods for quantifying spatial pattern in ecological data. Ecography 25: 578–600.
4. Rytkönen MJP (2004) Not all maps are equal: GIS and spatial analysis in epidemiology. International Journal of Circumpolar Health 63: 9–24.
5. Biggeri A, Dreassi E, Catelan D, Rinaldi L, Lagazio C, et al. (2006) Disease mapping in veterinary epidemiology: a Bayesian geostatistical approach. Statistical Methods in Medical Research 15: 337–352.
6. Kleinschmidt I, Pettifor A, Morris N, MacPhail C, Rees H (2007) Geographic distribution of human immunodeficiency virus in South Africa. The American Journal of Tropical Medicine and Hygiene 77: 1163–1169.
7. Vanhatalo J, Vehtari A (2007) Sparse log Gaussian processes via MCMC for spatial epidemiology. In: JMLR Workshop and Conference Proceedings. volume 1, pp. 73–89.
8. Hossain MM, Lawson AB (2009) Approximate methods in Bayesian point process spatial models. Computational Statistics & Data Analysis 53: 2831–2842.
9. Baddeley A, Berman M, Fisher NI, Hardegen A, Milne RK, et al. (2010) Spatial logistic regression and change-of-support in Poisson point processes. Electronic Journal of Statistics 4: 1151–1201.
10. Li Y, Brown P, Rue H, al Maini M, Fortin P (2012) Spatial modelling of lupus incidence over 40 years with changes in census areas. Journal of the Royal Statistical Society: Series C (Applied Statistics) 61: 99–115.
11. Meliker JR, Sloan CD (2011) Spatio-temporal epidemiology: Principles and opportunities. Spatial and Spatio-temporal Epidemiology 2: 1–9.
12. Rue H, Held L (2005) Gaussian Markov Random Fields: Theory and Applications, volume 104 of *Monographs on Statistics and Applied Probability*. Chapman & Hall, London.
13. Stein ML (1999) Interpolation of Spatial Data: Some Theory for Kriging. Chapter 2, Springer Verlag, New York.
14. Banerjee S, Gelfand AE, Carlin BP (2004) Hierarchical modeling and analysis for spatial data. Chapman and Hall/CRC, Boca Raton, Florida.
15. Besag J, York J, Mollié A (1991) Bayesian image restoration, with two applications in spatial statistics (with discussion). Annals of the Institute of Statistical Mathematics 43: 1–59.
16. Besag J, Green P, Higdon D, Mengersen K (1995) Bayesian computation and stochastic systems. Statistical Science 10: 3–41.
17. Clayton D, Bernardinelli L (1992) Bayesian methods for mapping disease risk. In: Elliott P, Cuzick J, English D, Stern R, editors, Geographical and Environmental Epidemiology: Methods for Small Area Studies, Oxford University Press. pp. 205–220.
18. Mollié A (1996) Bayesian mapping of disease. In: Gilks WR, Richardson S, Spiegelhalter DJ, editors, Markov Chain Monte Carlo in Practice, Chapman & Hall, London. pp. 359–379.
19. Wakefield JC, Best NG, Waller LA (2000) Bayesian approaches to disease mapping. In: Elliot P, Wakefield JC, Best NG, Briggs DJ, editors, Spatial Epidemiology: Methods and Applications, Oxford: Oxford University Press. pp. 104–127.
20. Rue H, Martino S, Chopin N (2009) Approximate Bayesian inference for latent Gaussian models by using integrated nested Laplace approximations. Journal of the Royal Statistical Society: Series B (Statistical Methodology) 71: 319–392.

21. Terzopoulos D (1988) The computation of visible-surface representations. IEEE Transactions on Pattern Analysis and Machine Intelligence 10: 417–438.
22. Handcock MS, Stein ML (1993) A Bayesian analysis of kriging. Technometrics 35: 403–410.
23. Minasny B, McBratney AB (2005) The Matérn function as a general model for soil variograms. Geoderma 128: 192–207.
24. Simpson D, Lindgren F, Rue H (2011) Fast approximate inference with INLA: the past, the present and the future. Arxiv preprint arXiv:11052982.
25. Sauleau EA, Mameli V, Musio M (2010) Using integrated nested Laplace approximation for modeling spatial healthcare utilization. Arxiv preprint arXiv:10063764.
26. Willgert K, Schroedle B, Schwermer H (2011) Spatial analysis of bluetongue cases and vaccination of Swiss cattle in 2008 and 2009. Geospatial Health 5: 227–237.
27. Schrödle B, Held L (2011) A primer on disease mapping and ecological regression using INLA. Computational Statistics 26: 241–258.
28. Schrödle B, Held L (2011) Spatio-temporal disease mapping using INLA. Environmetrics 22: 725–734.
29. Schrödle B, Held L, Riebler A, Danuser J (2011) Using integrated nested Laplace approximations for the evaluation of veterinary surveillance data from Switzerland: a case-study. Journal of the Royal Statistical Society: Series C (Applied Statistics) 60: 261–279.
30. Illian JB, Sørbye SH, Rue H, Hendrichsen DK (2012) Using INLA to fit a complex point process model with temporally varying effects - a case study. Journal of Environmental Statistics 3: 1–29.
31. Illian JB, Sørbye SH, Rue H (2012) A toolbox for fitting complex spatial point process models using integrated nested Laplace approximation (INLA). The Annals of Applied Statistics 6: 1499-1530.
32. Gneiting T, Raftery AE (2007) Strictly proper scoring rules, prediction, and estimation. Journal of the American Statistical Association 102: 359–378.
33. Czado C, Gneiting T, Held L (2009) Predictive model assessment for count data. Biometrics 65: 1254–1261.
34. Cuzick J, Edwards R (1990) Spatial clustering for inhomogeneous populations. Journal of the Royal Statistical Society Series B (Methodological) 52: 73–104.
35. Bivand R, Anselin L, Berke O, Bernat A, Carvalho M, et al. (2011) spdep: Spatial dependence: weighting schemes, statistics and models. R package version 05-31, URL http://CRAN R-project org/package= spdep.
36. Celeux G, Forbes F, Robert CP, Titterington DM (2006) Deviance information criteria for missing data models. Bayesian Analysis 1: 651–673.
37. Aitkin M (2010) Statistical Inference: An Integrated Bayesian/Likelihood Approach. Chapman & Hall/CRC.
38. Plummer M (2008) Penalized loss functions for Bayesian model comparison. Biostatistics 9: 523–539.
39. Spiegelhalter DJ, Best NG, Carlin BP, Linde AVD (2002) Bayesian measures of model complexity and fit. Journal of the Royal Statistical Society: Series B (Statistical Methodology) 64: 583–639.
40. Held L, Schrödle B, Rue H (2010) Posterior and cross-validatory predictive checks: a comparison of MCMC and INLA. In: Kneib T, Tutz G, editors, Statistical Modelling and Regression Structures–Festschrift in Honour of Ludwig Fahrmeir, Physica-Verlag HD, Dordrecht. pp. 91–110.
41. Roos M, Held L (2011) Sensitivity analysis in Bayesian generalized linear mixed models for binary data. Bayesian Analysis 6: 259–278.
42. Tango T (2000) A test for spatial disease clustering adjusted for multiple testing. Statistics in Medicine 19: 191–204.
43. Moran PAP (1950) Notes on continuous stochastic phenomena. Biometrika 37: 17–23.
44. Wakefield J (2007) Disease mapping and spatial regression with count data. Biostatistics 8: 158–83.
45. Cressie N, Chan NH (1989) Spatial modeling of regional variables. Journal of the American Statistical Association 84: 393–401.
46. Lee D (2011) A comparison of conditional autoregressive models used in Bayesian disease mapping. Spatial and Spatio-temporal Epidemiology 2: 79–89.

6

Climate Exposure of US National Parks in a New Era of Change

William B. Monahan*, Nicholas A. Fisichelli

National Park Service, Natural Resource Stewardship & Science, Fort Collins, Colorado, United States of America

Abstract

US national parks are challenged by climate and other forms of broad-scale environmental change that operate beyond administrative boundaries and in some instances are occurring at especially rapid rates. Here, we evaluate the climate change exposure of 289 natural resource parks administered by the US National Park Service (NPS), and ask which are presently (past 10 to 30 years) experiencing extreme (<5th percentile or >95th percentile) climates relative to their 1901–2012 historical range of variability (HRV). We consider parks in a landscape context (including surrounding 30 km) and evaluate both mean and inter-annual variation in 25 biologically relevant climate variables related to temperature, precipitation, frost and wet day frequencies, vapor pressure, cloud cover, and seasonality. We also consider sensitivity of findings to the moving time window of analysis (10, 20, and 30 year windows). Results show that parks are overwhelmingly at the extreme warm end of historical temperature distributions and this is true for several variables (e.g., annual mean temperature, minimum temperature of the coldest month, mean temperature of the warmest quarter). Precipitation and other moisture patterns are geographically more heterogeneous across parks and show greater variation among variables. Across climate variables, recent inter-annual variation is generally well within the range of variability observed since 1901. Moving window size has a measureable effect on these estimates, but parks with extreme climates also tend to exhibit low sensitivity to the time window of analysis. We highlight particular parks that illustrate different extremes and may facilitate understanding responses of park resources to ongoing climate change. We conclude with discussion of how results relate to anticipated future changes in climate, as well as how they can inform NPS and neighboring land management and planning in a new era of change.

Editor: Moncho Gomez-Gesteira, University of Vigo, Spain

Funding: Support for this study was provided by the National Park Service, and by the NASA Applied Sciences program through award number 10-BIOCLIM10-0034. The funders had no role in study design, data collection and analysis, decision to publish, or preparation of the manuscript.

Competing Interests: The authors have declared that no competing interests exist.

* Email: Bill_Monahan@nps.gov

Introduction

Recent scientific reviews of the US National Park Service (NPS) suggest the agency needs to manage its parks for changing climatic and ecological baselines in a landscape context [1–3]. These recommendations reiterate and heighten the imperative that it is no longer ecologically viable to manage resources solely within park boundaries, nor feasible to meet objectives based on pre-European settlement conditions. Broad-scale anthropogenic changes impact even the most remote areas with the highest levels of protection (e.g., federally designated wildernesses [4]). Climate change, land use change, pollution, and invasive species are pervasive and have additive and interactive effects on species and ecosystems [5], and they challenge the NPS to work under rapidly shifting conditions at broad spatial scales in order to protect park resources for future generations [6,7].

Climate change is a principal factor driving the imperative for progressive management aimed at achieving a specified 'desired future condition' (DFC) [8,9]. Magnitudes and rates of modeled future climate change suggest that temperatures in many regions of the globe may shift outside the envelope of historical (1860–2005) variability by mid-century [10], thus requiring DFCs to depart from past observations or records. Furthermore, climate change may already be affecting park natural resources. According to the Intergovernmental Panel on Climate Change (IPCC), globally each of the last three decades was increasingly warmer than any preceding decade since 1850, and – in the northern hemisphere –1983–2012 was possibly the warmest 30-year period of the last 1400 years [11]. In these situations, DFC management confronts existing 'retrospective' management practices that have for decades worked to preserve areas and resources within an 'historical range of variability' (HRV) [12]. For example, in a coastal park such as Point Reyes National Seashore [13], where the physical shoreline and intertidal wetlands are important resources, park managers may need to consider facilitating the inland migration of both features into new areas, rather than retain HRV by resisting sea level rise. Depending on the proximity of the current shoreline to the park's boundary, some of this DFC management may necessitate accepting or promoting resource migration beyond areas administered by the NPS. Due in part to logistical dilemmas posed by these possible futures, coupled with their novelty, there is often a desire to – where possible and as supported by current policy – manage under an existing, observed HRV, rather than an entirely new and uncertain set of conditions.

Recommendations to both scale up and consider changing baselines [1–3] encourage a thorough analysis of the exposure of individual parks and surrounding landscapes to climate change. Such insights are foundational to understanding which climate drivers are already statistically extreme relative to HRV, and whether ongoing climate change will further exacerbate (vs. alleviate) these conditions. Here, we evaluate the recent climate (past 10–30 years) of each US NPS natural resource park relative to its HRV across the past 112 years (1901–2012). We consider parks in a landscape context (park + surrounding 30 km) and evaluate both mean and inter-annual variation in major climate drivers over multiple climatically and management relevant time periods. Our analyses address three questions relevant to individual park management and NPS planning and policy: (i) relative to 1901–2012 HRV, do recent climate conditions NPS-wide tend to be unusually low or high on certain climate variables, (ii) how sensitive are these statistical distributions to the time window of analysis (10, 20, or 30 years), and (iii) which individual parks are climatically extreme relative to their HRV? We conclude with discussion of how results relate to anticipated future changes in climate and – by extension – how they may inform NPS climate adaptation.

Materials and Methods

Study Sites

Our analyses were performed for all natural resource parks considered by the NPS Inventory & Monitoring Program's landscape dynamics monitoring project, NPScape ($n = 289$), including a 30 km buffer around each park's boundary [14]. Hereafter these areas of analysis are referred to generally as 'parks'. The 30 km buffer was selected because of the relatively coarse spatial resolution of the climate data (see below), as well as to approximate the surrounding zone of ecological influence [15]. Natural resource parks range from approximately 14°S to 68°N latitude (spanning the Equator) and 115°W to 65°E longitude (spanning the International Date Line), hence the need to use global climate data as a single, common source for consistent coverage of all parks in this study.

Climate Data

Gridded climate data were obtained from the Climatic Research Unit (CRU) high-resolution time series version 3.21 (TS 3.21) [16]. CRU TS 3.21 data are globally available at 0.5 decimal degrees (approximately 3000 km^2 at the Equator and 2000 km^2 at 70°N latitude) for each month × year, 1901–2012. Although this dataset is spatially coarse, at least relative to the size of some parks (e.g., the smallest park with 30 km buffer encompasses roughly one grid cell), it offers for all NPS geographies the highest spatiotemporal resolution of observed climate over the 20th and 21st centuries. We selected for analysis monthly averages of daily temperature (minimum, maximum, mean; °C), percentage cloud cover, and vapor pressure (haP), as well as total monthly precipitation (mm) and monthly frequencies of frost days (a period of 24 hours in which the minimum temperature falls below 0°C) and wet days (rain days per month). We used the monthly climate variables to calculate a series of biologically meaningful variables, including 19 standard bioclimatic variables [17], as well as six other variables that relate to photosynthetic activity and plant growth [18,19] (Table 1).

Although correlations exist among some of these variables, we include the full suite in our analysis because the relevance of individual climate variables can vary dramatically by park resource (e.g., minimum winter temperature may influence a bird distributional limit, while maximum summer temperature may be a more important driver for particular mammals). As such, the elimination of certain variables due to high statistical correlations would adversely restrict the value of these analyses for park interpretation. Furthermore, the data presented here are percentiles calculated from HRV (see below), not the raw climate variables. Hence, whether variables are correlated is not simply a question of correlation at one point in time; it is a complex question of how sensitive correlations are across the entire 1901–2012 period of analysis, interacting with the moving time window of summarization.

Statistical Analysis

For each biologically relevant variable, over the entire time series (1901–2012), we used three moving windows (10, 20, and 30 years) to calculate a series of running means and standard deviations (SD). Hereafter we refer to these as 'moving window means' and 'moving window standard deviations'. These statistical distributions are what we use to estimate HRV. For example, with the 10 year window, we calculated the mean and SD for 103 windows (1901–1910, 1902–1911, ..., 2003–2012) to create the HRV distribution. The three windows were designed to encompass both near- and long-term management and planning considerations, as well as important climatic periods and cycles. In the NPS, many implementation plans have short horizons (<10 years), while strategic plans forecast out 10 to 20 years [20]. Traditional climate summaries span periods of 30 years [21], while major climate cycles affecting NPS geographies (e.g., Pacific Decadal Oscillation and North Atlantic Oscillation) tend to operate roughly on decadal to multi-decadal scales (e.g., over periods of 10, 20, or 30 years) [22].

For each variable, moving window size, and summary statistic (temporal mean and SD), we calculated the area-weighted mean for each park, based on the spatial intersect of park areas of analysis with the CRU lattice. Using processing logic contained in the NPScape climate grid analysis toolset [23], a resource designed by the NPS Inventory and Monitoring Program to facilitate common statistical analyses of gridded climate data, we then computed using R [24] the percentile of each most recent window (2003–2012, 1993–2012, 1983–2012) on the corresponding distribution of moving window values for each park (i.e., its HRV). In other words, we quantified recent conditions as the percentile of the most recent moving window value relative to the entire distribution of moving window values for that climate variable and window size (e.g., a 90th percentile for the most recent 10 year moving window of annual mean temperature signifies that the value of this climate variable during this time period was greater than 90% of all annual mean temperature 10 year moving window values). Hereafter we refer to these estimates as 'recent percentiles'.

We then averaged the recent percentiles of the most recent 10, 20, and 30 year moving windows and computed the maximum difference in percentile (max Δ) among windows; this resulted in – for each park, variable, and summary statistic – both an overall measure of recent climate change exposure with respect to HRV, and an estimate of how sensitive that measure was by comparing the different moving window sizes. Note that our measure of variability (moving window SD) does not reveal any information regarding extreme events, such as droughts, heat waves, and storms, which can also impact park resources [25].

An example of these steps with annual mean temperature (Bio1) is shown for Grand Canyon National Park (Figure 1). In this instance, the percentiles associated with the most recent mean temperature windows (computed over 10, 20, and 30 years) are all >95% (mean = 99%), showing that recent temperature has been at the extreme warm end of the entire distribution of moving windows (see location of red asterisks in boxplot distributions, Figure 1B). The max Δ for mean temperature is relatively small

Table 1. Biologically relevant variables considered in the analyses of climate change exposure of US national parks.

Code	Name
Bio1	Annual mean temperature
Bio2	Mean diurnal range (mean of monthly (max temp – min temp))
Bio3	Isothermality (Bio2/Bio7)
Bio4	Temperature seasonality (standard deviation)
Bio5	Maximum temperature of the warmest month
Bio6	Minimum temperature of the coldest month
Bio7	Temperature annual range (Bio5– Bio6)
Bio8	Mean temperature of the wettest quarter
Bio9	Mean temperature of the driest quarter
Bio10	Mean temperature of the warmest quarter
Bio11	Mean temperature of the coldest quarter
Bio12	Annual precipitation
Bio13	Precipitation of the wettest month
Bio14	Precipitation of the driest month
Bio15	Precipitation seasonality (coefficient of variation)
Bio16	Precipitation of the wettest quarter
Bio17	Precipitation of the driest quarter
Bio18	Precipitation of the warmest quarter
Bio19	Precipitation of the coldest quarter
Cld1	Mean annual percentage cloud cover
Cld4	Cloud seasonality (standard deviation)
Vap18	Vapor pressure of the warmest quarter
Wet12	Annual number of wet days
Wet18	Number of wet days of the warmest quarter
Frs12	Annual number frost days

(2.9%), confirming that recent conditions have been very warm across all three window sizes. The standard deviation of Bio1 (inter-annual variability over 10, 20, and 30 years) shows a more variable pattern among window sizes, with SD percentiles associated with the most recent windows varying between 25 and 80% (red asterisks, Figure 1D). Hence, estimates of recent inter-annual variability in annual mean temperature are sensitive to window size (max $\Delta = 55\%$).

Given the number and complexity of variables considered in our analysis, we performed two multivariate analyses in an effort to produce more integrated summaries of recent percentiles. Principal component analyses were used to identify a small set of uncorrelated axes for each combination of summary statistic (temporal mean and SD) and measure (mean percentile and max Δ percentile). We also conducted a threshold-based classification analysis of mean percentiles, on both the moving window mean and SD, to characterize which parks were extreme on their annual, monthly, or quarterly temperature (Bio1, 5, 6, 8–11) or precipitation (Bio12–14, 16–19) variables. In this classification analysis, as well as in other evaluations of individual climate variables, we define 'extreme' as either less than the 5th percentile (low) or greater than the 95th percentile (high), and all intervening percentiles as 'average'. We tallied the total number of temperature and precipitation variables that met either of these conditions and categorized each park as extreme low or high on temperature and precipitation if at least one variable in each class was low or high. Parks were considered 'mixed' if they possessed low and high estimates of temperature, or low and high estimates of precipitation. Then, considering only the low and high variables, we calculated the maximum value of their max Δ's to provide a maximum estimate of sensitivity to window size. Given the thresholds used to define 'extreme', the maximum possible value for max Δ is 15%.

For a subset of parks spanning the contiguous US, we conducted another sensitivity analysis using local weather station data to examine the degree to which CRU results differ from those obtained using direct climate observations. For local station data, we identified 18 United States Historical Climatology Network stations located within park boundaries that similarly span the 1901–2012 period of analysis [26]. We repeated the above threshold-based classification analysis and found strong correspondence between datasets. Agreement in classification of extreme warm temperatures based on moving window means occurred at 17 of 18 sites (94%); the station data identified all parks as extreme warm whereas the CRU data found one park that was not extreme. For precipitation moving window means, agreements in classification for extreme wet and extreme dry conditions were 83% and 94%, respectively. Considering extremes of both temperature and precipitation, the CRU and weather station datasets matched in moving window mean classifications at 14 of 18 sites (78%). With moving window standard deviations, agreements in temperature and precipitation classifications were on average 74%. These comparisons suggest that the CRU dataset likely provides reasonable climate estimates for parks in this

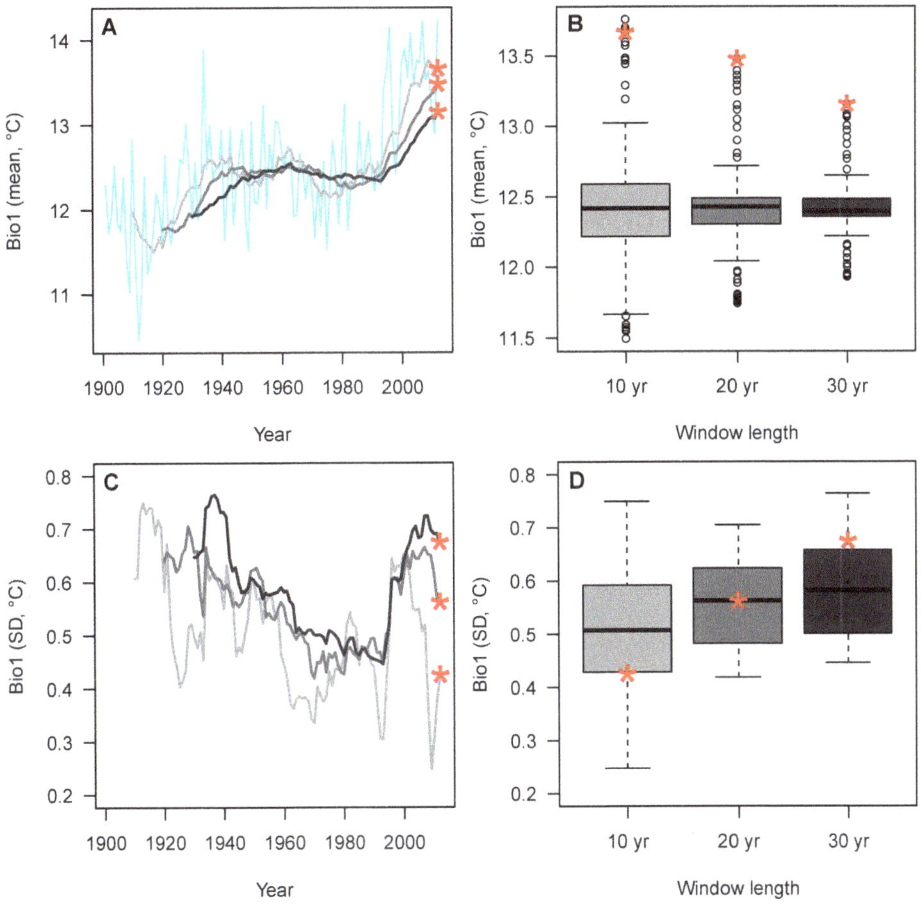

Figure 1. Example moving window time series (used to estimate HRV) and recent percentiles shown for Bio1 (annual mean temperature), Grand Canyon National Park (with 30 km buffer). A, B) Moving window means; C, D) Moving window standard deviations (SD). Three moving windows –10 years (light gray), 20 years (medium gray), and 30 years (dark gray) – are calculated from the annual time series (blue, A). Bio1 values for the most recent windows (2003–2012 (10 yr), 1993–2012 (20 yr), 1983–2012 (30 yr)) are indicated by the red asterisks. Boxes in B and D are the inter-quartile range (median = thick perpendicular line), dashed lines the outer tails (1.5 × inter-quartile range), and dots the outliers. Recent percentiles are calculated for the red asterisks shown in B and D.

analysis, especially considering how the two data sources measure climate at such different spatial scales (see discussion).

Lastly, as part of the CRU-based classification analysis, for each climate variable (Bio1, 5, 6, 8–14, 16–19) we used a Monte Carlo simulation to test whether the number of extreme parks was significantly different than random expectations. To conduct this test, we first randomly selected 100000 samples of size 289 from a uniform distribution (min = 0, max = 1) to develop a sampling distribution of the expected number of extreme parks under the null hypothesis (mean = 14.5, SD = 3.7). Using that sampling distribution, we then computed the probability of obtaining the observed number of extreme parks for each climate variable.

Results

Climatic Variation across the NPS

Parks vary considerably in how their recent climates compare to their 1901–2012 HRV, as estimated by mean percentiles for moving window means and standard deviations (Figure 2). For recent mean percentiles associated with moving window means (Figure 2A, white boxes), parks are most broadly distributed on precipitation variables (Bio12–19) and most skewed on temperature and cloud variables (Bio1, 6, 10, 11, Cld1, 4, Frs12). In particular, recent percentiles for moving window means associated with temperature variables tend to be quite high across most parks (e.g., annual mean temperature (Bio1) is higher than the 50th percentile in 99% of parks), while recent percentiles associated with the annual number of frost days (Frs12) are correspondingly quite low (Figure 2A). The mean diurnal temperature range (Bio2) is also low, suggesting a greater warming of nighttime low temperatures than of daytime highs.

For recent mean percentiles associated with moving window standard deviations (Figure 2A, gray boxes), parks are most skewed low on temperature variables (Bio3, 6, 7) and skewed high on the number of wet days (Wet12, 18) and vapor pressure (Vap18). Hence, recent patterns of inter-annual variability tend to be especially low or high on these variables. However, compared to moving window means, percentiles for moving window standard deviations are overall less skewed and more broadly distributed across parks.

Sensitivity to Moving Window Size

Moving window size has a considerable and consistent effect across most parks and variables on the maximum difference in recent percentile, as evident by inter-quartile ranges often extending to upwards of 30 to 40% (Figure 2B). In other words,

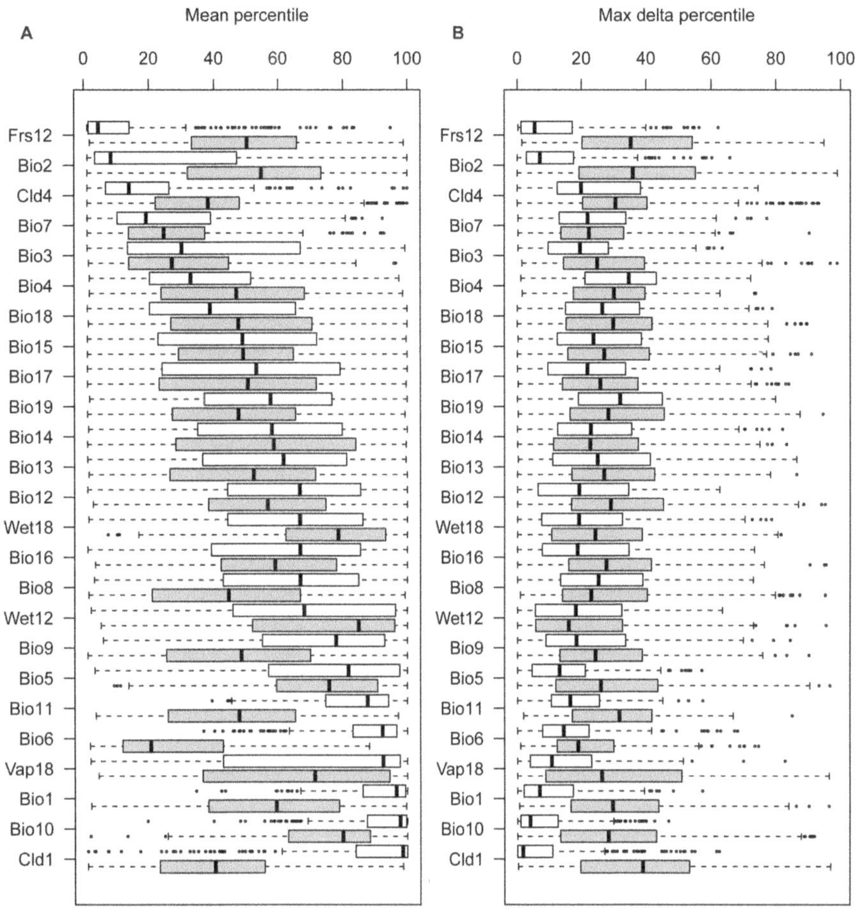

Figure 2. Distribution of recent climate percentiles from each park (with 30 km buffer) for 25 biologically relevant climate variables, showing both moving window means (white boxes) and moving window standard deviation (gray boxes), calculated for three moving windows (10, 20, 30 years). A) Mean percentile across windows; B) Maximum difference in percentile across windows. Boxes are the inter-quartile range (median = thick perpendicular line), dashed lines the outer tails (1.5× inter-quartile range), and dots the outliers. Climate variables are sorted based on median values of mean climate percentiles, thus for a large majority of parks, recent conditions include very low numbers of frost days (Frs12), low diurnal range (Bio2), very warm annual and summer temperatures (Bio1 and 10), and very high cloud cover (Cld1). See Table 1 for definitions of climate variables.

evaluating whether a park is presently low or high on its HRV can depend on the time window of analysis. In terms of moving window means, the same temperature and cloud variables with skewed recent mean percentile distributions (Figure 2A, white boxes) also tend to exhibit low maximum differences in recent percentile (Figure 2B, white boxes). Similar patterns are seen with moving window standard deviations (comparing gray boxes between Figures 2A and 2B).

Contrary to the example shown for Grand Canyon National Park in Figure 1, the maximum difference in recent percentile between moving windows is not always based on 10 and 30 year moving windows. Omitting ties (~10% of the park × variable combinations), 49% of the largest ranges in recent moving window means involve 10 and 30 years, 30% involve 10 and 20 years, and 21% involve 20 and 30 years. For moving window standard deviations, 46% of the largest ranges involve 10 and 30 years, 30% involve 10 and 20 years, and 24% involve 20 and 30 years. Hence, about half the observations do not follow a simple linear trend whereby change in the 10, 20, and 30 year windows is unidirectional.

Climatically Extreme vs. Average Parks

The mean and maximum difference in recent percentile for all parks and variables are given as a series of appendices for moving window means (Appendix S1: Bio1–6; Appendix S2: Bio7–11; Appendix S3: Bio12–19; Appendix S4: Cld1, 4, Vap18, Wet12, 18, Frs12) and standard deviations (Appendix S5: Bio1–6; Appendix S6: Bio7–11; Appendix S7: Bio12–19; Appendix S8: Cld1, 4, Vap18, Wet12, 18, Frs12).

Focusing on two common climate variables, annual mean temperature (Bio1) and annual precipitation (Bio12), geographic distributions of recent percentiles are illustrated for moving window means (Figure 3) and standard deviations (Figure 4). Considering moving window means for annual mean temperature (Figure 3A), practically all natural resource parks are warm with respect to their HRV (262 parks (91%) are greater than the 75th percentile; 158 parks (55%) are greater than the 95th percentile), with low sensitivity to moving window size (125 parks (43%) have a maximum difference in percentile that is less than 5%). This general pattern holds for parks from the Pacific Islands to Alaska to the East Coast of the US.

Corresponding results for annual precipitation (Figure 3B) are rather different: parks are heterogeneous with respect to where

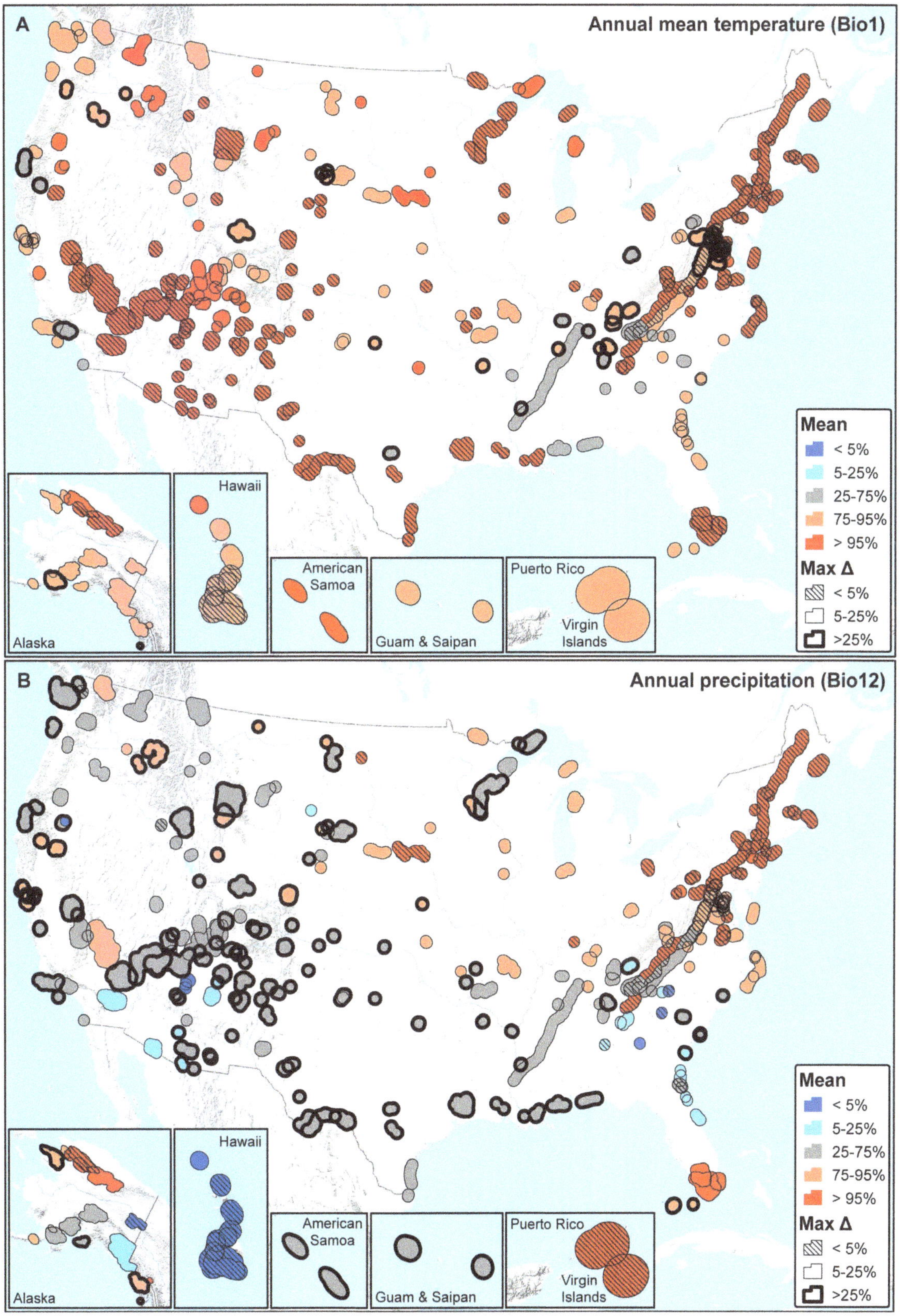

Figure 3. The average (Mean) and maximum difference (Max Δ) of recent percentiles calculated for moving window means. A) Annual mean temperature (Bio1); B) Annual precipitation (Bio12). Mean values provide an overall measure of recent (past 10, 20, and 30 year windows) climate change exposure with respect to 1901–2012 HRV, while the maximum difference measures sensitivity to moving window size (smaller values are less sensitive).

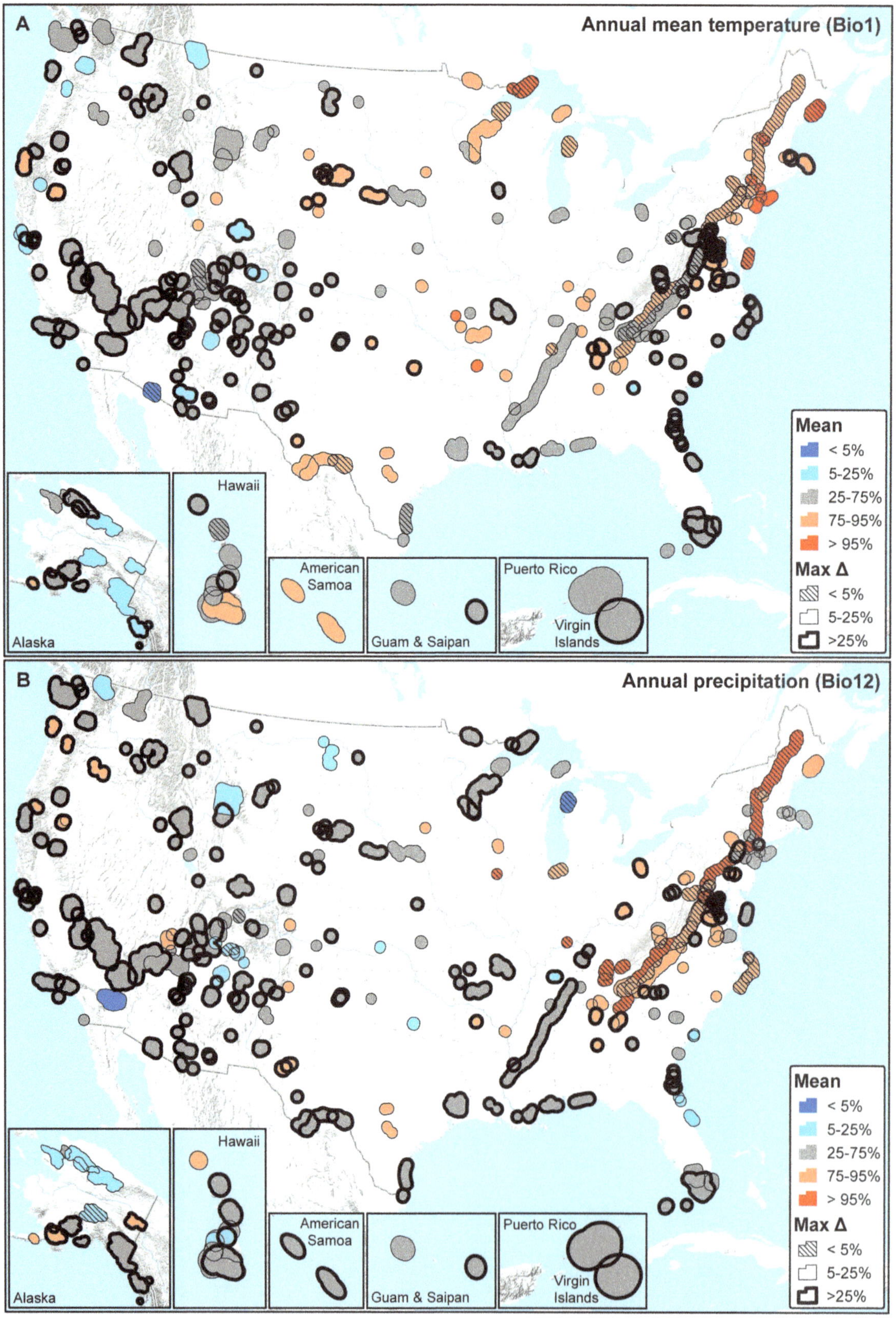

Figure 4. The average (Mean) and maximum difference (Max Δ) of recent percentiles calculated for moving window standard deviations. A) Annual mean temperature (Bio1); B) Annual precipitation (Bio12). Mean values provide an overall measure of recent (past 10, 20, and 30 year windows) climate change exposure with respect to 1901–2012 HRV, while the maximum difference measures sensitivity to moving window size (smaller values are less sensitive).

recent precipitation percentiles fall on their HRV and in many instances are very sensitive to moving window size (i.e., large max Δ). Furthermore, unlike annual mean temperature, most parks have average recent precipitation percentiles: 135 parks (47%) are within the 25th to 75th percentiles and 233 parks (81%) are within the 5th to 95th percentiles. Hence, only 56 parks (19%) are extreme ($<5^{th}$ percentile or $>95^{th}$ percentile) with respect to annual precipitation.

In terms of inter-annual variability, as measured by moving window standard deviations, recent percentiles for both annual mean temperature (Figure 4A) and annual precipitation (Figure 4B) also exhibit complex heterogeneous patterns that are highly sensitive to window size. Most parks have average recent percentiles for inter-annual variability. For annual mean temperature, 176 parks (61%) are within the 25th to 75th percentiles and 275 parks (95%) are within the 5th to 95th percentiles. Annual precipitation is similar: 188 parks (65%) are within the 25th to 75th percentiles and 277 parks (96%) are within the 5th to 95th percentiles. Hence, only about 10 parks are extreme with respect to either annual mean temperature or annual precipitation, and these parks differ between the two variables (Figure 4).

Principal component analyses on the data in Figure 2 reveal further that the percentiles encompass complex correlations that can only be partially captured by a handful of uncorrelated multivariate axes. Principal components 1 and 2 (PC1 and PC2) for the recent mean percentile associated with moving window means explain 38% of the variation (PC1: 21%; PC2: 17%) while those associated with moving window standard deviations explain 31% (PC1: 20%; PC2: 11%). Similarly, PC1 and PC2 for the maximum difference in percentile associated with moving window means explain 26% of the variation (PC1: 15%; PC2: 11%) while those associated with moving window standard deviations explain 21% (PC1: 12%; PC2: 9%). The top two components thus only capture a portion of the information contained by the original 25 variables and suggest strong heterogeneity in climate patterns across parks.

In the classification analysis, considering extreme climates ($<5^{th}$ percentile or $>95^{th}$ percentile) triggered by any one of the temperature (Bio1, 5, 6, 8–11) or precipitation (Bio12–14, 16–19) variables, most parks are presently dominated by extreme high temperatures (235 parks (81%) are higher than the 95th percentile), and – independent of temperature – some also by either extreme high precipitation (78 parks (27%) are higher than the 95th percentile) or low precipitation (43 parks (15%) are lower than the 5th percentile) (Figure 5A). Excluding the 30 parks (10%) that do not exhibit any extreme percentiles with respect to temperature or precipitation, the maximum difference in percentile across moving windows is less than 10% in 84% of the remaining parks (Figure 5A).

The number of parks that are extreme high or low on each temperature or precipitation variable considered in the classification analysis are reported in Table 2 and summarized below. For all temperature variables (Bio1, 5, 6, 8–11), a significantly large number of parks exhibit extreme high temperatures, while a significantly small number of parks exhibit extreme low temperatures (Table 2). The three most frequent extreme high temperature variables are Bio10 (170 parks, 59%), Bio1 (158 parks, 55%), and Bio6 (102 parks, 35%); numbers sum to more than 234 because parks could be extreme on more than one variable. Extreme low temperature is only documented for two variables (Bio5 and 8) in a total of 3 parks (1%). The number of parks with extreme high precipitation is significantly large for four out of 7 variables, e.g., Bio12 and 16 (44 parks each, 15%). Bio14 and 19 are associated with a significantly small number of parks with extreme low values (8 parks each, 3%). Two of the three 'mixed' category parks are explained by concurrent high and low extremes of precipitation. For example, Assateague Island National Seashore in Maryland has experienced extreme warm temperatures (Bio1, 5, 10), extreme wet conditions during the wettest quarter (Bio16), and extreme dry conditions during the driest month (Bio14). The third mixed category park, Santa Monica Mountains National Recreation Area in southern California, has experienced extreme cold conditions for maximum temperature of the warmest month (Bio5), and extreme warm conditions for minimum temperature of the coldest month (Bio6), resulting in an extreme low percentile for temperature annual range (Bio7).

Parks are quite heterogeneous in whether inter-annual variability is high ($>95^{th}$ percentile) or low ($<5^{th}$ percentile) on temperature or precipitation variables (Figure 5B). In terms of temperature, 43 parks (15%) exhibit extreme low inter-annual variability and 69 parks (24%) exhibit extreme high inter-annual variability. For precipitation, independent of temperature, 42 parks (15%) exhibit extreme low inter-annual variability and 76 parks (26%) exhibit extreme high inter-annual variability. Excluding the 108 parks (37%) that are not extreme with respect to either temperature or precipitation inter-annual variability, the maximum difference in percentile across moving windows is less than 10% in 91% of the remaining parks (Figure 5B).

A significantly small number of parks exhibit extreme low temperature inter-annual variability on four variables: Bio5 (0 parks), Bio1 and 10 (1 park each, <1%), and Bio11 (4 parks, 1%; Table 2). For high temperature inter-annual variability, multiple variables exhibit both significantly small and large numbers of extreme parks. For example, fewer parks than expected have high temperature inter-annual variability on Bio6 (0 parks) and Bio8 (5 parks, 2%), while a significantly large number of parks have high inter-annual variability on Bio5 (40 parks, 14%) and Bio10 (33 parks 11%). Patterns of inter-annual variability for precipitation are similar to temperature. A significantly small number of parks exhibit low precipitation inter-annual variability for Bio12 (2 parks, 1%), Bio13 (3 parks, 1%), and Bio16 (1 park, <1%). For high precipitation inter-annual variability, a significantly small number of parks are extreme for Bio19 (6 parks, 2%), while significantly large numbers of parks are extreme for Bio14 (32 parks, 11%) and Bio16 (24 parks, 8%). The 11 'mixed' category parks for inter-annual variability include 8 parks with high and low extremes of temperature and 3 parks with high and low extremes of precipitation.

Finally, we highlight parks from the classification analysis that exemplify particular cases of extreme temperature or precipitation and are supported by maximum differences in percentile that are less than 5%. Parks that exhibit extreme high temperatures and high temperature inter-annual variability include Mammoth Cave National Park in Kentucky and Appomattox Court House National Historical Park in Virginia. Examples of parks that also possess extreme high temperatures but low temperature inter-annual variability include Niobrara National Scenic River in Nebraska and Fort Larned National Historic Site in Kansas. Relative to HRV, pronounced warming characterizes these four parks, but they differ in terms of whether recent extreme warm temperatures have been more or less variable. In terms of precipitation, an example of an extreme wet park with high precipitation inter-annual variability is Chesapeake & Ohio Canal National Historical Park in Washington DC, Maryland, and West Virginia. At the other extreme, Pu'ukohola Heiau National Historic Site in Hawaii is extreme dry with low precipitation inter-annual variability. Parks that have been experiencing

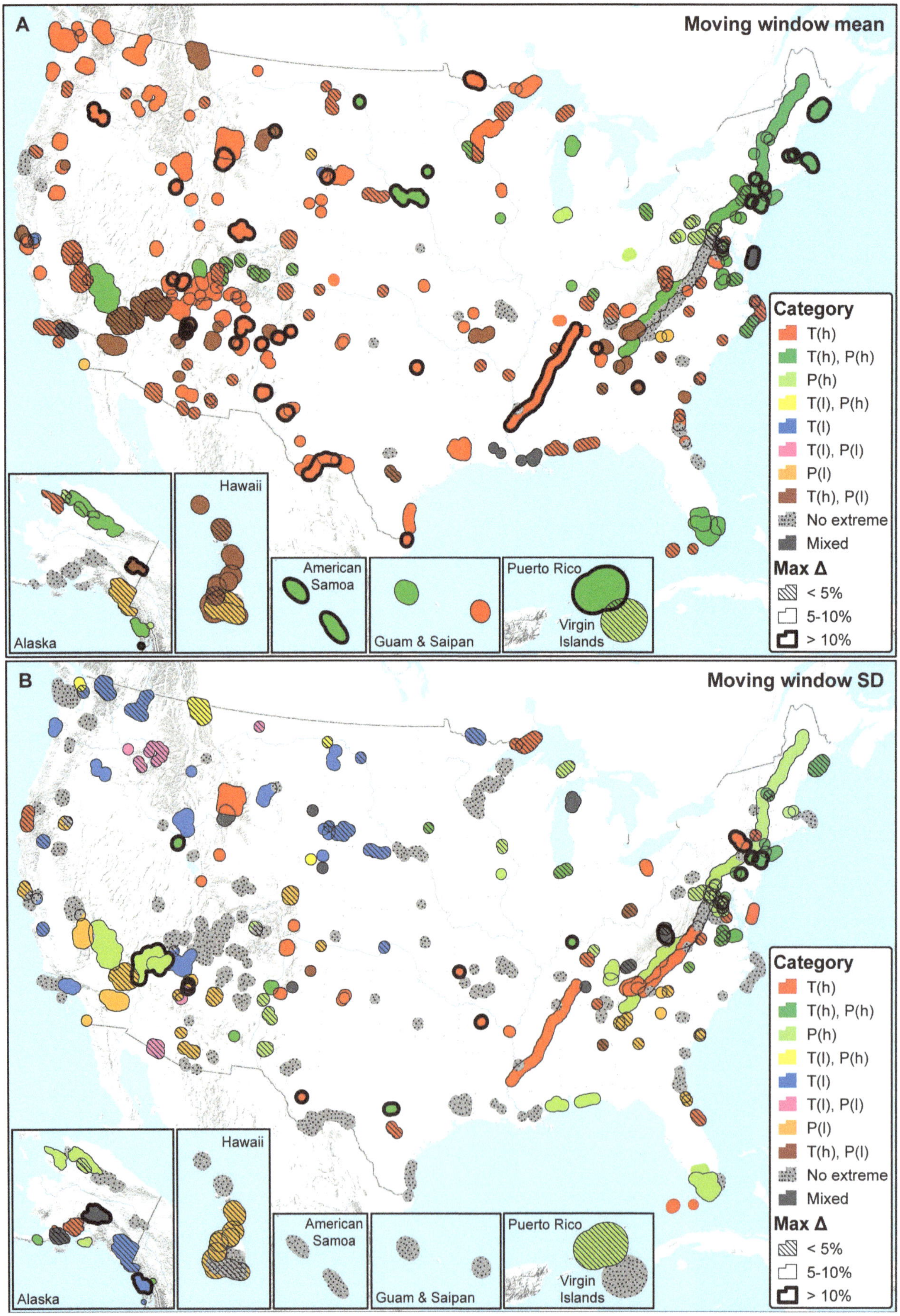
A
Moving window mean
Alaska
Hawaii
American Samoa
Guam & Saipan
Puerto Rico
Virgin Islands
Category
T(h)
T(h), P(h)
P(h)
T(l), P(h)
T(l)
T(l), P(l)
P(l)
T(h), P(l)
No extreme
Mixed
Max Δ
< 5%
5-10%
> 10%
B
Moving window SD
Alaska
Hawaii
American Samoa
Guam & Saipan
Puerto Rico
Virgin Islands
Category
T(h)
T(h), P(h)
P(h)
T(l), P(h)
T(l)
T(l), P(l)
P(l)
T(h), P(l)
No extreme
Mixed
Max Δ
< 5%
5-10%
> 10%

Figure 5. Summary of all temperature (Bio1, 5, 6, 8–11) and precipitation (Bio12–14, 16–19) variables with recent mean percentiles that are either less than the 5th percentile or greater than the 95th percentile (i.e., extreme on HRV). A) Moving window means; B) Moving window standard deviations (SD). T(l) = one or more temperature variables low (<5th percentile; 'cold' or decreased inter-annual variability). T(h) = one or more temperature variables high (>95th percentile; 'warm' or increased inter-annual variability). P(l) = one or more precipitation variables low (<5th percentile; 'dry' or decreased inter-annual variability). P(h) = one or more precipitation variables high (>95th percentile; 'wet' or increased inter-annual variability). Parks with both T(l) and T(h) or both P(l) and P(h) are symbolized as 'mixed'. Parks that fall between the 5th and 95th percentiles on all temperature and precipitation variables are symbolized as 'no extreme'. The maximum difference (Max Δ) in percentile was calculated only for temperature and precipitation variables that were extreme; maximum values are reported to represent maximum sensitivity. Max Δ is undefined in the case of parks symbolized as 'no extreme'; parks that fall into all other categories, and with a max Δ of 0, are also shown without outlines.

extreme warm and dry climates include Kalaupapa National Historical Park in Hawaii, Mojave National Preserve in southern California, and Lake Mead National Recreation Area in Nevada and Arizona. Parks that are extreme warm and wet include Cape Lookout National Seashore in North Carolina, Florissant Fossil Beds National Monument in Colorado, and Delaware Water Gap National Recreation Area in New Jersey and Pennsylvania. These examples represent different categories of extreme climate and may offer insights for understanding how park resources are responding to ongoing changes in climate.

Discussion

Natural resource parks in the US NPS exhibit complex patterns of exposure to climate change when their recent climates are evaluated relative to their 1901–2012 HRV. However, some regional patterns are clearly identified and corroborated by the literature: parks in the desert southwest are warmer and drier [27]; parks in Hawaii are warmer and drier [28]; parks in the northeast are warmer and wetter [29]; parks in the Midwest are warmer [30]; and parks in the southeast exhibit signs of the 'warming hole' [31]. Parks are overwhelmingly at the extreme warm end of historical temperature distributions and this is true for several variables, including annual mean temperature (Bio1), minimum temperature of the coldest month (Bio6), and mean temperature of the warmest quarter (Bio10). Precipitation patterns are more heterogeneous across parks and variables. For example, relative to HRV, precipitation may presently be high during the wettest month (Bio13) and low during the driest month (Bio14) (e.g., Jean Lafitte National Historical Park and Preserve in Louisiana). Climate variability did not show strong or consistent trends, i.e., recent inter-annual variability is generally within the range observed since 1901. Combined, these results suggest that many natural resource parks are already experiencing extreme climatic conditions with respect to their HRV, but these patterns vary across parks, climate variables, summary statistics (temporal mean or SD), and time windows of analysis.

Species within national parks are experiencing extreme climates and measurable plant and animal responses to recent climate change have already been documented [32–35]. The complexity of climate change shown here will likely be mirrored by species and other resources responding to different climate drivers. For example, temperate tree species in the Great Lakes region appear to be responding to summer temperature, while white-tailed deer are more sensitive to winter conditions [36,37]. Additionally,

Table 2. Monte Carlo simulations identifying significantly small (<9) or large (>20) numbers of parks with extreme climates (recent mean percentiles either <5th percentile or >95th percentile), reported for all temperature (Bio1, 5, 6, 8–11) and precipitation (Bio12–14, 16–19) variables considered in the classification analysis.

	Moving window mean		**Moving window SD**	
Variable	**<5%**	**>95%**	**<5%**	**>95%**
Bio1	0****	158****	1***	13
Bio5	2***	93****	0****	40****
Bio6	0****	102****	20	0****
Bio8	1***	39****	10	5**
Bio9	0****	54****	17	9
Bio10	0****	170****	1***	33****
Bio11	0****	58****	4**	1***
Bio12	12	44****	2***	10
Bio13	15	18	3**	10
Bio14	8*	30****	11	32****
Bio16	16	44****	1***	24**
Bio17	17	30****	12	13
Bio18	15	16	10	13
Bio19	8*	13	9	6*

*$P<0.05$,
**$P<0.01$,
***$P<0.001$,
****$P<0.0001$.

species adapted to more extreme precipitation regimes may show positive responses to warming temperatures, whereas mesic species may show increased stress [38]. These differential responses are likely to cause natural communities to disassemble and novel communities to form [39,40]. Climate complexity – as well as complex interactions between climate and other global change agents [5] – require that land managers possess tremendous scientific knowledge about the ecosystems under their jurisdiction, and develop coherent and achievable resource management goals.

Hansen et al. [9] identify HRV and DFC as different management philosophies necessitated by the degree to which past, present, and future ecological conditions overlap. Parks like Grand Canyon National Park, with natural resources that are sensitive to temperature (e.g., annual mean temperature, Figure 1), may require DFC management strategies because their temperature trajectories lead to a complete decoupling of past, present, and future conditions. Such a decision, however, should be resource specific, and especially in the case of culturally significant natural resources (e.g., cultural landscapes, seascapes, glaciers, and iconic organisms) may consider resistance as a viable, short-term management realization of DFC. For example, the General Sherman giant sequoia in Sequoia National Park is an iconic organism of cultural significance that – should it be threatened directly or indirectly by climate change through drought or fire – could receive intensive management intervention (i.e., it could be proactively maintained within its HRV). As climate shifts further outside of HRV bounds, resistance strategies will likely become less effective and extremely difficult management decisions – with input from the public and stakeholder groups – will be required. Indeed, because US national parks are for the benefit and enjoyment of the people [41] (the current motto is "Experience Your America"), many park resources will be subject to these decisions requiring compromise solutions that weigh results emerging from climate science against a suite of public values and socioeconomic considerations.

Estimates of future climate are – to varying degrees – uncertain, especially for particular variables considered in our analysis (e.g., cloud cover, vapor pressure) [42,43]. Some CRU variables (e.g., cloud cover, vapor pressure, number of frost and wet days) are also more uncertain than others (e.g., temperature and precipitation, used to derive bioclimatic variables), owing to a comparably sparse record of observation and the need to initially model and interpolate at a coarser resolution [16]. Nevertheless, even considering just temperature and precipitation, climate projections for the 21st century suggest many park geographies will become warmer and either drier or wetter [44]. Regionally, these predictions generally follow the same patterns and trends seen in our results. In other words, climate change is ongoing, and parks are already experiencing changes that can be documented without having to necessarily look out an additional 50 or 100 years. Importantly, future changes in temperature and precipitation will likely push many parks beyond the limits of their HRV. For example, recent mean temperature of the warmest quarter (Bio10) at Apostle Islands National Lakeshore in Wisconsin has already reached its 100th percentile, meaning that any continued increase in temperature will push the park higher than all warm quarter temperatures it has experienced since 1901. Similarly, many parks are already extreme dry or wet; if these observed extremes are followed by future changes in the same direction, then affected parks will experience precipitation regimes unlike any they have seen in over a century. Furthermore, changes in some climate variables may exacerbate or override shifts in other variables. For example, extreme warm temperatures will cause considerably drier conditions, even if precipitation amounts remain within HRV. Although it is reasonable to expect HRV to increase with time, and indeed there may be real biological, ecological, or cultural reasons to consider longer periods [12], other practical management and planning considerations affect the determination of a meaningful HRV baseline. The year 1901 predates the establishment of the NPS by over a decade [41] and now many parks are transcending all climates they have experienced during their entire tenure. In this sense, 1901–2012 represents a reasonable baseline, and HRV analyses based on it suggest that the NPS is entering a new era of change. Furthermore, given that recent temperatures may be extreme even compared to longer baselines [11], the present analyses likely offer conservative estimates of temperature extremes in parks.

Management and planning decisions will be greatly informed by an overall assessment of vulnerability that integrates the present climate change exposure analyses with others that consider both the sensitivity and adaptive capacity of park resources to climate change [9,45–48]. Beyond climate, additional change agents with additive and interactive effects (e.g., land use, pollution, biological invasions, fire) could also be profitably included as other analyses of exposure [49]. Here, a principal challenge will be developing vulnerability models that can be parameterized at the desired spatial and temporal scales of analysis. Given tradeoffs between the extent and resolution characteristics of scale (e.g., a long time series (extent) may be restricted to a monthly instead of daily time step (resolution)), vulnerability models will be subject to some of the classical model tradeoffs involving realism, precision, and generality [50]. Although there is an unstated desire in vulnerability assessments to maximize realism and precision at the cost of generality (i.e., collect site-level data and develop highly parameterized models that predict ecological complexity), we draw attention to the inability of these models to encompass broad geographic extents and suggest that vulnerability models maximizing generality and realism at the cost of precision stand to play an especially pivotal role in planning and policy.

Our measures of climate change exposure support broad-scale vulnerability assessments in the NPS, but they also help individual parks understand and interpret which of their major climate drivers are beginning to approach the limits of 1901–2012 HRV. Such insights facilitate discussion of how climate change may impact diverse park resources and values (e.g., natural resources, cultural resources, facilities, visitor experience, wilderness character), which are often sensitive to different measures of exposure, but they are not intended to replace more detailed analyses required for site-level management and planning within park boundaries. Many parks are in areas of complex topography and indeed this heterogeneity has important implications for climate adaptation [51,52]. We suggest that the present analyses may be used to prioritize and select climate variables for more park-specific analyses of meso- and topo-climatic exposure. This 'coarse filter' approach is valuable because it allows parks to initiate resource-specific assessments with *a priori* knowledge of how major climate drivers have changed in recent time. For resources influenced by climate variables that are presently well within the bounds of HRV, parks might decide that limited funding should be redirected to study other resources that are sensitive to variables approaching the limits of HRV. Parks might also decide to monitor variables with recent extreme percentiles to assess rates of change and detect when and where they become 'novel' with respect to 1901–2010 HRV. Beyond applications related to monitoring and management, results offer park interpreters scientific information that may be translated and communicated to the public.

In approaching the 100-year anniversary of its creation in 2016, the NPS is poised to enter its next century of natural resource stewardship and science. The new century brings new challenges in terms of stewarding park resources in the face of environmental drivers that operate beyond park boundaries. Climate change further challenges us to develop new, ecologically viable desired conditions to guide the preservation of park resources in this new era of change. While such challenges remain paramount, more integrative research and education in the climate and landscape arenas will contribute to solutions.

Supporting Information

Appendix S1 Mean percentile and maximum difference in percentile (in parentheses) for moving window means: annual mean temperature (Bio1), mean diurnal range (Bio2), isothermality (Bio3), temperature seasonality (Bio4), maximum temperature of the warmest month (Bio5), and minimum temperature of the coldest month (Bio6).

Appendix S2 Mean percentile and maximum difference in percentile (in parentheses) for moving window means: temperature annual range (Bio7), mean temperature of the wettest quarter (Bio8), mean temperature of the driest quarter (Bio9), mean temperature of the warmest quarter (Bio10), and mean temperature of the coldest quarter (Bio11).

Appendix S3 Mean percentile and maximum difference in percentile (in parentheses) for moving window means: annual precipitation (Bio12), precipitation of the wettest month (Bio13), precipitation of the driest month (Bio14), precipitation seasonality (Bio15), precipitation of the wettest quarter (Bio16), precipitation of the driest quarter (Bio17), precipitation of the warmest quarter (Bio18), and precipitation of the coldest quarter (Bio19).

Appendix S4 Mean percentile and maximum difference in percentile (in parentheses) for moving window means: mean annual percentage cloud cover (Cld1), cloud seasonality (Cld4), vapor pressure of the warmest quarter (Vap18), annual number of wet days (Wet12), number of wet days of the warmest quarter (Wet18), and annual number frost days (Frs12). NA's associated with Frs12 indicate that the variable is undefined (i.e., no frost days exist).

Appendix S5 Mean percentile and maximum difference in percentile (in parentheses) for moving window standard deviations: annual mean temperature (Bio1), mean diurnal range (Bio2), isothermality (Bio3), temperature seasonality (Bio4), maximum temperature of the warmest month (Bio5), and minimum temperature of the coldest month (Bio6).

Appendix S6 Mean percentile and maximum difference in percentile (in parentheses) for moving window standard deviations: temperature annual range (Bio7), mean temperature of the wettest quarter (Bio8), mean temperature of the driest quarter (Bio9), mean temperature of the warmest quarter (Bio10), and mean temperature of the coldest quarter (Bio11).

Appendix S7 Mean percentile and maximum difference in percentile (in parentheses) for moving window standard deviations: annual precipitation (Bio12), precipitation of the wettest month (Bio13), precipitation of the driest month (Bio14), precipitation seasonality (Bio15), precipitation of the wettest quarter (Bio16), precipitation of the driest quarter (Bio17), precipitation of the warmest quarter (Bio18), and precipitation of the coldest quarter (Bio19).

Appendix S8 Mean percentile and maximum difference in percentile (in parentheses) for moving window standard deviations: mean annual percentage cloud cover (Cld1), cloud seasonality (Cld4), vapor pressure of the warmest quarter (Vap18), annual number of wet days (Wet12), number of wet days of the warmest quarter (Wet18), and annual number frost days (Frs12). NA's associated with Frs12 indicate that the variable is undefined (i.e., no frost days exist).

Acknowledgments

This work was conducted in coordination with NPScape and the NASA-NPS Landscape Climate Change Vulnerability Project. We thank reviewers and colleagues who provided comments that greatly improved an earlier version of this manuscript: John Dennis, Brent Frakes, John Gross, Andy Hansen, Cat Hawkins Hoffman, Patrick Jantz, Jodi Norris, Tom Olliff, Gregor Schuurman, and Scott Zolkos.

Author Contributions

Conceived and designed the experiments: WBM NAF. Performed the experiments: WBM NAF. Analyzed the data: WBM NAF. Contributed reagents/materials/analysis tools: WBM NAF. Wrote the paper: WBM NAF.

References

1. National Parks Second Century Commission (2009) Advancing the National Park Idea: Science and Natural Resource Committee Report. National Parks Conservation Association. Available: http://www.npca.org/assets/pdf/Committee_Science.PDF. Accessed 2014 May.
2. National Academy of Public Administration (2010) Strengthening America's Best Idea: An Independent Review of the National Park Service's Natural Resource Stewardship and Science Directorate. Washington DC: National Academy of Public Administration. Available: http://www.napawash.org/wp-content/uploads/2010/08/09-05.pdf. Accessed 2014 May.
3. National Park System Advisory Board (2012) Revisiting Leopold: Resource Stewardship in the National Parks. Washington DC: National Park System Advisory Board. Available: http://www.nps.gov/calltoaction/PDF/LeopoldReport_2012.pdf. Accessed 2014 May.
4. Cole DN, Landers PB (1996) Threats to wilderness ecosystems: Impacts and research needs. Ecol Appl 6: 168–184.
5. Heller NE, Zavaleta ES (2009) Biodiversity management in the face of climate change: A review of 22 years of recommendations. Biol Conserv 142: 14–32.
6. Kareiva P, Enquist C, Johnson A, Julius S, Lawler J, et al. (2008) Synthesis and conclusions. In: Julius SH, West J, editors. Preliminary Review of Adaptation Options for Climate-Sensitive Ecosystems and Resources. Final Report, Synthesis and Assessment Product 4.4: US Climate Change Science Program and the Subcommittee on Global Change Research. 622–689.
7. National Park Service (2010) Climate Change Response Strategy. Fort Collins: National Park Service Climate Change Response Program. Available: http://www.nature.nps.gov/climatechange/docs/NPS_CCRS.pdf. Accessed 2014 May.

8. Harris JA, Hobbs RJ, Higgs E, Aronson J (2006) Ecological restoration and global climate change. Restor Ecol 14: 170–176.
9. Hansen AJ, Piekielek N, Davis D, Haas J, Theobald D, et al. (2014) Exposure of US National Parks to land use and climate change 1900–2100. Ecol Appl: 24: 484–502.
10. Mora C, Frazier AG, Longman RJ, Dacks RS, Walton MM, et al. (2013) The projected timing of climate departure from recent variability. Nature 502: 183–187.
11. Intergovernmental Panel on Climate Change (2013) Climate Change 2013– The Physical Science Basis – Summary for Policymakers. Switzerland: Intergovernmental Panel on Climate Change.
12. Keane RE, Hessburg PF, Landres PB, Swanson FJ (2009) The use of historical range and variability (HRV) in landscape management. Forest Ecol Manag 258: 1025–1037.
13. Hameed SO, Holzer KA, Doerr AN, Baty JH, Schwartz MW (2013) The value of a multi-faceted climate change vulnerability assessment to managing protected lands: Lessons from a case study in Point Reyes National Seashore. J Environ Manage 121: 37–47.
14. National Park Service (2011) NPScape NPS boundary-derived areas of analysis SOP: Park, 3 km, and 30 km areas of analysis. Fort Collins: Natural Resource Program Center. Available: from https://irma.nps.gov/App/Reference/Profile/2170522?lnv = True. Accessed 2014 May.
15. Hansen AJ, Davis CR, Piekielek N, Gross J, Theobald DM, et al. (2011) Delineating the ecosystems containing protected areas for monitoring and management. BioScience 61: 363–373.
16. Harris I, Jones PD, Osborn TJ, Lister DH (2014) Updated high-resolution grids of monthly climatic observations – the CRU TS3.10 Dataset. Int J Climatol 34: 623–642. doi:10.1002/joc.3711.
17. Hijmans R (2006) Bioclim-aml, version 2.3. Berkeley: Museum of Vertebrate Zoology. Available: http://www.worldclim.org/mkBCvars.aml. Accessed 2014 May.
18. Fritts HC (1976) Tree rings and climate. New York: Academic Press.
19. Lambers H, Chapin FS, Pons TL (1998) Plant Physiological Ecology. New York: Springer.
20. National Park Service (2004) Program Standards: Park Planning. Available: http://planning.nps.gov/document/aug2004standards.pdf. Accessed 2014 May.
21. MetOffice (2014) Climate averages. Available: http://www.metoffice.gov.uk/climate/uk/averages/. Accessed 2014 May.
22. Stenseth NC, Ottersen G, Hurrell JW, Mysterud A, Lima M, et al. (2003) Studying climate effects on ecology through the use of climate indices: The North Atlantic Oscillation, El Nino Southern Oscillation and beyond. P Roy Soc Lond B Bio 270: 2087–2096.
23. National Park Service (2013) NPScape Standard Operating Procedure: Climate Grid Analysis Toolset. Fort Collins: Natural Resource Stewardship and Science. Available: https://irma.nps.gov/App/Reference/Profile/2193159?lnv = True. Accessed 2014 May.
24. R Core Team (2014) R: A Language and Environment for Statistical Computing. R Foundation for Statistical Computing, Vienna, Austria. Available: http://www.R-project.org/. Accessed 2014 May.
25. Parmesan C, Root TL, Willig MR (2000) Impacts of extreme weather and climate on terrestrial biota. B Am Meteorol Soc 81: 443–450.
26. Menne MJ, Williams Jr CN, Vose RS (2013) United States Historical Climatology Network (USHCN) Version 2.5 Serial Monthly Dataset. Carbon Dioxide Information Analysis Center, Oak Ridge National Laboratory, Oak Ridge, Tennessee. Available: http://cdiac.ornl.gov/ftp/ushcn_v2.5_monthly/. Accessed 2014 May.
27. Garfin G, Franco G, Blanco H, Comrie A, Gonzalez P, et al. (2014) Chapter 20: Southwest. Climate Change Impacts in the United States: The Third National Climate Assessment. In: Melillo JM, Richmond TC, Yohe GW, editors. U.S. Global Change Research Program, 462–486. doi:10.7930/J08G8HMN.
28. Leong JA, Marra JJ, Finucane ML, Giambelluca T, Merrifield M, et al. (2014) Chapter 23: Hawai'i and U.S. Affiliated Pacific Islands. Climate Change Impacts in the United States: The Third National Climate Assessment. In: Melillo JM, Richmond TC, Yohe GW, editors. U.S. Global Change Research Program, 537–556. doi:10.7930/J0W66HPM.
29. Horton R, Yohe G, Easterling W, Kates R, Ruth M, et al. (2014) Chapter 16: Northeast. Climate Change Impacts in the United States: The Third National Climate Assessment. In: Melillo JM, Richmond TC, Yohe GW, editors. U.S. Global Change Research Program, 371–395. doi:10.7930/J0SF2T3P.
30. Pryor SC, Scavia D, Downer C, Gaden M, Iverson L, et al. (2014) Chapter 18: Midwest. Climate Change Impacts in the United States: The Third National Climate Assessment. In: Melillo JM, Richmond TC, Yohe GW, editors. U.S. Global Change Research Program, 418–440. doi:10.7930/J0J1012N.
31. Carter LM, Jones JW, Berry L, Burkett V, Murley JF, et al. (2014) Chapter 17: Southeast and the Caribbean. Climate Change Impacts in the United States: The Third National Climate Assessment. In: Melillo JM, Richmond TC, Yohe GW, editors. U.S. Global Change Research Program, 396–417. doi:10.7930/J0NP22CB.
32. Moritz C, Patton JL, Conroy CJ, Parra JL, White GC, et al. (2008) Impact of a century of climate change on small-mammal communities in Yosemite National Park, USA. Science 322: 261–264.
33. Tingley MW, Monahan WB, Beissinger SR, Moritz C (2009) Birds track their Grinnellian niche through a century of climate change. Proc Natl Acad Sci U S A 106: 19637–19643.
34. Rubidge EM, Monahan WB, Parra JL, Cameron SE, Brashares JS (2011) The role of climate, habitat, and species co-occurrence as drivers of change in small mammal distributions over the past century. Glob Chang Biol 17: 696–708.
35. Dolanc CR, Thorne JH, Safford HD (2013) Widespread shifts in the demographic structure of subalpine forests in the Sierra Nevada, California, 1934 to 2007. Global Ecol Biogeogr 22: 264–276.
36. Côté SD, Rooney TP, Tremblay JP, Dussault C, Waller DM (2004) Ecological impacts of deer overabundance. Annu Rev Ecol Evol Syst 35: 113–147.
37. Fisichelli NA, Frelich LE, Reich PB (2012) Sapling growth responses to warmer temperatures 'cooled' by browse pressure. Glob Change Biol 18: 3455–3463.
38. Knapp AK, Beier C, Briske DD, Classen AT, Luo Y, et al. (2008) Consequences of more extreme precipitation regimes for terrestrial ecosystems. BioScience 58: 811–821.
39. Hobbs RJ, Arico S, Aronson J, Baron JS, Bridgewater P, et al. (2006) Novel ecosystems: Theoretical and management aspects of the new ecological world order. Global Ecol Biogeogr 15: 1–7.
40. Hobbs RJ, Higgs E, Harris JA (2009) Novel ecosystems: Implications for conservation and restoration. Trends Ecol Evol 24: 599–605.
41. National Park Service Organic Act (1916) 16 USC 1–4; Aug. 25, 1916, ch. 408, 39 Stat. 535.
42. Rosenfeld D, Sherwood S, Wood R, Donner L (2014) Climate effects of aerosol-cloud interactions. Science 343: 379–380.
43. Sherwood SC, Bony S, Dufresne JL (2014) Spread in model climate sensitivity traced to atmospheric convective mixing. Nature 505: 37–42.
44. Walsh J, Wuebbles D, Hayhoe K, Kossin J, Kunkel K, et al. (2014) Chapter 2: Our Changing Climate. Climate Change Impacts in the United States: The Third National Climate Assessment. In: Melillo JM, Richmond TC, Yohe GW, editors. U.S. Global Change Research Program, 19–67. doi:10.7930/J0KW5CXT.
45. Dawson TP, Jackson ST, House JI, Prentice IC, Mace GM (2011) Beyond predictions: Biodiversity conservation in a changing climate. Science 332: 53–58.
46. Glick P, Stein BA, Edelson NA, editors (2011) Scanning the Conservation Horizon: A Guide to Climate Change Vulnerability Assessment. Washington DC: National Wildlife Federation.
47. Klausmeyer KR, Shaw MR, MacKenzie JB, Cameron DR (2011) Landscape-scale indicators of biodiversity's vulnerability to climate change. Ecosphere 2(8): art88. doi:10.1890/ES11-00044.1.
48. Magness DR, Morton JM, Huettmann F, Chapin FS III, McGuire AD (2011) A climate-change adaptation framework to reduce continental-scale vulnerability across conservation reserves. Ecosphere 2(10): 112. doi:10.1890/ES11-00200.1.
49. Fisichelli NA, Abella SR, Peters M, Krist FJ Jr (2014) Climate, trees, pests, and weeds: Change, uncertainty, and biotic stressors in eastern U.S. national park forests. Forest Ecol Manag 327: 31–39.
50. Levins R (1966) The strategy of model building in population biology. Am Sci 54: 421–431.
51. Loarie SR, Duffy PB, Hamilton H, Asner GP, Field CB, et al. (2009) The velocity of climate change. Nature 462: 1052–1055.
52. Ackerly DD, Loarie SR, Cornwell WK, Weiss SB, Hamilton H, et al. (2010) The geography of climate change: Implications for conservation biogeography. Divers Distrib 16: 476–487.

7

Ground Layer Plant Species Turnover and Beta Diversity in Southern-European Old-Growth Forests

Francesco Maria Sabatini[1]*, Sabina Burrascano[1], Hanna Tuomisto[2], Carlo Blasi[1]

1 Department of Environmental Biology, Sapienza, University of Rome, Rome, Italy, **2** Department of Biology, University of Turku, Turku, Finland

Abstract

Different assembly processes may simultaneously affect local-scale variation of species composition in temperate old-growth forests. Ground layer species diversity reflects chance colonization and persistence of low-dispersal species, as well as fine-scale environmental heterogeneity. The latter depends on both purely abiotic factors, such as soil properties and topography, and factors primarily determined by overstorey structure, such as light availability. Understanding the degree to which plant diversity in old-growth forests is associated with structural heterogeneity and/or to dispersal limitation will help assessing the effectiveness of silvicultural practices that recreate old-growth patterns and structures for the conservation or restoration of plant diversity. We used a nested sampling design to assess fine-scale species turnover, i.e. the proportion of species composition that changes among sampling units, across 11 beech-dominated old-growth forests in Southern Europe. For each stand, we also measured a wide range of environmental and structural variables that might explain ground layer species turnover. Our aim was to quantify the relative importance of dispersal limitation in comparison to that of stand structural heterogeneity while controlling for other sources of environmental heterogeneity. For this purpose, we used multiple regression on distance matrices at the within-stand extent, and mixed effect models at the extent of the whole dataset. Species turnover was best predicted by structural and environmental heterogeneity, especially by differences in light availability and in topsoil nutrient concentration and texture. Spatial distances were significant only in four out of eleven stands with a relatively low explanatory power. This suggests that structural heterogeneity is a more important driver of local-scale ground layer species turnover than dispersal limitation in southern European old-growth beech forests.

Editor: Ben Bond-Lamberty, DOE Pacific Northwest National Laboratory, United States of America

Funding: The original project was funded by two grants of Sapienza, University of Rome: 1. grant number C26N12ESSE - F.M. Sabatini (PI); 2. grant number C26A128K4A - C. Blasi (PI). The funders had no role in study design, data collection and analysis, decision to publish, or preparation of the manuscript.

Competing Interests: The authors have declared that no competing interests exist.

* E-mail: francescomaria.sabatini@uniroma1.it

Introduction

The composition of plant species assemblages varies in space and time as a result of the complex interplay among several structuring factors [1]. Spatial distribution of plant species depends on different mechanisms related to their tolerances to environmental factors, intra- and inter-specific interactions (e.g. dispersal, competition, herbivory, pathogens) and random variation. Given the theoretical and practical relevance of understanding these drivers, they have received much attention in recent years [2–7].

The relative importance of the mechanisms that generate floristic turnover can differ widely among systems as a consequence of habitat type, biogeographical context, plant life-history traits and other local contingencies [5,8,9]. The spatial scale of the study and the length of the relevant ecological gradients may also have an effect. When the sampled ecological gradient is long, i.e. when a wide range of environmental conditions are considered, resource-driven processes are expected to have a stronger influence on the community than when the gradient is truncated, since focusing on small geographic extents and relatively homogeneous habitats can lead to higher noise-to-signal ratios [10–13]. At local scales, therefore, floristic patterns may be expected to mostly reflect the effects of dispersal limitation and random survival [7,14]. This complexity has fostered a long-standing debate on the degree to which community assembly is a result of niche vs. neutral processes, which was revived after the publication of Hubbell's book on the neutral theory of biodiversity [15]. Most studies addressing the issue since then have found both neutral and niche processes to operate simultaneously [2,8,9,14,16].

In temperate forests, niche and neutral processes may concurrently control local-scale variation of ground layer species composition, although their relative roles may vary with successional stage. As late-successional stands develop, structural heterogeneity (vertical and horizontal) and the number and variety of ecological niches (e.g. microhabitat) increase [17–19]. The persistence of relatively stable ecological conditions for a long time (ecological continuity), may allow for the emergence of patterns due to the colonization and persistence of low-dispersal species dependent on very old, undisturbed forests [20]. Both the accumulation of microhabitats and the colonization of forest interior species depend on time since last disturbance, although it is not clear what their relative roles are in determining local-scale variation of ground layer composition.

An integrated approach is necessary to assess the individual contributions of structural heterogeneity and dispersal limitation to explaining species turnover, because most ecologically relevant variables are spatially autocorrelated [5,7]. Understorey plants can be affected by factors determined by overstorey structure, such as light availability, but also by purely abiotic factors, such as soil properties and topography. In principle, after accounting for all sources of environmental heterogeneity, the residual correlation between geographical distances and species turnover can be considered a proxy of spatially autocorrelated biological processes such as dispersal limitation or other biotic interactions [21].

The study of the mechanisms that cause biological diversity to accumulate and be maintained in natural ecosystems at different spatial scales and under different conditions is an essential foundation of management and conservation strategies implemented at the local, landscape and regional levels [3,22,23]. Old-growth forests represent a reference point for evaluating human impacts on other forest ecosystems and for improving current sustainable forest management practices. These forests have avoided severe disturbance for centuries, and species distribution can therefore be expected to reflect environmental gradients and demographic dynamics more clearly than in managed forests, where anthropogenic contingencies (such as land-use history and forest management) dominate [4,24–26]. Old-growth remnants, thus represent an ideal situation for investigating the dynamics and patterns of ground layer vegetation under natural conditions, although their relative scarcity in several biogeographic regions is a challenge for unravelling the drivers of community assembly. This is particularly true for Europe, where forests have been impacted by man for millennia and are likely to be, on average, more managed than forests in most parts of the world [24]. Understanding the degree to which plant diversity in old-growth forests is associated with structural heterogeneity and/or to dispersal limitation will help assessing the effectiveness of silvicultural practices that recreate old-growth patterns and structures for the conservation or restoration of plant diversity.

Here, we used a nested sampling design to study beta diversity and species turnover (i.e. the proportion of species composition that changes among sampling units (*sensu* Tuomisto [36,37]) across 11 of the best preserved beech forest stands with old-growth characteristics in Southern Europe. We primarily focused on ground layer flora because it hosts the vast majority of forest biodiversity [27], and plays an important role in the structure and functioning of forest ecosystems due to its influence on nutrient fluxes, tree regeneration, successional patterns and light regime [28,29].

We hypothesized that after accounting for the intrinsic heterogeneity due to abiotic factors, most of the variation in ground layer species turnover is related to stand structural heterogeneity. However, we also expected to observe a significant effect of dispersal limitation [7,14,23]. Finally, we hypothesized that the relative importance of different environmental and structural variables would vary across the stands according to their degree of fine-scale heterogeneity [10,11].

Materials and Methods

Study Sites

We selected 11 forests identified in the literature as being old-growth or having old-growth characteristics. These stands encompass four different countries in Southern Europe (Italy, Spain, Bosnia-Herzegovina and Montenegro) and are dominated by European beech (*Fagus sylvatica*); codominant species vary across stands and encompass silver fir (*Abies alba*), Turkish oak (*Quercus cerris*), or sessile oak (*Quercus petraea*). All study sites belong to the temperate oceanic bioclimate [30]. Eight out of 11 stands grow on sedimentary bedrock (limestone, marlstone, sandstone), two on eruptive rock, and one on metamorphic rock (Table 1). All necessary permits were obtained for the described study, which complied with all relevant regulations. Depending on the forest stand, the permits were issued by the competent National Parks' authorities, Italian State Forestry Corps, ARIF Puglia or by the Consejería de Agroganadería y Recursos Autóctonos, Gobierno del Principado de Asturias.

Sampling design

We used a nested sampling design. In each forest stand, we set a network of 25 sampling quadrats regularly distributed in a 1-ha square macroplot. Quadrats were 5 m×5 m and were located at the centres of a regular grid consisting of 20 m×20 m cells. For each quadrat we sampled vascular plant species composition, forest structure (live and deadwood) and environmental variables (topography, soil, light). The macroplots were positioned in the field in the same locations where previous research projects had investigated structural features of the stands [31,32].

The vegetation was divided into three layers: tree (height >3 m), shrub (1.3< height ≤3 m) and ground layer (height ≤1.3 m). We estimated plant cover using an ordinal cover class scale with class limits 0.5%, 1%, 2%, 5%, 10%, 15%, 20%, and thereafter every 10% up to 100% (each class includes its upper limit). Each plant species was assigned a cover value in each layer separately through visual estimation, and in addition we recorded the total cover of the tree and shrub layer.

For all trees with dbh >2.5 cm situated inside the quadrats, we recorded the species, diameter at breast height (dbh), and vitality class (1 = vigorous; 2 = living with dead parts; 3 = standing dead; 4 = broken above 1.3 m). This information was used to calculate total basal area of live trees within the quadrats. We also used a wedge prism to estimate the basal area per hectare on the basis of a variable radius plot centred on each quadrat. Hereafter, we will refer to the former measure as 'quadrat basal area' and to the latter as 'prism basal area'. For each quadrat we also measured the position, species and diameter of the four large live trees with dbh>40 cm closest to the quadrat centre. These data were used to calculate three indices of structural heterogeneity [33]. The first describes spatial distribution of large live trees (*Uniform Angle Index*, also known as Winkelmass Index), the second species clustering (*Species Mingling Index*) and the third diameter heterogeneity (*Modified Dbh Dominance Index* or DBHDM). The average of four readings of a hemispherical densiometer taken in the four cardinal directions was used to estimate the openness of the overstorey canopy cover ('canopy openness' hereafter).

Diffuse light transmittance of photosynthetically active radiation (μmol/m^2•sec, hereafter 'PAR') at 1 m height was measured using a *Licor Line Quantum Sensor - Li-191* under uniform sky conditions at three different times of day: morning (10–11 am), noon (12.30–1.30 pm) and afternoon (3–4 pm). In the stand Muniellos, all measurements were taken an hour later. Two measurements were taken within each quadrat, 1.5 m north and 1.5 m south of the quadrat centre, and their average was used in the analyses. An overall average of all six measurements per quadrat was also calculated.

For every deadwood piece of at least 10 cm length and 5 cm minimum diameter occurring in the quadrat, we measured the two end-diameters and length, identified the species when possible and assigned the piece to one of five decay classes [34]. Each quadrat was characterised by the total number of such deadwood pieces, their total volume (assuming a truncated cone shape for each piece), the number of different decay classes present, and the most advanced state of decay present.

The following micro-topographical variables were recorded for each quadrat: slope, aspect, topographical position (ridge top,

Table 1. Location and main features of the 11 stands with old-growth characteristics included in this study.

Forest Name	Location	Nation	Altitude	Average mean Temp. (°C)	Average Annual Prec. (mm)	Year since last human intervention	Dominant Species	Bedrock	Phytosociological type	Ref.
Abeti Soprani	Central Apennines	I	1250–1450	8.4	1124	30	AA, FS	Marly Limestone	*Pulmonario apenninae-Abietetum albae*	[31,61]
Biogradska Gora	Biogradska Gora NP	Mne	1300–1350	2–6	1750–2000	140	FS, AA	Eruptive rocks	*Abieti-Fagetum*	[32,62]
Val Cervara	Abruzzo, Lazio, Molise NP	I	1730–1830	7.2	1211	no ref	FS	Limestone	*Cardamino kitaibelii-Fagetum sylvaticae*	[18,31,61,63,64]
Monte Cimino	Central Italy	I	925–1053	14.3	1300	61	FS	Trachyte	*Allio pendulini- Fagetum sylvaticae*	[31,61]
Collemeluccio	Central Apennines	I	900–1000	9.2	960	50	AA, FS, QC	Marly limestone	*Pulmonario apenninae-Abietetum albae*	[31,61]
Fonte Novello	Gran Sasso, Monti della Laga NP	I	1340	10	1071	310	FS	Sandy marlstone	*Cardamino kitaibelii-Fagetum sylvaticae*	[31,61,65]
Gargano Pavari	Gargano NP	I	720–800	11.6	1041	56	FS	Limestone	*Anemono apenninae-Fagetum sylvaticae*	[31,61]
Monte di Mezzo	Central Apennines	I	950–1150	8.6	1022	55	FS, QC	Marly Limestone	*Anemono apenninae-Fagetum sylvaticae*	[31,61]
Muniellos	Asturias	E	1200–1250	13	1980	70	FS, QP	Quartzite, sandstone	*Blechno spicant-Fagetum sylvaticae*	[66,67]
Perucica	Sutjeska NP	BiH	1450–1500	8.6	1363	no ref	AA, FS, PA	Limestone, Sandstone	*Piceo-Abieti-Fagetum*	[32,56,68]
Sasso Fratino	Foreste Casentinesi NP	I	950–1050	9	1689	76	FS, AA	Sandstone, Marlstone	*Cardamino heptaphyllae-Fagetum sylvaticae; Aceri platanoidis-Fagetum sylvaticae*	[31,61,69,70]

Nations – I: Italy, E: Spain, BiH: Bosnia and Herzegovina, Mne: Montenegro. Dominant species – AA: *Abies alba*, FS: *Fagus sylvatica*, QC: *Quercus cerris*, QP: *Quercus petraea*, PA: *Picea abies*.

upper slope, mid slope, lower slope, valley, flat area), and percentage cover of outcropping rocks and stones. Data on slope and aspect, together with stand latitude, were used to calculate above-canopy potential annual direct incident solar radiation (MJ/cm^2•yr) [35].

A topsoil sample (a single core of the top 20 cm) was collected for each quadrat, and analysed in the laboratory for granulometry (Bouyoucos's densimetric method), pH (in distilled water), carbon concentration (Walkley and Black's method) and total nitrogen concentration (Kjeldahl's method). In the field we estimated litter depth and the percentage of ground covered by litter. We also measured soil volumetric water content (VWC, the ratio of water volume to soil volume) in each quadrat as the average of five measurements done with a soil moisture meter *Field Scout - TDR 100*.

For each stand, we noted the occurrence of different kinds of disturbance related to timber harvest (e.g. occurrence of man-made tree stumps), wild boar rooting, trampling and surface water soil erosion. For each kind of disturbance we subjectively assigned an ordinal value ranging from 0–3 (0 = none, 1 = low, 2 = moderate, 3 = severe). A synthetic index of disturbance was then calculated as the sum of the four individual indices.

Sampling was performed in late spring-early summer in years 2011–2012. Stands located at low altitude or with higher average temperature were sampled earlier during the growing season in order to sample all the stands in a comparable phenological stage.

Beta diversity

A wide range of different phenomena, including various kinds of heterogeneity, differentiation and rate of change, with or without reference to external explanatory factors are often referred to the term beta diversity [36–38]. To avoid confusion, here we use the term beta diversity only to refer to the effective number of compositionally distinct sampling units (which in our case corresponds to 5 m×5 m quadrats). This equals true beta diversity as defined by Tuomisto [36–39] on the basis of the foundation laid by Hill [40] and Jost [41,42]. When referring to the proportion of species composition that changes among sampling units, we use the term species turnover instead [37,38].

For each stand we calculated species diversity as follows [36,40]:

$$^{q}D = 1/\sqrt[q-1]{\sum_{i=1}^{S} p_i p_i^{q-1}} \tag{1}$$

where p_i is the proportional abundance of species i, S is the total number of species, and q is the order of the diversity. We calculated true diversity both with q = 0 and with q = 2. When q = 0, species abundances cancel out from the equation, so ^{0}D obtains the same numeric value as species richness. When q>1, abundant species are given more weight than implied by their proportional abundances, and at q = 2 the denominator of eq. (1) equals the original Simpson diversity index [41,42].

Species diversity observed at a particular spatial extent (in our case, within the 1-ha plot) can be thought to result from two independent factors observable at a smaller spatial grain, namely average species density within sampling units (the 5 m×5 m quadrats) and compositional heterogeneity among the sampling units. To quantify the relative roles of these, we followed Tuomisto [37] to perform a multiplicative partitioning of the total observed species diversity in each stand (gamma diversity): $^{q}D_{\gamma} = {}^{q}D_{\alpha} \times {}^{q}D_{\beta}$. We developed for this purpose an *ad hoc* R script that is provided in Appendix S1.

Multiple regression on dissimilarity matrices

There has been some controversy on whether beta diversity is better modelled with the raw-data or the distance approach [21,43], which partly stems from different definitions of beta diversity. Tuomisto and Ruokolainen [21] argued that both approaches can be justified in different situations, and that the choice of the statistical method depends on the ecological question of interest.

Here we were primarily interested in the relative contributions of environmental differences and dispersal limitation in explaining species turnover, which is a question spelled out in terms of distances (a level-3 question according to Tuomisto and Ruokolainen [21]). Therefore, we adopted a distance-based variation partitioning approach, i.e. we partitioned the variation of dissimilarity matrices, that are based on species abundance data, into fractions uniquely or jointly explained by a number of explanatory dissimilarity matrices [2,44,45].

Species turnover was calculated for sampling unit pairs with a dissimilarity measure that is a monotonic transformation of true beta diversity and can be thought of as a generalization of the Jaccard similarity index to abundance data [36,41]:

$$^{q}C_{\beta} = \frac{2}{^{q}D_{\beta}} - 1 \tag{2}$$

The complement of this index is a dissimilarity measure linearly related to proportional species turnover (*sensu* Tuomisto [37]). We used this measure with q = 2 to build a quadrat-to-quadrat dissimilarity matrix for each stand separately, based on ground layer species composition. Corresponding dissimilarity matrices were built based on each structural and environmental variable separately, using either the Euclidean distance (for quantitative variables, after standardization) or Gower dissimilarity (for ordinal, nominal and binary variables). Finally, we built a matrix of geographical distances between the centre-points of the quadrats.

Explanatory variables were organized into three sets: 1) forest structure, 2) abiotic environmental factors and 3) geographical distances (Table 2). Mean values and standard deviation of the original environmental variables in each stand are shown in table S1. Forest floor PAR was included among the structural variables, because it is mostly determined by canopy features. Set 3 contained a single distance matrix, that of linear spatial distances among quadrats. To check if species turnover is related to geographical distances in a non-linear manner, we ran the analyses also with distance matrices that had been log-transformed or rank-transformed, but the results were very similar and are not shown.

To quantify the variation explained by each set of explanatory variables in each stand, we first performed a preliminary selection of dissimilarity matrices that have a significant marginal effect on ground layer species turnover through simple Mantel tests (9,999 permutations). We then performed a separate Multiple Regression on Distance Matrices analysis (MRM [44,46]) for each of the three sets of explanatory variables, starting with all the variables that were individually significant and removing the non-significant ones through backward elimination. All explanatory variables that remained in any of the three models were then used to quantify the total variation explained by a full MRM model. Finally, we followed the standard decomposition techniques used in variation partitioning to extract 'unique' and 'shared' fractions of variance explained by the three sets of variables [2,45,47].

For each quantitative environmental and structural variable, we finally tested whether the marginal effect of its dissimilarity matrix

Table 2. Structural and environmental variables used to calculate explanatory dissimilarity matrices; sets and subsets in which the variables were included; type of variable (nominal: N, ordinal: O, quantitative: Q); and unit of measurement (range for ordinal variables).

Set	Subset	Variable used to calculate dissimilarity	Variable type	u.m.
1.Structure	1a.Composition	Tree richness	Q	n species
	1b. Live structure	Tree cover	Q	%
		Shrub cover	Q	%
		Developmental phase	O	1–5
		Basal Area (Prism)	Q	m^2/ha
		Stem density	Q	n/25 m^2
		Basal area (quadrat)	Q	m^2/25 m^2
		Canopy openness	Q	%
		Uniform Angle Index	Q	0–1
		Sp. Mingling Index	Q	0–1
		DBHDM Index	Q	0–1
		Distance closest large live tree	Q	M
	1c. Deadwood	Deadwood volume	Q	m^3/25 m^2
		Deadwood density	Q	n/25 m^2
		Max decay class	O	0–5
		Num. decay classes	Q	0–5
	1d. Ground layer transmitted Light	Morning PAR	Q	μmol/m^2•sec
		Noon PAR	Q	μmol/m^2•sec
		Afternoon PAR	Q	μmol/m^2•sec
		Average PAR	Q	μmol/m^2•sec
2. Abiotic Environment	2a. Topography	Slope	Q	°
		Folded aspect	Q	°
		Pot. solar irradiation	Q	ln(MJ/cm^2•yr)
		Topographic position	N	
		Slope position	N	
		Rock coverage	Q	%
		Stone coverage	Q	%
	2b. Soil	Soil pH	Q	
		Soil Organic Matter	Q	% (g/g)
		Soil tot. N	Q	% (g/g)
		C/N ratio	Q	%
		Soil stone content	O	0–3
		Coarse Sand %	Q	% (g/g)
		Medium Sand %	Q	% (g/g)
		Fine Sand %	Q	% (g/g)
		Silt %	Q	% (g/g)
		Clay %	Q	% (g/g)
		Texture	N	
		Soil water content	Q	%
		Litter cover	Q	%
		Litter depth	Q	cm
	2c. Disturbance	Disturbance-Trampling	O	0–3
		Disturbance-Rooting	O	0–3
		Disturbance-Timber harvest	O	0–3
		Disturbance-Water Erosion	O	0–3
3. Space	3a. Geographical location	X and Y coordinates	Q	m

was correlated (Pearson's ρ) with the standard deviation of that variable in different stands.

Multivariate dispersion around group centroids

Compositional heterogeneity can be quantified with the average dissimilarity of individual sampling units from their group centroid in a multivariate species space. This is known as multivariate dispersion [48]. When the same dissimilarity measure is used, multivariate dispersion is monotonically related to the mean of pairwise dissimilarities between sampling units [36]. We illustrated the dispersion of quadrats around their stand centroids in the space defined by the first two axes of a Principal Coordinates Analysis (PCoA) based on the quantitative Jaccard dissimilarity measure. Stand centroids were calculated as the median of the quadrats in the PCoA-derived species space [48]. We also passively projected stand-level environmental variables (Table 1) on the graph.

We used a linear mixed effect model to incorporate the across-stand variation into the estimation of the overall response of multivariate dispersion to different explanatory variables. Compared to MRM, this analysis uses random effects (i.e. stands) to separate between the within-stand and among-stand effects of variables. Pairwise dissimilarities cannot be used in a mixed effects model since they are not independent of each other. Although the distances to centroid are not completely independent either, non-independence is not a serious problem with reasonable sample sizes [48].

We included as fixed effects of the initial mixed effect model the multivariate dispersions based on compound dissimilarity matrix derived from eight subsets of explanatory variables, i.e. overstorey composition (subset 1a), live structure (1b), deadwood (1c), light availability (1d), topography (2a), soil (2b), disturbance (2c) as well as spatial distances (3, Table 2). Dissimilarities were mostly calculated using Gower dissimilarity, but for subset 1a we used the complement of the Jaccard index (presence-absence data) and for subset 3a we used the Euclidean distance. We also included in the fixed part of the initial full model the average stand altitude and climatic variables (annual average precipitation and temperature) to test whether these stand-level variables explained species turnover.

Model selection was performed following a standard protocol [49]. We started with a model containing all the explanatory variables; we then chose an appropriate random part comparing different models (e.g. random intercept or random intercept and slope) using REML (Restricted Maximum Likelihood) estimation and AICs. The fixed part of the model was selected through backward elimination of non-significant variables through likelihood ratio test and ML estimation. After model validation, two quadrats were eliminated as outliers from the model; both of them were characterized by extreme disturbance either due to wild boars or water erosion along a steep slope.

All statistical computations were performed in R 3.0.0 [50]. Multivariate dispersion from group centroid were calculated using 'vegan' package (version 2.0–7). Mixed effect models were computed in R using 'nlme' package (version 3.1–105).

Results

Ground layer diversity

A total of 238 species were found in the 11 stands (25 in the tree layer, 21 in the shrub layer and 231 in the ground layer), accounting for a total of 4,541 species•layer•quadrat observations. The stands differed both in species richness and in the evenness of the species abundances within them (Table 3). As a result, the

Table 3. Decomposition of true gamma diversity (total species diversity in a site) into true alpha diversity (mean species density per quadrat) and true beta diversity (effective number of compositionally distinct quadrats) in 11 old-growth beech forest stands in southern Europe.

	alpha	beta	gamma	alpha	beta	gamma
Abeti Soprani	19.04	3.26	62	9.17	2.63	24.13
Biogradska Gora	15.76	3.24	51	4.93	2.32	11.43
Valle Cervara	11.24	4.09	46	4.68	1.97	9.23
Monte Cimino	13.08	2.98	39	4.41	2.88	12.68
Collemeluccio	12.92	4.26	55	2.86	1.74	4.98
Fonte Novello	9.08	3.19	29	3.15	1.54	4.87
Gargano-Pavari	5.48	4.01	22	2.85	2.36	6.73
Monte di Mezzo	7.16	4.47	32	3.49	2.56	8.96
Muniellos	5.24	4.77	25	1.46	1.27	1.84
Perucica	15.96	2.32	37	3.45	1.24	4.28
Sassofratino	14.60	3.63	53	8.04	3.05	24.50

Diversities were quantified at q = 0 (which gives more weight to rare species) and q = 2 (which gives more weight to abundant species).

stands ranked differently in species diversity when species abundances were weighted in different ways and all diversity components were lower with q = 2 (which gives more weight to abundant species) than when q = 0.

Beta diversity ($^2D_\beta$) was positively related both to alpha diversity (Spearman's $\rho = 0.67$, $p = 0.023$) and to gamma diversity ($\rho = 0.80$, $p = 0.003$). None of the diversity measures was related to the values of the available stand-level descriptors (Table 1) or the means of most quadrat-level descriptors (Table 2). The exception was average PAR, for which there was a negative relationship with beta diversity ($\rho = -0.82$, $p = 0.002$). In other words, sites with more light were compositionally less heterogeneous. Furthermore, heterogeneity in the structural and environmental variables (as quantified with their standard deviations) were only related to the diversity components in two cases. Alpha diversity was positively related to heterogeneity in soil water content ($\rho = 0.62$, $p = 0.043$), and beta diversity was positively correlated with heterogeneity in soil C/N ratio ($\rho = 0.80$, $p = 0.003$).

Dependence of pairwise species turnover on forest structure, environmental variables and spatial distances

The results of multiple regression on dissimilarity matrices differed greatly among stands both in terms of explanatory power and in the explanatory variables retained in the final model (Fig. 1). Variables directly or indirectly related to light availability (i.e. tree layer cover, canopy openness and ground layer PAR at different times of the day) were significant in seven of the 11 stands, and so were variables related to site topography (e.g. potential solar irradiation, topographic position, and rock coverage).

Other structural variables that significantly explained ground layer species turnover were mostly related to basal area and stem density (three stands) and to deadwood density and decay class (four stands). Species turnover in the tree layer explained species turnover in the ground layer only in the two stands with the highest tree species richness. The most important soil variables were related to texture, organic matter and total nitrogen concentration. Disturbance was significant only in the stand

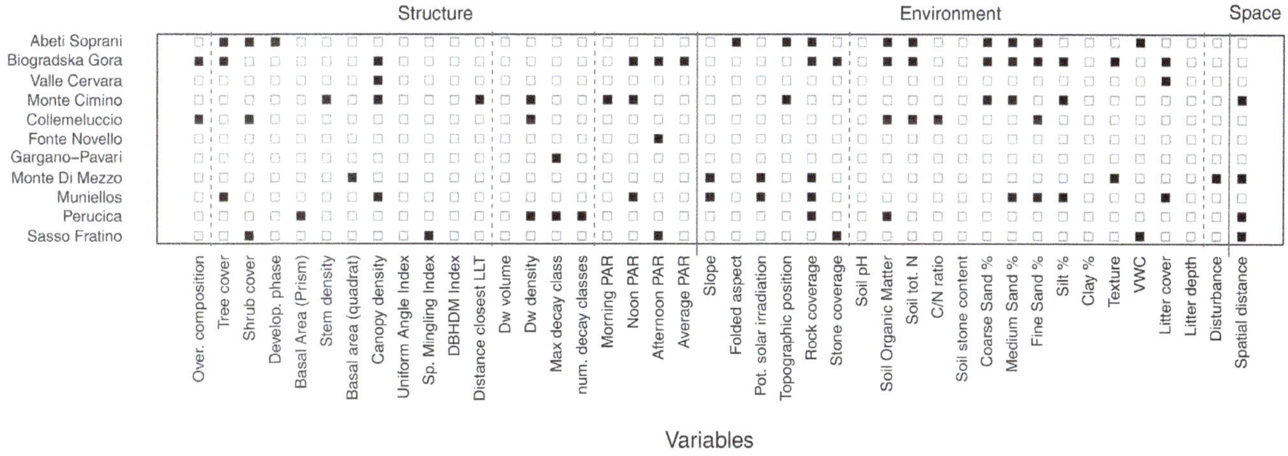

Figure 1. Significant Explanatory Variables. Variables that yield distance matrices with a significant ($p<0.05$) marginal effect (black quadrats) in mantel test (9,999 permutations) when modeling the ground layer plant species turnover in 11 old-growth beech forest stands in southern Europe.

'Monte di Mezzo'. Finally, between-quadrats spatial distance was significant in four stands.

The total variance in species turnover explained by the MRM models ranged from less than 2.0% (in stand 'Gargano-Pavari') to 44.8% (in 'Muniellos') with a median of 24.4% (Fig.2). Species turnover mainly responded to between-quadrats structural dissimilarities, which contributed to between 2% (in 'Gargano-Pavari') and 21.7% (in 'Biogradska Gora') of the explained variance, with a median of 13.0% (sum of the unique and shared fractions involving structural dissimilarities). The abiotic environmental dissimilarities contributed to between 0% (in 'Fonte Novello' and 'Gargano-Pavari') and 36.7% (in 'Muniellos') of the explained variance, with a median of 12.8% (Fig. 2). Between-quadrat spatial distances were significant only in four of the eleven stands, and contributed to between 0% and 8.1% (median $=0\%$) of the explained variance. The high total variance explained reported for three stands ('Muniellos', 'Biogradska Gora' and 'Abeti Soprani') was mostly related to their environmental heterogeneity (explaining respectively 36.7%, 30% and 25.1% of species turnover variation, Table 4), especially in soil and topographical features (Table S1 and S2).

The breakdown of variance into fractions explained by the different groups of variables are reported in Table 4 and Fig. 3. Three stands ('Monte di Mezzo', 'Perucica' and 'Sassofratino') returned negative shared fractions, which were very small except in 'Monte di Mezzo', where their magnitude was comparable to positive fractions.

Across the 11 stands, we found a significant correlation between explanatory variables' standard deviations (as an estimation of the underlying ecological gradient length) and the amount of variation explained by the corresponding dissimilarity matrices in MRM models in six cases only. These variables were: overstorey richness, canopy density, stone coverage, soil N content, soil silt fraction and water content (Table S3).

Mixed effect models – Multivariate dispersion

In the PCoA scatterplot based on understorey composition (Fig. 4) the first axis seems to be mainly related to altitude and time since last disturbance, and the second axis to the occurrence of conifers in the overstorey. Mean annual temperature and precipitation were also related to the second PCoA axis. The stands showing the largest dispersion of quadrats around the stand centroid in the multivariate species space (i.e. the highest species turnover between the quadrats and the stand centroid) were 'Monte Cimino', 'Sasso Fratino' and 'Gargano Pavari'. Those showing the smallest dispersion were 'Muniellos', 'Fonte Novello' and 'Perucica' (Table S2). With one exception, these were the same sites that showed the highest and lowest beta diversity at $q=2$, respectively. Indeed multivariate dispersion was highly correlated with $^{2}D_{\beta}$ (Pearson's $r=0.94$, $p<0.001$).

Multivariate compositional distance between a quadrat and its stand centroid was positively related to the corresponding differences in stand structure and soil properties (Table 5). The R^2 of the linear regression between the fitted and the observed values (a rough estimate of the variation explained by the model) was equal to 36.4%.

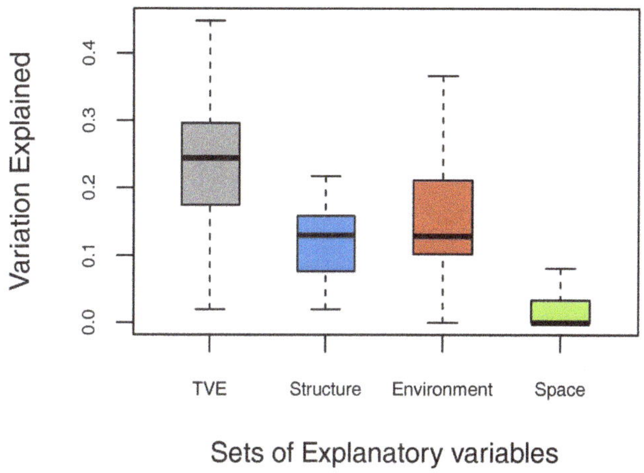

Figure 2. Variation Explained across stands. Boxplot showing the variability in how well ground layer species turnover was explained by multiple regression on distance matrices models (MRMs) in 11 old-growth beech stands in southern Europe. The first three boxplots correspond to models where explanatory variables are based on forest structural distances only, abiotic environmental distances only and spatial distances only. The last boxplot reports on the total variation explained (TVE) by the full models that include the explanatory variables of all three categories.

Table 4. Results of multiple regression on distance matrices analyses (MRMs) run on ground layer species turnover in 11 old-growth beech forest stands in southern Europe.

		Percentage of variation explained due to 'pure' effects and shared fractions						
	Full model	**Forest structure**	**Abiotic environment**	**Spatial**	**Str:Env| Spa**	**Str:Spa| Env**	**Env:Spa| Str**	**Str:Env:Spa**
Abeti Soprani	30.3	5.2	17.3	-	7.8	-	-	-
Biogradska Gora	37.0	7.0	15.3	-	14.7	-	-	-
Valle Cervara	15.6	5.9	7.8	-	2.0	-	-	-
Monte Cimino	19.3	7.0	4.4	1.0	5.4	0.6	0.6	0.2
Collemeluccio	24.4	11.6	9.7	-	3.1	-	-	-
Fonte Novello	7.3	7.3	-	-	-	-	-	-
Gargano-Pavari	2.0	2.0	-	-	-	-	-	-
Monte di Mezzo	28.9	10.0	20.7	10.1	−7.7	−8.4	−5.0	9.1
Muniellos	44.8	8.1	27.0	-	9.6	-	-	-
Perucica	28.5	9.2	8.8	3.5	2.4	1.8	3.3	-0.6
Sassofratino	21.0	7.9	1.4	1.0	7.4	−0.5	1.5	2.2

The full model includes all the explanatory variables that were retained after backward elimination in the models accounting for 1. forest structure, 2. abiotic environmental factors and 3. spatial distances. Str:Env|Spa = Joint effect of structural and abiotic environmental differences after controlling for the effect of spatial distance. Str:Env:Spa = Joint effect of all three types of dissimilarities.

Discussion

Structural heterogeneity rather than dispersal limitation is the main driver of floristic turnover

Our work provides insights into the mechanisms underlying ground layer species assembly in beech dominated old-growth forests. Environmental filtering, as determined by structural and environmental differences, explained a larger proportion of the variation in species turnover than dispersal limitation, as modelled by spatial distances. Within the stands, species turnover was mostly related to structural heterogeneity and, thereafter, to differences in abiotic environmental variables. The fraction of variation explained by spatial distances was on average very low, and not even statistically significant in most of the stands. Similar results were obtained when the 11 stands were analysed together using mixed effect models. In this case, the only significant predictors of species turnover were structural and soil heterogeneity. However, a large part of the variation in species turnover remained unexplained, which indicates that also other processes may be important, such as neutral dynamics.

When environmental factors are spatially autocorrelated, it is difficult to statistically separate the effects of dispersal limitation and environmental or habitat filtering [5,21,43]. A study performed in an old-growth temperate forest in Canada, which avoided this problem by using a sampling design that minimised the spatial autocorrelation of environmental variables, found environmental factors to be more important than dispersal-related mechanisms [6]. In our data, environmental and structural dissimilarities were only weakly or not at all correlated with spatial distances, and very little variance in species turnover was hence jointly explained by geographical and environmental distances (Table 4). The results of the mixed effect models further support the conclusion that dispersal limitation had only a small effect on ground layer species turnover in the southern European temperate forests we studied at the targeted spatial scales. Similar results have been found for tropical forests both at local and regional scales [12,51,52].

The low amount of variation explained by the spatial fraction may also be a spurious result due to the uneven number of explanatory variables among sets. Peres-Neto et al. [53] demonstrated for variation partitioning based on raw-data (e.g. RDA or CCA) that the estimation of the variation explained by a set should be corrected for the number of variables included in that set. Since no correction has been developed yet for MRM, we mitigated the effect of having uneven sets of explanatory variables by performing a preliminary selection of significant and non-collinear variables to be retained in each set.

In some stands, MRM returned negative values for the variation shared among sets of variables. This is a well-known problem of all variance partitioning techniques and has been attributed either to suppressor variables (i.e., a regressor having close to zero correlation with the response variable and a correlation with another regressor, which in turn is correlated with the response variables), or to two strongly correlated predictors with strong effects on the dependent variables of opposite signs [53]. However, since we performed both a preliminary and a backward selection of significant variables to be included in the full MRM, and thus minimized multicollinearity, we feel confident on the general validity of our results.

According to the mixed-effect model, within-stand species turnover was driven essentially by the heterogeneity in forest structure and soil variables, with no contribution from other variables such as deadwood, ground layer PAR, topography or within-stand spatial distances. The random intercept model we used supports the interpretation that the slope of the relationship between species turnover and soil or structural dissimilarity remains constant, but the amount of 'residual' species turnover between two quadrats with equal structural and soil conditions varies randomly across sites. This 'residual' species turnover is likely to include both locally important sources of environmental variability, not significant at the scale of the whole dataset, and other factors related to life-history traits of those species composing the ground layer assemblage. For instance, assemblages dominated by species with limited dispersal capabilities may have different 'residual' species turnover than those dominated by wind-dispersed species.

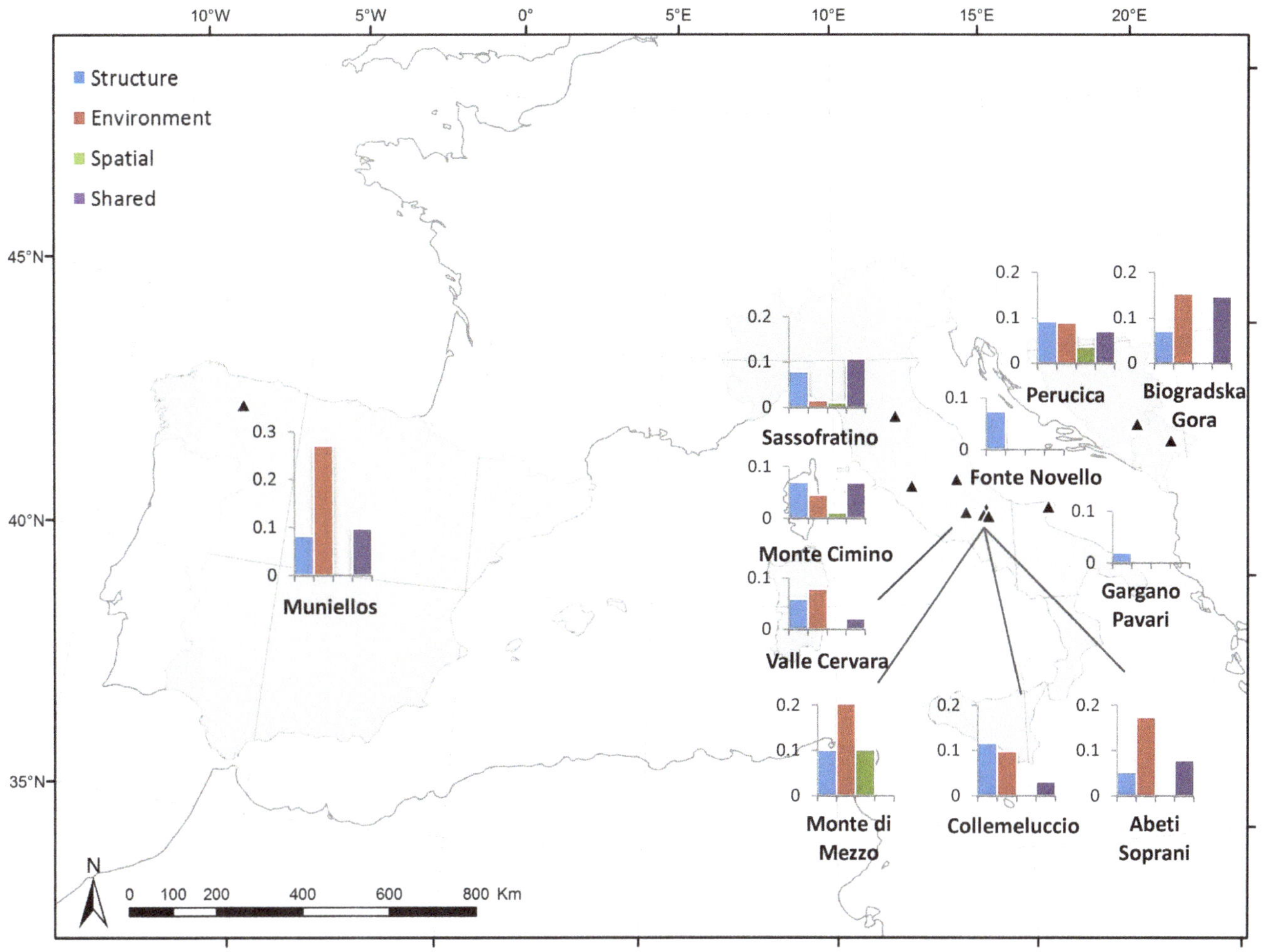

Figure 3. Fractions of Variation explained per stand. Fractions of variance in ground layer species turnover explained by multiple regression on distance matrices models (MRMs) in 11 old-growth beech forest stands. Bar charts report the relative explanatory power of forest structure, abiotic environment and spatial distance. Each pie-chart is placed next to the triangle that indicates the geographical location of the corresponding site. Negative shared fractions are not shown.

Length of the ecological gradients and relative roles of different environmental variables

The role of resource-driven niche processes and environmental variation is expected to be more limited at local than at broad spatial scales, because small areas usually harbour shorter ecological gradients than large areas do [10,54]. On the other hand, the effect of neutral processes, such as dispersal limitation, should be more easily detectable at small than at large geographical distances since it is related to the logarithm of distance [15,23]. However, the results of our study did not support the expectation that dispersal limitation dominates at local scales [14,55]. In contrast, structural and environmental heterogeneity always returned a higher amount of variation explained than spatial distances.

When considering our results, one should always keep in mind that the relative importance of different assembly processes critically varies with the spatial scale [9,16,55]. In this study, we only considered the within-stand component of ground layer spatial variation (encompassing a distance range of 20–150 m). The small extent (~1 ha) and grain (25 m^2) of our study design was aimed at matching the scale at which structural heterogeneity is created and maintained by gap dynamics [56], which is the dominant disturbance regime in southern European temperate old-growth beech forests [57]. Such a spatial scale is too fine to detect variations in the species pools due to biogeographical patterns, but probably too coarse to detect understorey plant autocorrelation occurring within typical neighbourhood sizes for plant populations [6].

The amount of ecological variability sampled in our macroplots varied among stands, also in relation to their developmental stage. Processes that produce horizontal spatial heterogeneity, such as gap development, are active throughout stand development. However, during the old-growth stage these processes increasingly generate within-stand structural heterogeneity [56,57]. Stands that developed into old-growth conditions a long time ago, are likely to be spatially more heterogeneous than those that only recently attained this condition, with repercussions on other ecological factors such as transmitted light distribution. Forest-floor light availability, as measured both directly (PAR) and using an indirect approach (tree cover, canopy openness) had a significant effect on ground layer species turnover except in four stands: 'Collemeluccio', 'Gargano-Pavari', 'Monte di Mezzo' and 'Perucica'. The first three of these stands were characterized by a

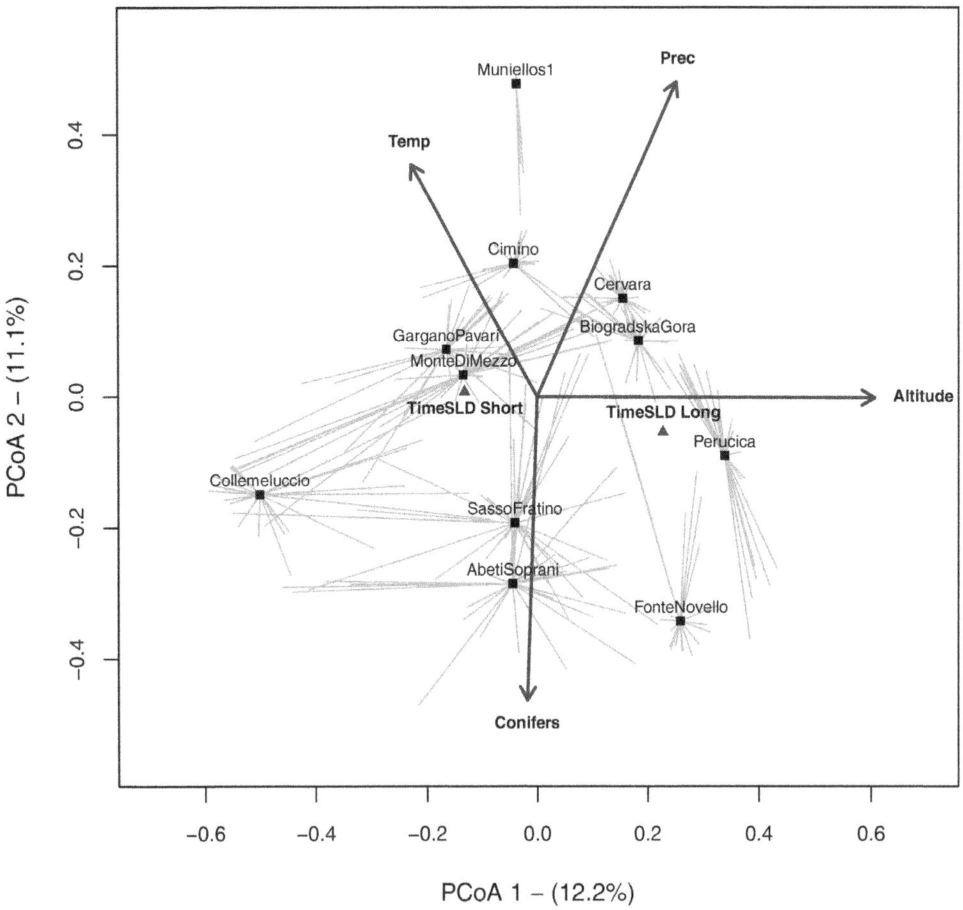

Figure 4. Understorey multivariate dispersion from stand centroids. Multivariate dispersion of quadrats (tips of gray lines) from stand centroids (black squares) in the multivariate space defined by the first two axes of a PCoA calculated on ground layer species composition in 11 old-growth beech forests of southern Europe. Site environmental variables were passively projected on the ordination to help the interpretation of the PCoA axes, either as vectors (quantitative variables) or black triangles (nominal variables). Temp: Average annual mean temperature; Prec: Average annual precipitation; Time SLD = Time Since Last Disturbance (Long >100 yrs, Short = <100 yrs).

Table 5. Output of the Mixed Effect Model on the response effect of within-stand structural and environmental dissimilarity and spatial distance on ground layer species turnover as modelled by the multivariate distance between quadrats and their stand centroids.

Linear mixed-effects model fit by REML

Random effects:		Formula: ~1 | stand			
	Plot	Residual			
StdDev:	0.116	0.167			
Fixed effects: Species Turnover ~ Live structure + Soil					
	Value	Std.Error	DF	t-value	p-value
(Intercept)	0.177	0.048	260	3.691	0.000
Live structure	0.772	0.197	260	3.919	0.000
Soil	0.690	0.216	260	3.199	0.002

A random intercept model with stand as the only random effect was selected as the best-fitting parsimonious model.

very continuous forest canopy with a scarce occurrence of canopy gaps, and thus a very short forest floor light gradient. Although these forests showed some old-growth features, they were probably still in a developmental stage in which gap-phase dynamics was only just starting, and high levels of horizontal diversification had not yet accumulated. 'Perucica' on the other hand, was a well preserved old-growth stand in which several gaps broke the horizontal continuity of the forest canopy, thanks to a relatively long period without human intervention. This was the only stand where ground layer species turnover was significantly related to basal area, probably in relation to its high within-stand variability. The relationships between basal area and ground layer species turnover are indirect, and are not limited to competition for light, but also for above- and below-ground growing space, given that basal area is a proxy for biomass. The relevance of structural variables, such as basal area and deadwood density, and the lack of a significant effect of light for this stand, support the hypothesis that the asymmetric competition exerted by the overstorey on ground layer flora is not only related to the shading effect [4].

Tree-layer composition was a significant predictor of species turnover only in two stands, both dominated by a mixture of broadleaved trees and conifers. It seems likely that, in those stands where the canopy composition was dominated by a single species

('Fonte Novello' and 'Valle Cervara'), compositional variation became uninformative. For the other stands, the compositional gradient may still be too short to be relevant, suggesting that overstorey species turnover may be more effective in predicting ground-layer species turnover at broader spatial scales. Indeed, when longer environmental gradients are considered, all components of the vegetation are likely to react to the environmental variation and hence become more strongly correlated [16,51,58].

The effect of top-soil heterogeneity on ground layer turnover was related either to soil carbon and nitrogen concentration or to soil texture. Soil C and N concentration were especially important in mixed stands, likely in relation to patchy accumulation of beech and fir litter, whose different nutrient contents and decomposition rates probably create a fine scale mosaic of heterogeneous topsoil parameters. Interestingly, in our study we did not observe any significant effect of topsoil pH variability on species turnover. In a study on the effect of fine-scale soil variability on plant distribution in German beech forests, very little variation was explained by pH alone, suggesting that the effect of pH may not have a substantial effect on ground layer species turnover at this scale [59].

Although we only recorded disturbance as ordinal indices, difference in disturbance level was a significant explanatory variable of species turnover in the stand 'Monte di Mezzo' which is a 300 ha forest patch heavily affected by the foraging activity of the wild boar. As in other natural reserves in central Italy, the population of wild boar has exploded in the recent decades, causing high pressure on ground-layer flora that accounts for most of the wild boar diet. Rooting activity is thus locally an important source of spatial variation in the ground-layer, both through direct herbivory and through its effect on soil features.

Our results partly supported the hypothesis that the relative importance of different environmental and structural variables depends on their degree of fine-scale heterogeneity. Only six variables showed a direct relationship between their standard deviation (as a measure of gradient length) and the fraction of variation explained by their dissimilarity matrices in MRM models (Table S3). However, all of these represent ecological factors generally thought to directly influence ground layer flora, namely light availability (canopy openness), soil water availability and nitrogen content, as well as competition by overstorey species. This is in agreement with the notion that the power of a statistical test to recognise an existing causal link between dependent and explanatory variables depends on the degree of heterogeneity in the explanatory variable. Although this result lends support to the interpretation that the observed relationships are indeed causal, our study was observational and not manipulative and may thus be vulnerable to potential (unknown) confounding factors.

Confounding factors may include unmeasured environmental variables, as well as the choice of inappropriate response models. These are usually identified as potential reasons accounting for the amount of 'unexplained variation' when performing variation partitioning [60]. In our study we tried to take into account as many sources of environmental heterogeneity that potentially affect ground-layer species turnover as possible. The stands in which MRM returned the highest amount of 'unexplained variation' were two monospecific beech stands ('Fonte Novello', 'Gargano-Pavari'), with a relatively uniform forest structure (continuous canopy cover with a low proportion of gaps) and little topographical heterogeneity (Table S1 and S2). Given the relatively homogeneous forest conditions, we expected these stands to return a beta-diversity value lower than average, but this was true only for 'Fonte Novello' (Table 3). In these stands some unmeasured abiotic factors, such as phosphorus or microelement concentrations in the soil, or depth of the water table, might have been particularly important. Unmeasured forms of natural disturbance (e.g. browsing by ungulates) might also be locally important in creating floristic patterns. We do not think, instead, that the amount of unexplained variation was related to lack-of-fit of the response model: given the local scale and the relatively short length of most environmental gradients within our study sites, the linear model underlying MRM appears justified (see also Figure S1).

We conclude that ground-layer species turnover strongly depends on niche-related processes, such as environmental filtering, which depend primarily on forest structural and environmental heterogeneity. Since old-growth forests represent only a very small percentage of forests in southern Europe, the need of developing and implementing silvicultural practices aimed at restoring old-growth structural attributes in regrowth and secondary forests is now widely recognized [25]. Based on the significant effects of structural heterogeneity on species turnover variation, our results suggest that silvicultural approaches that mimic the amount and patterns of structural heterogeneity observed in reference old-growth forests may have a positive effect on the conservation and restoration of high levels of ground layer species turnover in managed forests. However, conservation needs to pay attention also to the identity (life-history traits, conservation status) of the species to be conserved and not just aim at maximizing local heterogeneity [71]. Only with a complementary approach that includes the preservation of strict forest reserves, to ensure long-term ecological continuity of some reference forest stands, on the one hand, and a close-to-nature management of production forests on the other, we will be able to match both conservation goals and other social and economic needs.

Acknowledgments

We would like to thank Borja Jiménez-Alfaro, Emanuela Carli, Carmen Giancola, Eleonora Giarrizzo, Giorgia Martina, Manuel Palma, Danijela Stešević and Antonio Zoccola for help during the fieldwork, and Monica Zanini for the analysis of soil samples. We are grateful to Renzo Motta and Matteo Garbarino for logistic support. Further logistic assistance was provided by State Forestry Corps, 'Abruzzo, Lazio and Molise National Park', 'Gran Sasso and Monti della Laga National Park', 'ASBUC Intermesoli', 'Giardino della Flora Appenninica di Capracotta', 'Jardín Botánico Atlántico' (Gijón, Spain). The manuscript was partly prepared during FMS's stay at University of Turku.

Supporting Information

Figure S1 Relationships between understorey dissimilarities and a subset of environmental dissimilarities.

Appendix S1 An R script to perform true diversity decomposition.

Table S1 Averages and standard deviations of environmental variables.

Table S2 Stand average multivariate dispersion from group centroid.

Table S3 Correlation analysis between environmental and structural variables' standard deviations, and their marginal effect on ground layer species turnover.

References

1. Watt AS (1947) Pattern and process in the plant community. The Journal of Ecology 35: 1–22.
2. Tuomisto H, Ruokolainen K, Yli-Halla M (2003) Dispersal, environment, and floristic variation of western Amazonian forests. Science 299: 241–244.
3. Burrascano S, Anzellotti I, Carli E, Del Vico E, Facioni L, et al. (2013) Drivers of beta-diversity variation in Bromus erectus semi-natural dry grasslands. Appl Veg Sci 16: 404–416.
4. Burton JI, Mladenoff DJ, Clayton MK, Forrester JA (2011) The roles of environmental filtering and colonization in the fine-scale spatial patterning of ground-layer plant communities in north temperate deciduous forests. J Ecol 99: 764–776.
5. Dray S, Pélissier R, Couteron P, Fortin MJ, Legendre P, et al. (2012) Community ecology in the age of multivariate multiscale spatial analysis. Ecology 82: 257–275.
6. Gilbert B, Lechowicz MJ (2004) Neutrality, Niches, and Dispersal in a Temperate Forest Understory. Proc Natl Acad Sci U S A 101: 7651–7656.
7. Laliberte E, Paquette A, Legendre P, Bouchard A (2009) Assessing the scale-specific importance of niches and other spatial processes on beta diversity: a case study from a temperate forest. Oecologia 159: 377–388.
8. Cottenie K (2005) Integrating environmental and spatial processes in ecological community dynamics. Ecol Lett 8: 1175–1182.
9. Myers JA, Chase JM, Jiménez I, Jørgensen PM, Araujo-Murakami A, et al. (2013) Beta-diversity in temperate and tropical forests reflects dissimilar mechanisms of community assembly. Ecol Lett 16: 151–157.
10. Ruokolainen K, Linna A, Tuomisto H (1997) Use of Melastomataceae and pteridophytes for revealing phytogeographical patterns in Amazonian rain forests. J Trop Ecol 13: 243–256.
11. Peet R (1978) Forest vegetation of the Colorado Front Range: Patterns of species diversity. Vegetatio 37: 65–78.
12. Jones MM, Tuomisto H, Clark DB, Olivas P (2006) Effects of mesoscale environmental heterogeneity and dispersal limitation on floristic variation in rain forest ferns. J Ecol 94: 181–195.
13. Karst J, Gilbert B, Lechowicz MJ (2005) Fern community assembly: the roles of chance and the environment at local and intermediate scales. Ecology 86: 2473–2486.
14. Gravel D, Canham CD, Beaudet M, Messier C (2006) Reconciling niche and neutrality: the continuum hypothesis. Ecol Lett 9: 399–409.
15. Hubbell SP (2001) The Unified Neutral Theory of Biodiversity and Biogeography. Princeton, New Jersey, USA: Princeton University Press. 375 p.
16. Jones MM, Ferrier S, Condit R, Manion G, Aguilar S, et al. (2013) Strong congruence in tree and fern community turnover in response to soils and climate in central Panama. J Ecol 101: 506–516.
17. Winter S, Moller GC (2008) Microhabitats in lowland beech forests as monitoring tool for nature conservation. For Ecol Manag 255: 1251–1261.
18. Burrascano S, Rosati L, Blasi C (2009) Plant species diversity in Mediterranean old-growth forests: A case study from central Italy. Plant Biosyst 143: 190–200.
19. Saitta, Bernicchia A, Gorjon SP, Altobelli E, Granito VM, et al. (2012) Biodiversity of wood-decay fungi in Italy. Plant Biosyst 146: 488–488.
20. Norden B, Appelqvist T (2001) Conceptual problems of Ecological Continuity and its bioindicators. Biodivers Conserv 10: 779–791.
21. Tuomisto H, Ruokolainen K (2006) Analyzing or explaining beta diversity? Understanding the targets of different methods of analysis. Ecology 87: 2697–2708.
22. López-Martínez JO, Hernández-Stefanoni JL, Dupuy JM, Meave JA (2013) Partitioning the variation of woody plant β-diversity in a landscape of secondary tropical dry forests across spatial scales. J Veg Sci 24: 33–45.
23. Nekola JC, White PS (1999) The distance decay of similarity in biogeography and ecology. J Biogeogr 26: 867–878.
24. Burrascano S, Keeton WS, Sabatini FM, Blasi C (2013) Commonality and Variability in the Structural Attributes of Moist Temperate Old-growth Forests: a Global Review. For Ecol Manag 291: 458–479.
25. Bauhus J, Puettmann K, Messier C (2009) Silviculture for old-growth attributes. For Ecol Manag 258: 525–537.
26. Capotorti G, Zavattero L, Anzellotti I, Burrascano S, Frondoni R, et al. (2012) Do National Parks play an active role in conserving the natural capital of Italy? Plant Biosyst 146: 258–265.
27. Gilliam FS (2007) The ecological significance of the herbaceous layer in temperate forest ecosystems. Bioscience 57: 845–858.
28. Larsen JB (1995) Ecological stability of forests and sustainable silviculture. For Ecol Manag 73: 85–96.
29. Royo AA, Carson WP (2006) On the formation of dense understory layers in forests worldwide: consequences and implications for forest dynamics, biodiversity, and succession. Can J For Res 36: 1345–1362.
30. Rivas-Martinez S, Rivas-Saenz S (2009) Worldwide Bioclimatic Classification System, 1996–2009. Spain, http://www.globalbioclimatics.org: Phytosociological Research Center. Accessed 2013 Mar 1.
31. Calamini G, Maltoni A, Travaglini D, Iovino F, Nicolaci A, et al. (2011) Stand structure attributes in potential old-growth forests in the Apennines, Italy. Italia Forestale e Montana 66: 365–381.
32. Motta R, Bjelanovic I, Borgogno Mondino E, Curovic M, Garbarino M, et al. (*in press*) Analysis of the spatio-temporal dynamics of mixed beechsilver fir-Norway spruce old-growth forests of Biogradska Gora (Montenegro) and Perucica (Bosnia and Herzegovina). Plant Biosyst.
33. Aguirre O, Hui GY, von Gadow K, Jimenez J (2003) An analysis of spatial forest structure using neighbourhood-based variables. For Ecol Manag 183: 137–145.
34. Hunter ML (1990) Wildlife, forests and forestry: principles of managing forests for biological diversity. Englewood Cliffs: Prentice Hall.
35. McCune B, Keon D (2002) Equations for potential annual direct incident radiation and heat load. J Veg Sci 13: 603–606.
36. Tuomisto H (2010) A diversity of beta diversities: straightening up a concept gone awry. Part 2. Quantifying beta diversity and related phenomena. Ecography 33: 23–45.
37. Tuomisto H (2010) A diversity of beta diversities: straightening up a concept gone awry. Part 1. Defining beta diversity as a function of alpha and gamma diversity. Ecography 33: 2–22.
38. Tuomisto H (2011) Commentary: do we have a consistent terminology for species diversity? Yes, if we choose to use it. Oecologia 167: 903–911.
39. Tuomisto H (2010) A consistent terminology for quantifying species diversity? Yes, it does exist. Oecologia 164: 853–860.
40. Hill MO (1973) Diversity and Evenness: A Unifying Notation and Its Consequences. Ecology 54: 427–432.
41. Jost L (2006) Entropy and diversity. Oikos 113: 363–375.
42. Jost L (2007) Partitioning diversity into independent alpha and beta components. Ecology 88: 2427–2439.
43. Legendre P, Borcard D, Peres-Neto PR (2005) Analyzing beta diversity: Partitioning the spatial variation of community composition data. Ecol Monogr 75: 435–450.
44. Lichstein JW (2007) Multiple regression on distance matrices: a multivariate spatial analysis tool. Plant Ecol 188: 117–131.
45. Duivenvoorden JF, Svenning JC, Wright SJ (2002) Ecology - Beta diversity in tropical forests. Science 295: 636–637.
46. Legendre P, Lapointe FJ, Casgrain P (1994) Modeling brain evolution from behavior - A permutational regression approach. Evolution 48: 1487–1499.
47. Whittaker J (1984) Model Interpretation from the Additive Elements of the Likelihood Function. Journal of the Royal Statistical Society Series C (Applied Statistics) 33: 52–64.
48. Anderson MJ (2006) Distance-Based Tests for Homogeneity of Multivariate Dispersions. Biometrics 62: 245–253.
49. Zuur AF, Ieno EN, Walker NJ, Saveliev AA, Smith GM (2009) Mixed Effect Models and Extensions in Ecology with R. Berlin: Springer-Verlag.
50. R Development Core Team (2011) R: A language and environment for statistical computing Vienna, Austria: R Foundation for Statistical Computing.
51. Ruokolainen K, Tuomisto H, Macia MJ, Higgins MA, Yli-Halla M (2007) Are floristic and edaphic patterns in Amazonian rain forests congruent for trees, pteridophytes and Melastomataceae? J Trop Ecol 23: 13–25.
52. Poulsen AD, Tuomisto H, Balslev H (2006) Edaphic and Floristic Variation within a 1-ha Plot of Lowland Amazonian Rain Forest1. Biotropica 38: 468–478.
53. Peres-Neto PR, Legendre P, Dray S, Borcard D (2006) Variation partitioning of species data matrices: Estimation and comparison of fractions. Ecology 87: 2614–2625.
54. MacArthur RH, Wilson EO (1967) The theory of island biogeography. Princeton, NJ: Princeton University Press.
55. Guèze M, Paneque-Gálvez J, Luz AC, Pino J, Orta-Martínez M, et al. (2013) Determinants of tree species turnover in a southern Amazonian rain forest. J Veg Sci 24: 284–295.
56. Nagel TA, Svoboda M (2008) Gap disturbance regime in an old-growth Fagus-Abies forest in the Dinaric Mountains, Bosnia-Herzegovina. Can J For Res 38: 2728–2737.
57. Rugani T, Diaci J, Hladnik D (2013) Gap Dynamics and Structure of Two Old-Growth Beech Forest Remnants in Slovenia. PLoS ONE 8: e52641.
58. Burrascano S, Sabatini FM, Blasi C (2011) Testing indicators of sustainable forest management on understorey composition and diversity in southern Italy through variation partitioning. Plant Ecol 212: 829–841.
59. Bruelheide H, Udelhoven P (2005) Correspondence of the fine-scale spatial variation in soil chemistry and the herb layer vegetation in beech forests. For Ecol Manag 210: 205–223.

Author Contributions

Conceived and designed the experiments: FMS SB CB. Performed the experiments: FMS SB. Analyzed the data: FMS HT. Wrote the paper: FMS SB HT CB.

60. Borcard D, Legendre P, Drapeau P (1992) Partialling out the Spatial Component of Ecological Variation. Ecology 73: 1045–1055.
61. Lombardi F, Chirici G, Marchetti M, Tognetti R, Lasserre B, et al. (2010) Deadwood in forest stands close to old-growthness under Mediterranean conditions in the Italian Peninsula. Italia Forestale e Montana 65: 481–504.
62. Čurović M, Stešević D, Medarević M, Cvjetićanin R, Pantić D, et al. (2011) Ecological and structural characteristics of monodominant montane beech forests in the national park Biogradska Gora, Montenegro. Arch Biol Sci 63: 429–440.
63. Burrascano S, Lombardi F, Marchetti M (2008) Old-growth forest structure and deadwood: Are they indicators of plant species composition? A case study from central Italy. Plant Biosyst 142: 313–323.
64. Piovesan G, Di Filippo A, Alessandrini A, Biondi F, Schirone B (2005) Structure, dynamics and dendroecology of an old-growth Fagus forest in the Apennines. J Veg Sci 16: 13–28.
65. Sabatini FM, Burrascano S, Blasi C (2010) Niche heterogeneity and old-growth conservation value. Italia Forestale e Montana 65: 621–636.
66. Sabatini FM, Jiménez-Alfaro B, Burrascano S, Blasi C, (*in press*) Drivers of herb-layer species diversity in two unmanaged temperate forests in northern Spain. Community Ecol.
67. Fernández Prieto JA, Bueno Sánchez A (1992) A New Classification of the Forests of the Muniellos Biological Reserve in Northwest Spain. Vegetatio 102: 33–46.
68. Nagel TA, Svoboda M, Rugani T, Diaci J (2010) Gap regeneration and replacement patterns in an old-growth Fagus-Abies forest of Bosnia-Herzegovina. Plant Ecol 208: 307–318.
69. Bianchi L, Bottacci A, Calamini G, Maltoni A, Mariotti B, et al. (2011) Structure and dynamics of a beech forest in a fully protected area in the northern Apennines (Sasso Fratino, Italy). iForest - Biogeosciences and Forestry 4: 136–144.
70. Travaglini D, Paffetti D, Bianchi L, Bottacci A, Bottalico F, et al. (2012) Characterization, structure and genetic dating of an old-growth beech-fir forest in the northern Apennines (Italy). Plant Biosyst 146: 175–188.
71. Sabatini FM, Burton JI, Scheller RM, Amatangelo KL, Mladenoff DJ (*in press*) Functional diversity of ground-layer plant communities in old-growth and managed northern hardwood forests. Appl Veg Sci. doi: 10.1111/avsc.12083

8

Assessing Species Distribution Using Google Street View: A Pilot Study with the Pine Processionary Moth

Jérôme Rousselet[1]ⓢ, Charles-Edouard Imbert[1]ⓢ, Anissa Dekri[1], Jacques Garcia[1], Francis Goussard[1], Bruno Vincent[1], Olivier Denux[1], Christelle Robinet[1], Franck Dorkeld[2], Alain Roques[1], Jean-Pierre Rossi[2]*

1 Unité de Recherche 633 Zoologie Forestière, Institut National Recherche Agronomique, Centre d'Orléans, France, **2** Unité Mixte de Recherche 1062 Centre de Biologie pour la Gestion des Populations, Institut National Recherche Agronomique, Centre de Montpellier, France

Abstract

Mapping species spatial distribution using spatial inference and prediction requires a lot of data. Occurrence data are generally not easily available from the literature and are very time-consuming to collect in the field. For that reason, we designed a survey to explore to which extent large-scale databases such as Google maps and Google street view could be used to derive valid occurrence data. We worked with the Pine Processionary Moth (PPM) *Thaumetopoea pityocampa* because the larvae of that moth build silk nests that are easily visible. The presence of the species at one location can therefore be inferred from visual records derived from the panoramic views available from Google street view. We designed a standardized procedure allowing evaluating the presence of the PPM on a sampling grid covering the landscape under study. The outputs were compared to field data. We investigated two landscapes using grids of different extent and mesh size. Data derived from Google street view were highly similar to field data in the large-scale analysis based on a square grid with a mesh of 16 km (96% of matching records). Using a 2 km mesh size led to a strong divergence between field and Google-derived data (46% of matching records). We conclude that Google database might provide useful occurrence data for mapping the distribution of species which presence can be visually evaluated such as the PPM. However, the accuracy of the output strongly depends on the spatial scales considered and on the sampling grid used. Other factors such as the coverage of Google street view network with regards to sampling grid size and the spatial distribution of host trees with regards to road network may also be determinant.

Editor: Jon Moen, Umea University, Sweden

Funding: Research was funded by the French National Agency for Research (project URTICLIM) and the région Centre (project ADRIEN). The funders had no role in study design, data collection and analysis, decision to publish, or preparation of the manuscript.

Competing Interests: The authors have declared that no competing interests exist.

* E-mail: rossi@supagro.inra.fr

ⓢ These authors contributed equally to this work.

Introduction

Globalisation and one of its most publicized consequences, the species redistribution, has received a considerable attention during the last decades [1–3]. A lot of research efforts has been directed towards unravelling ecological processes implied in the spread of species [2,4]. In that context, GISs have proven particularly precious to monitor species spread [5]. There is a concomitant increasing demand for data documenting the spatial distribution of species for different objectives such as monitoring and modelling species range expansion [6,7], anticipating future distributions and devising control strategies [8–11], studying mechanisms at work with species dispersal and the relationships with landscape composition and physiognomy [12]. Unfortunately, the amount of data available is limited for a majority of taxa and geographical regions. The consequences of data scarcity are dramatic for example in the field of species distribution modelling [13,14] where collecting occurrence data is not always easy, time consuming, and is often non-environmentally friendly (because of the gas emission of vehicles used for the survey). As a result, updating or completing existing data sets is difficult albeit it is the very first step of ecological analysis and modelling.

During the last decade, geospatial data have become increasingly accessible with the advent of new mapping technologies such as Google Earth that offers free satellite imagery and aerial photos of most of earth's land surface. Google Earth has been used in several research areas that require mapping technology such as human or animal health [15,16], conservation biology [17,18] or biodiversity assessment [19]. A new level of spatial information has been recently reached with the development of Google street view (GSV) in 2007 [20]. This new technology provides panoramic imagery captured in hundreds of cities in different countries around the world. It corresponds to an unprecedented amount of information at street-level scale. Not only dedicated to cities and urban areas, GSV documents rural areas and unpopulated places. GSV is based on the idea of operating numerous data-collection vehicles around the world. Each vehicle is equipped with camera and GPS, and records images while driving paved and unpaved roads. Resulting data are processed and served via the Internet [20]. Street imagery consists of detailed views allowing users to navigating and exploring streets and cities [21]. The aim of the present study was to explore how the GSV technology could be helpful to ecological research in documenting the geographical distribution of species. Recent studies have shown that the GSV imagery could be used to depict and audit neighborhood environments in the framework of social science [22] and preventive medicine [23] but to our knowledge, no ecological application has been published so far.

We assessed the presence of an insect species by roadside sampling [24] based on GSV imagery and compared the outputs with independent field data. We selected the Pine Processionary Moth (PPM) (*Thaumetopoea pityocampa* Den. & Schiff., Lepidoptera, Notodontidae) as the biological model for our survey because it is a good example of expanding species that offers various advantages with regard to our aims. The PPM is a pine defoliator occurring on various tree species of the genera *Pinus* and *Cedrus* that are used in forestry or as ornamental trees in urban and rural areas throughout Europe. PPM larvae build white winter nests that are easily discernable and thus provide unambiguous indication of species presence since no other organism produce similar structures in these tree species at that time of the year. The nests are spatially aggregated [25] and exhibit a strong edge effect with considerably higher densities at stand edges [26] or along host tree-lined streets. This characteristic is invaluable with regards to roadside sampling. We sampled the subject species in two sampling areas using both field sampling and visual examination of GSV imagery and worked at two resolutions (*i.e.* grains) in order to test possible scale effects. Attention was paid to the coverage of the GSV database because it could constitute an important cause of discrepancies between field and *in silico* data sets. This study is a first step towards new methodologies for monitoring species geographical distribution across large spatial scales making use of the ever-increasing amount of data available through the Internet.

Materials and Methods

Model species

The PPM is a common defoliator occurring on various native and exotic conifer species throughout southern Europe and Mediterranean countries, where it is the most important pine and cedar defoliator. Its preferred host, *Pinus nigra*, as well as several other potential host tree species have been used for both large-scale afforestation and ornamental plantations and are thus widely distributed. The PPM range is largely controlled by the minimum winter temperatures [27,28]. A recent study has revealed that the PPM geographical range is expanding both northward and in altitude [29], probably in relation to climate changes [11,27,30,31]. Adult emergence occurs during summer depending on local climatic conditions. Soon after emergence, adults mate and females select a host tree and lay one egg batch on the host tree needles. Hatching occurs roughly one month later *i.e.* from August to September in our study area. Larvae are gregarious, feed on pine needles, and build a silk nest [32,33]. The first two instars build small temporary silk nests only detectable from nearby the host tree. From the third instar on, larvae built a definitive nest in which they will develop during autumn and winter. The winter nest is white and shiny due to newly produced silk. The pupation procession (which gave its name to the species) is the migration of larvae into the soil where they pupate until the following summer. It occurs at the end of winter or in early spring according to climatic and meteorological conditions. Empty nests turn to brownish and deteriorate. They have usually disappeared the following year when next PPM generation starts to build new nests.

Ethics statements

All the data used in the present work were collected on the public thoroughfare and thus did not require specific permissions according to the French law (Arrêt n. 516 du 7 mai 2004 Cour de cassation – Assemblée plénière). The present survey did not involve endangered or protected species.

Sampling zones and grid resolution

We surveyed the PPM spatial distribution in two sampling areas of contrasted extent (Figure 1). A large sampling area (46 848 km^2) was designed so as to cover the Région Centre in France which northern and eastern parts have been recently or are currently being colonized by the PPM. A second, smaller sampling area (22×22 km =484 km^2) was located within the former at the north of the Beauce area. Each sampling area was discretized into a set of sampling cells which size defined the grain or resolution of the survey. The large sampling area was discretized into 183 cells of 16×16 km size (Figure 1). This sampling zone will be thereafter referred to as LG (large grid). Similarly, the Beauce sampling window was discretized into a set of 121 cells of 2×2 km size and the resulting sampling grid will be referred to as SG (small grid) (Figure 1).

Field sampling protocols

The PPM was sampled within each individual cell throughout each sampling grid according to a protocol defined to monitor the species range expansion towards northern France [30,34] on the basis of nest road sampling. Each cell is visited by car and PPM host trees are observed by eye and with binoculars (when necessary) from the road and public land. When a nest is observed, it is georeferenced, the species is considered as present in the cell and then the neighbouring cells are prospected. When all the paved and unpaved roads practicable by car within a cell have been visited without detecting a nest, the PPM is considered as absent. The geographic coordinates of the tree hosting the observed nest are recorded by GPS (or the location of this tree is mapped onto a georeferenced aerial photo using PhotoExplor-eur or Arpentgis mobile in case of distant observation).

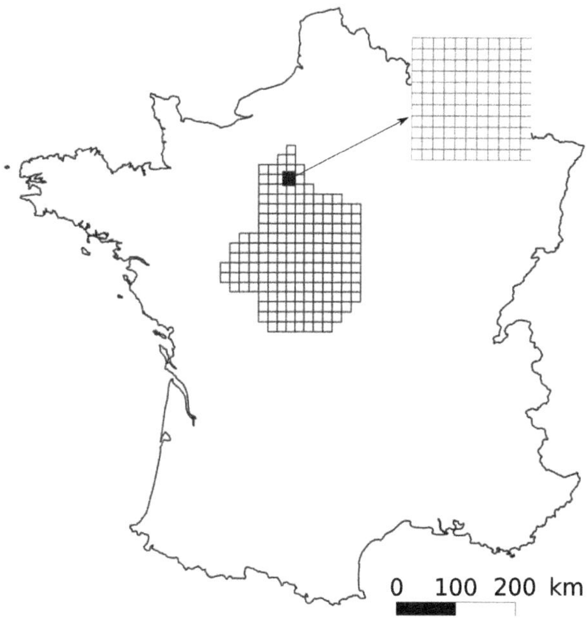

Figure 1. Pine processionary moth sampling grids. A large sampling grid covering the administrative region called "Région Centre" (46 848 km^2) in France was investigated. A second, smaller (121 km^2), sampling grid was nested within the former.

PPM presence-absence data in both grids were collected between 2007 and 2009 following the protocol described above. Note that a grid cell where the PPM was sighted on one sampling occasion is assumed to remain colonized the following years.

Sampling based on GSV

For each cell, an operator virtually drove along the roads available in the GSV database. He performed *in silico* roadside sampling by visually analyzing the available panoramic views. Figure 2 shows different pictures of PPM nests, infested trees and several trees located along streets in the region of Orléans, France, as they could be observed using GSV. When different views of the same place were available, all the points of view were explored. This work was realized in September and October 2011. As soon as a nest was detected, the cell was considered colonised by the PPM and the operator switched to another cell.

When a nest was difficult to identify, several views with different zoom levels and different viewing angles were examined. The spatial coordinates of each nest were recorded. Four kinds of cells were thus considered: cells without road covered by GSV at the date of the observation; cells containing at least one nest identifiable without ambiguity ("presence" cell); cells with a putative nest but with a persistent doubt ("indistinct" cells); cells without nest detected ("absence" cells).

Spatial coverage of Google roads and GSV

For the SG, we encountered several cells with no available GSV data and a high discrepancy between field and *in silico* data (see results). We therefore assessed the structure of the road network covered by GSV to examine its spatial variability and determine if it could explain the divergences between data sets. We developed a java script to collect information from the Google API (code available in Text S1), we then derived maps of roads included in the GSV database on a regular grid of 250 by 250 m mesh. The coordinates of each point were used to retrieve the nearest road and the nearest road available in the GSV database. We additionally computed an index to quantify the coverage by GSV within each sampling cell using the following procedure: cells were rasterized at the resolution of 250 by 250 m using the R package raster [35]. We then computed the proportion of pixels corresponding to road covered in the GSV database.

Figure 2. Pictures showing the pine processionary moth silk nest and different examples of infested trees located along streets in the region of Orléans, France. A. Winter silk nest. B. Host tree infested by several PPM colonies. C to E. Infested trees located along traffic lanes. F. Picture taken from within a car. All host trees are black pines (*Pinus nigra*) except in C. where black pines and scots pines (*P. sylvestris*) are present. All photos by J. Rousselet.

Comparing field and Google-derived data sets

Both zones led to two grids of PPM occurrences corresponding to field and Google-derived data. This was considered as a two-class prediction problem *i.e.* binary classification. Cells in which the PPM was observed formed the "positive" class while cells where the PPM was absent formed the "negative" class. We computed the four possible outcomes of that binary classifier, namely the true positive (TP), the true negative (TN), the false positive (FP) and the false negative (FN) [36], assuming that field sampling gave true data. TP corresponded to cases where the PPM was observed both in the field and from Google database. TN corresponded to absence in the field and in the Google derived data. FP corresponded to cells where the PPM was not observed in the field but was present according to Google data. Finally FN corresponded to field observations associated to absence according to Google data. TN, FN, TP and FP formed the confusion matrix reported in the results section (Tables 1 and 2). The sensitivity (rate of TP) and the specificity (rate of TN) were used to measure the proportion of good predictions derived from Google data in the case of cells where PPM is present and absent respectively. The sensitivity and the specificity were estimated as TP/P and TN/N where P and N are the total number of positive and negative cases respectively. We computed the accuracy or the rate of good predictions as (TP+TN)/(P+N). We additionally computed the Phi/Matthews correlation coefficient as a measure of discrete covariation between field and Google derived data [37].

This coefficient ranges from 1 to +1 with 1 indicating a perfect prediction, 0 a random prediction and negative values a worse than random prediction. All computations were done using the R statistical software [38] and the R package ROCR [37].

Results

Large sampling grid

The field data collected for the large scale survey showed that the PPM was present all over the study area: there were 178 cells classified as "presenc" and 5 as "absence". Google-derived data

Table 1. Confusion matrix for the pine processionnary moth field data (true class) and Google-derived data (hypothetized class) in the large-scale study grid (LG).

		field data	
		presence	absence
Google	presence	TP = 165	FP = 0
	absence	FN = 13	TN = 5

Table 2. Confusion matrix for the pine processionnary moth field data (true class) and Google-derived data (hypothetized class) in the small-scale study grid (SG).

		field data	
		presence	**absence**
Google	presence	TP = 3	FP = 0
	absence	FN = 63	TN = 49

Of the 121 sampled cells, 6 were removed from the analysis because no GSV data were available for comparison with field data.

provided slightly different values with 165 cells classified as "presence" and 17 as "absence". One cell was reported as "indistinct" because the pictures did not allow ascertaining the presence of PPM nests (Figure 3). In that case, we assigned the status of absence to the cell when computing sensitivity and other indices. Off the 183 cells, 165 were true positives and 5 were true negatives (Table 1). The number of false negatives was 13 and there was no case of false positive (Table 1). This led to very high values of the sensitivity (true positive rate) and specificity (true negative rate), which were respectively 0.927 and 1. Both field and Google-derived data are reported in Figure 3. Overall, Google-derived data showed a good agreement with field data as revealed by an accuracy of 0.929 and a Matthews correlation coefficient of 0.507.

Small sampling grid

The field data collected over the SG consisted in 70 and 51 cells classified as "presence" and "absence" respectively. Google-derived data comprised 3 cells classified as "presence", 109 cells where the PPM was absent, 3 indistinct cells, and 6 cells for which no GSV data were available (Figure 4). As in the case of the large-scale survey, the "indistinct" cells were classified as "absence" and we did not account for the cells with no data in the computation of the statistics.

There were 6 true positives and 49 true negatives while the number of false negatives was 63 and there was no case of false positive (Table 2). As a consequence, these values led to a low sensitivity of 0.045 while the specificity was 1. The Google-derived data showed a fairly low agreement with field data as the accuracy was 0.452 and the Matthews correlation coefficient was 0.141.

Linking GSV coverage and the PPM sampling

The coverage of the network was expressed as the proportion of pixels corresponding to roads within each sampling cell. Overall, this network covered the whole sampling area, with higher coverage around the main towns like in the bottom right part of the map that corresponds to the surroundings of the city of Chartres. The importance of the GSV coverage was not related to the number of true positive, false negative or true negative cases as shown in Figure 5. Note that the percentage used here is not the proportion of roads available within GSV but the amount of pixels (250 by 250 m) corresponding to roads available within GSV.

Discussion

Spatial scale issues: resolution

The present study showed that the data derived from GSV imagery were good surrogates for field data when assessing the spatial distribution of the PPM. By comparing Google-derived and field data, however, we showed that the resolution of the survey was critical in that fine scale *i.e.* high resolution sampling failed to properly describe the actual distribution (*viz* as perceived with field sampling). Scale effects are well known in ecology [39] and in our case they are directly linked to the resolution or grain, *i.e.* to the mesh of the sampling grid (Figure 1). It is analogous to the size of the sampling unit, which has been shown to strongly affect our perception of spatial variability [40,41] both in intensity and range.

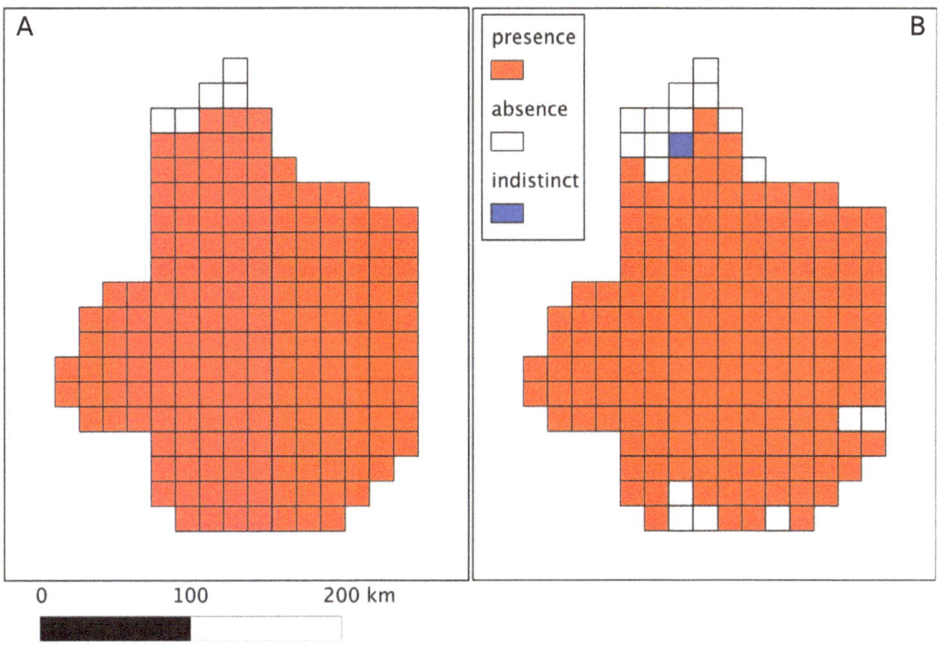

Figure 3. Large-scale study of the pine processionnary moth in France. A. Field data B. Google street view derived data. The sampling units are cells of 16 km×16 km.

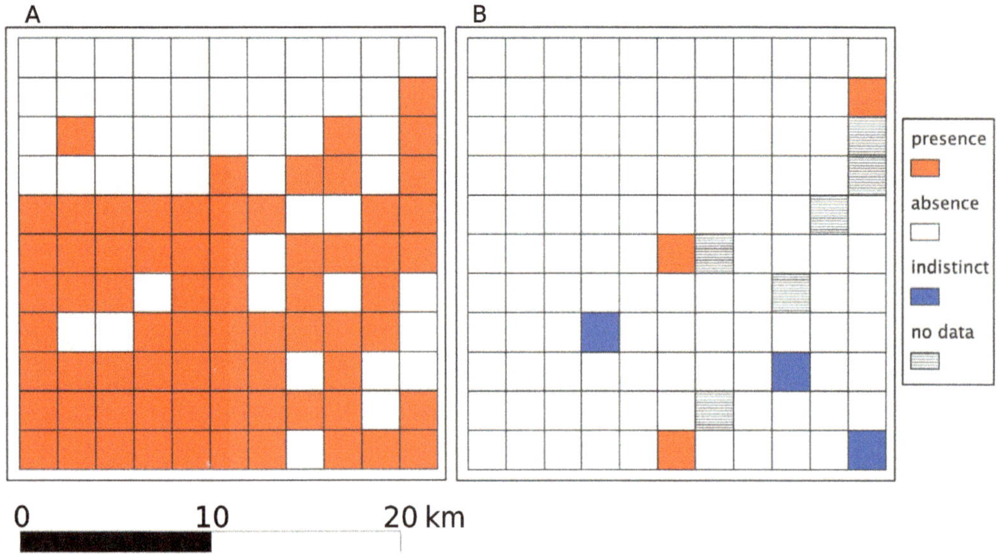

Figure 4. Small-scale study of the pine processionary moth in France. A. Field data B. Google street view derived data. The sampling units are cells of 2 km×2 km.

Google-derived data depend on the density of the GSV coverage which parameter in turn depends on the size of the sampling units. Large sampling units such as squares of 16×16 km used in the LG survey are more likely to include a large amount of roads covered in the GSV system which implies more chance to properly spot PPM nests when present. Decreasing the grain amounts to increasing the sampling effect with lower level of information per unit and ultimately no information in some locations where no street views are available (we reported 6 such cases in the SG). Such effect may become less meaningful in the future, as the density of the Google coverage will be progressively enhanced.

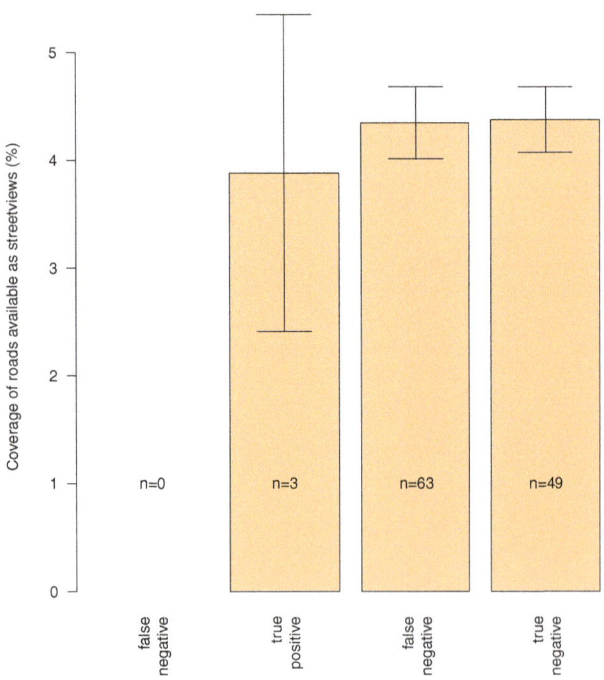

Figure 5. Relationships between the Google street view coverage (%) and the counts of true and false positives, and true and false negatives in the SG.

The grain also affects the variability of the date of the Street views picture, which may be a central parameter for some species as the PPM (see below).

Date of pictures: seasonality and other sources of uncertainty

Some organisms may concentrate or be visible at particular season or times of the day and this defines the periods when census becomes possible. Because we used the Google views (pictures) as a data source, we depend on the date at which pictures where taken. This is of a particular importance in the case of the PPM because nests can mostly be observed during winter and deteriorates after larvae migration into the soil. More generally, using GSV for species assessment should be approached with caution when species census is seasonal. As evoked above, the uncertainty introduced by processing pictures taken at various dates along the year is increasing as the grain is decreasing. It is important to highlight the value of the Google data for future research. Google database is updated and this growing body of data will constitute a tremendous amount of information in the next decades allowing for example researchers to perform retrospective analyses of species expansion processes and biological invasions. Chen et al. [42] recently published the results of a study assessing 100 years of environmental changes in Western China based on the comparison of modern versus 100-years old pictures of 250 localities from Western China. They showed how this type of data could help detecting and characterising changes in vegetation, landscape and more globally biodiversity. We believe that GSV may contribute to such type of retrospective surveys with unprecedented power and resolution.

Potential of GSV regarding species monitoring

As underlined above, only species visible from photographs or associated with conspicuous sign of presence can be surveyed using GSV imagery. This encompasses various organisms that alter significantly and specifically the colour or shape of trees as well as tree and plant species that colonize road edges or can be reliably sighted from roads. With that regards, the PPM constitutes a good biological model since its winter nests are white and easily visible in the tree foliage. In addition, PPM spatial aggregation at the edges of forest plots or along roads [26] increases the probability that the species be spotted if present in a given area. The present study focused on presence/absence data and the potential use of GSV for density census has not been considered. This question is more complicated and requires an estimate of the prospected surface [43]. Although GSV initially focused on city streets it was quickly extended to peri-urban areas and is now increasingly available for rural areas including large agricultural regions with low population densities. This is the case for different parts of our study region and in particular the Beauce area (Figure 4). The results reported in the present paper show that the GSV system, in its present form, provides a visual overview of urban, peri-urban and rural streets network allowing a proper assessment of the PPM regional distribution when assessed using large sampling window (here, 16×16 km). It is likely that the value of the GSV system will increase in the future as the density of the network will increase and that, for that reason, studies based on finer resolution will become possible.

In fact, the better representativeness of urban areas in GSV may constitute an opportune bias. Populated places often constitute major points of establishment for exotic organisms [44] from which invasive species might disperse towards other areas using different types of corridors [45]. Species monitoring in urban areas using GSV may prove useful to identify spots of invasive species, their dispersal pathways and the potential landscape features that slow down or speed up their dispersal.

Conclusions

The present survey illustrated how the GSV imagery could be used to perform *in silico* sampling of species occurrences. It must be emphasized that only organisms that can be reliably detected by road sampling can be assessed using GSV. The case of the PPM is straightforward as this species produces easily visible winter nests but many other organisms might require important calibration efforts. The scale issue deserves to be considered carefully and it must be noted that the method might perform poorly at small grains (high resolution). Although the ever-increasing coverage of GSV system should improve the method performances, we have no clue as to when and where small-scale sampling would become accurate. With only one case study, we obviously lack hindsight to advertise the technique for general use but our results show that it has some promise for future use, at least with species easily observed by means of road sampling such as the PPM.

Acknowledgments

We would like to thank Franois-Xavier Saintonge, Laurence Bouhot-Delduc, Hubert Pauly and Morgane Goudet of the DSF (Forest Health Department of the French Ministry of Agriculture) and Carole Kerdelhué (INRA) for helpful comments.

Author Contributions

Conceived and designed the experiments: JR CEI JPR. Performed the experiments: JR AD JG FG BV OD AR. Analyzed the data: JPR. Contributed reagents/materials/analysis tools: FD. Wrote the paper: JPR JR CEI CR AR.

References

1. Lin W, Zhou G, Cheng X, Xu R (2007) Fast economic development accelerates biological invasions in China. PLoS ONE 2: e1208.
2. Lockwood JL, Hoopes MF, Marchetti MP, editors (2007). Invasion ecology. Oxford: Blackwell Publishing. 304 p.
3. Perrings C, Mooney H, Williamson M, editors. (2010) Bioinvasions and globalization. Ecology, economics, management, and policy. Oxford: Oxford University Press. 267 p.
4. Thomas CD, Ohlemüller R (2010) Climate change and species distributions: An alien future? In: Perrings C, Mooney H, Williamson M, editors. Bioinvasions and globalization. Ecology, economics, management, and policy. Oxford: Oxford University Press. 19–29.
5. Clay SA, editor. (2011) GIS applications in agriculture. Volume Three: Invasive species. Boca Raton: CRC Press. 418 p.
6. Brooker RW, Travis JMJ, Clark EJ, Dytham C (2007) Modelling species' range shifts in a changing climate: The impacts of biotic interactions, dispersal distance and the rate of climate change. J Theor Biol 245: 59–65.
7. Rey O, Estoup A, Vonshak M, Loiseau A, Blanchet S, et al. (2012) Where do adaptive shifts occur during invasion? A multidisciplinary approach to unravelling cold adaptation in a tropical ant species invading the Mediterranean area. Ecol Lett 15: 1266–1275.
8. Franklin J (2009) Mapping species distributions: spatial inference and prediction. Cambridge: Cambridge University Press. 320 p.
9. Horn A, Kerdelhué C, Lieutier F, Rossi J-P (2012) Predicting the distribution of the two bark beetles *Tomicus destruens* and *Tomicus piniperda* in Europe and the Mediterranean region. Agric Forest Entomol 14: 358366.
10. Mulñoz A-R, Real R (2006) Assessing the potential range expansion of the exotic monk parakeet in Spain. Divers Distrib 12: 656–665.
11. Robinet C, Baier P, Pennerstorfer J, Schopf A, Roques A (2007) Modelling the effects of climate change on the potential feeding activity of *Thaumetopoea pityocampa* (Den. & Schiff.) (Lep., Notodontidae) in France. Global Ecol Biogeogr 16: 460–471.
12. Hill JK, Collingham YC, Thomas CD, Blakeley DS, Fox R, et al. (2001) Impacts of landscape structure on butterfly range expansion. Ecol Lett 4: 313–321.
13. Hernandez PA, Graham CH, Master LL, Albert DL (2006) The effect of sample size and species characteristics on performance of different species distribution modeling methods. Ecography 29: 773–785.
14. Wisz MS, Hijmans RJ, Li J, Peterson AT, Graham CH, et al. (2008) Effects of sample size on the performance of species distribution models. Divers Distrib 14: 763–773.
15. Carvalho LFR, de Melo CB, McManus C, Haddad JPA (2012) Use of satellite images for geographical localization of livestock holdings in Brazil. Prev Vet Med 103: 74–77.
16. Chang AY, Parrales ME, Jimenez J, Sobieszczyk ME, Hammer SM, et al. (2009) Combining Google Earth and GIS mapping technologies in a dengue surveillance system for developing countries. Int J Health Geogr 8: 49.
17. Asner GP, Rudel TK, Aide TM, Defries R, Emerson R (2009) A contemporary assessment of change in humid tropical forests. Conserv Biol 23: 1386–1395.
18. Giri C, Ochieng E, Tieszen LL, Zhu Z, Singh A, et al. (2011) Status and distribution of mangrove forests of the world using earth observation satellite data. Global Ecol Biogeogr 20: 154–159.
19. Guralnick RP, Hill AW, Lane M (2007) Towards a collaborative, global infrastructure for biodiversity assessment. Ecol Lett 10: 663–672.
20. Anguelov D, Dulong C, Filip D, Frueh C, Lafon S, et al. (2010) Google street view: Capturing the world at street level. Computer 43: 32–38.
21. Frome A, Cheung G, Abdulkader A, Zennaro M, Bissacco A, et al. (2009) Large-scale privacy protection in Google street view. In: Proc. 12th IEEE Intl Conf. Computer Vision (ICCV 09), IEEE Press 2373–2380.
22. Odgers CL, Caspi A, Bates CJ, Sampson RJ, Moffitt TE (2012) Systematic social observation of childrens neighborhoods using Google street view: a reliable and cost-effective method. J Child Psychol Psychiatry 53: 1009–1017.
23. Rundle AG, Bader MDM, Richards CA, Neckerman KM, Teitler JO (2011) Using Google street view to audit neighborhood environments. Am J Prev Med 40: 94–100.
24. Samalens J-C, Rossi J-P, Guyon D, Van Halder I, Menassieu P, et al. (2007) Adaptive roadside sampling for bark beetle damage assessment. Forest Ecol Manage 253: 177–187.
25. Arnaldo PS, Torres LM (2005) Spatial distribution and sampling of *Thaumetopoea pityocampa* (Den. & Schiff.) (Lep. Thaumetopoeidea) populations on *Pinus pinaster* Ait. in Montesinho, N. Portugal. Forest Ecol Manage 210: 1–7.
26. Samalens J-C, Rossi J-P (2011) Does landscape composition alter the spatiotemporal distribution of the pine processionary moth in a pine plantation forest? Popul Ecol 53: 287–296.

27. Battisti A, Stastny M, Netherer S, Robinet C, Schopf A, et al. (2005) Expansion of geographic range in the pine processionary moth caused by increased winter temperatures. Ecol Applic 15: 2084–2096.
28. Huchon H, Démolin G (1970) La bioécologie de la processionnaire du pin: dispersion potentielle-dispersion actuelle. Rev Forest Fr 22: 220–233.
29. Battisti A, Stastny M, Buffo E, Larsson S (2006) A rapid altitudinal range expansion in the pine processionary moth produced by the 2003 climatic anomaly. Global Change Biol 12: 662–671.
30. Robinet C, Rousselet J, Imbert C-E, Sauvard D, Garcia J, et al. (2010) Le réchauffement climatique et le transport accidentel par l'homme responsables de l'expansion de la chenille processionnaire du pin. Forêt Wallonne 108: 19–27.
31. Rosenzweig C, Casassa G, Karoly D, Imeson A, Liu C, et al. (2007) Assessment of observed changes and responses in natural and managed systems. In: Parry ML, Canziani OF, Palutikof JP, van der Linden PJ, Hanson CE, editors. Climate change 2007: Impacts, adaptation and vulnerability. Contribution of working group II to the fourth assessment report of the intergovernmental panel on climate change. Cambridge: Cambridge University Press. 79–131.
32. Démolin G (1969) Comportement des adultes de *Thaumetopoea pityocampa* Schiff.: dispersion spatiale, importance écologique. Ann Sci Forest 26: 81–102.
33. Geri C, Millier C, Xeuxet D (1985) Mesure des populations de processionnaire du pin (*Thaumetopoea pityocampa* Schiff – Lépidoptère Thaumetopoeidae) au Mont-Ventoux. Ann Sci Forest 42: 143–184.
34. Roques L, Soubeyrand S, Rousselet J (2011) A statistical-reaction-diffusion approach for analyzing expansion processes. J Theor Biol 274: 43–51.
35. Hijmans RJ, van Etten J (2012) raster: Geographic data analysis and modeling. R package version 2.0-41. Available: http://CRAN.R-project.org/package=raster. Accessed: 10 Sep 2013.
36. Fawcett T (2006) An introduction to ROC analysis. Pattern Recog Lett 27: 861874.
37. Sing T, Sander O, Beerenwinkel N, Lengauer T (2005) ROCR: visualizing classifier performance in R. Bioinformatics 21: 3940–3941.
38. R Development Core Team (2013) R: A Language and Environment for Statistical Computing. Vienna, Austria: R Foundation for Statistical Computing. Available: http://CRAN.R-project.org/. Accessed: 10 Sep 2013.
39. Levin SA (1992) The problem of pattern and scale in ecology: The Robert H. MacArthur award lecture. Ecology 73: 1943–1967.
40. Bellehumeur C, Legendre P, Marcotte D (1997) Variance and spatial scales in a tropical rain forest: Changing the size of sampling units. Plant Ecol 130: 89–98.
41. Rossi J-P, Nuutinen V (2004) The effect of sampling unit size on the perception of the spatial pattern of earthworm (*Lumbricus terrestris* L.) middens. Appl Soil Ecol 27: 189–196.
42. Chen H, Yin K, Wang H, Zhong S, Wu N, et al. (2011) Detecting one-hundred-year environmental changes in western China using seven-year repeat photography. PLoS ONE 6: e25008.
43. Southwood TRE, Henderson PA (2000) Ecological methods. Oxford: Blackwell Science. 575 p.
44. Smith RM, Baker RHA, Malumphy CP, Hockland S, Hammon RP, et al. (2007) Recent non-native invertebrate plant pest establishments in Great Britain: origins, pathways, and trends. Agric Forest Entomol 9: 307–326.
45. Säumel I, Kowarik I (2010) Urban rivers as dispersal corridors for primarily wind-dispersed invasive tree species. Landscape Urban Plann 94: 244–249.

9

Fish with Chips: Tracking Reef Fish Movements to Evaluate Size and Connectivity of Caribbean Marine Protected Areas

Simon J. Pittman[1,2,3]*, Mark E. Monaco[1], Alan M. Friedlander[4], Bryan Legare[2,5], Richard S. Nemeth[2], Matthew S. Kendall[1], Matthew Poti[1], Randall D. Clark[1], Lisa M. Wedding[1,6], Chris Caldow[1]

1 U.S. National Oceanic and Atmospheric Administration, Center for Coastal Monitoring and Assessment, Silver Spring, Maryland, United States of America, **2** Center for Marine and Environmental Studies, University of the Virgin Islands, St. Thomas, Virgin Islands, U.S. Virgin Islands, **3** Centre for Marine and Coastal Policy Research, The Marine Institute, Marine Building, Plymouth University, Plymouth, Devon, United Kingdom, **4** Department of Biology, Fisheries Ecology Research Laboratory, University of Hawai'i at Mānoa, Honolulu, Hawai'i, United States of America, **5** Texas Parks and Wildlife, Dickinson Marine Lab, Dickinson, Texas, United States of America, **6** Center for Ocean Solutions, Woods Institute for the Environment, Stanford University, Monterey, California, United States of America

Abstract

Coral reefs and associated fish populations have experienced rapid decline in the Caribbean region and marine protected areas (MPAs) have been widely implemented to address this decline. The performance of no-take MPAs (i.e., marine reserves) for protecting and rebuilding fish populations is influenced by the movement of animals within and across their boundaries. Very little is known about Caribbean reef fish movements creating a critical knowledge gap that can impede effective MPA design, performance and evaluation. Using miniature implanted acoustic transmitters and a fixed acoustic receiver array, we address three key questions: How far can reef fish move? Does connectivity exist between adjacent MPAs? Does existing MPA size match the spatial scale of reef fish movements? We show that many reef fishes are capable of traveling far greater distances and in shorter duration than was previously known. Across the Puerto Rican Shelf, more than half of our 163 tagged fish (18 species of 10 families) moved distances greater than 1 km with three fish moving more than 10 km in a single day and a quarter spending time outside of MPAs. We provide direct evidence of ecological connectivity across a network of MPAs, including estimated movements of more than 40 km connecting a nearshore MPA with a shelf-edge spawning aggregation. Most tagged fish showed high fidelity to MPAs, but also spent time outside MPAs, potentially contributing to spillover. Three-quarters of our fish were capable of traveling distances that would take them beyond the protection offered by at least 40–64% of the existing eastern Caribbean MPAs. We recommend that key species movement patterns be used to inform and evaluate MPA functionality and design, particularly size and shape. A re-scaling of our perception of Caribbean reef fish mobility and habitat use is imperative, with important implications for ecology and management effectiveness.

Editor: Christopher J. Fulton, The Australian National University, Australia

Funding: SJP, MEM, AMF and LMW were supported by U.S. National Park Service, NOAA Center for Coastal Monitoring and Assessment (CCMA) and NOAA Coral Reef Conservation Program. BL was supported by University of the Virgin Islands. RSN was supported by the National Science Foundation's Virgin Islands Experimental Program to Stimulate Competitive Research (0814417) and the University of the Virgin Islands Center for Marine and Environmental Studies (CMES). The UVI acoustic array was supported by grants from Puerto Rico Sea Grant (R-31-1-06), NOAA Saltonstall-Kennedy (NA09NMF4270068), NOAA cooperative agreement (NA08NOS4260348), and USGS State Partnership Program (07ERAG0078). The funders had no role in study design, data collection and analysis, decision to publish, or preparation of the manuscript.

Competing Interests: The authors have declared that no competing interests exist.

* E-mail: simon.pittman@noaa.gov

Introduction

In the past 50 years, coral reef ecosystems of the Caribbean Sea have experienced considerable region-wide declines in the abundance and body size of many fished species and in the quality of their habitat [1,2,3]. In response, marine protected areas (MPAs), including areas closed to fishing, have been established throughout the region [4]. However, MPAs often fail to reach their full potential as a consequence of factors such as illegal harvesting, regulations that legally allow detrimental harvesting, or emigration of animals outside boundaries because of continuous habitat or inadequate size of the MPA [42]. In the Caribbean and elsewhere, MPA placement, size and shape, typically have been delineated without knowledge of the movements of the animals they are intended to protect. This is a widespread practice for coral reef-centered MPAs where placement and shape are generally designed using relatively static biophysical features of the seafloor and with political and socio-economic considerations rather than criteria based on functional ecology [32,33,43,44]. Research on the optimal design and connectivity of MPAs and networks of MPAs has focused primarily on models of larval dispersal [5,13,37]. It is now increasingly acknowledged, however, that adult and juvenile movements have a considerable influence on the functioning of MPAs [7,34,40,41]. Spatially-explicit information on the movement patterns of fishes, including home range size, relocation movements and distances traveled on migrations, can play a central role to support the ecologically meaningful design of MPAs [5,6,10,45]. For highly mobile fishes, protection could be sub-optimal if MPAs are too small relative to the routine space use patterns of fishes, or if placed in locations that offer incomplete protection for animals when undertaking critical migrations [5,6,7,13]. Modeling studies suggest that even limited adult dispersal can have a large impact on MPA functioning [44]. Conversely, emigration from protected areas can contribute to "spillover" of fish into the neighboring fishery, an important fisheries management goal [8]. Clearly, knowledge of the spatial scale and patterns of fish movements across the seascape can provide direct evidence of ecological connectivity between individual MPAs, as well as data for the evaluation of ecological coherence within an MPA network.

Understanding movement patterns has profound implications for ecology too. Organism movement behavior is a fundamental ecological process that can be used to identify habitat use patterns, investigate bioenergetics, energy flows and to select the appropriate spatial (and temporal) scale(s) at which to study species ecology [9,11]. An organism-centered approach, that provides explicit spatial and temporal data on organism movements, can help ecologists formulate research questions and design studies that measure ecological patterns and processes at ecologically meaningful scales [9]. When analyzed together with maps of benthic composition, or seafloor topographic structure, movement data can identify critical pathways, or 'blue corridors', connecting patch types in the seascape [9,10,11,12]. Very little is known, however, about the spatial scale of adult fish movements, travel times and locations visited by Caribbean fishes during their life span [7,9].

Fish assemblages of Caribbean coral reef ecosystems are exceptionally diverse, with many species using multiple habitat types (i.e., seagrasses, mangroves, coral reefs) through daily home range movements, ontogenetic habitat shifts and spawning and non-spawning migrations [9,14]. As such, the spatial domain of routine movements, sometimes referred to as the ecological neighborhood [9,10], will differ between species, individuals and life-stages [6,14,15]. However, direct and spatially explicit evidence for the distances traveled and timing of movements by fishes associated with Caribbean coral reef ecosystems is scarce [16]. Studies have either focused on fine-scale movements for single species and families, usually confined geographically to a single embayment or MPA [19,20], or large-bodied species of high conservation or fishery concern, such as groupers, revealing long (>30–220 km) distance movements to spawning aggregations [35,36,49]. Relative to the fish life span, movement observations have been short (days-weeks) potentially resulting in an underestimate of the spatial scale of movement [7].

The most effective technique for gathering direct evidence of movements in time and space for highly mobile marine animals, such as fish, is to tag them with miniature acoustic transmitters [9,21]. Advances in microelectronics and a demand for ever smaller tags has resulted in the development of miniaturized underwater acoustic transmitters small enough to be harmlessly implanted in small (<20 cm) fishes, yet with sufficient battery life to emit a unique identification code every minute for a year or more. These coded transmitters are detected by underwater acoustic receivers that can be strategically moored above the seafloor to form an array of listening stations [22].

A lack of information on fish movements has been identified as a critical knowledge gap in the application of MPAs as a management tool [17,18,34]. To address the knowledge gap for the U.S Caribbean we utilized an array of 225 underwater acoustic receivers to track fish movements between MPAs and across the contiguous insular shelf of the U.S. Virgin Islands and southeastern Puerto Rico. The array known as the U.S. Caribbean Acoustic Network (USCAN), formed through a consortium of collaborating scientists from multiple institutions, was one of the largest acoustic fish telemetry networks in the world. Tracking movements of highly mobile organisms with an acoustic array allowed us, from afar, to locate fish and then map points of presence along their movement pathways as they swam freely across the seascape.

Our primary research questions were: 1.) How far do fish move and how long does it take them to move those distances? 2.) Are MPAs connected by fish movements and if so then how often do fish cross MPA boundaries? and 3.) Are MPAs in the eastern Caribbean of sufficient size to encompass the movements of reef fish?

Materials and Methods

We tagged 184 individual fish with surgically implanted miniature acoustic transmitters (see File S1) representing 19 species from 10 different families commonly occurring in Caribbean coral reef ecosystems (Table 1) [23]. Tagged fishes included ecologically and economically important species targeted in the Virgin Islands fishery. Fish sizes, measured as total body length (TL), ranged from 19 to 70 cm (mean = 30.3 cm). The largest individual tagged was a 70 cm TL nurse shark (*Ginglymostoma cirratum*) and the smallest was a blue tang (*Acanthurus coeruleus*) of 19 cm TL. Five species of snapper (Lutjanidae) accounted for 31% of all individuals tagged and three species of grunt (Haemulidae) accounted for 23% of the sample. At the species level, bluestriped grunt (*Haemulon sciurus*) comprised 22% of all individuals tagged, lane snapper (*Lutjanus synagris*) 21%, bar jack (*Caranx ruber*) 11%, and saucereye porgy (*Calamus calamus*) 10%.

Fish were captured, tagged and released between July 2006 and July 2008 within the Virgin Islands National Park (VINP) under research permit (# 118794) provided by the National Park Service of the U.S. Department of Interior. Fish locations were recorded for a total of 1,244 days (approx. 3 year and 4 months) calculated as the time elapsed between the first and last detection on a receiver. During that time, a total of 2,848,192 detections were recorded from 163 individual fish of 18 species (21 individuals were not detected after release).

Experimental procedures were approved by the Office of Research Compliance Animal Welfare and Biosafety Program of the University of Hawaii and the Animal Care and Use Committee. Permission to capture fish and conduct experiments including tagging and tracking inside waters managed by the U.S. Department of Interior were provided by the National Park Service under permit NPS PIMS Project #118794 Ecological linkages between the Virgin Islands Coral Reef National Monument and the Virgin Islands National Park.

Acoustic array

The underwater acoustic monitoring system consisted of an array of Vemco™ VR2/VR2W acoustic receivers (340 mm long 960 mm diameter, weight in water 300 g) capable of detecting coded transmitter signals when located within a 300 m minimum radius of the receiver. Receivers were deployed on polypropylene lines or vinyl coated stainless steel wire attached to the seabed with sand screws or concrete blocks at depths of between 1 and 38 meters. The array was positioned close to the edges of nearshore coral reefs and across the insular shelf to create both an alongshore

Table 1. Fish species tagged with ultrasonic transmitters and their biological characteristics and summary information on the duration of tracking and least cost distance between farthest receivers with detections.

Family	Scientific name	Common name	Number tagged	Trophic group	Total length range (cm)	Max. days tracked	Max inter-receiver distance (range in km)
Haemulidae	*Haemulon sciurus*	Bluestriped grunt	40	Invertivore	24–30.6	930	0.4–13.5
Lutjanidae	*Lutjanus synagris*	Lane snapper	38	Inv/Pisc	20–36	722	0.5–11.5
Carangidae	*Caranx ruber*	Bar jack	21	Piscivore	29.8–47	329	0.58–13.6
Sparidae	*Calamus calamus*	Saucereye porgy	19	Invertivore	21.3–35	180	0.47–15.3
Lutjanidae	*Ocyurus chrysurus*	Yellowtail snapper	14	Invertivore	22.5–38	333	0.3–16.1
Lutjanidae	*Lutjanus analis*	Mutton snapper	12	Invertivore	31–65	784	1.1–42.2
Holocentridae	*Holocentrus adscensionis*	Longjaw squirrelfish	9	Invertivore	26–29	348	1.5–5.1
Acanthuridae	*Acanthurus coeruleus*	Blue tang	5	Herbivore	19–24	106	2–8.5
Balistidae	*Balistes vetula*	Queen triggerfish	5	Invertivore	29–39	38	0.58–1.5
Lutjanidae	*Lutjanus griseus*	Gray snapper	5	Invertivore	25.2–35.4	657	1–3.5
Mullidae	*Mulloidichthys martinicus*	Yellow goatfish	3	Invertivore	31–32.3	34	0.4–4.8
Acanthuridae	*Acanthurus chirurgus*	Doctorfish	2	Herbivore	19.2–23.9	85	1–3.6
Serranidae	*Epinephelus guttatus*	Red hind	2	Inv/Pisc	29.5–36.5	402	6.3
Ginglymostomatidae	*Ginglymostoma cirratum*	Nurse shark	2	Inv/Pisc	55–70	233	0.7
Haemulidae	*Haemulon flavolineatum*	French grunt	2	Invertivore	20–21.5	1	0
Haemulidae	*Haemulon plumierii*	White grunt	2	Invertivore	25–31.5	332	6.2
Lutjanidae	*Lutjanus apodus*	Schoolmaster	1	Inv/Pisc	27	382	5.1
Lutjanidae	*Lutjanus jocu*	Dog snapper	1	Inv/Pisc	41.4	311	12.8
Mullidae	*Pseudupeneus maculatus*	Spotted goatfish	1	Invertivore	27	0	0

Note that these estimates represent the minimum known cross-boundary movements. We except that the maximum distance estimates are likely to be limited by the array configuration and that many movements are undetected, but this only increases the probability that distances traveled could be even greater than highlighted here rather than less extensive.

and an inshore-offshore corridor of listening stations (Fig. 1). Receivers were located both inside and outside four well-established, permanent and seasonal marine protected areas of the U.S. Virgin Islands: Virgin Islands Coral Reef National Monument (VICR) (approx. 51.5 km^2); the Virgin Islands National Park (VINP) (approx. 22.9 km^2); Hind Bank Marine Conservation District (MCD) (approx. 41 km^2), and Grammanik Bank (GB) (approx. 4.5 km^2), the latter two being sites of regionally important multispecies fish spawning aggregations. VICR and MCD are both no-take MPAs and GB is a seasonal closure for a range of spawning species [50]. Our array also provided information on trans-boundary movements from waters managed by U.S. Federal Government to waters managed by the Governments of Puerto Rico and U.S. Virgin Islands. Receivers were maintained and data managed and shared by a consortium of institutions known collectively as the U.S. Caribbean Acoustic Network (USCAN) comprising the National Oceanic and Atmospheric Administration's (NOAA) National Centers for Coastal Ocean Science (NCCOS) and NOAA National Marine Fisheries Service (NMFS), together with the University of the Virgin Island's Center for Marine and Environmental Studies. When operating simultaneously, this network was able to detect tagged fish movements between receivers across distances of between 300 m (closest receivers) to 103.3 km (farthest receivers) across the insular shelf. Although the integrated data from the entire array network provided greater coverage than could any of the individual project arrays, some spatial clustering of receivers resulted from arrays configured to address specific research questions.

Calculating least-cost pathways between acoustic receivers

To quantify the spatial domain of movements for each fish, we calculated between- receiver distance for all receivers that detected a fish and reported the distance between the pair of receivers that were farthest apart, together with shortest time elapsed between those detections. The distances we present here provide estimates of maximum known movement capability to define a spatial domain of space use for fishes. We select these data to address specific questions about the spatio-temporal scale of movements and we do not herein address detailed habitat use patterns. Rather than calculating straight line distances between each pair of receivers, some of which would have transected land or deep waters beyond the shelf edge, we quantified distances from modeled least-cost pathways (LCP) conditioned to follow suitable habitat for reef fish (i.e. colonized hardbottom habitat and shallow water non-reef habitat [sand]) while avoiding less suitable or unsuitable environments (i.e., land and deep water beyond the shelf edge) (see File S1). Tracking studies and underwater observations provide evidence that fish follow coral reefs to maximize access to food and minimize predation risk, whereas land and deep water and sometimes large expanses of sand can function as barriers to movement [7,24,39]. Non-reef sandy areas were weighted with relatively low cost such that pathways could cross sand to connect patches of more preferred (lowest cost) coral reef patches. Parameterization of LCP models was based on the known daytime occurrence of species over shallow water coral reef ecosystems where a strong dependence exists for colonized

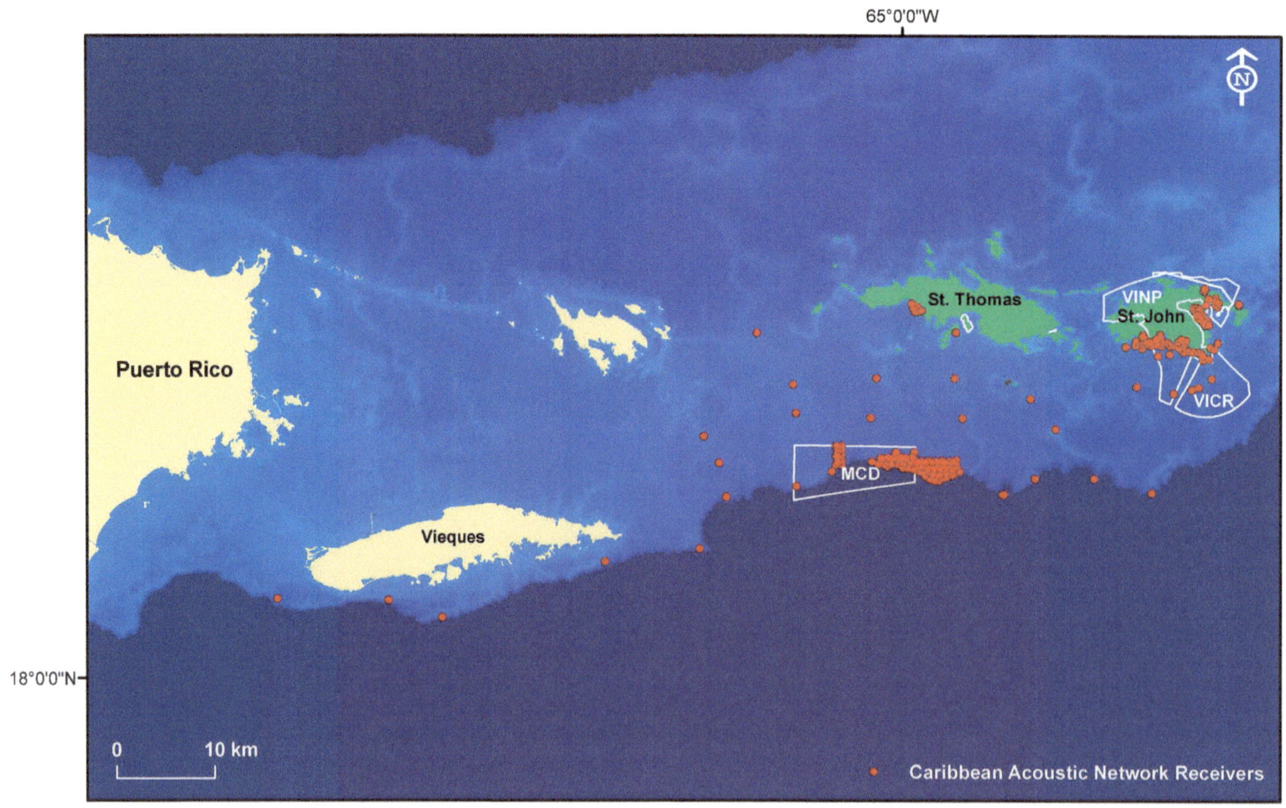

Figure 1. Location of underwater acoustic receivers in the U.S. Caribbean Acoustic Network (USCAN) on the southern insular shelf of Puerto Rico and U.S. Virgin Islands. MPA boundaries are shown for the Virgin Islands National Park (VINP), Virgin Islands Coral Reef Monument (VICR) and the Marine Conservation District (MCD).

carbonate seafloor which provides appropriate refuge from predation and potential food resources for the traveling fish.

Quantifying movements across MPA boundaries

To examine movements relative to MPA boundaries, we calculated the number of times each fish crossed a management boundary including excursions to unprotected areas outside of MPAs. Fish that did not cross a boundary and appeared to remain within the VINP during the tracking period were also quantified. Trans-boundary movements were recorded between the Virgin Islands National Park (limited fishing restrictions); Virgin Islands Coral Reef National Monument (no-take); Hind Bank Marine Conservation District (no-take spawning aggregation); Grammanik Bank (seasonal closure spawning aggregation/pelagic fishing only) and non-MPA waters. Detections from receivers that were located too close to a boundary to provide high certainty that a fish had crossed a boundary were removed from this part of the analysis. This resulted in removal of seven receivers along the VINP-VICR boundary and a large spatial gap in detection, therefore, our estimates of boundary crossings should be interpreted as a minimum of crossings.

Estimating the proportion of tracked time fishes spend inside and outside MPAs

To examine the presence and residence times of individual fish inside and outside MPAs, a Fidelity Index was calculated as a relative estimate of minimum residency. Individuals were considered present if two or more detections occurred on at least one receiver within a given day. Fishes were excluded from analysis where less than a total of ten detections were recorded across the monitoring period or where presence was recorded for only one day or less. The monitoring period for each fish was defined as the number of days between the tagging event and the last detection. Hourly fish presence was calculated by pooling detections into hourly increments (24 per day), such that if a fish was detected in an MPA or outside between 00:00 and 00:59 of any hour then it was considered present. If a fish was present in two zones in the same hour it received a value of 1 hour for each zone.

The proportion of time a fish was observed during the study was calculated as the hours present in all zones divided by the total monitoring period. To understand the preference for each zone (i.e. VICR, VINP and unprotected area), a location-specific Fidelity Index (FI) was quantified for each fish as the sum of hours in each zone divided by the total hours present. The proportion of time represented in the FI ranged from 0 (no fidelity) to 1 (high fidelity).

Due to uncertainty in fish locations during periods without detections, the FI is best interpreted as the minimum amount of time a fish is present in a zone and can be used to understand the relative preference between zones, rather than absolute time spent in each zone.

Quantifying spatial dimensions of eastern Caribbean MPAs

To calculate the spatial dimensions of MPAs (n = 220) throughout the Leeward and Windward Islands (U.S. Virgin Islands to Aruba, Bonaire and Curacao) we applied a GIS tool called Minimum Bounding Geometry using ArcGIS v10.1 (ESRI Inc, 2012) to quantify the length of the long axis (referred to as MPA length) and the perpendicular axis (referred to as MPA width) for each MPA polygon. To estimate the greatest distance that a fish inside an MPA would need to travel to reach the MPA boundary, we calculated the farthest-edge to centroid distance (the *apothem* for regular polygons) for each MPA polygon by halving the length and width dimensions. Spatial data on MPAs was provided by The Nature Conservancy compiled for the Caribbean Marine Protected Areas Managers Network and Forum (CaMPAM). The MPA database was in development at the time of writing and therefore incomplete and the regulations associated with each MPA were not available to determine the proportion of MPAs designated as no-take marine reserves. Because of differences in the way that fish pathways and MPA axes were measured we do not statistically compare fish distances traveled versus MPA length and width, but instead visually approximate the potential scale mismatch between spatial domains of MPAs and fish movement capability. In addition, differences in seascape configuration between individual MPAs preclude any further comparison.

Results and Discussion

How far can Caribbean reef fishes move?

Our long-term acoustic tracking data extend considerably the known movement abilities for Caribbean reef fish (Fig. 2). The mean tracking duration calculated from all tagged individuals was 164 days (±15.6 SE), with a maximum detection period of 967 days for a bluestriped grunt (*Haemulon sciurus*), exceeding the manufacturer's estimate of battery life for the transmitters. Other long duration tracking periods were 722 days for a lane snapper (*L. synagris*); 784 days for a mutton snapper (*L. analis*); 657 days for a gray snapper (*L. griseus*) and 402 days for a red hind (*Epinephelus guttatus*) (Table 1).

From a total of 163 fish tracked, the estimated mean between-receiver distance traveled was 3.8 km (median = 2.1 km). Seventy-five percent of fish undertook long range movements greater than 1 km and 33% more than 5 km. Twenty-eight individual fish traveled more than 1 km between two receivers in a single 24 hour period. Three fishes traveled 14.9 km (*Calamus calamus*, 24.5 cm TL); 11.7 km (*Haemulon sciurus*, 30 cm TL); and 10.6 km (*Lutjanus synagris*, 30.4 cm TL) in a single day. At the family level, grunts (haemulids), snappers (lutjanids), jacks (carangids), and porgies (sparids) showed the greatest long-range movements. Unexpectedly high mobility, however, was also detected for several species that are usually perceived as exhibiting relatively low mobility highlighting how little is known about reef fish movements. For instance, during a detection period of 348 days, a longjaw squirrelfish (*Holocentrus adscensionis*) was detected on receivers separated by an estimated least-cost path distance of 5.1 km. In addition, we include information on species with little previous information on distance travelled, including an individual blue tang (*Acanthurus coeruleus*) with maximum estimated distance of 8.5 km; a yellow goatfish (*Mulloidichthys martinicus*) 4.8 km and 1.5 km for a Queen triggerfish (*Balistes vetula*). In contrast, eleven fish were only detected within 500 m of their first detection point during an average period of 40 days of tracking (range = 1–167). Longer range movements, however, may have occurred out of range of our receiver array. For example, in a conventional tag-recapture study red hind migrated from 5 to 33 km from spawning aggregation sites to home range sites [51,52]. The modeled estimates of movement distances should be interpreted as best-available information on the spatial dimensions of the activity spaces for individual fish and an underestimate of the true distances moved within the life span due to spatial bias in the detectability of fish inherent with a fixed acoustic array. Furthermore, quantification of model uncertainty for least cost paths modeled on minimum cost for movements across a benthic cost surface requires further validation with detailed animal movement trajectories [48].

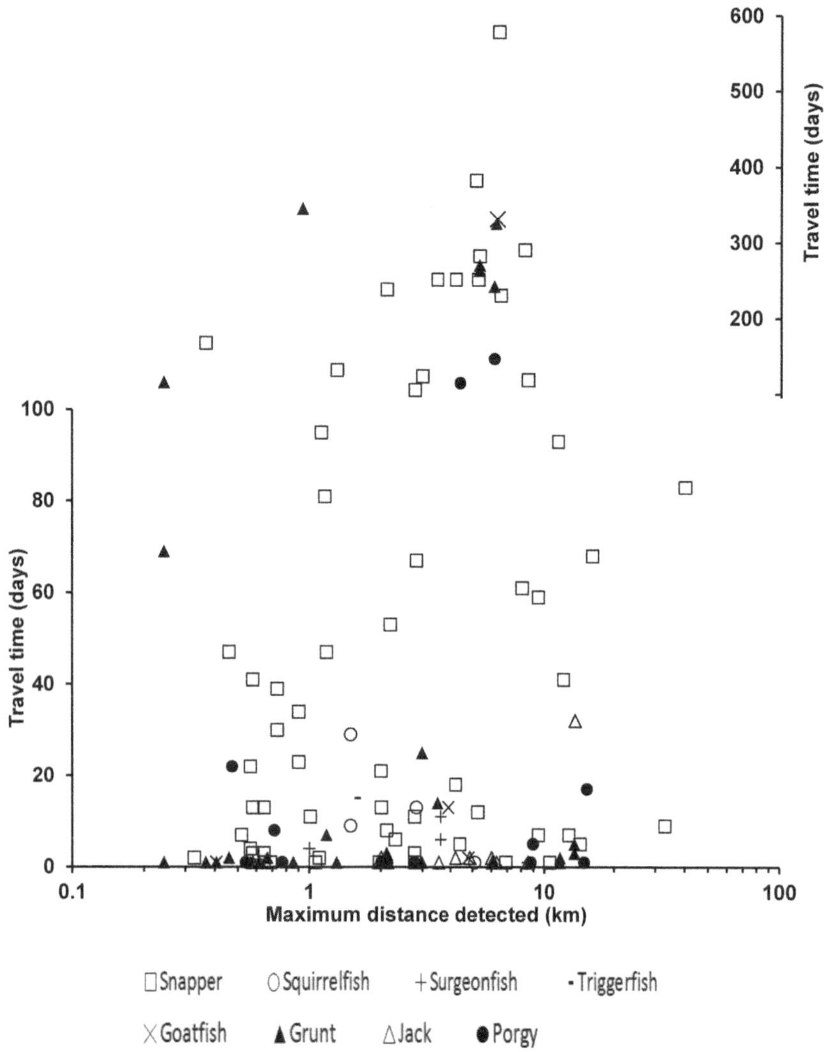

Figure 2. Distance and travel time between farthest receivers visited by acoustically tagged fish based on the time stamps of detections on the Caribbean acoustic array. Symbols identify fish family for0 each individual fish data point.

Association between fish body size and spatial scale of movements

Several reviews of home range movements have found large-bodied reef fishes often have larger home ranges than smaller bodied fishes [6,17]. After outliers were removed (fish detected only on a single receiver and the shark), our data demonstrate that body size (TL) and maximum distance of detection are positively and significantly correlated ($r = 0.48$, $p = <0.05$). Several large-bodied snapper which undertook long-distance migrations to shelf-edge spawning sites strengthened the positive correlation, but the correlation was also decoupled by high variability among individual fish together with highly variable tracking duration (Table 1). These differences in mobility are likely to be related to morphological and life-history characteristics, as well as the distribution of habitat structure across the region they inhabit. High behavioral polymorphism in movement patterns among fish species, life stages and individuals of the same species presents a major challenge in MPA network design, because over time unequal protection will exert a selection pressure for certain life history strategies [5]. In general, larger-bodied reef fishes are highly mobile, such as many large snappers, groupers and jacks that may undertake ontogenetic habitat shifts and spawning migrations across the shelf [38]. These broad-scale movements are thought to have a large impact on MPA functioning [5,10,17]. For example, Edgar et al. [42] found that MPA size was important for predicting biomass of jacks (Carangidae), a highly mobile family, whereby smaller MPAs correlated with lower biomass which was thought to result from increased time spent outside of park boundaries.

As often is the case with remote sensing approaches like acoustic telemetry, we are unable to provide information on the drivers of the movements that we report here. With a diverse range of species and families comes a diverse range of movement strategies. Some movements will be ontogenetic habitat shifts, others will be nomadic movements to find more optimal habitat, or enlargements of the home range, or very directional migratory movements to spawning aggregations. To adequately protect a range of species within an MPA or network of MPAs, the movement patterns for adult fish from a range of fished species needs to be considered in the MPA network design process

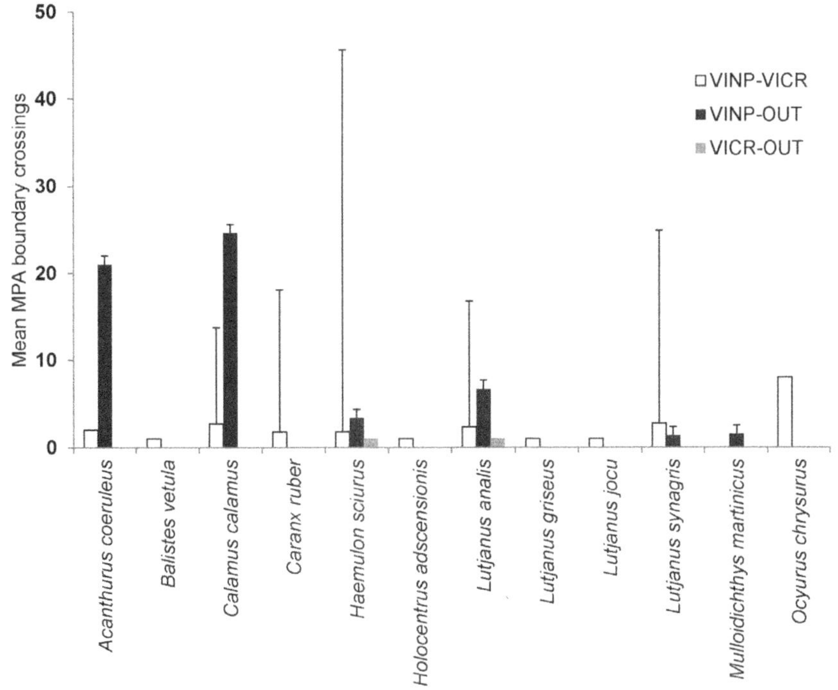

Figure 3. Mean (± SE) number of times individuals of a fish species crossed a management boundary (VINP – Virgin Islands National Park; VICR – Virgin Islands Coral Reef National Monument; Out – outside MPA) in the U.S.Virgin Islands (based on receiver detections) during the study period. Individuals from 12 of 18 species were confirmed to have crossed a management boundary.

[7,9,10]. Although our study provides new information for several species, longer duration continuous tracking of multiple individuals at different life stages, together with high resolution mapping of the seafloor will help to better understand the movement ecology of fishes associated with Caribbean coral reefs.

Connectivity within an MPA network

A total of 278 confirmed boundary crossings were recorded during the study period by 12 of the 18 species. Twenty-five percent of individuals crossed one or more MPA boundaries in the U.S. Virgin Islands network of MPA's. Five percent (7 individuals of 4 species) crossed two or more different management boundaries. Sixteen percent (24 individuals of 11 species) crossed the boundary between the VINP and the VICR. The highest number of crossings (222 crossings from 21 individuals and 11 species) occurred between VINP and areas outside of any MPA, with only two crossings from VICR to outside areas. Fifty-four crossings (24 individuals of 11 species) were recorded between VINP and VICR confirming the existing of ecological connectivity between adjacent MPA units.

Blue tang (*A. coeruleus*) bluestripped grunts (*H. sciurus*), four snapper species (*Lutjanus synagris, L. griseus, L. analis and Ocyurus chrysurus*), a jack species (*Caranx ruber*) and porgies (*Calamus calamus*) undertook most boundary crossings (Fig. 3). Many of these boundary crossings probably occurred because routine daily home range areas straddled the boundary of two adjacent MPAs or during a spawning migration. For instance, distinct day and night activity spaces sometimes occur in geographically discrete locations that can involve movements from deep to shallow habitat or from reef to seagrass beds [15,20]. From a management perspective this means that many of the tagged fishes spent time in management units providing different levels of protection from fishing, with uncertain effects on MPA performance. The siting of a fully protected MPA adjoining a partially protected MPA could result in a net flow of organisms into the fully protected MPA, especially if that MPA provides higher habitat quality. Furthermore, connectivity, particularly for herbivorous fishes, has been found to boost MPA performance and increase reef resilience [40,46,47], presenting additional ecological consideration for the relevance of fish movements in the MPA design process.

One individual mutton snapper traveled from the nearshore seascapes of VINP to the shelf edge (Grammanik Bank MPA) and back through non-MPA waters providing new direct evidence of ecological connectivity between multiple neighboring MPAs and between near-shore protected areas and known shelf edge spawning aggregations (Fig. 4). The modeled path distance between farthest receivers for this long-distance migration was estimated at 40.2 km. Our study has demonstrated that fish tagged within MPAs in the U.S. Virgin Islands spend time (and use resources) under differing management regimes that offer different levels of protection from fishing. Furthermore, we provide confirmation that fishes tagged in protected areas also spend time in areas open to fishing. Therefore we demonstrate that fish do swim in and out of MPAs in the U.S. Virgin Islands. Even if only temporary excursions to areas outside MPAs, these fishes contribute to a 'spillover' process influencing biodiversity, productivity and ecological functions in neighboring unprotected areas [25,26]. Although we conclude that ecological connectivity exists between MPAs in study area, a wide range of reef associated fish also undertake movements that are broader than the dimensions of the existing MPAs. Our sample is biased to the more highly mobile species of the reef fish community (snapper, porgies, jacks, grunts), and more work is required to include other key fish families not sampled or under-represented here, such as parrotfishes, wrasses and grouper.

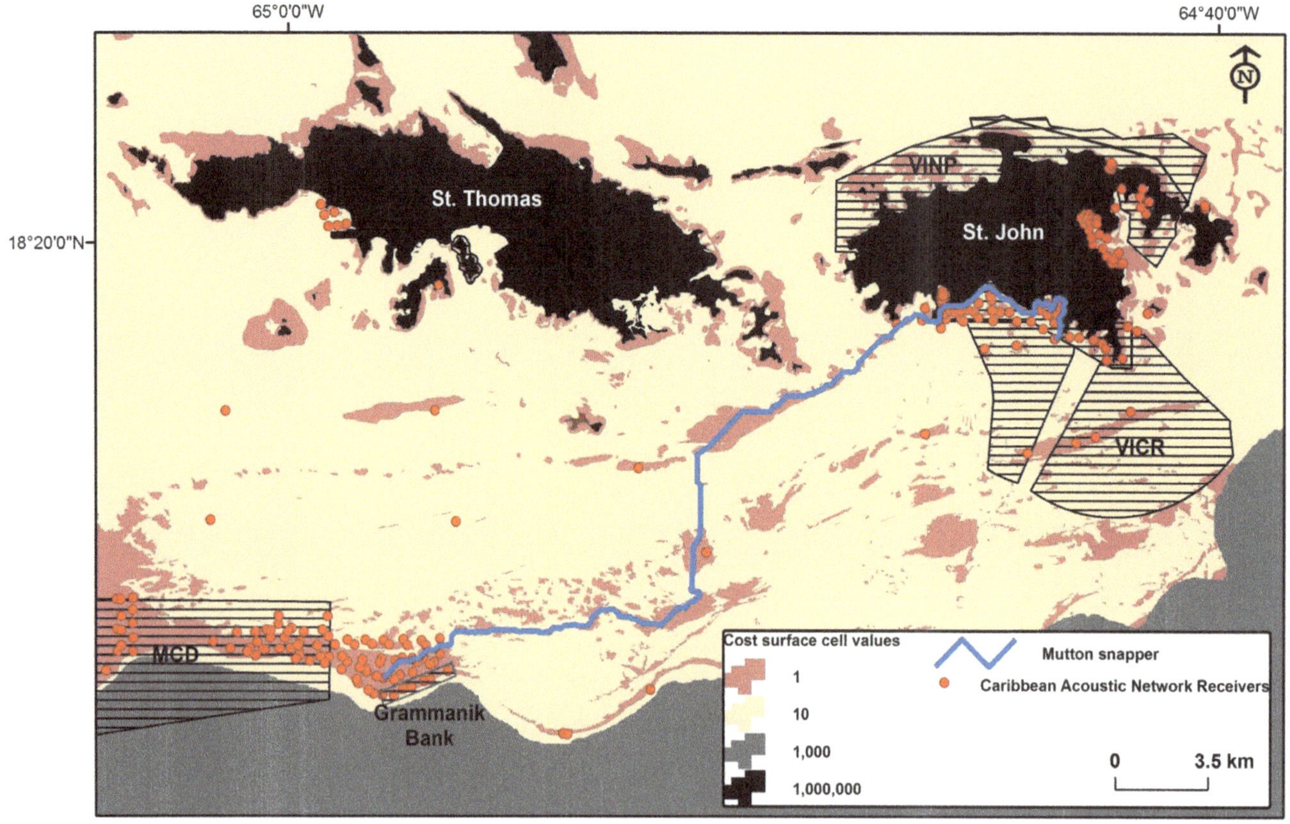

Figure 4. Cost surface for the U.S. Virgin Islands based on relative environmental suitability for shallow-water reef fish. One modeled least-cost path is shown for an individual mutton snapper (*Lutjanus analis*) that moved a maximum of 40.2 km between two receivers (Virgin Islands National Park [VINP] to shelf edge spawning aggregation). The path was parameterized to preferentially follow wherever possible the low cost coral reef to achieve a minimum accumulative travel cost between the two farthest receivers with detections.

Time spent by fishes inside and outside MPAs in the U.S. Virgin Islands

Monitoring periods ranged from 1 to 967 days with high variance within and between species (Table 2). Few fish were detected for over 50% of the hours available during their monitoring period. French grunt (n = 1) and yellow goatfish (n = 3) exhibited the highest percentage time observed (81 & 64±.12%), followed by red hind (63%), and doctorfish (52±0.09%). The majority of fishes (7 out of 18 species) were present only 20-50% of the time and schoolmaster, nurse shark and dog snapper less than 20% of the time. Tagged fish exhibited high fidelity to VINP wherein they were initially captured and tagged. VINP represented over 70% of the time observed for 17 of the 18 fish species tagged. Fourteen percent of fishes (21 individuals of 6 species) tagged in VINP spent time outside of MPAs. Eight species spent 5–30% of their time inside VICR, with seven species not present at all and one species, queen triggerfish, present 61% of the time in VICR. Relatively little time was spent by fishes in the unprotected area outside of MPAs, with the highest being bar jacks (12%) and mutton snapper (5%).

Regardless of some extensive long distance movements and many shorter duration movements outside MPAs, the high fidelity to VINP and VICR of fishes tagged inside VINP indicates that on the south shore of St. John, these MPAs can be managed with confidence that many reef-associated fishes will likely experience the consequences of management actions. Most individuals, however, were also detected in the unprotected zone at some point during their tracking period. The time spent outside MPAs has greater uncertainty due to the inherent physical gaps in detection as a result of the design of our acoustic array. Our estimates of time spent outside MPAs should therefore be interpreted as a minimum estimate. Temporal gaps in detection during the total period of the study were most likely because of: 1) fish spending time in areas outside of the listening range of the receivers (further than 300 m or under coral heads or in noisy shallow water); or 2) fish moved outside of the study area and could have been outside of MPAs. Although unaccounted for in this study, it is possible that fish movements in and around highly structured coral reefs interfered with acoustic detectability and could explain the majority of unobserved time for reef-associated species such as grunts, squirrelfish, nurse shark and small snapper (i.e., schoolmaster, lane snapper). From a management efficacy perspective, however, if U.S. Virgin Islands MPAs can offer sufficient protection from fishing, and the boundaries and spawning aggregations are not heavily fished, then the high residence times recorded here suggest that these MPAs would offer adequate protection for many fishes.

Potential scale mismatch with Caribbean MPAs

According to the 2005 World Database on Protected Areas an estimated 40% of the protected areas in the global network of coral reef MPAs were smaller than 2 km^2 [12]. Based on measurements of the spatial dimensions of 220 MPAs in the eastern Caribbean we reveal here that 85% are less than 5 km

Table 2. Summary data of percent time observed by species (total hours observed/total hours monitored (i.e., first to last detection) and the Fidelity Index score for VICR, VINP and non-MPA waters.

Family	Scientific name	Percent time observed (hrs present/hrs available)	Average Fidelity Index (± Stdev)		
			VICR	VINP	Outside MPA
Haemulidae	*Haemulon sciurus*	34±0.32	0.12±0.26	0.91±0.19	0.03±0.14
Lutjanidae	*Lutjanus synagris*	37±0.3	0.11±0.18	0.9±0.17	0.01±0.03
Sparidae	*Caranx ruber*	36±0.26	0.20±.25	0.85±0.23	0.02±0.06
Carangidae	*Calamus calamus*	49±0.31	0.06±0.11	0.91±0.2	0.12±0.25
Lutjanidae	*Ocyurus chrysurus*	41±0.3	0.01±0.03	0.98±0.05	0.02±0.08
Lutjanidae	*Lutjanus analis*	47±0.24	0.03±0.07	0.93±0.15	0.05±0.14
Holocentridae	*Holocentrus adscensionis*	38±0.38	0.14±0.37	0.93±0.19	0±0
Acanthuridae	*Acanthurus coeruleus*	39±0.27	0.01±0.02	0.99±0.01	0.01±0.02
Balistidae	*Balistes vetula*	30±0.24	0.61±0.54	0.39±0.54	0±0
Lutjanidae	*Lutjanus griseus*	39±0.17	0.33±0.4	0.7±0.35	0±0
Mullidae	*Mulloidichthys martinicus*	64±0.12	0.14±0.24	0.87±0.2	0.01±0.02
Acanthuridae	*Acanthurus chirurgus*	52±0.09	0±0	1±0	0±0
Serranidae	*Epinephelus guttatus*	63	0	1	0
Rhincodontidae	*Ginglymostoma cirratum*	22	0	1	0
Haemulidae	*Haemulon flavolineatum*	81	0	1	0
Haemulidae	*Haemulon plumierii*	13	0	1	0
Lutjanidae	*Lutjanus apodus*	0	0	1	0
Lutjanidae	*Lutjanus jocu*	11	0.06	0.94	0.04

The proportion of time represented in the FI ranged from 0 (no fidelity) to 1 (high fidelity).

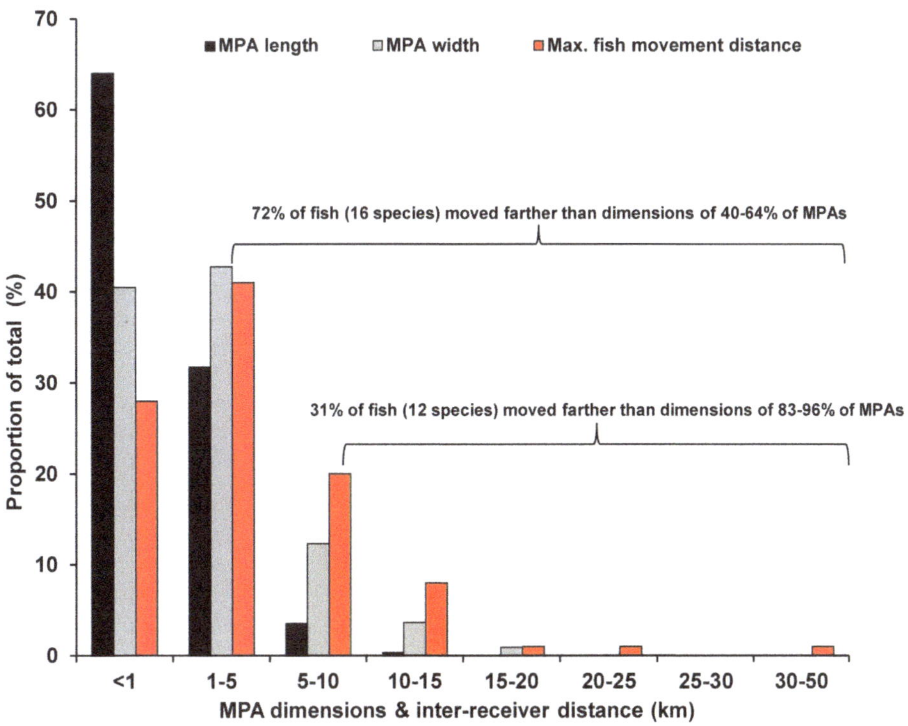

Figure 5. Histograms for the maximum spatial extent of movements for all tagged fish (n = 163) based on the least-cost pathway distance between the two farthest receivers with detections in the U.S. Caribbean; and the spatial dimensions of existing MPAs (n = 220) in the eastern Caribbean (only Leeward and Windward Islands).

wide (median 1.35 km; range 0.04–21) and 70% are less than 5 km long (median 2.5 km; range 0.13–47.2 km). MPAs tend to be longer (along shore) than they are wide (shore to shelf) because they are sited primarily to protect the near-shore marine environments. Our comparison of distances moved by fish versus MPA size across the eastern Caribbean region revealed that 74% of our tracked fish (from 16 of 18 species) traversed distances (i.e. >1 km) greater than the dimensions of 40–64% of existing MPAs (Fig. 5). Twenty-eight percent of our fish (from 12 of 18 species) traversed distances broader than the dimensions of 69–85% of MPAs, emphasizing a potentially problematic scale mismatch between MPA design and the spatial scales of fish movements.

Many of the challenges encountered in managing populations of highly mobile animals occur because of a mismatch between the scale of management and the scale(s) of the ecological processes being managed [9,10,16]. Movement studies and data such as seafloor maps can be used to assess if an MPA encompasses the routine and critical spaces (i.e., nursery, spawning) used during the life-history of highly mobile marine fauna. Re-scaling MPAs to more fully encompass ecological neighborhoods for a group of target species could make major advances in achieving conservation goals. Alternatively, if MPA boundaries can be delineated along physical features (i.e., expanses of unsuitable habitat) that function as natural barriers to fish movement then perhaps even relatively small marine reserves would offer enhanced protection through de-facto containment of reef-associated fish [41,42,44]. On small tropical islands, designing MPAs that extend across the insular shelf would better protect inshore-offshore migration corridors, thus encompassing the critical habitats within the life-cycle. For fishes vulnerable to fishing, but with well-defined linear movement routes, design of MPAs that include key migration pathways, so called 'blue corridors', connecting essential habitat areas will likely result in greater protection. Terrestrial conservation planning for highly mobile populations routinely consider connectivity and incorporate movement corridors in protected area design, although efficacy of corridors is not well tested [27].

To some extent, the expansion of two no-take MPAs in the U.S. Virgin Islands in 2003 to encompass some of the adjacent deeper water coral reefs, improved the likelihood that fishes undertaking onshore-offshore ontogenetic shifts or migrations will receive protection from fishing. Further research is needed to examine the shape of MPAs relative to their placement across the shelf in order to assess the level of protection provided to fishes that undergo cross-shelf movements (i.e. nearshore reefs to deeper offshore waters) through ontogenetic habitat shifts and spawning migrations. Conversely, if the objective is to enhance spillover then MPA placement in an area where the habitat inside is spatially connected to suitable habitat outside would enhance emigration of fish biomass [19]. Little is known about the influence of MPA placement and shape relative to seascape patterning, fish population response and MPA performance. Comparative studies of fish movements across multiple MPAs of different sizes protecting different spatial configurations of coral reef are required to address this knowledge gap.

It is important to note that our database of Caribbean MPAs did not indicate which MPAs (or portions of MPAs) were no-take areas/fishery closures. We also did not account for the level of protection offered by individual MPAs or the fact that MPAs could be adjacent or overlapping another MPA therefore providing greater spatial coverage in protection. This level of information has not yet been compiled for the entire Caribbean region, but it is likely that most MPAs are multi-use and open to fishing. No-take areas in the Caribbean are typically smaller than multi-use MPAs due to the socio-economic impact of restricting fishing on small island developing states. If true, this would mean that the scale mismatch between no-take areas and the spatial scale of fish movement could be an even greater disparity than highlighted here possibly resulting in insufficient protection to achieve the expected role in replenishing fished populations and rebuilding sustainable local fisheries.

Implications for ecologists

Rarely do fish ecologists measure habitat structure at spatial scales relevant to organism movements [9]. Based on the spatial movements revealed by our study, sampling and quantifying habitat correlates only with conventional survey techniques such as underwater transect lines and quadrats (10 s–100 s m^2) appears inadequate. Such fine-scale measurements provide only a snapshot of habitat association and potentially result in identification of misleading environmental drivers. In light of the greater mobility of reef fish demonstrated here and an acknowledgement that many fish use mosaics of habitat types, we advocate that the quantification of fish-habitat requires a conceptual shift to a multi-scale, multi-habitat, organism-centered approach analogous to the recent growth of landscape ecology to study mobile terrestrial fauna [28–30]. Several guiding principles that exist at the core of landscape ecology have made major contributions to terrestrial landscape planning and conservation, but in marine systems our understanding is still in its infancy. For example, (1) environmental heterogeneity exists at multiple spatial scales to which organisms respond differently and at different scales; (2) connectivity is an important ecological pattern and process; (3) patch boundaries/edges influence ecological processes. In landscape ecology, the organism's activity space helps us to anchor the scale selection to a known and functionally meaningful spatiotemporal scale that can then form the focal scale in a hierarchical approach [9]. The integration of concepts from behavioral ecology and landscape ecology represents a new frontier in marine ecology and offer many opportunities for scientific progress together with providing support for information-led design in marine management [31]. The ecological importance of adult movements is receiving renewed attention and it has been argued that information on adult-mediated population connectivity is critical to understanding the complex dynamics of connectivity between subpopulations, patterns of recruitment and the evolution of population structuring [37]. If movements of adult fishes play a large role in regulating regional patterns in productivity then this has important implications for fisheries management and the design and functioning of place-based strategies such as MPAs. Technological advances in microelectronics for telemetry, spatial analytical techniques and marine remote sensing now offer an unprecedented opportunity to address major knowledge gaps in marine animal movements for a wide range of mobile marine organisms.

Acknowledgments

The research was coordinated by NOAA National Centers for Coastal Ocean Science (NCCOS) in collaboration with the National Park Service. We thank S. Hitt and J. Blondeau for assisting with field work; R. Hill (NOAA National Marine Fisheries Service, Galveston), G. Skomel (Division of Marine Fisheries, Massachusetts State Government), B. DeAngelis (The Nature Conservancy) M. Shivji (Nova Southeastern

University, Florida) and B. Weatherbee (University of Rhode Island) for sharing receiver data, and J. Knowles of The Nature Conservancy for the Caribbean MPA data. Thank you to three anonymous reviewers for helping to refine the manuscript.

Author Contributions

Conceived and designed the experiments: MEM AMF MSK RDC CC. Performed the experiments: SJP MEM AMF BL RSN MSK RDC. Analyzed the data: SJP BL LMW. Contributed reagents/materials/analysis tools: SJP MEM RSN MP. Wrote the paper: SJP BL.

References

1. Alvarez-Filip L, Dulvy NK, Gill JA, Cote IM, Watkinson AR (2009) Flattening of Caribbean coral reefs: region-wide declines in architectural complexity. Proc R Soc B 276: 3019–3025.
2. Paddack MJ, Reynolds JD, Aguilar C, Appeldoorn RS, Beets J, et al. (2009) Recent region-wide declines in Caribbean reef fish abundance. Curr Biol 19: 590–595.
3. Stallings CD (2009) Fishery-independent data reveal negative effect of human population density on Caribbean predatory fish communities. PLoS One 4(5): e5333.
4. Guarderas AP, Hacker SD, Lubchenco J (2008) Current status of marine protected areas in Latin America and the Caribbean. Conserv Biol 22: 1630–1640.
5. Grüss A, Kaplan DM, Guénette S, Roberts CM, Botsford LW (2011) Consequences of adult and juvenile movement for marine protected areas. Biol Conserv 144(2): 692–702.
6. Kramer DL, Chapman MR (1999) Implications of fish home range size and relocation for marine reserve function. Env Biol Fish 55: 65–79.
7. Meyer CG, Papastamatiou YP, Clark TB (2010) Differential movement patterns and site fidelity among trophic groups of reef fishes in a Hawaiian marine protected area. Mar Biol 157: 1499–1511.
8. Alcala AC, Russ GR, Maypa AP, Calumpong HP (2005) A long-term, spatially replicated experimental test of the effect of marine reserves on local fish yields. Can J Fish Aquat Sci 62: 98–108.
9. Pittman SJ, McAlpine CA (2003) Movements of marine fish and decapod crustaceans: Process, theory and application. Adv Mar Biol 44: 205–294.
10. Palumbi SR (2004) Marine reserves and ocean neighborhoods: The spatial scale of marine populations and their management. Annu Rev Enviro Res 29: 31–68.
11. Nathan R, Getz WM, Revilla E, Holyoak M, Kadmon R, et al. (2008) A movement ecology paradigm for unifying organismal movement research. PNAS 105: 19052–19059.
12. Beck MW, Heck KL Jr., Able KW, Childers DL, Eggleston DB, et al. (2001) The identification, conservation, and management of estuarine and marine nurseries for fish and invertebrates. Bioscience 51(8): 633–641.
13. Gaines SD, Lester SE, Grorud-Colvert K, Costello C, Pollnac R (2010) Evolving science of marine reserves: New developments and emerging research frontiers. PNAS 107: 18251–18255.
14. Chapman MR, Kramer DL (2000) Movement of post-settlement fishes within and among fringing coral reefs in Barbados. Env Biol Fish 57: 11–24.
15. Hitt S, Pittman SJ, Brown KA (2011) Tracking and mapping sun-synchronous migrations and diel space use patterns of *Haemulon sciurus* and *Lutjanus apodus* in the U.S. Virgin Islands. Environ Biol Fish 92(4): 525–538.
16. Sale PF (1998) Appropriate spatial scales for studies of reef-fish ecology. Aust J Ecol 23 (3): 202–208.
17. Sale PF, Cowen RK, Danilowicz BS, Jones GP, Kritzer JP, et al. (2005) Critical science gaps impede use of no-take fishery reserves. TREE 20: 74–80.
18. Mora C, Andréfouët S, Costello MJ, Kranenburg C, Rollo A, et al. (2006) Ecology. Coral reefs and the global network of Marine Protected Areas. Science 312: 1750–1751.
19. Popple ID, Hunte W (2005) Movement patterns of *Cephalopholis cruentata* in a marine reserve in St Lucia, WI, obtained from ultrasonic telemetry. J Fish Biol 67: 981–992.
20. Hitt S, Pittman SJ, Nemeth RS (2011) Diel movements of fish linked to benthic seascape structure in a Caribbean coral reef ecosystem. Mar Ecol Prog Ser 427: 275–291.
21. Heupel MR, Semmens JM, Hobday AJ (2006) Automated acoustic tracking of aquatic animals: scales, design, and deployment of listening station arrays. Mar Fresh Res 57: 1–13.
22. Grothues TM (2009) A review of acoustic telemetry technology and a perspective on diversification in coastal tracking arrays. Meth.Techn. Fish Biol. Fisheries 9: 77–90.
23. Monaco ME, Friedlander A, Hile SD, Pittman SJ, Boulon RH (2009) The coupling of St. John, US Virgin Islands marine protected areas based on reef fish habitat affinities and movements across management boundaries. Proceedings 11th International Coral Reef Symposium, Ft. Lauderdale, Florida, 7–11 July 2008.
24. Turgeon K, Robillard A, Grégoire J, Duclos V, Kramer DL (2010) Functional connectivity from a reef fish perspective: Behavioral tactics for moving in a fragmented landscape. Ecology 91(11): 3332–3342.
25. Russ GR, Alcala AC (2004) Marine reserve benefits local fisheries. Ecol Appl 14: 597–606.
26. Russ GR, Alcala AC (2011) Enhanced biodiversity beyond marine reserve boundaries: the cup spillith over. Ecol Appl 21 (1): 241–250.
27. Rosenberg DK, Noon BR, Meslow EC (1997) Biological corridors: form, function, and efficacy. BioScience 47 (10): 667–687.
28. Levin SA (1992) The problem of pattern and scale in ecology: The Robert H. MacArthur Award Lecture. Ecology 73: 1943–1967.
29. Pittman SJ, Kneib RT, Simenstad C (2011) Practicing coastal seascape ecology. Mar Ecol Prog Ser 427: 187–190.
30. Boström C, Pittman SJ, Simenstad C, Kneib RT (2011) Seascape ecology of coastal biogenic habitats: advances, gaps and challenges. Mar Ecol Prog Ser 427: 191–217.
31. Grober-Dunsmore R, Pittman SJ, Caldow C, Kendall MA, Fraser T (2009) A landscape ecology approach for the study of ecological connectivity across tropical marine seascapes. Chapter 14, p493–529. In (Nagelkerken ed.) Ecological Connectivity of Coral Reef Ecosystems. Springer.
32. Roberts CM (2000) Selecting marine reserve locations: optimality versus opportunism. Bull Mar Sci 66(3): 581–592.
33. Roberts CM, Andelman S, Branch G, Bustamante RH, Castilla JC, et al. (2003) Ecological criteria for evaluating candidate sites for marine reserves. Ecol Appl 13: 199–214.
34. Botsford LW, Micheli F, Hastings A (2003) Principles for the design of marine reserves. Ecol Appl 13(Supp. 1): 25–31.
35. Pina-Amargós F, González-Sansón G (2009) Movement patterns of goliath grouper *Epinephelus itajara* around southeast Cuba: implications for conservation. Endang Spec Res 7(3): 243–247.
36. Bolden SK (2000) Long-distance movement of a Nassau grouper (*Epinephelus striatus*) to a spawning aggregation in the central Bahamas. Fish Bull 98(3): 642–645.
37. Frisk MG, Jordaan A, Miller TJ (2013) Moving beyond the current paradigm in marine population connectivity: are adults the missing link? Fish Fish. doi: 10.1111/faf.12014.
38. Kobara S, Heyman WD, Pittman SJ, Nemeth RS (2013) Biogeography of transient reef-fish spawning aggregations in the Caribbean: a synthesis for future research and management. Oceanogr Mar Biol 51: 281–326.
39. Lee KA, Huveneers C, Macdonald T, Harcourt RG (2014) Size isn't everything: movements, home range, and habitat preferences of eastern blue gropers (*Achoerodus viridis*) demonstrate the efficacy of a small marine reserve. Aquatic Conserv: doi: 10.1002/aqc.243.
40. Mumby PJ (2006) Connectivity of reef fish between mangroves and coral reefs: algorithms for the design of marine reserves at seascape scales. Biol Conserv 128(2): 215–222.
41. Roberts CM, Sargant H (2002) Fishery benefits of fully protected marine reserves: why habitat and behavior are important. Nat Res Model 15(4): 487–507.
42. Edgar GJ, Stuart-Smith RD, Willis TJ, Kininmonth S, Baker SC, et al. (2014) Global conservation outcomes depend on marine protected areas with five key features. Nature 506(7487): 216–220.
43. Devillers R, Pressey RL, Grech A, Kittinger JN, Edgar GJ, et al. (2014) Reinventing residual reserves in the sea: are we favouring ease of establishment over need for protection? Aquat Conserv. DOI: 10.1002/aqc.2445.
44. Gaines SD, White C, Carr MH, Palumbi SR (2010) Designing marine reserve networks for both conservation and fisheries management. PNAS 107(43): 18286–18293.
45. Moffitt EA, Botsford LW, Kaplan DM, O'Farrell MR (2009) Marine reserve networks for species that move within a home range. Ecol Appl 19: 1835–1847.
46. Olds AD, Connolly RM, Pitt KA, Maxwell PS (2012) Habitat connectivity improves reserve performance. Cons Letters 5: 56–63.
47. Olds AD, Albert S, Maxwell PS, Pitt KA, Connolly RM (2013) Mangrove-reef connectivity promotes the effectiveness of marine reserves across the western Pacific. Global Ecol and Biogeogr 22: 1040–1049.
48. Sawyer SC, Epps CW, Brashares JS (2011) Placing linkages among fragmented habitats: do least-cost models reflect how animals use landscapes? J Appl Ecol 48(3): 668–678.
49. Nemeth RS (2009) Chapter 4: Dynamics of reef fish and decapod crustacean spawning aggregations: underlying mechanisms, habitat linkages and trophic interactions. p 73–134. In: I Nagelkerken, ed. Ecological connectivity among tropical coastal ecosystems. Springer, Netherlands.
50. Nemeth RS, Kadison E, Herzlieb S, Blondeau J, Whiteman E (2006) Status of a yellowfin grouper (*Mycteroperca venenosa*) spawning aggregation in the U.S. Virgin Islands with notes on other species. Proc. 57th Gulf Carib Fish Inst 57: 543–558.
51. Nemeth RS (2005) Population characteristics of a recovering U.S. Virgin Islands red hind spawning aggregation following protection. Mar Ecol Prog Ser 286: 81–97.
52. Nemeth RS, Blondeau J, Herzlieb S, Kadison E (2007) Spatial and temporal patterns of movement and migration at spawning aggregations of red hind, *Epinephelus guttatus*, in the U.S.Virgin Islands. Env Biol Fish 78(4): 365–381.

10

Ecological Conditions Favoring Budding in Colonial Organisms under Environmental Disturbance

Mayuko Nakamaru[1]*, Takenori Takada[2], Akiko Ohtsuki[1,3], Sayaki U. Suzuki[1,3¤], Kanan Miura[1], Kazuki Tsuji[4]

1 Department of Value and Decision Science, Tokyo Institute of Technology, Tokyo, Japan, 2 Graduate School of Environmental Earth Science, Hokkaido University, Sapporo, Japan, 3 The Graduate University for Advanced Studies, Kanagawa, Japan, 4 Faculty of Agriculture, University of the Ryukyus, Okinawa, Japan

Abstract

Dispersal is a topic of great interest in ecology. Many organisms adopt one of two distinct dispersal tactics at reproduction: the production of small offspring that can disperse over long distances (such as seeds and spawned eggs), or budding. The latter is observed in some colonial organisms, such as clonal plants, corals and ants, in which (super)organisms split their body into components of relatively large size that disperse to a short distance. Contrary to the common dispersal viewpoint, short-dispersal colonial organisms often flourish even in environments with frequent disturbances. In this paper, we investigate the conditions that favor budding over long-distance dispersal of small offspring, focusing on the life history of the colony growth and the colony division ratio. These conditions are the relatively high mortality of very small colonies, logistic growth, the ability of dispersers to peacefully seek and settle unoccupied spaces, and small spatial scale of environmental disturbance. If these conditions hold, budding is advantageous even when environmental disturbance is frequent. These results suggest that the demography or life history of the colony underlies the behaviors of the colonial organisms.

Editor: Ahmed Moustafa, American University in Cairo, Egypt

Funding: This work is supported by JSPS KAKENHI Grant Number 21770076 (M.N.), 23370006 (T.T.), 25340115 (T.T.), 20033015 (K.T.), 21247006 (K.T.), and Inamori Foundation (M.N.). The funders had no role in study design, data collection and analysis, decision to publish, or preparation of the manuscript.

Competing Interests: The authors have declared that no competing interests exist.

* E-mail: nakamaru@valdes.titech.ac.jp

¤ Current address: National Agricultural Research Center, Ibaraki, Japan

Introduction

Dispersal distance varies widely among organisms. Dispersal determines the genetic structure of the meta-population and influences the evolution of individual characteristics such as mortality, reproductive efforts, sex allocation and altruism [1–3]. Since dispersal is a costly activity, why organism disperse has become an important issue in evolutionary ecology [4,5]. In the pioneering meta-population model of Hamilton and May (1977) it was shown that dispersal can evolve even when the dispersal cost is so high that nearly all dispersers are destined to die. Subsequent studies identified specific factors that favor dispersal over non-dispersal, such as the avoidance of competition among relatives and inbreeding [1,2]. Disturbances such as predation, drought and flooding are regarded as important general triggers of dispersal, encouraging organisms to hedge the risks [6].

However, many organisms do not disperse long distances. For example, some organisms often adopt an "budding" strategy, where a colony splits into two or more parts which subsequently move away from each other. The budding by clones, buds, or propagules has been studied theoretically and experimentally from the viewpoint of the evolution of altruism and sex ratio [7–9]. However, those studies did not focus on the difference in dispersal distance, while in many real organisms the budding strategy is typically characterized by shorter-range dispersal in comparison to non-budding dispersal. Contrary to the conventional view, organisms reproducing by budding often survive in environments undergoing frequent disturbance. For example, many invasive clonal plants that are specialists of disturbed habitats predominantly use vegetative reproduction, despite retaining the seed-producing capability that would enhance their spread [10,11]. New colonies of branching gorgonian coral (*Plexaurella* sp.) exposed to waves are predominantly founded by fragments of broken branches, rather than by inseminated gametes that can migrate long distances [12]. In some invasive ants, such as *Linepithema humile* (the Argentine ant) and *Wasmannia auropunctata* (the tiny fire ant), winged queens do not engage in nuptial (dispersal) flights. Instead they mate within or near natal nests, and new colonies are founded by colony-splitting (fission or budding), in which queens search for a new nest site on foot, accompanied by some workers [13,14]. These invaders are recognized as specialists of disturbed habitats [13,15,16]. However, the ecological conditions favoring budding remain obscure.

All of the above examples (clonal plants, corals and ants) are colonial or super organisms that must maintain a spatially fixed life following initial settlement. Assuming such colonial life, Nakamaru et al. (2007) pioneered the investigation of ecological conditions favoring budding under disturbance in a spatially explicit computational model (lattice model) [17]. They identified an important trade-off between dispersal distance and offspring size (table 1); small reproductive units, such as plant seeds, spawned coral eggs and winged ant queens, travel long-distance, whereas large units, such as the propagules of clonal plants, broken branches of corals and dependently-founded ant colonies remain within their local environment. This difference in dispersal affects the survival and growth of offspring; small offspring suffer low survival and reach maturity comparatively slowly, while large offspring (buds) benefit from high survival and rapid growth to maturity. Nakamaru et al. (2007) concluded that the short distance dispersal strategy (budding) is advantageous over long distance dispersal (non-budding) under conditions of (i) very high mortality of a small-sized colony, or (ii) relatively frequent environmental disturbance over a small spatial scale.

However, Nakamaru et al. did not clarify the effect of spatial structure and the life history on the above-mentioned advantage. For each colony, they specifically assumed that growth is logistic and that death rate is a negative exponential function of colony size. However, other survivorship patterns and other growth functions are possible in nature. The authors failed in their quest for a mathematical model that could accommodate relatively complicated assumptions and therefore could not obtain analytical results. These drawbacks have precluded a precise understanding of the adaptive significance of budding. To more generally understand the ecological conditions that induce budding dispersal, here we investigate the characteristics of an organism's life history functions, such as survivorship and colony growth, that favor budding dispersal under disturbance.

We adopt the general view that life history parameters (survival, growth and reproduction functions) directly determine the evolved dispersal strategy [18,19]. Environmental conditions, such as spatial structure and disturbance, influence the above parameters in a complicated manner through environment vs. phenotype interactions. In this paper we make a simple generalized assumption that is applicable to colonial organisms adopting similar resource allocation strategies. We first focus on colonial life history using a model without spatial structure, called the baseline model. This model assumes discrete colony size, assigning different death and growth probabilities to each colony size. Under this assumption, we can construct discrete equations or a matrix model in the baseline model that permit exact solutions. We clarify the effects of life histories or demographics (such as survivorship and growth rate) on the competition between the two strategies, i.e. budding vs. non-budding. We also examine the conditions under which budding is advantageous over non-budding when realistic and possible assumptions are added to the baseline model.

Methods and Results

Model 1: The baseline model in the completely mixed population

It is common that an individual is assumed to be a basic unit in the agent-based simulation model. In contrast, here the basic unit in our agent-based simulation model is the clonal (super) organism or the colony of individuals, and the model is called the colony-based model. Gardner and Grafen (2009) proved that there is no mathematical difference between an individual organism and a clonal group if social groups have no intracolonial conflicts [20].

We consider only non-spatial structures or completely mixed populations, ignoring the effect of dispersal distance. Instead, we aim to clarify the effect of colonial life history patterns on the population dynamics. The colony size-structured matrix model assumes four colony sizes, designated size 1, size 2, size 3 and size 4. The first three denote immature colonies, while size 4 represents the mature colony. Life history pattern is described by two parameters; the baseline death probability and the growth probability of the colony (Fig. 1 A). The baseline death probability of a colony of size i is defined as d_i ($0 \leq d_i \leq 1$, $i = 1, 2, 3$ and 4.). An additional death probability, independent of the baseline death, is imposed by disturbance. The survival probability of a colony of size i is $1-d_i$, defined as p_i ($0 \leq p_i \leq 1$, $i = 1, 2, 3$ and 4). Let g_i be the probability that a colony of size i expands to size i+1 ($0 \leq g_i \leq 1$, $i = 1, 2, 3$). When a colony of size i survives with probability p_i, it expands to size i+1 with probability g_i ($i = 1, 2, 3$); otherwise it remains at its present size with probability $1-g_i$. When the colony has reached size 4, it can divide into two colonies, one of which will colonize an empty site. Therefore the chance of division is proportional to the density of empty sites in the population multiplied by the division rate, h, where h controls the division speed. A size 4 colony deprived of the opportunity to divide will survive with probability p_4.

The colony divides by one of two strategies: designated 2:2 division and 1:3 division (Fig.1 B and C). A size 4 colony adopting the 2:2 division strategy divides into two colonies of size 2, one of which moves to an empty site. Alternatively, a size 4 colony adopting the 1:3 division strategy divides into two unequally sized colonies (of size 1 and size 3), the smaller of which moves to an empty site. This situation mimics nature; if a biological colony splits into two equal parts, one of the daughter colonies moves to a new site close to the natal colony; if the colony is unequally split, the smaller colony disperses far from the natal colony. The 2:2 and 1:3 division strategies correspond to budding and seed-like dispersal, respectively. To purely clarify the effect of division ratio on the population dynamics, we ignore spatial structure and distance.

The size-structured matrix model for colonies adopting the 2:2 division strategy in the population is constructed as

$$x(t) = S(t)x(t), \quad (1)$$

Table 1. Life history characteristics of long and short dispersal units.

Dispersal units	Seeds, spawned eggs, and independently founding queens	vegetative propagule, broken coral braches, or dependently-founded ant colonies
Dispersal distance	long	short
Size	small	large
Time to reproduction	long	short
Number	many	few
Survival rate	low	high

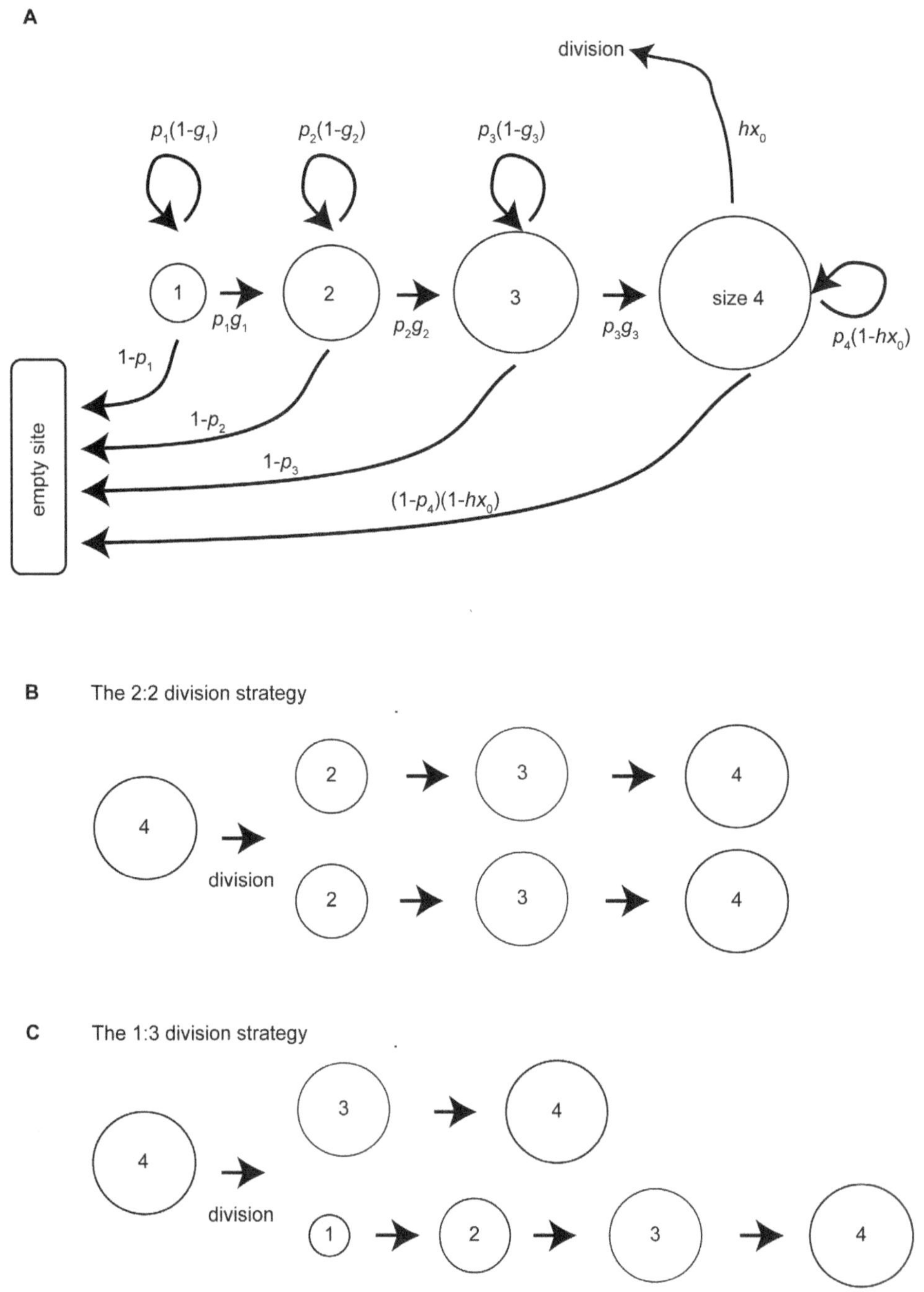

Figure 1. Life history of a colony and its division strategies. (A) Life history of a colony. (B) The 2:2 division strategy, in which a mature (size 4) always divides into two size 2 daughter colonies. The new colonies grow to maturity. (C) The 1:3 division strategy, in which a mature colony always divides into a small (size 1) and a large (size 3) daughter colony. The size 1 matures slowly via size 2 and size 3, while the size 3 colony matures rapidly.

where $x(t) = (x_0(t), x_1(t), x_2(t), x_3(t), x_4(t))'$ and

$$S(t) = \begin{bmatrix} 1-hx_4(t) & 1-p_1 & 1-p_2 & 1-p_3 & (1-p_4)(1-hx_0(t)) \\ 0 & p_1(1-g_1) & 0 & 0 & 0 \\ hx_4(t) & p_1g_1 & p_2(1-g_2) & 0 & hx_0(t) \\ 0 & 0 & p_2g_2 & p_3(1-g_3) & 0 \\ 0 & 0 & 0 & p_3g_3 & p_4(1-hx_0(t)) \end{bmatrix}.$$

Let $x_0(t)$ be the density of empty sites at time t and $x_i(t)$ ($i = 1, 2, 3$ and 4) be the density of colonies of size i adopting the 2:2 division strategy at time t, subject to $\Sigma_i\, x_i(t) = 1$. The values of $hx_4(t)$ in element (3, 1) and $hx_0(t)$ in element (3, 5) show that, when a size 4 colony divides, one of the two resulting size 2 colonies disperses to an empty site while the other remains in its natal habitat.

Similarly, the size-structured matrix model for colonies adopting the 1:3 division strategy in the population is constructed as

$$y(t) = L(t)y(t), \tag{2}$$

where $y(t) = (y_0(t), y_1(t), y_2(t), y_3(t), y_4(t))'$ and

$$L(t)=\begin{bmatrix} 1-hy_4(t) & 1-p_1 & 1-p_2 & 1-p_3 & (1-p_4)(1-hy_0(t)) \\ hy_4(t) & p_1(1-g_1) & 0 & 0 & 0 \\ 0 & p_1g_1 & p_2(1-g_2) & 0 & 0 \\ 0 & 0 & p_2g_2 & p_3(1-g_3) & hy_0(t) \\ 0 & 0 & 0 & p_3g_3 & p_4(1-hy_0(t)) \end{bmatrix}.$$

Let $y_0(t)$ be the density of empty sites at time t and $y_i(t)$ ($i = 1, 2, 3$ and 4) be the density of colonies of size i adopting the 1:3 division strategy at time t, subject to $\Sigma_i y_i(t) = 1$. When a size 4 colony divides, the probability of the resulting size 1 colony successfully migrating to an empty site is determined by $hy_4(t)$ in element (2, 1), while the probability that the size 3 colony remains in its natal habitat is $hy_0(t)$ in element (4, 5).

The dynamics of competition between colonies adopting 1:3 and 2:2 division strategies can be described using the competition matrix, $F(w(t))$:

$$w(t+1) = F(w(t))w(t), \tag{3}$$

in which the vector $w(t)$ consists of $x_i(t)$ and $y_i(t)$ ($i = 1, 2, 3, 4$) and $z_0(t)$ which is the density of empty sites, and the matrix F represents the colony size transition. Detailed information about Eq. (3) is given in Appendix S4.

Results when all growth probabilities are one

Mathematical results of the colony size-structured matrix model. Here we consider the simplest case in which all growth probabilities are one ($g_i = 1$). By local stability analysis of Eq. (1), the equilibrium point where colonies adopting the 2:2 division strategy become extinct ($x_0{}^*(t) = 1.0$) is locally unstable when

$$2p_2p_3 > p_4 + (1-p_4)/h, \tag{4}$$

otherwise the equilibrium point $x_0{}^*(t) = 1.0$ is locally stable (see appendix S1 and S2-(i)). This result implies that colonies adopting the 2:2 division strategy could stably survive in the local environment under conditions satisfying the inequality (4). Inequality (4) is called the viability condition of the 2:2 division strategy. When $p_1 \le p_2 \le p_3 \le p_4$, inequality (4) is locally stable at the equilibrium $x_1{}^*(t)+x_2{}^*(t)+x_3{}^*(t)+x_4{}^*(t)>0$ and $x_0{}^*(t)<1.0$ (see appendix S1 and S2-(ii)).

Similarly, by local stability analysis of Eq. (2), the equilibrium point where colonies adopting the 1:3 division strategy become extinct ($y_0{}^*(t) = 1.0$) is locally unstable when

$$(1+p_1p_2)p_3 > p_4 + (1-p_4)/h. \tag{5}$$

Under this condition, colonies adopting the 1:3 division strategy can stably survive; hence, inequality (5) is called the viability condition of the 1:3 division strategy (appendix S1 and S3-(i)). When $p_1 \le p_2 \le p_3 \le p_4$, inequality (5) defines the locally stable condition at equilibrium ($y_1{}^*(t) + y_2{}^*(t) + y_3{}^*(t) + y_4{}^*(t) > 0$ and $y_0{}^*(t) < 1.0$) (see appendix S1 and S3-(ii)).

We also analyze the matrix model in which two strategies compete for empty sites (see appendix S4). The notation of parameters and variables is unaltered, apart from the density of empty sites at time t, denoted $z_0(t)$. As shown in appendix S4-(ii), the equilibrium point of the pure 2:2 division strategy, ($x_1{}^*(t)$ + $x_2{}^*(t) + x_3{}^*(t) + x_4{}^*(t) > 0$ and $y_i{}^*(t) = 0$ ($i = 1$–4)), is locally stable when inequalities (4) and (6) are satisfied, where (6) is given by

$$2p_2 > 1 + p_1p_2. \tag{6}$$

Appendix S4-(iii) also shows that the equilibrium point of the pure 1:3 division strategy, ($y_1{}^*(t) + y_2{}^*(t) + y_3{}^*(t) + y_4{}^*(t) > 0$ and $x_i{}^*(t) = 0$ ($i = 1$–4)), is locally stable when inequalities (5) and (7) are satisfied, where (7) is given by

$$2p_2 < 1 + p_1p_2. \tag{7}$$

Inequalities (6) and (7) are called the competition conditions.

These analyses indicates that the 2:2 division strategy wins against the 1:3 division strategy when inequalities (4) and (6) are satisfied, while upholding inequalities (5) and (7) favors the 1:3 division strategy. Inequalities (4)–(7) indicate that the division rate h does not affect the competition condition, but does affect the viability conditions. Under the viability conditions, small h restrains colony survivorship. The competition conditions are influenced by the survival probabilities of colonies of size 1 and 2 (see inequalities (6) and (7)), but not by those of older colonies.

Equation (6) can be rewritten as $p_2 > 1/(2-p_1) \ge 1/2$, indicating that colonies adopting the 2:2 division strategy can overcome their 1:3 counterparts when p_2 exceeds 0.5. Inequality (7) can be rewritten as $p_2 < 1/(2-p_1) \le 1$. Equations (6) and (7) also indicate that relatively high p_2 can favor colonies adopting the 2:2 division strategy.

When $d_1 = d_2 = d_3 = d_4$, i.e. the dynamics are size-independent, inequality (7) always holds. This result implies that, in size-independent dynamics, the 1:3 division strategy is always advantageous over the 2:2 division strategy.

To investigate whether inequalities (4) – (7) hold for any values of p_1, p_2, p_3 and p_4, we conduct the numerical simulations of Eq. (D1) presented in appendix S4, and also run the colony-based simulations.

Simulation results of competition for empty sites between two reproductive strategies (2:2 division and the 1:3 division). We conducted colony-based simulations based on the assumptions in the previous section. We assume 2,500 sites, either unoccupied or occupied by colonies adopting one of two reproductive strategies. Each parameter set was run 100 times.

Three forms of the baseline death probability function are considered in the simulations. Death type (i) assumes that $d_2 = d_3 = d_4$. If d_1 is higher than the other rates, this function is similar to the exponential decreasing function adopted in Nakamaru et al. [17]. In death type (ii), the death probability function is a linear function of size, and defining d_1 and d_4 automatically determines d_2 and d_3. In death type (iii), we assume that $d_1 = d_2 = d_3$. If $d_1 = d_4$, the four death probabilities are identical. In this scenario, we can investigate the size-independent ecological dynamics and compare them with the size-dependent dynamics.

To understand the effect of colony division and death probability on the ecological dynamics described by Eqs. (1), (2) and (D1), the theoretical results (computed from inequalities (4) – (7)) and the simulation results are plotted in Fig. 2. We now compare the results of inequalities (4) – (7) with those of the colony-based simulations and the numerical calculations of eq. (3), and show that these approaches are consistent. Figure 2 A shows how the outcomes are affected by two death probabilities (d_1 and d_4) in death type (i). Each parameter set was tested in 100

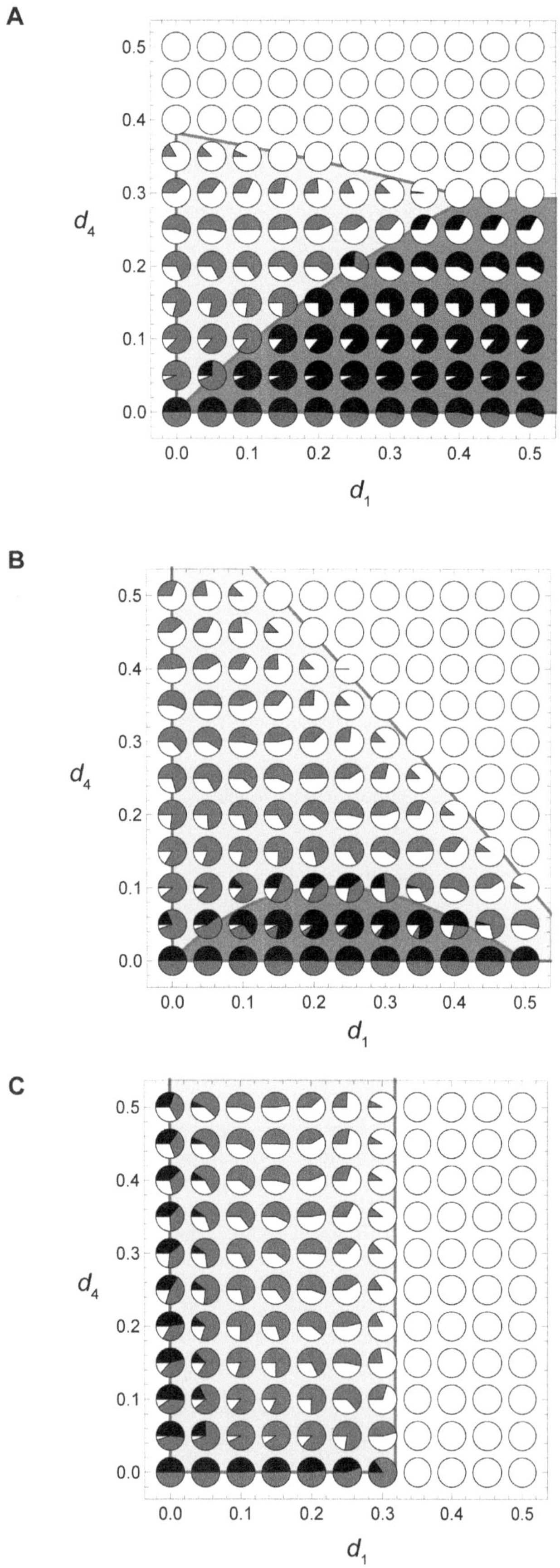

Figure 2. The d_1-d_4 graphs of the theoretical and simulated solutions in a completely mixed population. The horizontal and vertical axes represent d_1 and d_4, respectively The lines in the graphs delineate the viability and competition conditions in the completely mixed population. Theoretical analysis predicts that the 2:2 division strategy wins in the dark gray area, while the 1:3 division strategy wins in the light gray area. Each pie chart shows the average densities of sites occupied by either of the two strategies, as well as the empty sites (open circles) after 10,000 iterations in 100 simulation runs. The black and gray portions of the filled circles indicate the average density of sites occupied by the 2:2 and 1:3 division strategies, respectively. In $d_4 = 0.0$, matured colonies adopting both strategies ultimately survive and neutrally coexist. Initial densities are $z_0 = 0.2$ and $x_1 = x_2 = x_3 = x_4 = y_1 = y_2 = y_3 = y_4 = 0.1$. Other parameters are: 2,500 sites in the population, $u = 0$ and $h = 1$. (A) Death probability function assumes that $d_2 = d_3 = d_4$. (B) Death probability function is linear. (C) Death probability function assumes that $d_1 = d_2 = d_3$.

simulation runs. Since the simulated and theoretical outcomes are consistent, we conclude that inequalities (4) – (7) can predict the simulated ecological dynamics even though $p_1 \leq p_2 \leq p_3 \leq p_4$ is violated (see appendix S2). When inequality (7) is satisfied (roughly, $d_1 \leq d_4$), colonies adopting the 1:3 division strategy dominate the population. Conversely, if inequality (6) is upheld (roughly, $d_1 > d_4$), the population is dominated by colonies undergoing 2:2 division. When d_4 is high or neither viability condition is satisfied, size 4 colonies die out before dispersing their daughter colonies and all colonies become extinct.

Figure 2 B shows the simulation outcomes in the d_1- d_4 graph in death type (ii). In parameter sets satisfying $d_1 > d_4$ and small d_4, colonies adopting the 2:2 division strategy dominate the population in some cases. When d_1 or d_4 is high, all colonies become extinct. Otherwise, the population is dominated by colonies adopting the 1:3 division strategy. As also shown in Fig. 2 B, inequalities (4) – (7) can predict the simulated ecological dynamics even when $p_1 \leq p_2 \leq p_3 \leq p_4$ is violated (See appendix S2). In death type (iii), colonies adopting the 2:2 division strategy are always overwhelmed by those undertaking the 1:3 division strategy (figure 2 C).

Figure 2 displays the outcomes of special cases of the death functions. One fixed ($d_4 = 0.2$) and three variable (d_1, d_2 and d_3) death probabilities were assumed in the additional simulations (figures are not shown). The outcome of this scenario also support inequalities (4) – (7), in which the competition between the two strategies depends on both d_1 and d_2, and colony survivorship is affected by all death probabilities. The high d_1 ensures that colonies adopting the 2:2 division strategy ultimately win, while low d_2 favors the 2:2 division strategy.

We conclude that the 2:2 division strategy is advantageous over the 1:3 strategy when the death probability of size 1 colonies is relatively higher than that of size 2 colonies (see Eq. (6) and Fig. 2). Otherwise, the 1:3 strategy is favorable. The reasons for these findings are now discussed: Fig. 1 C shows that the 1:3 strategy divides a mature colony into two daughter colonies of unequal size (1 and 3). The size 1 colony matures much more slowly than the size 3 colony, which rapidly (by comparison) reaches size 4 and divides. By contrast, the 2:2 strategy divides a mature colony into two half-sized colonies (figure 1 B). A size 2 colony matures more rapidly than a size 1 colony but less rapidly than one of size 3. Therefore, if all colonies share the same death probability, the 1:3 division strategy is advantageous over the 2:2 strategy because the probability of reproduction is higher for size 3 colonies than for size 2 colonies (see Fig. 1 B and C).

Now consider the scenario in which the death probability of a size 1 colony is prohibitively high. Once colonies adopting the 1:3

division strategy have divided, the size 1 daughter colonies die out, while the size 3 colonies survive and mature. Colonies adopting the 2:2 division strategy are not affected by the small survival chance of size 1 colonies, and both daughter colonies will likely survive.

Next, we assign a high death probability to a size 2 colony, and observe the effects on the ecological dynamics. Figure 1 B indicates that colonies adopting the 2:2 division strategy can scarcely increase their numbers because most of the daughter colonies dies out before reaching maturity. Meanwhile, although new size 1 colonies also die out, the size 3 colonies can survive to maturity. Therefore, imposing a high death probability on size 2 colonies favors colonies adopting the 1:3 division strategy. This discussion is supported by the competition conditions (Eqs. (6) and (7)).

The results when all growth probabilities are less than one

The following analysis concerns the effect of growth probability on the ecological dynamics. Applying the methods of appendices S1–S3 to Eqs. (1) and (2), the viability condition of the 2:2 and 1:3 division strategies are respectively obtained as

$$2p_2p_3g_2g_3>(p_4+(1-p_4)/h)(1-p_2(1-g_2))(1-p_3(1-g_3)), \qquad (8)$$

and $(p_1p_2g_1g_2+(1-p_1(1-g_1))(1-p_2(1-g_2)))\,p_3g_3$
$$>(p_4+(1-p_4)/h)\,(1-p_1(1-g_1))\,(1-p_2(1-g_2))(1-p_3(1-g_3)). \qquad (9)$$

Applying the methods of appendix S4 to Eq. (3) and to the general equation (D1), we can calculate the conditions under which the 2:2 division strategy wins against the 1:3 strategy when inequalities (8) and (10) are satisfied, where (10) is given by

$$2p_2g_2(1-p_1(1-g_1)) > (1-p_1(1-g_1))(1-p_2(1-g_2)) + p_1p_2g_1g_2. \qquad (10)$$

On the other hand, if inequalities (9) and (11) hold, the 1:3 division strategy can defeat the 2:2 strategy, where (11) is given by

$$2p_2g_2(1-p_1(1-g_1)) < (1-p_1(1-g_1))(1-p_2(1-g_2)) + p_1p_2g_1g_2. \qquad (11)$$

Inequalities (10) and (11) are called the competition conditions.

The competition conditions are independent of the division rate h, but h does affect the viability conditions. In essence, small h restrains colony survival. The competition conditions are influenced by the survival and growth probabilities of size 1 and 2 colonies (see also inequalities (6) and (7)), but not by those of older colonies.

When $d_1=d_2=d_3=d_4$ and $g_1=g_2=g_3$, interpreted as no colony dynamics, inequality (11) is always upheld. This result implies that, in static colonies, the 1:3 division strategy always confers survival at the expense of the 2:2 division strategy.

We now compare the results of inequalities (8)-(11) with those of the colony-based simulations and the numerical calculations of eq. (3), and show that these approaches are consistent. First, we examine the effect of growth probabilities on the system using the outcomes derived from inequalities (8) – (11). Figure 3 shows the effect of g_2 and g_1 ($=g_3$) on the competition and viability conditions defined in inequalities (8) – (11). If $g_2>g_1=g_3$, the fastest growing colonies are those increasing from size 2 to 3. This scenario mimics logistic growth. When $d_1=0.35>d_2=d_3=d_4=0.15$, and also $g_1=g_2=g_3=1$ (see Figures 2 A and 3 A), the population is dominated by colonies adopting the 2:2 division strategy. When

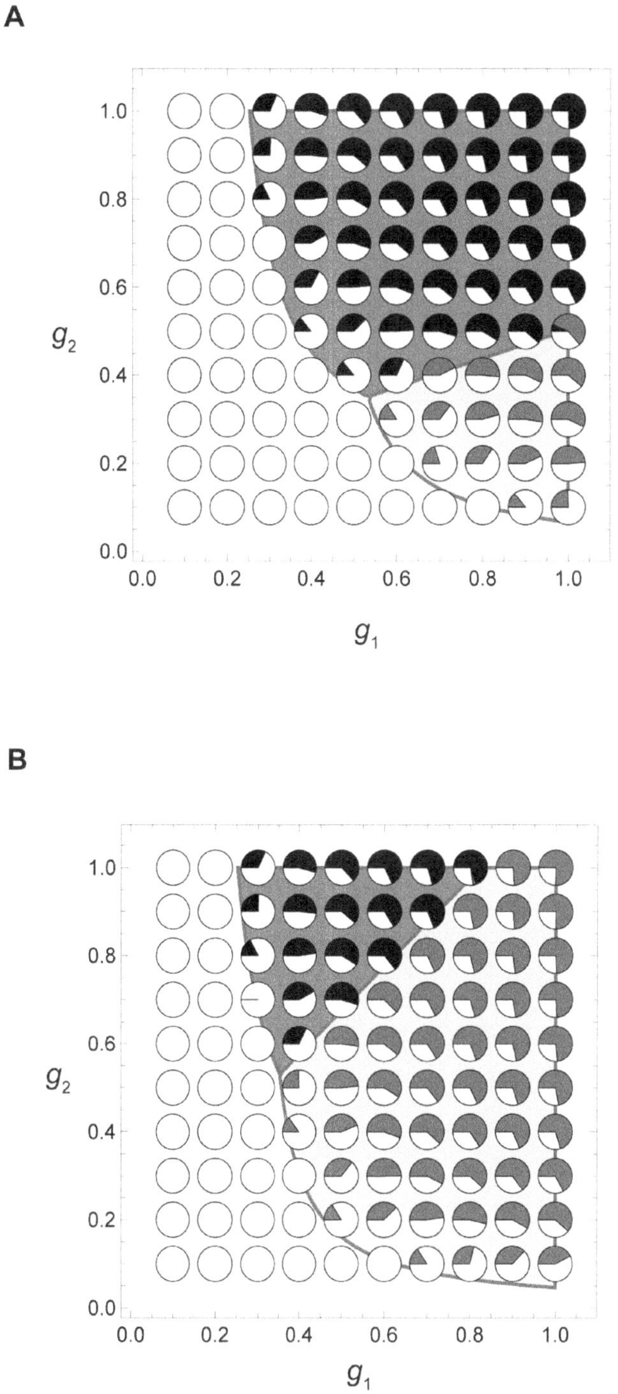

Figure 3. The g_1-g_2 plot of simulated and theoretical results in a completely mixed population. Simulations are iterated 6,250 times. Parameters are $d_2=d_3=d_4=0.15$, $h=1$, $u=0$. The horizontal and vertical axes represent g_1 ($=g_3$) and g_2, respectively. Initial densities are $z_0=0.6$ and $x_1=x_2=x_3=x_4=y_1=y_2=y_3=y_4=0.05$. More information is provided in the caption to Fig. 2. (A) $d_1=0.35$. (B) $d_1=0.15$.

$g_2>g_1=g_3$, colonies adopting the 2:2 division strategy can always defeat their unequally dividing counterparts. Conversely, lower g_2 favors colonies adopting the 1:3 division strategy. When $d_1=d_2=d_3=d_4=0.15$, and also $g_1=g_2=g_3=1$, the 1:3 division strategy becomes favorable (see figure 3 B). When $g_2<g_1=g_3$, colonies adopting the 1:3 division strategy outcompete those

adopting the 2:2 division strategy. The reverse occurs when $g_2 > g_1 = g_3$, especially when g_2 is high. According to Figure 3, low and high g_2 promote the 1:3 division strategy and the 2:2 division strategy, respectively. These results are supported by the simulation outcomes.

Model 2 and results: Direct competition between a dispersing colony and settled colonies

Thus far, the model has assumed no direct competition between colonies; dispersing colonies avoid sites occupied by existing colonies, and seek unoccupied sites in which to settle. However, direct competition is the norm in nature. In this section, we examine the effect of direct competition imposed by dispersing colonies settling at new sites.

Competition is introduced to the baseline model with $g_i = 1$. Once a colony has divided into two, one of the daughter colonies migrates from the natal colony site to a new site chosen randomly from the site population. If the site is empty, the dispersing colony settles there. If the site is occupied by another colony, the dispersing colony competes with the settled colony for the space. Let w_{ij} be the probability that a dispersing colony of size i invades and annihilates a settled colony of size j. We investigate three types of direct competition: (i) $w_{ij} = 0.0$, (ii) $w_{ij} = 1.0$, (iii) $w_{ij} = 1$ if $i > j$, $w_{ij} = 0.5$ if $i = j$, and $w_{ij} = 0.0$ if $i < j$. In situation (i), the settled colony can overcome the dispersing colony regardless of its size, and the dispersing colony dies out. In (ii), the dispersing colony always combats the settled colony, the settled colony dies out, and the dispersing colony occupies the site. Situation (iii) implies that a larger colony always overtakes a smaller one.

If all colonies in the population adopt the 2:2 division strategy, the matrix construct is

$$x(t) = S_1(t)x(t), \tag{12}$$

where $x(t) = (x_0(t), x_1(t), x_2(t), x_3(t), x_4(t))^t$ and

$$S_1(t) = \begin{bmatrix} 1-hx_4(t) & 1-p_1 & 1-p_2 & 1-p_3 & (1-p_4)(1-h) \\ 0 & -hx_4(t)w_{21} & 0 & 0 & 0 \\ hx_4(t) & p_1+hx_4(t)w_{21} & 0 & hx_4(t)w_{23} & h+hx_4(t)w_{24} \\ 0 & 0 & p_2 & -hx_4(t)w_{23} & 0 \\ 0 & 0 & 0 & p_3 & p_4(1-h)-hx_4(t)w_{24} \end{bmatrix}.$$

At equilibrium, $x^* = (1, 0, 0, 0, 0)^t$ is locally unstable when $2p_2p_3 > p_4 + (1-p_4)/h$ regardless of w_{ij}, which corresponds to inequality (4).

Similarly, if all colonies in the population adopt the 1:3 division strategy, the matrix construct is

$$y(t) = L_1(t)y(t), \tag{13}$$

where $y(t) = (y_0(t), y_1(t), y_2(t), y_3(t), y_4(t))^t$ and

$$L_1(t) = \begin{bmatrix} 1-hy_4(t) & 1-p_1 & 1-p_2 & 1-p_3 & (1-p_4)(1-h) \\ hy_4(t) & 0 & hy_4(t)w_{12} & hy_4(t)w_{13} & hy_4(t)w_{14} \\ 0 & p_1 & -hy_4(t)w_{12} & 0 & 0 \\ 0 & 0 & p_2 & -hy_4(t)w_{13} & h \\ 0 & 0 & 0 & p_3 & p_4(1-h)-hy_4(t)w_{14} \end{bmatrix}.$$

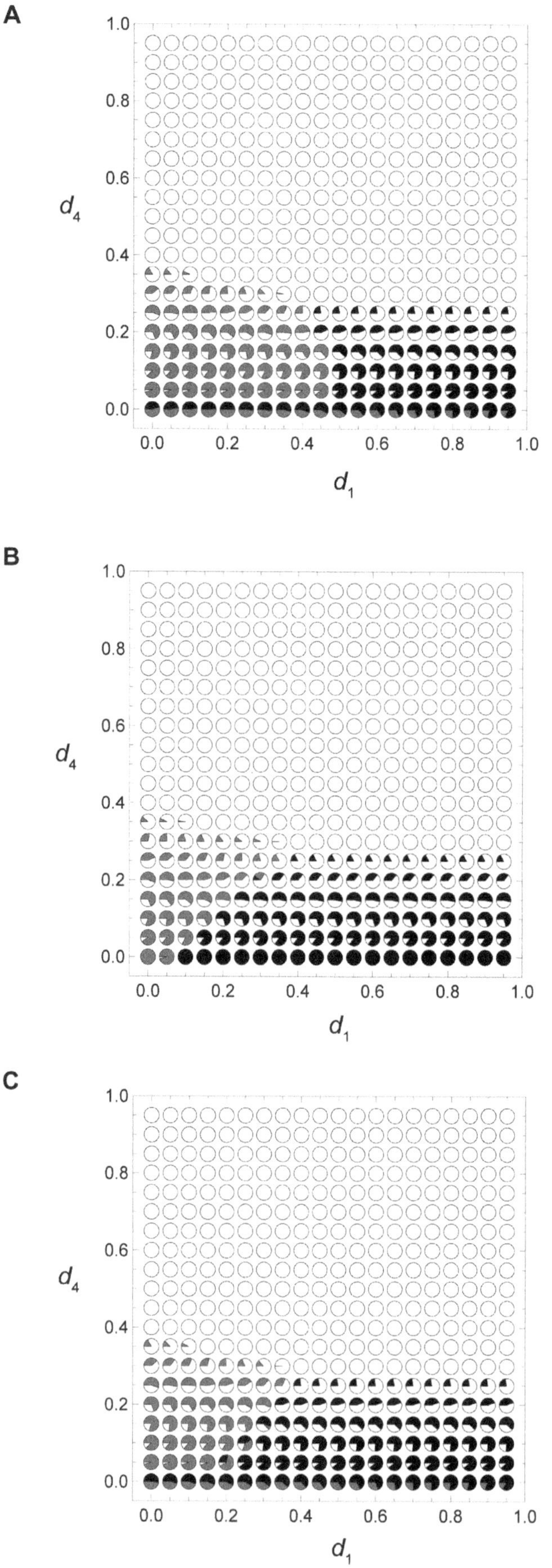

Figure 4. The d_1-d_4 graph of the colony-based simulations in which a dispersing colony competes with settled colonies. The horizontal and vertical axes represent d_1 and d_4, respectively. Death type (i), in which $d_2 = d_3 = d_4$, is assumed. Each pie chart shows the average densities of sites occupied by either of the two strategies, as well as the empty sites (open circles) after 5,000 iterations of (A) and (B) and after 20,000 iterations of (C) in 100 simulations. The black and gray portions of each circle indicate the average densities of sites occupied by colonies undertaking the 2:2 division strategy and the 1:3 division strategy, respectively. The open portion represents the average density of empty sites. Initial densities are $z_0 = 0.6$ and $x_1 = x_2 = x_3 = x_4 = y_1 = y_2 = y_3 = y_4 = 0.05$. The other parameters are: 2,500 sites, $u = 0$ and $h = 1$. (A) Direct competition (i), in which $w_{ij} = 0.0$. (B) Direct competition (ii), in which $w_{ij} = 1.0$. (C) Direct competition (iii), in which a bigger colony always combats a smaller one.

At equilibrium, $y^* = (1, 0, 0, 0, 0)^t$ is locally unstable when $(1+p_1p_2)p_3 > p_4 + (1-p_4)/h$ regardless of w_{ij}, which corresponds to inequality (5).

Interestingly, the baseline model with $g_i = 1$ and the direct competition model share the same local stability condition; an equilibrium in which all colonies become extinct. This arises because a newly-divided dispersing colony encountering many empty sites at equilibrium will more likely settle at a site not already occupied. In this situation, direct competition exerts little influence on the population dynamics.

To investigate the effect of direct competition among colonies adopting either reproductive strategy, we run the colony-based simulations under death type (i), which most strongly favors the 2:2 division strategy among the three death categories. As shown in Figure 4, among the three categories of direct competition, the 2:2 division strategy is most favored in category (ii).

Next, we discuss why direct competition category (ii), rather than category (i), confers advantage to the 2:2 division strategy, focusing on the disadvantage of the 1:3 division strategy in this scenario. The 1:3 division strategy disperses a size 1 colony, while the larger colony remains at the site. Recall that, in this strategy, a size 3 colony matures more rapidly than a colony of size 1. In direct competition category (ii), a size 3 colony dies out, whereas a size 1 colony disperses and settles at a new site. Since a size 1 colony matures more slowly than a size 2 colony, the 2:2 division strategy is advantageous over the 1:3 division strategy. Conversely, in direct competition category (i), a size 3 colony can survive but a size 1 colony cannot colonize a new site. In this scenario, the 1:3 division strategy is preferable.

The assumptions of the baseline model and direct competition category (i) are similar in that dispersing colonies can settle only at empty sites. However the two scenarios generate completely different outcomes (see Figures 2 A and 4 A). In Figure 2 A, the 2:2 division strategy is more advantageous than in Figure 4 A, because dispersing colonies always die out under direct competition (i), unless they find an unoccupied site to colonize. Conversely, in the baseline model, colonies do not disperse unless empty sites are available. Although the dispersing colonies die out, the 1:3 division strategy confers less damage than the 2:2 division strategy because the size 3 colonies remaining at the natal site after division compensate for the loss of the smallest colonies in direct competition category (i). Populations of colonies adopting the 2:2 division strategy lose their dispersing size 2 members, which is more damaging to the population dynamics than the loss of size 1 colonies. Therefore the 2:2 division strategy cannot outcompete the 1:3 division strategy in this scenario.

Although direct competition (ii) most favors the 2:2 division strategy among the three direct competitions, it disadvantages the 2:2 division strategy relative to the baseline model with $g_i = 1$ (See Figures 2 A and 4 B). Therefore we conclude that the 2:2 division strategy is advantageous if newly-divided colonies disperse solely to empty sites and never directly compete with colonies to claim an occupied site. In the following section, we investigate the effect of spatial structure and environmental disturbance on the population dynamics by imposing spatial structure on the baseline model.

Model 3 and results: Spatial structure and environmental disturbance

In a previous section, we found that imposing death type (i) and the logistic growth of colonies favor the 2:2 division strategy over the 1:3 strategy when dispersing colonies seek and settle at unoccupied spaces in a completely mixed population. In this section, we examine the environmental factors that favor the 2:2 division strategy over the 1:3 division strategy. To this end, we use the lattice-structured population as a spatial structure, assuming death type (i) and that a dispersing colony seeks and settles at an unoccupied site. Although the 2:2 division strategy is advantageous when $g_2 > g_1$ and g_3, here we set $g_1 = g_2 = g_3 = 1$ for simplicity. We then run simulations to investigate how the population dynamics are affected by spatial structure and environmental disturbance.

We first present the assumptions of the spatial structured model without environmental disturbance. We use a lattice model as a spatial structure model. Each colony distributed in each lattice site, possesses two traits; colony division and dispersal distance. Each colony adopts either the 2:2 or the 1:3 division strategy (Figure 1), and disperses either short- or long-distance.

This setup yields four possible division/dispersal strategies: 1:3 long, 1:3 short, 2:2 long and 2:2 short. The 1:3 long strategy disperses the size 1 colony to a random site in the lattice-structured population. The 1:3 short strategy displaces the size 1 colony to an empty site among the nearest neighbors. The 2:2 long disperses one of the size 2 colonies to a random site in the lattice-structured population. The 2:2 short strategy and deposits one of the divided size 2 colonies in a nearest-neighbor empty site. Among these strategies, the 1:3 long and 2:2 short are most commonly found in nature. Thus, we focus on the competition between the 1:3 long and 2:2 short strategies in the lattice-structured population.

Simulated results in the absence of environmental disturbance. First, to clarify the effect of the lattice-structured population on the competition between the two strategies, we assume no environmental disturbance. Computer simulations were conducted using death type (i) and $g_1 = g_2 = g_3 = 1$. The outcomes of the simulations are presented in Figure 5. When the colony is static ($d_1 = d_2 = d_3 = d_4$), and the death probabilities relatively low, the 1:3 long overwhelms 2:2 short. Colonies adopting the 2:2 short strategy can outcompete their 1:3 long counterparts if the death probability of size 1 colonies is sufficiently high. The 1:3 long strategy is favorable over the 2:2 short because the latter strategy places a daughter colony only at a nearest neighbor site. Thus, the 2:2 short strategy cannot search for unoccupied sites when the local colony density is high, and colonies adopting this strategy are prevented from increasing their number. On the other hand, the 1:3 long strategy disperses a new colony to a random site, releasing it from local density effects. Provided that empty sites exist in the population, colonies adopting the 1:3 long strategy can sustain population growth. We also conducted simulations in which death probabilities d_1, d_2 and d_3 were variable while d_4 was fixed at 0.2. The outcomes of these simulations are consistent with other results (see Figure S1).

When all of g_1, g_2, and g_3 are less than one, the lattice-structured population displays logistic growth, which favors the 2:2 short strategy over the 1:3 long (see Figure S2).

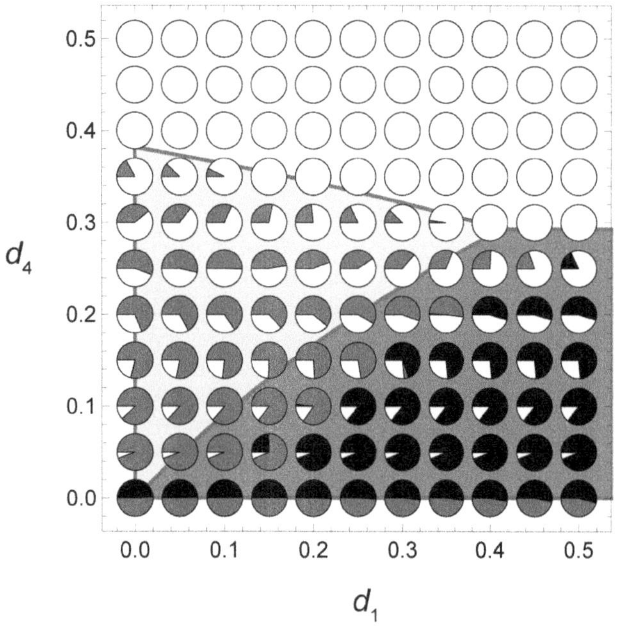

Figure 5. The d_1-d_4 graph of the two-dimensional lattice-structured population. The horizontal and vertical axes represent d_1 and d_4, respectively. Each circle shows the average densities of sites occupied by either of the two strategies, as well as the empty sites (open circles) after 10,000 iterations in 100 simulations on the two-dimensional lattice. The black and gray portions of each circle indicate the average density of the sites occupied by colonies undertaking the 2:2 division and 1:3 division strategies, respectively. The open portion represents the average density of the empty sites. The lines in the graphs delineate the theoretical solutions in a completely mixed population. In the dark gray area, the 2:2 division strategy theoretically wins over the 1:3 strategy. In the light gray area, the predicted winning strategy is 1:3. In $d_4 = 0.0$, mature colonies always survive and a neutral relationship develops between the strategies, which then coexist. Initial densities are $z_0 = 0.2$ and $x_1 = x_2 = x_3 = x_4 = y_1 = y_2 = y_3 = y_4 = 0.1$. The other parameters are: 2,500 lattice sites, $u = 0$ and $h = 1$. The death probability assumes death type (i).

Furthermore, if competition is introduced among all four division/dispersal strategies, colonies adopting either short-distance strategy rapidly become extinct because competition strongly favors long dispersal distance. After just a few iterations, the population comprises colonies adopting the 1:3 long strategy and the 2:2 long strategy. Since the 1:3 and 2:2 long-dispersing colonies in the lattice-structured population behave identically to their counterparts adopting the 1:3 and the 2:2 division strategies in the completely mixed population, the final competition outcomes among all four division/dispersal strategies are those of the 1:3 and the 2:2 division strategies in the completely mixed population (see Figure S3).

Simulated results in the presence of environmental disturbance. We now introduce two parameters of environmental disturbance, u and q, defined as the frequency of disturbance and the disturbance scale, respectively [17]. These parameters are independent of the baseline natural death. q can be interpreted as the degree of spatial autocorrelation. The probability that a colony of size i is removed from a site by environmental disturbance (leaving the previously occupied site empty) is defined as u_i ($0.0 \leq u_i \leq 1.0$). Disturbance is assumed independent of colony size and density ($u_1 = u_2 = u_3 = u_4 = u$). A disturbance kills a colony occupying a lattice site with probability u. The disturbance then spreads to a nearest neighbor site (selected randomly among eight neighbors) with probability q and kills the residential colony (if present). If the site is unoccupied, it is merely disturbed. The disturbance is repeated with probability q until halted with probability $1 - q$. For example, if u is low and q is high, the lattice contains a few large or clustered disturbed areas. If u is high and q is low, many small areas are disturbed in the lattice.

Figure 6 shows the effect of the disturbance parameters u and q on the competition between 1:3 long and 2:2 short when $g_1 = g_2 = g_3 = 1$. This simulation assumes death type (i). As demonstrated in Figure 6, low q and relatively high u confer an advantage to the 2:2 short strategy when d_1 is higher and the difference between d_1 and d_2 ($= d_3 = d_4$) is larger (see figure 6 C, D and E). Increasing u and q favors the 1:3 long strategy at certain death probabilities favoring the 2:2 short strategy in the absence of disturbance. These outcomes are supported by standard dispersal theory. The simulation outcomes are influenced by disturbance when the growth probabilities are less than 1 (see figure S4). According to this figure, the 2:2 short strategy is advantageous over 1:3 long when size 2 colonies grow rapidly (high g_2) relative to g_1 (which is fixed; see figure S4 A–C and D–F). Figure S4 A and D, together with Figure S4 B and E, indicate that the 2:2 short strategy is advantageous when g_1 is low relative to g_2.

Discussion and Conclusions

We examined the ecological conditions under which budding can be advantageous over migration in colonial organisms. The main conclusions are summarized as follows: (1) when the life history parameters are independent of colony size, budding colonies are overwhelmed by dispersive colonies. (2) When the death probability of the smallest colonies is relatively higher than that of more mature colonies, survival favors the budding strategy. The death function can be modeled as a negative exponential function of the continuous lifespan. (3) Budding wins against dispersal when the growth rates of size 1 and size 3 colonies are lower than that of size 2 colonies. In this scenario, colonies undergo logistic growth, which ensures their survival if they adopt the budding strategy. (4) When a newly-divided dispersing colony does not compete with a settled colony for space, but instead colonizes open spaces, the budding strategy is advantageous over dispersal. This corresponds to the situation in which founding colonies leave their natal site, and seek and settle at a nearby open space.

We then examine the effects of spatial structure and environmental disturbance on the population dynamics. Relative to dispersal, the budding strategy proved less favorable in spatial structured populations than in completely mixed populations. This occurs because long distance dispersal prevents local overcrowding. In the event of environmental disturbance, long distance dispersal confers a survival advantage over budding species because migrating species can escape the risk. However, when the scale of environmental disturbance is small, budding can retain an advantage even under relatively high frequency of environmental disturbance.

Our current paper roughly supports the main results presented in Nakamaru *et al.* (2007). Among the various growth and death functions, logistic colony growth (corresponding to g_1, $g_3 < g_2$) and an exponentially decreasing death function (corresponding to $d_1 >> d_2$, d_3, d_4) favor budding over dispersal. However, the assumptions of our current model differ from those in Nakamaru et al (2007). For example, the previous study assumed continuous colony size, and that each colony obeys logistic growth and an exponentially decreasing death function. Here, to more easily investigate the effect of different growth and death functions on the

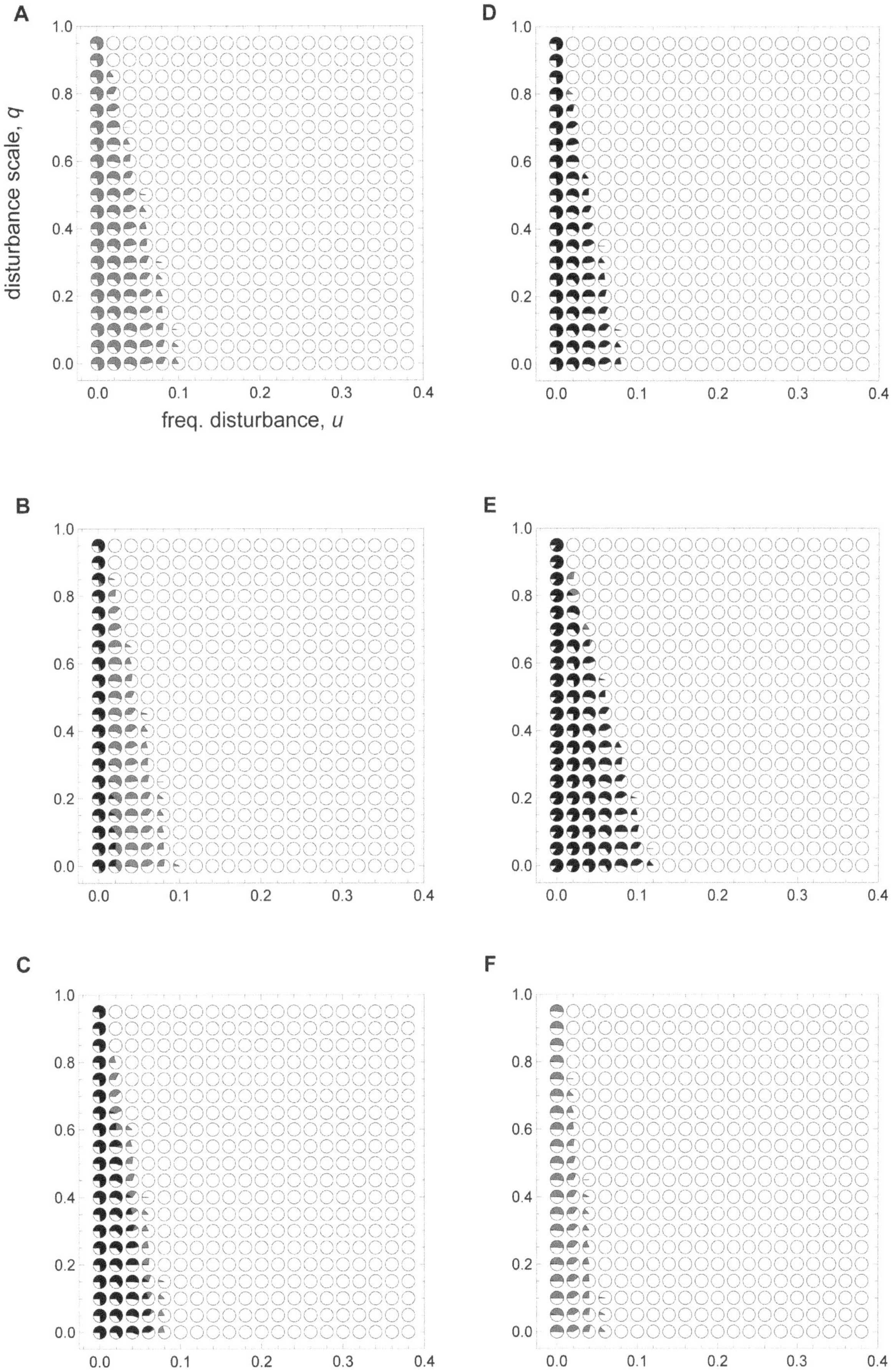
A
B
C
D
E
F
disturbance scale, q
freq. disturbance, u
0.0
0.2
0.4
0.6
0.8
1.0
0.0
0.1
0.2
0.3
0.4

Figure 6. The u-q graph of colony-based simulations in the two-dimensional lattice-structured population after 5,000 iterations. Details are provided in the caption to Figure 5. The death probability assumes death type (i). (A) $d_1 = 0.25$ and $d_2 = 0.15$, (B) $d_1 = 0.3$ and $d_2 = 0.15$, (C) $d_1 = 0.35$ and $d_2 = 0.15$, (D) $d_1 = 0.4$ and $d_2 = 0.15$, (E) $d_1 = 0.35$ and $d_2 = 0.1$ and (F) $d_1 = 0.35$ and $d_2 = 0.2$. Other parameters are: $g_1 = g_2 = g_3 = 1$ and $h = 1$.

evolutionary population dynamics, we discretized the colony size. Despite these differences in model assumptions, the results are consistent between the two studies, which strengthens our conclusions.

Budding likely evolves when very small individuals (or colonies) undergo logistic growth and are subject to relatively high mortality. Furthermore, our models suggest that budding is favored when dispersing offspring (colonies) can seek unoccupied spaces in which to settle without inciting conflict. Ant or coral colonies (or colonies of ramets in plants) usually exhibit type III survivorship curves (negative exponential mortality functions) [21–23] and logistic growth [21,24–26]. Furthermore, ants are known to seek suitable empty sites when their colony disperses [27,28]. Thus, these organisms are naturally predisposed to budding. Although our results also suggest that environmental disturbance promotes long-dispersing by its risk hedging effect and discourages budding, the benefits of budding may outweigh the disadvantage of short-distance dispersal in some of the above mentioned organisms (some corals, plants and ants) that reproduce by budding in frequently disturbed environments.

Rigorous empirical testing of our hypothesis is limited by the difficulty of distinguishing the baseline death and growth rates from the changes in those parameters induced by disturbance. However, we consider that our model predictions could be tested on realized average death and growth rates that include both baseline and additional (or reduced) rates caused by disturbance. As already mentioned, life history parameters such as survival and growth functions appear to directly determine which strategy should evolve. If the average death and growth rates of fission-dividing colonial organisms fulfill the predicted conditions favoring the 2:2 division strategy described in Figure 2 and Eqs. (4) and (6), our models would be supported. The life history of species or closely related species adopting both budding and non-budding tactics must be investigated in empirical studies.

Previous studies on the evolution of body-condition dependency in dispersal have not directly addressed the budding phenomenon [29–31]. Instead, these models assume semelparity or an annual lifecycle. Budding is possible only when a parent (or part of the parent's body) survives after reproduction, coexists with the offspring (perennial) and re-reproduces (iteroparity). Therefore, our results are not directly comparable with the results of these studies.

We now discuss the factors that should be incorporated into our current model to render it more realistic. We assumed that a colony reproduces by dividing in two. We also assumed that (not always, however) large offspring disperse over a smaller distance (size-dispersal distance trade off). However, another reproductive constraint has been reported in the literature; the trade-off between size and number of offspring (e.g. [32]). We consider that the size-number trade-off is a sensible proposition when the parental effort or total parental resource that can be allocated to reproduction is constant. For generality in our current study, the total reproductive effort was assumed labile (for instance, the dispersive strategy (3:1 strategy) invests more resource in survival than budding, given that the dispersed portion is the offspring). We now introduce the 2:1:1: division strategy, in which a size 4 colony divides into one size 2 colony and two size 1 colonies that then disperse, and the 1:1:1:1 division strategy, in which a size 4 colony divides into four size 1 colonies. We constrained the reproduction effort to be constant (mimicking a size–number trade off) in the competition between 2:1:1 vs. 2:2 strategies. However, biological strategies might be more precisely described by 2:1:1 and 1:1:1:1 strategies; for example; plants may release many seeds and ant colonies disperse many winged queens simultaneously. We have partially tackled this situation. In well-mixed populations, the local stable conditions of the 1:1:1:1 and the 2:1:1 division strategies are $4p_1p_2p_3 > p_4 + (1-p_4)/h$ and $p_2p_3(1+2p_1) > p_4 + (1-p_4)/h$, respectively (method presented in the Appendices). The left hand side of both inequalities specifies the probability that dividing colonies mature to size 4. This result implies that (i) budding wins against the other three strategies when the survival probability of size 1 colonies (p_1) is very small, and (ii) if the converse is true (p_1 is high), the 1:1:1:1 division strategy may be favored. The latter condition might simulate the evolution of semelparity. However, the strongest strategy when members of the population adopt all four strategies has yet to be elucidated.Furthermore, one must consider relaxing the assumption on the timing of reproduction. Size 4 colonies only reproduce in the current model. If we assume that size 2 and 3 colonies can reproduce, we can analyze the conditions under which annual or perennial life cycles evolve (regarding the reproductive strategy of size 2 colonies as annual semelparity).

The current model does not include colony size effects in disturbance. For instance, if smaller colonies are more vulnerable to environmental disturbance than larger ones, the budding might be more advantageous over dispersal than is predicted in this paper. Finally, our study considers neither explicit competition among individuals within each colony nor direct interference induced by competition between neighboring colonies. In future work, we will incorporate these factors into a more realistic model.

Budding has been considered to play a key role in evolutionary ecology. Theoretically, budding increases the genetic variance among groups and/or the genetic relatedness within groups, thus promoting the evolution of group-advantageous traits such as altruism and female-biased sex allocation [33–35]. On the other hand, restricted migration (which frequently accompanies fission) intensifies the local competition among related individuals and hinders the evolution of altruism, despite the apparent enhancement of genetic relatedness of interactants [36–38]. However, recent theoretical and empirical studies suggest that budding can solve this "viscous population dilemma" and enable the evolution of altruism, because budding increases the number of more related individuals in the background local population [7,8,39–45]. All of the above-mentioned theoretical studies focus on the effect of budding on evolved social traits. Thus, budding is assumed as the given condition, without considering the environments in which budding evolves. The latter problem is the focus of our current paper. On the other hand, the ecological life history models introduced here neglect issues such as relatedness and intra-colonial conflicts. Therefore, our current study is complementary to existing studies. In future studies, we hope to incorporate relatedness and intra-colonial conflicts into the model scheme, to better understand the ecological drivers of coevolving life history and social traits.

Supporting Information

Figure S1 Effect of life history parameters on competition between the two strategies in the two-dimensional lattice. Black spheres indicate that the 2:2 short strategy overcomes the 1:3 long strategy, while gray spheres indicate the opposite. Where no spheres are presented, colonies have become extinct. The volume of each sphere denotes the average density of the strategy after 1,000 iterations. This graph shows the effect of imposing the death probabilities d_1, d_2, d_3. The parameters are $d_4=0.2$, $g_1=g_2=g_3=1$, $h=1$. Initial densities are $z_0=0.6$ and $x_1=x_2=x_3=x_4=y_1=y_2=y_3=y_4=0.05$.

Figure S2 Effect of life history parameters on competition between the two strategies in the two-dimensional lattice. Black spheres indicate that the 2:2 short strategy overcomes the 1:3 long strategy, while gray spheres indicate the opposite. Where no spheres are presented, the colonies have become extinct. The volume of each sphere denotes the average density of the strategy after 6,250 iterations. This graph illustrates the effect of growth probabilities g_1, g_2, and g_3. Other parameters are: $d_1=0.35$, $d_2=d_3=d_4=0.15$. Initial densities are $z_0=0.6$ and $x_1=x_2=x_3=x_4=y_1=y_2=y_3=y_4=0.05$.

Figure S3 Competition among four possible strategies in a structured population residing on a two dimensional lattice, after 10,000 iterations. The parameters are $g_2=g_3$, $d_1=0.35$, $d_2=d_3=d_4=0.15$ and $h=1$. The initial population density of each strategy is the same and the initial density of vacant sites is 0.6.

Figure S4 The *u*-*q* graph of colony-based simulations in the two-dimensional lattice structured population after 6,250 iterations. Details are provided in the caption to Figure 5. (A) $g_1=0.8$ and $g_2=0.8$, (B) $g_1=0.8$ and $g_2=0.6$, (C) $g_1=0.8$ and $g_2=0.4$, (D) $g_1=0.6$ and $g_2=0.8$, (E) $g_1=0.6$ and $g_2=0.6$ and (F) $g_1=0.6$ and $g_2=0.4$. Other parameters are: $d_1=0.35$, $d_2=d_3=d_4=0.15$, $h=1$ and $g_3=0.9$. Initial density of each size of each strategy is identical, and the initial density of vacant sites is 0.6.

Appendix S1 Corollary of Jury's criterion.

Appendix S2 Local stability analysis of positive equilibrium on the 2:2 division strategy.

Appendix S3 Local stability analysis of positive equilibrium on the 1:3 division strategy.

Appendix S4 Local stability analysis in competitive system between two strategies.

Acknowledgments

We thank Evan Economo for his helpful comments.

Author Contributions

Conceived and designed the experiments: MN. Performed the experiments: AO SUS KM. Analyzed the data: AO SUS KM MN. Wrote the paper: MN TT KT. Made the mathematical model: MN. Proved the mathematical model: TT. Analyzed the mathematical model: TT MN.

References

1. Clobert J, Danchin E, Dhonde AA, Nichols JD (2001) Dispersal. Oxford: Oxford University Press.
2. Levin SA, Muller-Landau HC, Nathan R, Chave J (2003) The ecology and evolution of seed dispersal: A theoretical perspective. Annu Rev Ecol Syst 34: 575–604.
3. Clobert J, Baguette M, Benton TG, Bullock JM (2012) Dispersal ecology and evolution. Oxford: Oxford University Press.
4. Wynee-Edwards V (1962) Animal dispersion in relation to social behaviour. New York: Hafaer.
5. Williams GC (1966) Adaptation and natural selection. Princeton: Princeton University Press.
6. Ronce O, Olivieri I, Clobert J, Danchin E (2001) Perspectives on the study of dispersal evolution. In: Clobert J, Danchin E, Dhonde AA, Nichols JD, editors. Dispersal. Oxford: Oxford University Press. pp. 341–357.
7. Gardner A, West SA (2006) Demography, altruism, and the benefits of budding. J Evol Biol 19: 1707–1716.
8. Kummerli R, Gardner A, West SA, Griffin AS (2009) Limited dispersal, budding dispersal, and cooperation: an experimental study. Evolution 63: 939–949.
9. Gardner A, Arce A, Alpedrinha J (2009) Budding dispersal and the sex ratio. Journal of Evolutionary Biology 22: 1036–1045.
10. Liu J, Dong M, Miao SL, Li ZY, Song MH, et al. (2006) Invasive alien plants in China: role of clonality and geographical origin. Biol Invasions 8: 1461–1470.
11. Honnay O, Jacquemyn H (2010) Clonal plants: beyond the patterns—ecological and evolutionary dynamics of asexual reproduction. Evol Ecol 24: 1393–1397.
12. Coffroth MA, Lasker HR (1998) Population structure of a clonal gorgonian coral: The interplay between clonal reproduction and disturbance. Evolution 52: 379–393.
13. Passera L (1994) Characteristics of tramp species. In: William DF, editor. Exotic ants: biology, impact, and control of introduced species. Colorado: Westview Press. pp. 23–43.
14. Holway DA, Lach L, Suarez AV, Tsutsui ND, Case TJ (2002) The causes and consequences of ant invasions. Annu Rev Ecol Syst 33: 181–233.
15. Carpintero S, Reyes-Lopez J, de Reyna LA (2003) Impact of human dwelling on the distribution of the exotic Argentine ant: a case study in the Doñana National Park. Biol Conserv 115: 279–289.
16. Tsuji K (2010) Unicolonial ants: loss of colony identity. In: Breed M, Moore J, editors. Encyclopedia of Animal Behavior. Oxford: Academic Press. pp. 469–473.
17. Nakamaru M, Beppu Y, Tsuji K (2007) Does disturbance favor dispersal? An analysis of ant migration using the colony-based lattice model. J Theor Biol 248: 288–300.
18. Caswell H (2000) Matrix population models 2nd edition. Sunderland: Sinauer Associates, Inc., Publishers.
19. Stearns SC (1992) The Evolution of Life Histories. Oxford: Oxford University Press.
20. Gardner A, Grafen A (2009) Capturing the superorganism: a formal theory of group adaptation. J Evol Biol 22: 659–671.
21. Hölldobler B, Wilson EO (1990) The ants. Cambridge: Belknap Press of Harvard University Press.
22. Babcock RC (1991) Comparative demography of three species of scleractinian corals using age- and size-dependent classifications. Ecol Monogr 61: 225–244.
23. Harper JL (1977) Population Biology of Plants. Waltham: Academic Press.
24. Yamaguchi M (1983) Growth data analysis in the reef-building coral {Pocillopora} {damicornis} (Linnaeus). Galaxea 2: 21–27
25. van Woesik R, Sakai K, Ganase A, Loya Y (2011) Revisiting the winners and the losers a decade after coral bleaching. Mar Ecol Prog Ser 434: 67–76.
26. Suzuki JI, Hucthings MJ (1997) Interactions between shoots in clonal plants and the effects of stored resources on the structure of shoot population. In: de Kroon H, van Groenendael J, editors. The Ecology and Evolution of Clonal Plants. Leiden: Backhuys Publishers.
27. Stroeymeyt N, Giurfa M, Franks NR (2010) Improving decision speed, accuracy and group cohesion through early information gathering in house-hunting ants. PLoS One 5: e13059.
28. Cronin AL (2012) Consensus decision making in the ant Myrmecina nipponica: house-hunters combine pheromone trails with quorum responses. Anim Behav 84: 1243–1251.
29. Gyllenberg M, Kisdi E, Utz M (2008) Evolution of condition-dependent dispersal under kin competition. J Math Biol 57: 285–307.
30. Gyllenberg M, Kisdi E, Utz M (2011) Body condition dependent dispersal in a heterogeneous environment. Theor Popul Biol 79: 139–154.
31. Gyllenberg M, Kisdi E, Utz M (2011) Variability within families and the evolution of body-condition-dependent dispersal. J Biol Dyn 5: 191–211.
32. Smith CC, Fretwell SD (1974) The optimal balance between size and number of offspring. Am Nat 108: 499–506.
33. Slatkin M, Wade MJ (1978) Group selection on a quantitative character. Proc Natl Acad Sci U S A 75: 3531–3534.

34. Goodnight CJ, Stevens L (1997) Experimental studies of group selection: what do they tell us about group selection in nature? Am Nat 150: S59–S79.
35. Ono S, Misawa K, Tsuji K (2003) Effect of Group Selection on the Evolution of Altruistic Behavior. J Theor Biol 220: 55–66.
36. Wilson DS, Pollock GB, Dugatkin LA (1992) Can sltruism evolve in purely viscous populations? Evol Ecol 6: 331–341.
37. Taylor PD (1992) Altruism in viscous populations - an inclusive fitness model. Evol Ecol 6: 352–356.
38. Taylor PD (1992) Inclusive fitness in a homogeneous environment. Proc Biol Sci 249: 299–302.
39. Sharp SP, Simeoni M, Hatchwell BJ (2008) Dispersal of sibling coalitions promotes helping among immigrants in a cooperatively breeding bird. Proc Biol Sci 275: 2125–2130.
40. Slatkin M (1977) Gene flow and genetic drift in a species subject to frequent local extinctions. Theor Popul Biol 12: 253–262.
41. Goodnight KT (1992) The effect of stochastic variation on kin selection in a budding-viscous population. Am Nat 140: 1028–1040.
42. Lehmann L, Perrin N, Rpusset F (2006) Population demography and the evolution of helping behaviors. Evolution 60: 1137–1151.
43. Bradley BJ, Doran-Sheehy DM, Vigilant L (2007) Potential for female kin associations in wild western gorillas despite female dispersal. Proc Biol Sci 274: 2179–2185.
44. Heinsohn R, Dunn P, Leggs S, Double M (2000) Coalitions of relatives and reproductive skew in cooperatively breeding white-winged choughs. Proc Biol Sci 267: 243–249.
45. Williams DA, Rabenold KN (2004) Male-biased dispersal, female philopatry, and routes to fitness in a social corvid. J Anim Ecol 74: 150–159.

11

On a Dhole Trail: Examining Ecological and Anthropogenic Correlates of Dhole Habitat Occupancy in the Western Ghats of India

Arjun Srivathsa[1,2*], Krithi K. Karanth[2,3], Devcharan Jathanna[2], N. Samba Kumar[2,4], K. Ullas Karanth[2,3,5]

1 Post-Graduate Programme in Wildlife Biology and Conservation, Wildlife Conservation Society, India Program, National Centre for Biological Sciences (Tata Institute of Fundamental Research), Bangalore, India, **2** Centre for Wildlife Studies, Bangalore, India, **3** Wildlife Conservation Society, Bronx, New York, United States of America, **4** Wildlife Conservation Society, India Program, Bangalore, India, **5** Nicholas School of Environment, Duke University, Durham, North Carolina, United States of America

Abstract

Although they play a critical role in shaping ecological communities, many threatened predator species are data-deficient. The Dhole *Cuon alpinus* is one such rare canid with a global population thought to be <2500 wild individuals. We assessed habitat occupancy patterns of dholes in the Western Ghats of Karnataka, India, to understand ecological and anthropogenic determinants of their distribution and habitat-use. We conducted spatially replicated detection/non-detection surveys of dhole signs along forest trails at two appropriate scales: the entire landscape and a single wildlife reserve. Landscape-scale habitat occupancy was assessed across 38,728 km^2 surveying 206 grid cells of 188-km^2 each. Finer scale habitat-use within 935 km^2 Bandipur Reserve was studied surveying 92 grid cells of 13-km^2 km each. We analyzed the resulting data of dhole signs using likelihood-based habitat occupancy models. The models explicitly addressed the problematic issue of imperfect detection of dhole signs during field surveys as well as potential spatial auto-correlation between sign detections made on adjacent trail segments. We show that traditional 'presence versus absence' analyses underestimated dhole habitat occupancy by 60% or 8682 km^2 [naïve = 0.27; $\hat{\psi}_L$(SE) = 0.68 (0.08)] in the landscape. Addressing imperfect sign detections by estimating detection probabilities [$\hat{p}_{t(L)}$ (SE) = 0.12 (0.11)] was critical for reliable estimation. Similar underestimation occurred while estimating habitat-use probability at reserve-scale [naïve = 0.39; $\hat{\Psi}_s$(SE) = 0.71 (0.06)]. At landscape scale, relative abundance of principal ungulate prey primarily influenced dhole habitat occupancy. Habitat-use within a reserve, however, was predominantly and negatively influenced by anthropogenic disturbance. Our results are the first rigorous assessment of dhole occupancy at multiple spatial scales with potential conservation value. The approach used in this study has potential utility for cost-effectively assessing spatial distribution and habitat-use in other species, landscapes and reserves.

Editor: John Goodrich, Panthera United States of America

Funding: Department of Science and Technology, Government of India. (URL: http://www.dst.gov.in). Save the Tiger Fund of National Fish and Wildlife Foundation (URL: http://www.nfwf.org/Pages/default.aspx). Rhino & Tiger Conservation Fund of US Fish and Wildlife Service, Washington, DC. (URL: http://www.fws.gov/international/wildlife-without-borders/rhino-and-tiger-conservation-fund.html). The funders had no role in study design, data collection and analysis, decision to publish, or preparation of the manuscript.

Competing Interests: Co-author K. Ullas Karanth is a PLOS ONE editorial board member. This does not alter the authors' adherence to PLOS ONE Editorial policies and criteria.

* E-mail: arjuns.cws@gmail.com

Introduction

Large carnivores are highly threatened across the world [1] and populations face risks of local extinction because they are wide-ranging and occur at relatively low densities [2–4]. Wide-ranging carnivores are often found across diverse habitats such as forests, open grasslands, cultivated lands and agricultural landscapes. Therefore, understanding carnivore distributions at the spatial scale of large landscapes is crucial for identifying sites for targeted conservation efforts [5–8]. Even within designated wildlife reserves, effective conservation requires data on how carnivores use different habitat types in a matrix. Furthermore, a full understanding of large-scale carnivore distribution as well as patterns of habitat-use within wildlife reserves can be gained only by identifying ecological and anthropogenic factors that are key determinants of habitat occupancy.

Wild canids, in particular, have large geographic ranges with interspersion of human-modified landscapes. Consequently, dependence on common prey resources has led to conflict, persecution by humans and spread of disease from domestic animals [9,10]. The Asiatic Wild Dog or Dhole *Cuon alpinus*, Pallas 1811, is the only Asian wild canid that primarily inhabits forested areas. Dholes are among the top social predators of large ungulates in tropical forests [11–13]. Their numbers have significantly declined and trace populations are now largely restricted to forested areas [14]. In India, dholes were considered vermin and bounty-hunted to the verge of extinction before they received legal protection in 1972 [14,15]. They have been extirpated from 60% of their former range in the last century due to human persecution and loss of forest cover, and now occur primarily in protected wildlife reserves embedded within larger multiple-use landscapes [16,17].

Although historically a widespread species, current subjective assessments suggest that <2500 individuals of dholes may survive globally [14]. Despite their endangered status [14], dholes are the least studied social large carnivores. Previous studies of dholes have mostly focused on diet profiles [18–21] and behavior through *ad libitum* data collection methods [22–24]. This dearth of quantitative studies of their population ecology hinders conservation. Therefore, an assessment of dhole distribution and habitat-use is important in elucidating ecological drivers of their present occurrence and for measuring the viability of remaining populations and habitats.

A multi-scale analysis of spatial distribution is important for conservation in order to isolate scale-dependent ecological processes [25]. For example, assessment of landscape-level distribution of dholes could provide information on meta-population structure, population source-sites, functional corridors, landscape connectivity and other threats, which are useful for regional conservation planning. In areas where dholes are known to occur, such as individual wildlife reserves, patterns of habitat-use can provide insights on ecological drivers of local densities and efficacy of management interventions.

Because dhole signs such as tracks and scats are relatively easy to identify, we designed rigorous sign surveys under an occupancy-modeling framework. We examined dhole occupancy at two distinct spatial scales identified as relevant for understanding dhole ecology and addressing conservation needs. At landscape-scale, we surveyed a 38, 728 km^2 area along the Western Ghats (Karnataka, India) to examine dhole distribution patterns. In order to assess patterns of fine-scale habitat-use, we chose a subset of the larger landscape, and surveyed a 935-km^2 area in a single wildlife reserve.

Animal distribution patterns across space are typically non-uniform as a consequence of varying habitat characteristics [26,27]. Variations in distribution patterns can be estimated when data from well-designed field surveys are confronted with ecologically relevant predictors and analyzed through parsimonious modeling [28]. Factors such as abundance of prey species, land cover type, anthropogenic disturbances and protection efforts may influence spatial distribution of carnivores [29–31]. We hypothesized that for dholes, site-specific probabilities of occupancy at both scales would be influenced by a combination of ecological and anthropogenic factors.

Prey densities are fundamental determinants of carnivore densities [32,33]. We measured relative abundances of all prey species (gaur *Bos gaurus*, sambar *Rusa unicolor*, chital *Axis axis*, wild pig *Sus scrofa* and muntjac *Muntiacus muntjak*) that make up >90% of biomass in dhole diet [18,19,23,24]. However, our interest was in identifying the influence of preferred prey species abundance on distribution of dholes. Based on previous diet studies, we predicted that chital and sambar abundance would positively influence dhole occupancy at both spatial scales. However, anecdotal records also indicate that dholes avoid human settlements and presence. We therefore also predicted that anthropogenic disturbance would negatively influence dhole distribution and habitat-use.

This is the first systematic application of robust occupancy models to quantitatively assess dhole distribution and habitat-use patterns anywhere in their range.

Materials and Methods

Ethics Statement

The study was conducted in the protected areas and adjoining forests of Western Ghats within the state of Karnataka. The Karnataka State Forest Department provided necessary research permits for the study. Since the methods used were non-invasive and relied completely on recording indirect signs of animals, animal care and use committee approval was not required.

Study Area

A countrywide occupancy-based questionnaire survey of large mammal distributions in India showed dhole presence in the Western Ghats, Central India and North-east India [16]. Our study was conducted in the central part of the Western Ghats, located in the state of Karnataka (Fig. 1).

We assessed landscape-scale dhole occupancy across a 38,728 km^2 area in Karnataka's Western Ghats. The study area consists of semi-evergreen, tropical moist-deciduous, tropical dry-deciduous and dry-deciduous forests with substantial anthropogenic modifications, creating a heterogeneous vegetation matrix [34]. There are 16 protected reserves in the landscape, encompassing an area of *c.* 8700 km^2. The 21,176 km^2 of forested areas support a diverse assemblage of wild ungulates such as the chital, four-horned antelope *Tertracerus quadricornis*, gaur, mouse deer *Moschiola indica*, muntjac, sambar and wild pig. The landscape also has populations of dhole and its co-predators, the tiger *Panthera tigris* and the leopard *Panthera pardus* [33].

To examine fine-scale habitat-use by dholes, we chose Bandipur National Park, a 935-km^2 subset of the Western Ghats landscape in Karnataka. This protected area predominantly has tropical moist and dry deciduous forests, with some areas degraded to scrub due to human impacts [35]. It supports a density of 35.2/100 km^2 medium to large sized prey species [33]. It is contiguous with Wayanad Wildlife Sanctuary and Mudumalai Wildlife Sanctuary to its south and Nagarahole National Park to its northwestern sides (Fig. 2).

Occupancy Modeling

Although conventional radio-telemetry can generate reliable data on large carnivore habitat-use, its application is limited because it is expensive, labor-intensive and involves the difficult tasks of safe capture, immobilization and handling of elusive animals. Due to these difficulties, emphasis has increasingly shifted to collection of non-invasive data such as indirect signs, camera-trap photographs, fecal DNA, etc. to estimate distribution and abundances [36–39].

From such data, estimates of space-use at the scale of patch or habitat are sometimes obtained using Resource Selection Functions (RSF) or Resource Selection Probability Functions (RSPF). Weaknesses of the approach include the need for some estimate of resource 'availability' which is often difficult to obtain in practice [48]. Moreover, since such functions do not take variable detectability into account, it usually results in imprecise estimates and flawed inferences on species-habitat relationships [40,41]. Other 'predictive' distribution approaches such a Habitat Suitability Models (HSM) and Ecological Niche Factor Analysis [42] that rely on presence-only data introduce biases in estimates because of imperfect detection issues, unequal sampling effort and unaccounted true absence of species [43,44].

On the other hand, occupancy modeling permits using detection/non-detection data of animal signs, while explicitly addressing the issue of imperfect detection [26,28,45]. This method works well for studies of carnivores that generally occur at low local densities and large spatial scales [31,46,47]. Recent occupancy models can also yield reliable estimates of probability of habitat-use at finer spatial scales [48–50]. The stronger inference derived from occupancy models arises from their ability to make full use of available information, decomposing true absence from non-detection, within a probabilistic framework [48,51,52].

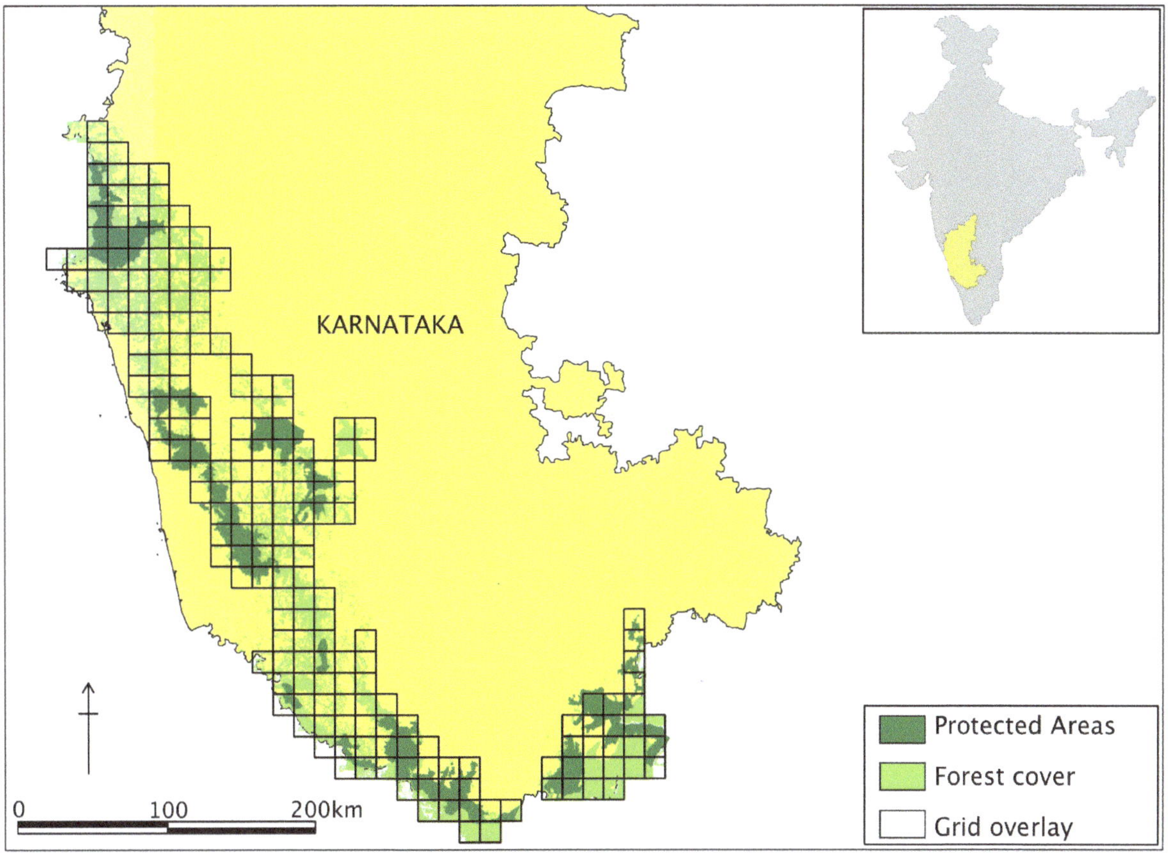

Figure 1. Study area map of the Western Ghats landscape in Karnataka. Study area and survey design used for the landscape scale habitat occupancy of dholes in the Western Ghats, Karnataka State, India (2006–2007). The map shows overall forest cover, protected wildlife reserves with superimposition of 188 km^2-grid-array. Inset: location of the study area in India.

Field Survey Methods

Dholes use forest roads and trails extensively for hunting, movement, scent-marking and defecation at latrine sites [18,19,23]. Therefore, surveying for their signs such as tracks and scat-piles along roads/trails improves detection rates and sample sizes. To optimize spatial coverage and logistical feasibility, we implemented occupancy survey design using spatial rather than temporal replicates, surveying consecutive segments along forest roads and trails [53,54]. A team of three surveyors skilled in mammal sign recognition surveyed forest roads and trails to detect signs of dholes. Only fresh, correctly identifiable signs were recorded to avoid biases arising from misclassifications or sign decay [55,56].

Landscape-scale survey design. We adapted the occupancy sampling design implemented for surveys of tigers in the same landscape [31,54]. Field surveys were carried out between February 2006–May 2006 and December 2006–May 2007 (during the dry season) maintaining uniform detection conditions with an assumption that dhole distribution did not change over the larger landscape in this short period.

The study area was gridded with 206 square cells of 188-km^2 size each to estimate true habitat occupancy (proportion of area occupied by dholes). Since we aimed at estimating the proportion of total area occupied by dholes at large scale, it was necessary that each grid (sample unit) was larger than the maximum home range size of dhole packs in the area. We relied on home range sizes estimated from sighting records of known individual packs in earlier studies [23,24] as well as radio-telemetry based estimates from other landscapes [57], setting cell sizes much larger than these estimates (range 20–105 km^2). We standardized the sampling effort per cell such that walk effort was ~40 km for a cell that was fully occupied by dhole habitat. The effort reduced in proportion to available dhole habitat as measured by extent of forest cover.

Detections of dhole signs were recorded as either '1' (detected) or '0' (undetected) on each successive 100-meter long trail/road segment. For analyses these detection/non-detection data were later aggregated at 1-km segments. The number of spatial replicates we sampled per cell varied from 2 to 42. The survey effort invested was 2021 man-days with 4174 km distance walked, yielding a total of 278 detections of dhole signs (211 scat deposit locations and 67 track sets). In terms of the number of replicates, as well as signs detected, the survey generated data appropriate for formal occupancy analyses.

Reserve-scale survey design. Within the larger landscape, at the scale of a single reserve field surveys were conducted during the dry season (January 2012–April 2012) in Bandipur Tiger Reserve. A grid with 92 cells of 13 km^2 each was overlaid on the area. The estimated dhole home-range size of *c.* 20–40 km^2 in the dry season [23,24] was larger than the cell size with the objective being measuring patterns of 'habitat-use probability' rather than proportion of habitat occupied (true habitat occupancy). Our attempt was to maximize spatial coverage and to discriminate between more and less intensively used areas by dhole within a single reserve. Therefore, the entire network of dirt roads within

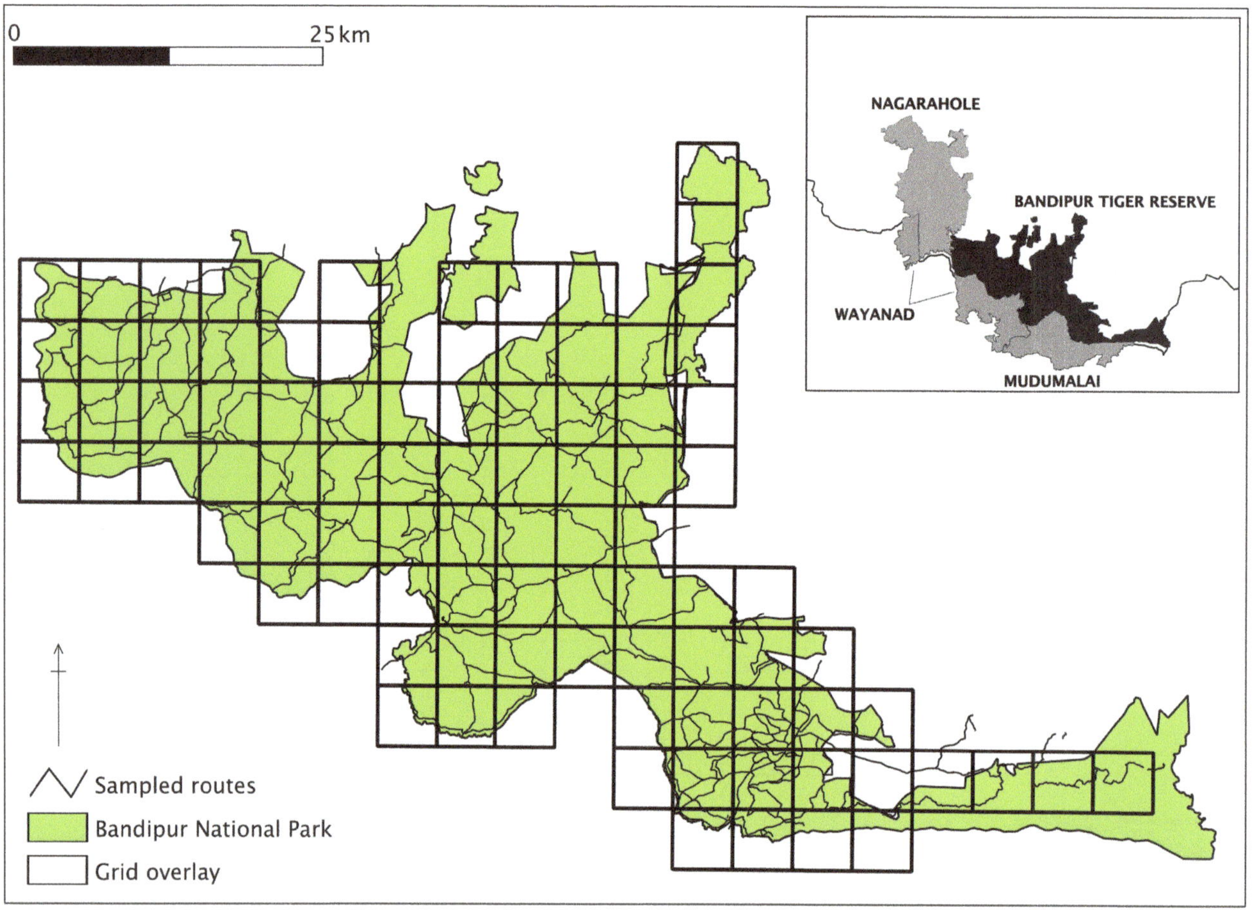

Figure 2. Study area map of Bandipur Tiger Reserve. Study area and survey design for Bandipur Tiger Reserve, India (2012) showing protected area boundary, forest road sign-survey routes and 13-km^2-grid array. Inset: location of the study area and adjoining protected areas.

Bandipur was sampled intensively (Fig. 2). Detection/non-detection data were recorded from consecutive 100-meter segments along forest-roads. The number of replicates per cell varied from 8 to 254 100-m segments. We surveyed 92 such small cells with a total of 730 km of walk effort, resulting in 235 detections of dhole signs (57 fresh scat deposits and 178 track sets).

Analytical Approach

The standard occupancy model of MacKenzie et al. [28] describes the likelihood of observing a detection history for a site (detections and non-detections) as a function of the occurrence probability (occupancy, Ψ) and detection probability (detectability, p). Maximum likelihood methods may be used to estimate these parameters [58].

We used an extension to the standard model above, developed by Hines et al. [54] that uses spatial replicates instead of temporal ones, and explicitly accounts for spatial auto-correlation of detections on contiguous replicates. This model has been applied successfully for surveys of other carnivores (see Karanth et al. [31] and Sunarto et al. [59] for tigers; Thorn et al. [60] for brown hyaenas *Hyaena brunnea*). The key parameters estimated by the Hines et al. [54] spatial dependence model are:

$\boldsymbol{\Psi}$ - probability of dhole presence in a site

$\boldsymbol{\theta^0}$ - probability of dhole presence in a replicate conditional on absence in the previous replicate

$\boldsymbol{\theta^1}$ - probability of dhole presence in a replicate conditional on presence in the previous replicate

$\boldsymbol{p_t}$ - probability of detecting dhole sign in a replicate conditional on presence in the replicate

The detection parameters $\boldsymbol{\theta^0}$ and $\boldsymbol{\theta^1}$ express the magnitude of spatial dependence between contiguous replicates ($\boldsymbol{\theta^0} = \boldsymbol{\theta^1}$ signifies complete independence of replicates). We use the subscripts 'L' and 's' with each of these parameters to indicate two spatial scales. Ψ_L refers to probability of dhole occurrence at landscape-scale and Ψ_s refers to probability of habitat-use by dholes at reserve-scale.

Covariates for distribution and habitat-use

Detections of signs of principal ungulate prey species were recorded on spatial replicates within each sampled grid cell. Because direct observations of anthropogenic disturbances are rare and difficult to quantify, we used indirect evidence, such as signs of livestock presence (e.g. tracks, dung, pellets, etc.) as reasonable surrogates of human disturbance factors (see Karanth et al. [31]). These covariates were quantified as a ratio of number replicates with indirect signs to the total number of replicates in each site.

We modeled site-specific probabilities of dhole occupancy Ψ_i as linear functions of the above mentioned covariates using a logit link function [28,40]:

$$Logit\Psi_i = \beta_0 + \beta_1 x_1 + \beta_2 x_2 + \ldots\ldots + \beta_n x_n$$

Table 1. Comparison between the standard occupancy model [28] and the model using spatial replicates with potential auto-correlation of sign detections among them [54] used to generate estimates of dhole habitat occupancy at landscape scale (Western Ghats, India, 2006–2007) and habitat-use at reserve scale (Bandipur Tiger Reserve, India, 2012).

	Model	AIC	ΔAIC	AIC weight	Model Likelihood	K*	Deviance
Landscape-scale occupancy	Ψ_L (.) θ^0_L (.) θ^1_L (.) $p_{t(L)}$ (.)	975.47	0	1	1	5	965.47
	Ψ_L (.) $p_{(L)}$ (.)	1005.91	30.44	0	0	3	999.91
Reserve-scale occupancy	Ψ_S (.) θ^0_S (.) θ^1_S (.) $p_{t(s)}$ (.)	991.27	0	1	1	5	981.27
	Ψ_S (.) p_S (.)	1508.67	517.4	0	0	3	1502.67

*K = number of parameters.

where β_i refers to the magnitude of influence of individual ecological or anthropogenic covariate x_i. All the site-specific covariates measured as proportions were converted to scaled and centered values [61]. We used model selection tests based on Akaike information criterion (AIC) implemented in program PRESENCE version 5.7 [62] to rank competing covariate models [63].

Results

Occupancy models to address spatial auto-correlation

When estimating occupancy at both landscape and reserve scales, we first compared the standard MacKenzie et al. [28] model with the Hines et al. [54] model that explicitly addresses the likely spatial auto-correlation of sign detections made along spatial replicates. These models fit the data better based on AIC values, than the standard model that assumes such sign detections are independent events (Table 1). Parameter estimates showed lack of independence strongly at both scales (Table 2).

Landscape-scale Occupancy

Detection probability of dhole signs. We initially modeled the replicate-level detection probability ($p_{t(L)}$) as a function of abundance of preferred prey (chital, sambar) and human disturbance (livestock presence) as plausible ecological and anthropogenic covariates, while maintaining a global model for the occupancy parameter [Ψ_L(*chital+sambar+livestock*)]. We estimated cell-specific detection probability $p_{t(L)}$ at landscape-scale as a function of combined abundance of all prey species (*chital+sambar+gaur+pig+muntjac*). We also used human disturbance (livestock sign abundance) as an additive variable expected to negatively influence dhole occupancy. The best fit model (AIC weight = 0.60) showed that $p_{t(L)}$ was a function of combined abundance of all prey. We used this model for all subsequent analyses (Table 3).

Probability of occurrence. We tested six models with different sets of plausible covariates against the basic model [here the basic model refers to Ψ_L (.), θ^0_L (.), θ^1_L (.), $p_{t(L)}$ (*allprey*)]. We used Akaike model weights to assess to strength of evidence in favor of each of these seven models (Table 4). Four models that include abundance of chital, sambar as positive influences and livestock signs as negative variables ranked higher in the candidate set (ΔAIC<2.0). We derived the maximum likelihood parameter estimates for occupancy at landscape-scale (Ψ_L) from models with Akaike weight >0.01 using model averaging [63].

Influence of covariates on dhole distribution. We had predicted that variations in distribution at the landscape-scale would be a function of ecological and anthropogenic variables. We further hypothesized that abundance of wild ungulate prey would positively influence dhole distribution, whereas human disturbance (indicated by livestock presence) would have a negative influence. Therefore, we examined the untransformed β coefficient values of these variables. The magnitude and direction (sign) on the β coefficient values for chital and sambar were positive, as we had expected (Table 5).

Human disturbance negatively influenced dhole distributions as indicated by the negative sign on β values. The model-specific β coefficients for all covariates at landscape-scale, along with model-averaged values are presented in Table 5. We summed the Akaike weights of key covariates in order to examine the relative influence of each covariate. Chital abundance was found to have the highest influence on landscape-scale occupancy (summed Akaike weight = 0.90).

Parameter estimates of dhole distribution. The final parameter estimates of dhole occupancy and replicate-level

Table 2. Estimates of dhole habitat occupancy at landscape-scale (Western Ghats, India, 2006–2007) and habitat-use at reserve-scale (Bandipur) generated using spatially replicated sign surveys under the model [54] incorporating spatial auto-correlation of sign detections.

	Naïve	Ψ (SE)	θ^0 (SE)	θ^1 (SE)	p_t (SE)
Landscape-scale occupancy	0.27	0.75 (0.32)	0.03 (0.09)	0.88 (0.04)	0.19 (0.03)
Reserve-scale occupancy	0.39	0.71 (0.12)	0.01 (0.00)	0.75 (0.04)	0.78 (0.06)

Please see Methods section for parameter descriptions.
Footnote: Here the parameters Ψ and p_t are estimated only for the basic model with no covariates. Final estimates of these parameters were derived from covariate models.

detection probabilities were derived from model averaging across the covariate models. The model-averaged estimate of probability of landscape-scale occupancy was $\hat{\Psi}_L$ (SE) = 0.67 (0.08).

We used the mean estimate of probability of occupancy at large scale to arrive at total proportion of the landscape occupied by dholes ($\hat{\Psi}_L$). For this, the cell-specific values of occupancy were obtained as a weighted product of probability of occupancy ($\hat{\Psi}_i$) and proportional area of forest cover in each cell (a_i/21,176).

$$\hat{\hat{\psi}}_L = \frac{\sum_{i=1}^{206} a_i \hat{\Psi}_i}{21{,}176}$$

The conventional presence-absence approach indicated dhole occupancy in the Western Ghats landscape to be 27% (*c.* 5,700 km^2). After accounting for imperfect detection [$\hat{p}_{t(L)}$(SE) = 0.12 (0.11)], Markovian dependence of replicates, and model–averaging the occupancy parameter, we estimate that dholes occupied 14,185 km^2 (68%) of the landscape [$\hat{\hat{\psi}}_L$(SE) = 0.68 (0.08)], which shows that the presence-versus-absence approach severely underestimated true occupancy (Fig. 3).

Reserve-scale Occupancy

Detection probability of dhole signs. At the reserve-scale, replicate level detectability was estimated from intensive surveys of the entire road-route network of Bandipur. This yielded a substantially high estimate of detection probability [$\hat{p}_{t(s)}$ (SE) = 0.78 (0.06)] even for the basic model [Ψ_s (.), θ^0_s (.), θ^1_s (.), $p_{t(s)}$ (.)]. Therefore we used constant detectability $p_{t(s)}$ (.) in modeling probability of habitat-use (Ψ_s).

Probability of habitat-use. We chose seven plausible covariate models and tested them against the basic model [here, the basic model refers to Ψ_s (.), θ^0_s (.), θ^1_s (.), $p_{t(s)}$ (.)] (Table 6). Highly cross-correlated variables were not included in the same model (for example: Pearson correlation coefficient $r_{chital \times sambar}$ = 0.52). We note that at this scale, the basic model without any covariates showed the best fit. However, the difference among AIC values across all models was low (ΔAIC<2.52) with all competing models having Akaike weights >0.01. Therefore, at this scale also we performed model averaging to estimate occupancy, defined as probability of habitat-use (Ψ_s) within the reserve.

Influence of covariates on dhole habitat-use. We had hypothesized that even at the reserve scale, abundance of preferred ungulate prey species would positively influence dhole habitat-use, and human disturbance would be a negative influence. Therefore, we examined the untransformed β coefficient values of these variables. At this scale, however, only sambar abundance positively influenced dhole habitat-use but chital abundance as well as that of all prey combined had little effect.

Human disturbance negatively influenced probability of habitat-use as shown by the negative sign on β values. The model-specific β coefficients for covariates at reserve-scale, along with model-averaged values are in Table 7. As in the case of the large-scale, we summed the Akaike weights of key variables in order to examine the relative influence of each covariate. Dhole habitat-use pattern within the reserve was most significantly influenced by human disturbance (summed Akaike weight = 0.42).

Parameter estimates of dhole habitat-use. The naïve 'presence-versus-absence' estimate suggested that dholes used about 39% of Bandipur National Park. We estimated true probability of use to be $\hat{\Psi}_s$ (SE) = 0.71 (0.06) and detectability $\hat{p}_{t(s)}$ (SE) = 0.78 (0.06). Site-specific variations in Ψ_s calculated from averaging across all covariate models, show a matrix of high and low habitat-use probabilities (Fig. 4). These interspersed high

Table 3. Results of comparisons to select models for estimating probability of detecting dhole signs p_t on 1-km long spatial replicates used in the field survey conducted at landscape scale in the Western Ghats landscape, India (2006–2007), under the constant global model for dhole occurrence [Ψ_L (*chital+sambar+livestock)*].

Model	AIC	ΔAIC	AIC weight	Model Likelihood	K*	Deviance
Ψ_L (global) θ^0_L (.) θ^1_L (.) $p_{t(L)}$ (allprey)	966.33	0	0.6031	1	6	954.33
Ψ_L (global) θ^0_L (.) θ^1_L(.) $p_{t(L)}$(allprey+livestock)	967.21	0.88	0.3884	0.644	7	953.21
Ψ_L (global) θ^0_L (.) θ^1_L (.) $p_{t(L)}$ (.)	975.47	9.14	0.0062	0.0104	5	965.47
Ψ_L (global) θ^0_L (.) θ^1_L (.) $p_{t(L)}$ (livestock)	977.47	11.14	0.0023	0.0038	6	965.47

*K = number of parameters.

Table 4. Model comparisons to identify ecological and anthropogenic habitat covariates influencing dhole distribution Ψ_L at landscape-scale (Western Ghats, India, 2006–2007) from spatially replicated sign surveys during 2006–2007.

Model	AIC	ΔAIC	AIC weight	Model Likelihood	K*	Deviance
Ψ_L (chital+livestock) θ^0_L (.) θ^1_L (.) $p_{t(L)}$ (allprey)	961.98	0	0.2958	1	8	945.98
Ψ_L (chital) θ^0_L (.) θ^1_L (.) $p_{t(L)}$ (allprey)	962.02	0.04	0.2899	0.9802	7	948.02
Ψ_L (chital+sambar) θ^0_L (.) θ^1_L (.) $p_{t(L)}$ (allprey)	962.89	0.91	0.1876	0.6344	8	946.89
Ψ_L (chital+sambar+livestock) θ^0_L (.) θ^1_L (.) $p_{t(L)}$ (allprey)	963.62	1.64	0.1303	0.4404	9	945.62
Ψ_L (.) θ^0_L (.) θ^1_L (.) $p_{t(L)}$ (allprey)	966.33	4.35	0.0336	0.1136	6	954.33
Ψ_L (livestock) θ^0_L (.) θ^1_L (.) $p_{t(L)}$ (allprey)	966.44	4.46	0.0318	0.1075	7	952.44
Ψ_L (sambar) θ^0_L (.) θ^1_L (.) $p_{t(L)}$ (allprey)	966.49	4.51	0.031	0.1049	7	952.49

*K = number of parameters.

and low habitat-use patches suggest that even within an occupied wildlife reserve, dholes selectively use certain habitats more than others.

Discussion

Methodological Issues

Animal sign surveys are efficient because they do not require technically trained field personnel or advanced equipment. Consequently, they have wide applicability in assessing animal distributions at large spatial scales [28,37]. Surveys using an occupancy modeling framework can estimate and predict species distributions at single population or meta-population levels, scaling up from individual reserves to larger landscapes or regions.

In our study, as expected, a traditional presence-versus-absence approach underestimated dhole occupancy by 8682 km^2 (60%). Analysis under the approach used by Hines et al. [54] and Karanth et al. [31] for surveying tigers overcame this bias and generated a reasonable estimate of landscape occupancy at 68% (14,185 km^2). The models we used partitioned the observation process [52] into components of detection probability at the replicate level and cell level, thus fully addressing the twin problems of imperfect detection as well as autocorrelation of sign detections (Table 2). The superiority of this approach was clearly demonstrated at both reserve scale and landscape scale, with the models clearly fitting the data better (Tables 1 and 2). We believe this approach is superior to other predictive distribution assessments that use only data on presence [44,64,75], because it uses information built into absence data that is traditionally discarded. Furthermore, our study adapted the temporal replication-based model to one that is based on spatial replications, which is logistically more efficient in practice [31].

Although absolute abundance is a more sensitive metric for managing endangered wildlife populations, this parameter is difficult and expensive to derive at large spatial scales [51,65]. Since dholes do not have distinct morphological traits (body/pelage markings), generating simultaneous estimates of distributions at large scales and habitat-use patterns at reserve scale may be of great value because abundance estimation methods that rely on individual identification from photographic captures [66] cannot be used. In the future, we anticipate development of models that can integrate landscape-scale occupancy data with reserve-scale dhole abundances derived using potentially useful methods such as genetic capture-recapture sampling [38]. While we have tried to reduce biases in our estimates from misclassification of indirect signs (incorrect identification of tracks and scats from other wild carnivores or feral/domestic dogs), we do not discount the possibility of potential errors. Recent developments in occupancy modeling have addressed these issues [55], though

Table 5. Estimates of β coefficient values for different individual habitat covariates hypothesized to influence dhole distribution Ψ_L at landscape scale (Western Ghats, India, 2006–2007).

Model	$\hat{\beta}_{Intercept}$ (SE)	$\hat{\beta}_{chital}$ (SE)	$\hat{\beta}_{sambar}$ (SE)	$\hat{\beta}_{livestock}$ (SE)
Ψ_L (chital+livestock) θ^0_L (.) θ^1_L (.) $p_{t(L)}$ (allprey)	1.38 (1.11)	4.52 (3.01)	-	−1.18 (0.94)
Ψ_L (chital) θ^0_L (.) θ^1_L (.) $p_{t(L)}$ (allprey)	1.26 (1.04)	4.05 (2.81)	-	-
Ψ_L (chital+sambar) θ^0_L (.) θ^1_L (.) $p_{t(L)}$ (allprey)	1.57 (1.30)	4.84 (3.41)	0.99 (1.29)	-
Ψ_L (chital+sambar+livestock) θ^0_L (.) θ^1_L (.) $p_{t(L)}$ (allprey)	1.50 (1.17)	4.85 (3.20)	0.56 (1.06)	−1.00 (0.94)
Ψ_L (.) θ^0_L (.) θ^1_L (.) $p_{t(L)}$ (allprey)	1.21 (1.46)	-	-	-
Ψ_L (livestock) θ^0_L (.) θ^1_L (.) $p_{t(L)}$ (allprey)	1.50 (2.11)	-	-	−1.63 (2.30)
Ψ_L (sambar) θ^0_L (.) θ^1_L (.) $p_{t(L)}$ (allprey)	2.27 (4.11)	-	4.47 (9.48)	-
Model averaged values	**1.42 (1.28)**	**4.05 (2.8)**	**0.40 (0.73)**	**−0.53 (0.56)**

These include model-averaged β estimates with unconditional standard errors.

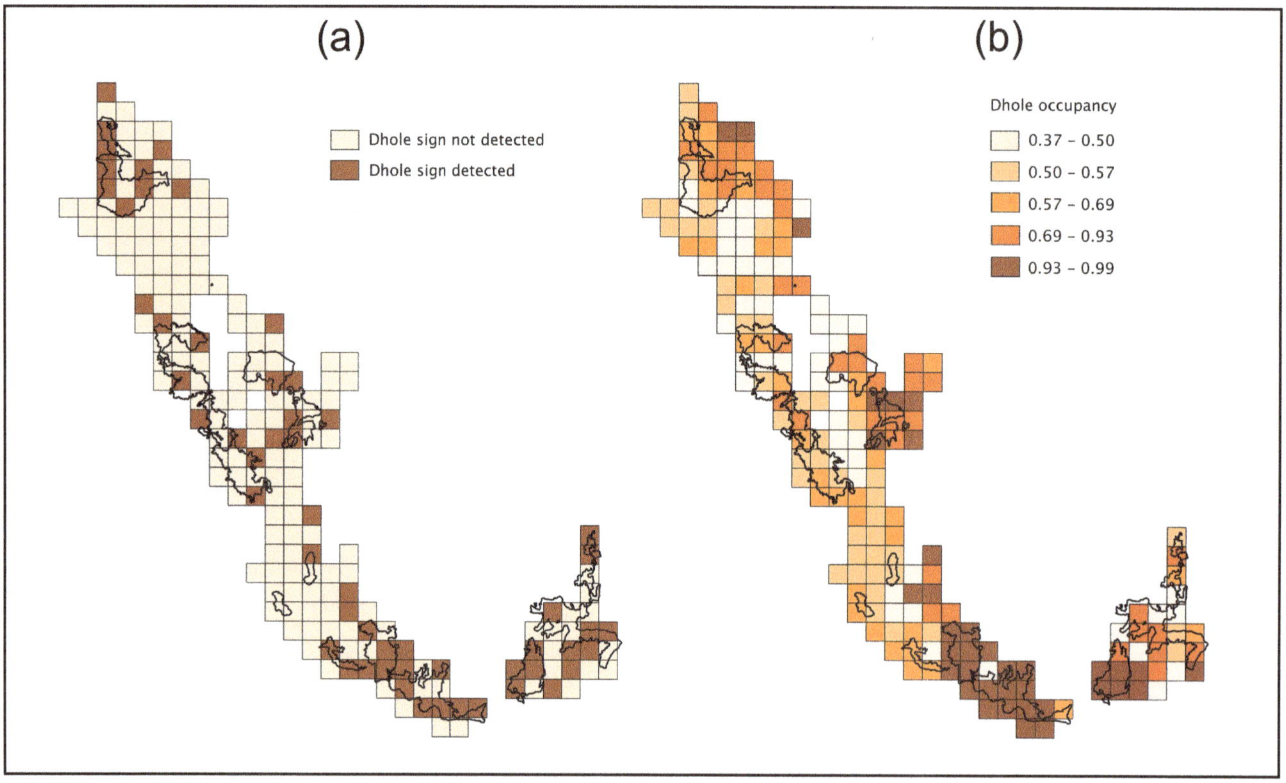

Figure 3. Dhole distribution patterns. Patterns of landscape-scale occupancy: dhole distribution in the Western Ghats of Karnataka, India (2006–2007). (a) Naïve estimate and (b) estimated probabilities of occupancy.

currently there are no methods that integrate spatial dependence models [54] with false-positive error models.

Ecological interactions, habitat characteristics and dhole occupancy

The interactions between a species and its habitat(s) may occur at multiple scales [67]. Our results reaffirm the importance of choosing the appropriate spatial scale for assessing species-habitat relationships. At the landscape scale, surveys of tigers that used cells (sites) much larger than expected home range size yielded reliable inferences on true occupancy [31,54] and possibly even relative abundance [68]. In our study, at the landscape-scale we could reliably estimate proportion of potential habitat occupied by dholes. The influence of ecological and anthropogenic covariates at the cell (site) level on dhole distribution was generally in conformity with our predictions (Table 4). Abundance of chital emerged as the most important ecological driver in determining dhole distribution. However, distribution of chital in the landscape was limited to deciduous forests with flatter terrain [69, this study]. Wherever chital were absent, sambar appears to be principal prey species influencing dhole occurrence (Table 5).

'Probability of occupancy' (using large cells) is not a very useful metric at the level of individual reserves where dholes are present. The parameter 'probability of habitat-use', which measures intensity with which dholes use different areas within a reserve (Bandipur) is more useful for identifying important habitats for dholes. Estimates of probability of habitat-use in our study showed how factors influencing species-habitat relationships are scale dependent. For example, chital abundance, a key covariate in determining distribution of dholes at landscape scale, does not appear to affect habitat-use patterns of dholes at reserve scale, as does sambar abundance (Table 7). We note that chital occur in large herds (up to 81 animals/group) compared to sambar (up to 6 animals/group) [70]. Because chital occur in almost every site (13 km^2 cell in our study), the relationship between their abundance and habitat occupancy breaks down at smaller scales [71–73]. Chital attain high densities in all surveyed cells within Bandipur [74] and therefore their abundance does not appear to influence fine-scale habitat-use by dholes. We further examined this issue *post hoc* by including combined abundance of all ungulate prey as a covariate in the reserve-scale model comparisons (Table 6). The combined prey abundance also had no influence on patterns of habitat-use by dholes (Table 7), confirming that cell-specific abundance of select prey species is a strictly scale-dependent factor in determining dhole occupancy. On the other hand, anthropogenic disturbance (typified by livestock presence in our study) clearly and negatively influences dhole distribution at landscape scale (Table 5) as well as dhole habitat-use at the reserve scale (Table 7).

Carnivore densities and other demographic parameters are chiefly influenced by prey abundance [32,33,75–77]. Our results highlight the importance of medium to large-sized prey for driving dhole occupancy across the landscape, in concordance with other studies of dhole feeding ecology in the Western Ghats and across Asia [12,13,19,20,78].

Conservation Implications

Landscape-scale estimates of dhole occupancy presented in this paper show that dholes are not strictly confined to protected wildlife reserves as is generally perceived (Fig. 3). Dholes occupy about 14,185 km^2 (68%) of the Karnataka Western Ghats

Table 6. Model comparisons to identify ecological and anthropogenic habitat covariates influencing dhole habitat-use Ψ_s at reserve scale (Bandipur Tiger Reserve, India) from spatially replicated sign surveys during January-April 2012.

Model	AIC	ΔAIC	AIC weight	Model Likelihood	K*	Deviance
Ψ_s (.) θ^0_s (.) θ^1_s (.) $p_{t(s)}$ (.)	991.27	0	0.2406	1	5	981.27
Ψ_s (livestock) θ^0_s (.) θ^1_s (.) $p_{t(s)}$ (.)	991.81	0.54	0.1837	0.7634	6	979.81
Ψ_s (sambar) θ^0_s (.) θ^1_s (.) $p_{t(s)}$ (.)	992.20	0.93	0.1511	0.6281	6	980.20
Ψ_s (sambar+livestock) θ^0_s (.) θ^1_s (.) $p_{t(s)}$ (.)	993.02	1.75	0.1003	0.4169	7	979.02
Ψ_s (allprey) θ^0_s (.) θ^1_s (.) $p_{t(s)}$ (.)	993.12	1.85	0.0954	0.3965	6	981.12
Ψ_s (chital) θ^0_s (.) θ^1_s (.) $p_{t(s)}$ (.)	993.23	1.96	0.0903	0.3753	6	981.23
Ψ_s (allprey+livestock) θ^0_s (.) θ^1_s (.) $p_{t(s)}$ (.)	993.73	2.46	0.0703	0.2923	7	979.73
Ψ_s (chital+livestock) θ^0_s (.) θ^1_s (.) $p_{t(s)}$ (.)	993.79	2.52	0.0682	0.2837	7	979.79

*K = number of parameters.

landscape, of which only *c.* 8700 km^2 (41%) is in protected wildlife reserves.

A meta-analysis by Woodroffe and Ginsberg [79] predicted 723 km^2 as the lower reserve size threshold for persistence of dhole populations in India. With the average reserve size being 570 km^2 in our study area, dhole populations seem to thrive in smaller reserves or even outside the reserve system, in multi-use forests. This is possibly enabled by larger populations in reserves like Bandipur-Nagarahole that serve as 'sources' in sustaining a dhole meta-population in the Western Ghats landscape of Karnataka. Meta-populations comprising of multiple small populations of animals are more likely to persist in landscapes with multiple high quality habitat patches with good connectivity [80] as in the Western Ghats. Therefore, retention of landscape connectivity beyond well-protected wildlife reserves should receive greater attention in this region.

Unlike tigers and leopards in this landscape, dholes are not involved in significant levels of conflict with humans [81] or targeted for illegal trade of body parts [82]. However, their persistence outside protected reserves is problematic, because of depleted prey densities [16,33] and potential risk of disease from large populations of semi-feral dogs and cats [83]. The Indian government is currently promoting voluntary village relocations from protected wildlife habitats as a strategy for conflict mitigation and improving social welfare [84,85]. This could be a very effective indirect tool for ensuring dhole population viability. However, with rapid economic growth in the Western Ghats, many infrastructure projects involving highways, pipelines, dams, canals and power lines are likely to increase habitat fragmentation and subsequently impact dhole populations. We believe that our results may be very useful to mitigate the impact of such projects, if considered seriously while planning regional infrastructure development. Voluntary relocation projects and establishment of ecologically sensitive zones [86] may also benefit from considering spatial distribution patterns of dhole populations reported in this study.

At the level of a single reserve (Bandipur), the available prey abundance appears to be more than adequate to support current densities of dholes [33,74]. Abundance index of chital and combined prey species measured in our study show presence in almost all grids with very high relative abundance across grids. Current management practices include habitat manipulations in the form of creating water holes and artificially increasing forage availability to increase herbivore populations. Such targeted

Figure 4. Dhole habitat-use patterns. Patterns of reserve-scale occupancy: habitat-use by dholes in Bandipur Tiger Reserve, India (2012). (a) Naïve estimate from presence-versus-absence approach and (b) estimated probabilities of occupancy.

Table 7. Estimates of β coefficient values for different individual habitat covariates hypothesized to influence dhole habitat-use Ψ_s at reserve scale (Bandipur Tiger Reserve, India, 2012).

Model	$\hat{\beta}_{Intercept}$ (SE)	$\hat{\beta}_{chital}$ (SE)	$\hat{\beta}_{sambar}$ (SE)	$\hat{\beta}_{allprey}$ (SE)	$\hat{\beta}_{livestock}$ (SE)
Ψ_s (.) θ^0_s (.) θ^1_s (.) $p_{t(s)}$ (.)	0.88 (0.57)	-	-	-	-
Ψ_s (livestock) θ^0_s (.) θ^1_s (.) $p_{t(s)}$ (.)	0.85 (0.58)	-	-	-	−0.96 (0.76)
Ψ_s (sambar) θ^0_s (.) θ^1_s (.) $p_{t(s)}$ (.)	1.11 (0.89)	-	1.35 (1.83)	-	-
Ψ_s (sambar+livestock) θ^0_s (.) θ^1_s (.) $p_{t(s)}$ (.)	1.00 (0.75)	-	1.09 (1.53)	-	−0.93 (0.78)
Ψ_s (allprey) θ^0_s (.) θ^1_s (.) $p_{t(s)}$ (.)	0.92 (0.61)	-	-	0.29 (0.78)	-
Ψ_s (chital) θ^0_s (.) θ^1_s (.) $p_{t(s)}$ (.)	0.90 (0.60)	0.17 (0.92)	-	-	-
Ψ_s (allprey+livestock) θ^0_s (.) θ^1_s (.) $p_{t(s)}$ (.)	0.88 (0.62)	-	-	0.22 (0.80)	−0.94 (0.76)
Ψ_s (chital+livestock) θ^0_s (.) θ^1_s (.) $p_{t(s)}$ (.)	0.89 (0.61)	0.12 (0.93)	-	-	−0.96 (0.76)
Model averaged values	**0.96 (0.68)**	**0.02 (0.15)**	**0.32 (0.50)**	**0.05 (0.16)**	**−0.43 (0.42)**

These include model-averaged β estimates with unconditional standard errors.

habitat manipulations to further increase prey abundance, particularly that of chital which are at excessive densities (>50–100 animals/km^2) in some locations [74], therefore seem unnecessary. In view of the significant negative influence of human disturbance on dhole occupancy, in the landscape as well as within the designated wildlife reserve, conservation resources should be focused on strengthening protection and patrolling to reduce such impacts.

With major declines in carnivore numbers globally and the data-deficient status of most dhole populations [1,3], our findings have potential utility for conservation of dholes across their geographic range. While we reliably and rigorously identified key ecological drivers of dhole occupancy at multiple spatial scales, we believe our approach has wider application for assessing distribution and habitat-use patterns in other rare, elusive and wide-ranging carnivore species.

Acknowledgments

We acknowledge Wildlife Conservation Society (New York), Wildlife Conservation Society-India, Centre for Wildlife Studies and the National Centre for Biological Sciences for supporting the research. We are grateful to the Karnataka State Forest Department for providing research permits to carry out the study. We thank Arjun M. Gopalaswamy for analytical support, Dr. Ajith Kumar for useful discussions and Srinivas Vaidyanathan for assistance with GIS analysis. We thank all the research assistants, field assistants and volunteers who helped in collecting and processing field data.

Author Contributions

Conceived and designed the experiments: AS KKK NSK KUK. Performed the experiments: AS NSK. Analyzed the data: AS. Contributed reagents/materials/analysis tools: AS KKK DJ NSK KUK. Wrote the paper: AS KKK KUK.

References

1. Ripple WJ, Estes JA, Beschta RL, Wilmers CC, Ritchie EG, et al. (2014) Status and ecological effects of the world's largest carnivores. Science 343: 1241484–1241484. DOI: 10.1126/science.1241484.
2. Ceballos G, Ehrlich PR (2002) Mammal population losses and the extinction crisis. Science 296: 904–907. DOI: 10.1126/science.1069349.
3. Cardillo M, Purvis A, Bielby J, Mace GM, Sechrest W, et al. (2004) Human population density and extinction risk in the world's carnivores. PLoS Biol 2: 0909–0914. DOI: 10.1371/journal.pbio.0020197.
4. Ceballos G, Ehrlich PR, Soberón J, Salazar I, Fay JP (2005) Global mammal conservation: what must we manage? Science 309: 603–607. DOI: 10.1126/science.1114015.
5. Wikramanayake ED, Dinerstein E, Robinson JG, Karanth U, Rabinowitz A, et al. (1998) An ecology-based method for defining priorities for large mammal conservation: The tiger as case study. Conserv Biol 12: 865–878. DOI: 10.1111/j.1755-263X.2010.00162.x.
6. Ceballos G, Ehrlich PR (2006) Global mammal distributions, biodiversity hotspots, and conservation. Proc Natl Acad Sci USA 103: 19374–19379. DOI: 10.1073/pnas.0609334103.
7. Morrison JC, Sechrest W, Dinerstein E, Wilcove DS, Lamoreux JF (2007) Persistence of large mammal faunas as indicators of global human impacts. J Mammal 88: 1363–1380. DOI: 10.1644/06-MAMM-A-124R2.1.
8. Athreya V, Odden M, Linnell JDC, Krishnaswamy J, Karanth KU (2013) Big cats in our backyards: persistence of large carnivores in a human dominated landscape in India. PLoS One 8: e57872. DOI: 10.1371/journal.pone.0057872.
9. Treves A, Karanth KU (2003) Human-carnivore conflict and perspectives on carnivore management worldwide. Conserv Biol 17: 1491–1499. DOI: 10.1111/j.1523-1739.2003.00059.x.
10. Sillero-Zubiri C, Hoffmann M, Macdonald DW (2004) Canids: Foxes, Wolves, Jackals and Dogs. Status survey and conservation action plan. IUCN, Gland, Switzerland and Cambridge, United Kingdom.
11. Karanth KU, Sunquist ME (2000) Behavioural correlates of predation by tiger (*Panthera tigris*), leopard (*Panthera pardus*) and dhole (*Cuon alpinus*) in Nagarahole, India. J Zool 250: 255–265. DOI: 10.1111/j.1469-7998.2000.tb01076.x.
12. Grassman LI, Tewes ME, Silvy NJ, Kreetiyutanont K (2005) Spatial ecology and diet of the dhole *Cuon alpinus* (Canidae, Carnivora) in north central Thailand. Mammalia 69: 11–20. DOI: 10.1515/mamm.2005.002.
13. Kamler JF, Johnson A, Vongkhamheng C, Bousa A (2012) The diet, prey selection, and activity of dholes (*Cuon alpinus*) in northern Laos. J Mammal 93: 627–633. DOI: 10.1644/11-MAMM-A-241.1.
14. Durbin LS, Hedges S, Duckworth W, Tyson M, Lyenga A, Venkataraman A (2008) Dhole *Cuon alpinus*. In: IUCN 2010.IUCN Red List of Threatened Species, Gland, Switzerland. Accessed from http://www.iucnredlist.org/apps/redlist/details/5953/0. Accessed 2014 January 30.
15. Cohen JA (1978) *Cuon alpinus*. Mamm Species 100: 1–3.
16. Karanth KK, Nichols JD, Hines JE, Karanth KU, Christensen NL (2009) Patterns and determinants of mammal species occurrence in India. J Appl Ecol: 1189–1200. DOI:10.1111/j.1365-2664.2009.01710.x.
17. Karanth KK, Nichols JD, Karanth KU, Hines JE, Christensen NL (2010) The shrinking ark: patterns of large mammal extinctions in India. Proc Biol Sci 277: 1971–1979. DOI: 10.1098/rspb.2010.0171.
18. Karanth KU, Sunquist ME (1995) Prey selection by tiger, leopard and dhole in tropical forests. J Anim Ecol 64: 439.
19. Andheria AP, Karanth KU, Kumar NS (2007) Diet and prey profiles of three sympatric large carnivores in Bandipur Tiger Reserve, India. J Zool 273: 169–175. DOI: 10.1111/j.1469-7998.2007.00310.x.
20. Kawanishi K, Sunquist ME (2008) Food habits and activity patterns of the Asiatic golden cat (*Catopuma temminckii*) and dhole (*Cuon alpinus*) in a primary rainforest of Peninsular Malaysia. Mammal Study 33: 173–177. DOI: http://dx.doi.org/10.3106/1348-6160-33.4.173.
21. Thinley P, Kamler JF, Wang SW, Lham K, Stenkewitz U, et al. (2011) Seasonal diet of dholes (*Cuon alpinus*) in northwestern Bhutan. Mamm Biol 76: 518–520. DOI: 10.1016/j.mambio.2011.02.003.

22. Johnsingh AJT (1982) Reproduction and social behaviour of the dhole, *Cuon alpinus* (Canidae). J Zool 198: 443–463.
23. Venkataraman AB, Arumugam R, Sukumar R (1995) The foraging ecology of dhole (*Cuon alpinus*) in Mudumalai Sanctuary, southern India. J Zool 237: 543–561. DOI: 10.1111/j.1469-7998.1995.tb05014.x.
24. Karanth KU, Sunquist ME (2000) Behavioural correlates of predation by tiger (*Panthera tigris*), leopard (*Panthera pardus*) and dhole (*Cuon alpinus*) in Nagarahole, India. J Zool 250: 255–265. DOI: 10.1017/S0952836900002119.
25. Hartley S, Kunin WE (2003) Scale dependency of rarity, extinction risk, and conservation priority. Conserv Biol 17: 1559–1570. DOI: 10.1111/j.1523-1739.2003.00015.x.
26. MacKenzie DI, Nichols JD, Hines JE, Knutson MG, Franklin AB (2003) Estimating site occupancy, colonization, and local extinction when a species is detected imperfectly. Ecology 84: 2200–2207. DOI: 10.1890/02-3090.
27. Royle JA (2006) Site occupancy models with heterogeneous detection probabilities. Biometrics 62: 97–102. DOI: 10.1111/j.1541-0420.2005.00439.x.
28. MacKenzie DI, Nichols JD, Lachman GB, Droege S, Royle JA, et al. (2002) Estimating site occupancy rates when detection probabilities are less than one. Ecology 83: 2248–2255. DOI: 10.1890/0012-9658(2002)083[2248:ESORWD]2.0.CO;2.
29. Linkie M, Chapron G, Martyr DJ, Holden J, Leader Williams N (2006) Assessing the viability of tiger subpopulations in a fragmented landscape. J Appl Ecol 43: 576–586. DOI: 10.1111/j.1365-2664.2006.01153.x.
30. Lyra-Jorge M, Ciocheti G, Pivello V (2008) Carnivore mammals in a fragmented landscape in northeast of Sao Paulo State, Brazil. Biodivers Conserv 17: 1573–1580. DOI: 10.1007/s10344-009-0324-x.
31. Karanth KU, Gopalaswamy AM, Kumar NS, Vaidyanathan S, Nichols JD, et al. (2011) Monitoring carnivore populations at the landscape scale: occupancy modelling of tigers from sign surveys. J Appl Ecol 48: 1048–1056. DOI: 10.1111/j.1365-2664.2011.02002.x.
32. Carbone C, Gittleman JL (2002) A common rule for the scaling of carnivore density. Science 295: 2273–2276. DOI: 10.1126/science.1067994.
33. Karanth KU, Nichols JD, Kumar NS, Link WA, Hines JE (2004) Tigers and their prey: Predicting carnivore densities from prey abundance. PNAS 101: 4854–4858. DOI: 10.1073/pnas.0306210101.
34. Das A, Krishnaswamy J, Bawa KS, Kiran MC, Srinivas V, Kumar NS, et al. (2006) Prioritization of conservation areas in the Western Ghats, India. Biological Conservation, 133, 16–31. DOI: 10.1016/j.biocon.2006.05.023.
35. Kodandapani N, Cochrane MA, Sukumar R (2008) A comparative analysis of spatial, temporal, and ecological characteristics of forest fires in seasonally dry tropical ecosystems in the Western Ghats, India. For Ecol Manage 256: 607–617. DOI: 10.1016/j.foreco.2008.05.006.
36. Karanth KU, Nichols JD (1998) Estimation of tiger densities in India using photographic captures and recaptures. Ecology 79: 2852–2862. DOI: 10.1890/0012-9658(1998)079[2852:EOTDII]2.0.CO;2.
37. Stanley TR, Royle JA (2005) Estimating site occupancy and abundance using indirect detection indices. J Wildl Manage 69: 874–883. DOI: 10.2193/0022-541X(2005)069[0874:ESOAAU]2.0.CO;2.
38. Mondol S, Karanth KU, Kumar NS, Gopalaswamy A, Andheria A, et al. (2009) Evaluation of non-invasive genetic sampling methods for estimating tiger population size. Biol Conserv 142: 2350–2360. DOI: 10.1016/j.biocon.2009.05.014.
39. Long RA, Donovan TM, MacKay P, Zielinski WJ, Buzas JS (2011) Predicting carnivore occurrence with noninvasive surveys and occupancy modeling. Landsc Ecol 26: 327–340. DOI: 10.1007/s10980-010-9547-1.
40. Tyre AJ, Tenhumberg B, Field SA, Niejalke D, Parris K, et al. (2003) Improving precision and reducing bias in biological surveys: Estimating false-negative error rates. Ecol Appl 13: 1790–1801. DOI: 10.1890/02-5078.
41. Gu W, Swihart RK (2004) Absent or undetected? Effects of non-detection of species occurrence on wildlife–habitat models. Biol Conserv 116: 195–203. DOI: 10.1016/S0006-3207(03)00190-3.
42. Hirzel AH, Hausser J, Chessel D, Perrin N (2002) Ecological-Niche Factor Analysis: How to compute habitat-suitability maps without absence data? Ecology 83: 2027–2036. DOI: 10.1890/0012-9658(2002)083[2027:ENFAHT]2.0.CO;2.
43. Jones JPG (2011) Monitoring species abundance and distribution at the landscape scale. J Appl Ecol 48: 9–13. DOI: 10.1111/j.1365-2664.2010.01917.x.
44. Yackulic CB, Chandler R, Zipkin EF, Royle JA, Nichols JD, et al. (2013) Presence-only modelling using MAXENT: when can we trust the inferences? Methods Ecol Evol 4: 236–243. DOI: 10.1111/2041-210x.12004.
45. O'Connell AF, Talancy NW, Bailey LL, Sauer JR, Cook R, et al. (2006) Estimating site occupancy and detection probability parameters for meso- and large mammals in a coastal ecosystem. J Wildl Manage 70: 1625–1633. DOI: 10.2193/0022-541X(2006)70[1625:ESOADP]2.0.CO;2.
46. Sargeant GA, Sovada MA, Slivinski MA, Johnson DH (2005) Markov Chain Monte-Carlo estimation of species distributions: A case study of the swift fox in western Kansas. J Wildl Manage 69: 483–497. DOI: 10.2193/0022-541X(2005)069[0483:MCMCEO]2.0.CO;2.
47. Magoun A, Ray J, Johnson D, Valkenburg P, Dawson F, et al. (2007) Modeling wolverine occurrence using aerial surveys of tracks in snow. J Wildl Manage 71: 2221–2229. DOI: 10.2193/2006-372.
48. Mackenzie DI (2006) Modeling the Probability of Resource Use: The effect of, and dealing with, detecting a species imperfectly. J Wildl Manage 70: 367–374. DOI: 10.2193/0022-541X(2006)70[367:MTPORU]2.0.CO;2.
49. Nichols J, Bailey L, O'Connell Jr A, Talancy N, Campbell Grant E, et al. (2008) Multi-scale occupancy estimation and modelling using multiple detection methods. J Appl Ecol 45: 1321–1329. DOI: 10.1111/j.1365-2664.2008.01509.x.
50. Mordecai RS, Mattsson BJ, Tzilkowski CJ, Cooper RJ (2011) Addressing challenges when studying mobile or episodic species: hierarchical Bayes estimation of occupancy and use. J Appl Ecol 48: 56–66. DOI: 10.1111/j.1365-2664.2010.01921.x.
51. Williams BK, Nichols JD, Conroy MJ (2002). Analysis and management of animal populations. California: Academic Press, San Diego.
52. Royle JA, Dorazio RM (2008). Hierarchical modeling and inference in ecology: The analysis of data from populations, metapopulations, and communities. Academic Press, San Diego, California.
53. Kendall WL, White GC (2009) A cautionary note on substituting spatial subunits for repeated temporal sampling in studies of site occupancy. J Appl Ecol 46: 1182–1188. DOI: 10.1111/j.1365-2664.2009.01732.x.
54. Hines JE, Nichols JD, Royle JA, MacKenzie DI, Gopalaswamy AM, et al. (2010) Tigers on trails: occupancy modeling for cluster sampling. Ecol Appl 20: 1456–1466. DOI: 10.1890/09-0321.1.
55. Miller DA, Nichols JD, McClintock BT, Campbell Grant EH, Bailey LL, et al. (2011) Improving occupancy estimation when two types of observational error occur: non-detection and species misidentification. Ecology 92: 1422–1428.
56. Rhodes JR, Lunney D, Moon C, Matthews A, McAlpine CA (2011) The consequences of using indirect signs that decay to determine species' occupancy. Ecography (Cop) 34: 141–150. DOI: 10.1111/j.1600-0587.2010.05908.x.
57. Acharya BB, Johnsingh AJT, Sankar K (2010) Dhole telemetry studies in Pench Tiger Reserve, central India. Telem Wildl Sci 13: 69–79.
58. MacKenzie DI, Nichols JD, Royle JA, Pollock KH, Hines JE, et al. (2006) Occupancy estimation and modeling: Inferring patterns and dynamics of species occurrence. Elsevier, San Diego.
59. Sunarto S, Kelly MJ, Parakkasi K, Klenzendorf S, Septayuda E, et al. (2012) Tigers need cover: Multi-scale occupancy study of the big cat in Sumatran forest and plantation landscapes. PLoS One 7: e30859. DOI: 10.1371/journal.pone.0030859.
60. Thorn M, Green M, Bateman PW, Waite S, Scott DM (2011) Brown hyaenas on roads: Estimating carnivore occupancy and abundance using spatially auto-correlated sign survey replicates. Biol Conserv 144: 1799–1807.
61. Gelman A (2008) Scaling regression inputs by dividing by two standard deviations: 2865–2873. DOI:10.1002/sim.
62. Hines JE (2006) PRESENCE – Software to estimate patch occupancy and related parameters. Version 5.7. USGS-PWRC. Accessed from http://www.mbr-pwrc.usgs.gov/software/presence.html. Accessed 2013 September 10.
63. Burnham KP, Anderson DR (2002) Model selection and multimodel inference: A practical information-theoretic approach. New York: Springer-Verlag.
64. Jenks KE, Kitamura S, Lynam AJ, Ngoprasert D, Chutipong W, et al. (2012) Mapping the distribution of dholes, *Cuon alpinus* (Canidae, Carnivora), in Thailand. Mammalia 76: 175–184.
65. Witmer GW (2005) Wildlife population monitoring: some practical considerations. Wildl Res 32: 259–263.
66. O'Connell AF, Nichols JD, Karanth KU (2011) Camera traps in animal ecology: Methods and analyses. Tokyo: Springer.
67. Elith J, Leathwick JR (2009) Species distribution models: ecological explanation and prediction across space and time. Annu Rev Ecol Evol Syst 40: 677–697. DOI: 10.1146/annurev.ecolsys.110308.120159.
68. Wibisono HT, Linkie M, Guillera-Arroita G, Smith JA, Sunarto, et al. (2011) Population status of a cryptic top predator: an island-wide assessment of tigers in Sumatran rainforests. PLoS One 6: e25931. DOI: 10.1371/journal.pone.0025931.
69. Karanth KU (1986) Status of wildlife and habitat conservation in Karnataka. J - Bombay Nat Hist Soc 83: 166–179.
70. Karanth KU, Sunquist ME (1992) Population structure, density and biomass of large herbivores in the tropical forests of Nagarahole, India. J Trop Ecol 8: 21–35.
71. Gaston KJ, Blackburn TM, Greenwood JJD, Gregory RD, Quinn RM, et al. (2000) Abundance - occupancy relationships. J Appl Ecol 37: 39–59.
72. Freckleton RP, Gill JA, Noble D, Watkinson AR (2005) Large-scale population dynamics, abundance-occupancy relationships and the scaling from local to regional population size. J Anim Ecol 74: 353–364. DOI: 10.1111/j.1365-2656.2005.00931.x.
73. He F, Condit R (2007) The distribution of species: occupancy, scale, and rarity. In: Storch D, Marquet PA, Brown JH, editors. Scaling biodiversity. Cambridge University Press. 32–50 p.
74. Kumar NS (2010) PhD thesis. Assessment of distribution and abundance of ungulate prey using spatial models in Nagarahole and Bandipur Tiger Reserves, India. Manipal University, India.
75. Royle JA, Chandler RB, Yackulic C, Nichols JD (2012) Likelihood analysis of species occurrence probability from presence-only data for modelling species distributions. Methods Ecol Evol 3: 545–554. DOI: 10.1111/j.2041-210X.2011.00182.x.
76. Trainor AM, Schmitz OJ, Ivan J, Shenk TM (2013) Enhancing species distribution modeling by characterizing predator-prey interactions. Ecol Appl: 130710123503006. DOI: 10.1890/13-0336.1.

77. Williams TM, Estes JA, Doak DF, Springer AM (2004) Killer appetites: Assessing the role of predators in ecological communities. Ecology 85: 3373–3384.
78. Wang SW, MacDonald DW (2009) Feeding habits and niche partitioning in a predator guild composed of tigers, leopards and dholes in a temperate ecosystem in central Bhutan. J Zool 277: 275–283. DOI: 10.1111/j.1469-7998.2008.00537.x.
79. Woodroffe R, Ginsberg J (1998) Edge effects and the extinction of populations inside protected areas. Science 280: 2126–2128.
80. Robert A (2009) The effects of spatially correlated perturbations and habitat configuration on metapopulation persistence. Oikos 118: 1590–1600. DOI: 10.1111/j.1600-0706.2009.17818.x.
81. Karanth KK, Gopalaswamy AM, Prasad PK, Dasgupta S (2013) Patterns of human–wildlife conflicts and compensation: Insights from Western Ghats protected areas. Biol Conserv 166: 175–185.
82. Velho N, Karanth KK, Laurance WF (2012) Hunting: A serious and understudied threat in India, a globally significant conservation region. Biol Conserv 148: 210–215.
83. Woodroffe R, Cleaveland S, Courtenay O, Laurenson MK, Artois M (2004) Infectious disease in the management and conservation of wild canids. In: Macdonald DW, Sillero-Zubiri C, editors. The biology and conservation of wild canids. Oxford University Press. pp. 123–142.
84. Narain S, Panwar HS, Gadgil M, Thapar V, Singh S (2005) Joining the dots: The report of the Tiger Task Force. Project Tiger Directorate, Union Ministry of Environment, Government of India, New Delhi.
85. Rangarajan M, Shahabuddin G (2006) Displacement and relocation from protected areas: Towards a biological and historical synthesis. Conservation and Society 4: 359–378.
86. DeFries R, Karanth KK, Pareeth S (2010) Interactions between protected areas and their surroundings in human-dominated tropical landscapes. Biol Conserv 143: 2870–2880.

12

The Effects of Spatial Scale on Breakdown of Leaves in a Tropical Watershed

Renan S. Rezende[1,2]*, Mauricio M. Petrucio[1], José F. Gonçalves, Jr.[2]

1 Graduate Program in Ecology of Department of Ecology and Zoology, Institute of Biological Sciences, Federal University of Santa Catarina, Florianópolis, Santa Catarina, Brazil, **2** Department of Ecology, Institute of Biology, University of Brasília, Brasília, Federal District, Brazil

Abstract

The objective was to assess the effects of natural variation in the physical structure of the environment on biological communities and on the processing of *Eucalyptus cloeziana* and *Inga laurina* and to identify the controlling factors at different scales along stream order gradients. The study area consisted of 14 sampling sites distributed within a tropical watershed (1st, 2nd, 3rd and 4th order streams replicated in 4 sub-basins). Our samples consisted of 3 g of leaves of *E. cloeziana* (high-quality) and *I. laurina* (low-quality) placed in 252 bags with 10mm mesh (measured by the chemical composition of the detritus). Four samples of each leaf type were collected periodically (three times) over a period of 75–125 days and washed on a sieve to separate the invertebrates. A series of leaf disks were cut to determine ash-free dry mass, polyphenol, lignin, cellulose, total microbial biomass and fungal biomass, and the remaining material was oven-dried to determine the dry weight. We performed analyses within and between spatial scales (regional and local) to assess which watershed scale was the more import determinant of the leaf breakdown rate (*k*). The microbial and shredder were most influenced at the local scale (stream order). Shredders were influenced by microorganisms, with stronger interactions between them than were found to drive the *k* at the local scale. Moreover, differences in the overall *k* and abiotic variables were more strongly influenced at the regional scale (sub-basin), showing that the study scale alters the response of the studied variables. We found higher *k* values at higher values of water velocity, dissolved oxygen and temperature, all of which accelerate biological metabolism in response to variations on the regional scale. Watersheds with warmer microclimates and streams with higher nutrient levels and oxygen could be accelerating the ecosystem metabolism, independent of the detritus quality.

Editor: Kerrie Swadling, University of Tasmania, Australia

Funding: The sources of funding that have supported the work were PROCAD-NF/CAPES (No. 173/2010), PNADB/Edital 17/2009/CAPES (No. 1098/2010), PELD/CNPq (No. 558233/2009-0), FAPEMIG (No. APQ-00274-12), Federal University of Santa Catarina (UFSC) and funds of University of Brasília (UnB - DPP; No. 121366/2011). The funders had no role in study design, data collection and analysis, decision to publish, or preparation of the manuscript.

Competing Interests: The authors have declared that no competing interests exist.

* E-mail: renanrezende30@gmail.com

Introduction

The characteristics of lotic ecosystems show natural patterns along an upstream-downstream gradient due to variations in geomorphology and topography in the watershed [1,2]. Natural changes along this gradient (upstream to downstream) include an increase in the dimensions of the stream (width), changes (increases or decreases) in the velocity of the water, and openings in the riparian canopy that allow greater light penetration [3,4]. Greater luminosity increases the temperature and accelerates photosynthetic production and autotrophic metabolism [3]. Along this gradient, therefore, the relative abundance of micro-organisms increases but that of invertebrates decreases (primarily shredders of organic matter), decreases from the headwaters to the downstream sections [1]. These natural changes also modify the energy input and cycling of organic matter in space and time [1,5]. Senescent leaves are an important source of nutrients and food resources in heterotrophic metabolic environments, primarily in headwaters and small streams [6,7]. However, this material is mineralized and available for use by primary producers and other trophic levels after its decomposition. As a result, leaf breakdown is a key process in lotic ecosystems [6,8]. Leaf breakdown can be influenced by many factors, such as physical and chemical variables (associated with water and detritus) and the activities of communities of decomposers (micro-organisms and aquatic invertebrates) [6–13].

The study of leaf breakdown at the scale of a watershed allows us to observe emerging patterns and identify certain factors that structure the ecosystems at different scales [14,15]. It is evident that a series of successively smaller and nested geomorphologic units can have various patterns and structures depending on the scale that is being analyzed [14,16]. These patterns and structures can be observed in riffles and pools within continuous stretches, which are nested within large rivers that make up a watershed [15,17,18]. Studies that address only one scale are subject to problems because certain variables are measured directly in small areas or across short time intervals, whereas few can be measured at fine resolution over large areas [17,18]. In addition, changes at smaller scales are not maintained at larger scales [14]. Therefore, the issue is that unless patterns are consistent at all scales, the findings at one scale cannot be extrapolated to yield accurate predictions at other scales. Accordingly, tests at multiple scales are needed for confident extrapolation. From this perspective, the evaluation of leaf breakdown at different scales enables the development of an integrated vision of the landscape during this important ecological process [19,20].

The streams in a watershed can be considered within a hierarchical framework that presents organized view of spatial and temporal variations among and within stream systems along the "riverscapes" [21–23]. Therefore, several studies have examined leaf processing at large spatial scales across biomes [24,25], latitudinal gradients [24], and altitudinal gradients [2] and influences of land use [26,27]. Moreover, several recurrent topics emerge from considerations of several spatial scales. These topics include the relative importance of fungi and invertebrates [28], the use of bioindicators [20] and the hierarchical nature of lotic ecosystems [15]. Studies assessing allochthonous leaf breakdown at a watershed scale are rare worldwide, but they have been performed in temperate systems [15,20,23,29].

In tropical streams, individual riffles or short stream reaches continue to be the most frequent sites for studies of leaf processing based on the traditional conceptual model [30,31]. Several factors are known to cause variation in the rates of processing within and among tropical stream reaches [13,32]. These factors include the effects of species mixing [33,34], litter quality [35,36], micro-organism communities [37,38], invertebrate communities [39–41], detritivores and shredders [10,42] and seasonal effects [8,43,44]. However, systematic assessments of variability in allochthonous leaf breakdown rates across multiple spatial scales using the watershed as the sampling unit have not been performed in tropical stream systems. This study could help to answer important questions, such as "how does spatial structure influence ecosystem function and how do we integrate within and between spatial scales to assess function", suggested by Sutherland et al. [45] as one of 100 fundamental ecological questions.

Based on the premise that leaf breakdown is the result of the activity of decomposer organisms and the physical and chemical processes occurring in the stream water, which vary along the scale investigated in the study scale [6,12,13], the following hypotheses were tested in this study: (i) natural differences in the physical nature of the stream (increasing canopy opening, water velocity, temperature and nutrient concentrations) accelerate biological metabolism and leaf breakdown from upstream to downstream; (ii) shredders decrease and micro-organisms increase in importance from upstream to downstream; and (iii) differences in the overall *k* values will be more clearly understandable (strongly explained) at an increased spatial scale. The objective of the study was to assess the natural effects of variation in the physical environment on biological communities and the leaf breakdown rates of *Eucalyptus cloeziana F. Muell* and *Inga laurina Sw. Willd* and to identify the controlling factors at different scales along the stream order gradient.

Methods

The Study System

The study area consisted of 14 sampling sites distributed along the Gama-Cabeça do Veado watershed, a part of the Federal District in west central Brazil, comprising 1st, 2nd, 3rd and 4th order streams replicated in 4 sub-basins (Figure 1). The area includes important waterways that form the Paraná basin, a part of the Cerrado biome (Brazilian Savannah). The climate is tropical and has distinct rainy-hot (October to April) and cold-dry seasons (May to September). The mean annual temperature is 20°C, and the altitude varies between 1025 and 1150 m above sea level. The study area included three conservation units protecting the entire watershed and all sampling sites studied (Ecological Station of the University of Brasília, Ecological Reserve of IBGE and Ecological Station of the Botanical Gardens of Brasília). The study was approved by Ministry of Environment of Brazil through the System of Information and Authorization on Biodiversity (SISBIO) for activities with scientific purpose (code: 39629-1), and also was approved by the Scientific and Technical Council of the Ecological Station of the University of Brasília (code: 05–12), IBGE Ecological Reserve (code: 54 PC - PAD 1) and Botanical Gardens of Brasília (code: 13/2011).

Procedures

The experiment was conducted from June through October 2011 (the winter season), a period of zero rainfall, resulting in a high level of homogeneity in the physical and chemical features of the environment. Based on their chemical composition, we selected leaves from two species for use as detritus. The leaves of an exotic species (*Eucalyptus cloeziana F. Muell*) were used to represent high-quality detritus. The planting of *Eucalyptus* monocultures in place of the native vegetation has potential repercussions for stream basins. The area in which the study was conducted does not contain *Eucalyptus* plantations. However, the substitution of this monoculture for native vegetation has occurred in neighboring basins, where this monoculture has expanded, as it has throughout Brazil, e.g., to supply charcoal for steelmaking and pulp for papermaking [35]. The low-quality detritus from a native species (abundant in riparian vegetation; *Inga laurina Sw. Willd*) was also used in the study, as this species best represents the plants of the Cerrado.

The leave The two types of leaves were chemically characterized by the mean values of total polyphenols (22.80 ± 2.5; 18.29 ± 1.8 mg/g^{-1}), total tannic acids (0.003 ± 0.0002; 0.002 ± 0.0004 mg/g^{-1}), lignin (42.61 ± 0.7; 45.94 ± 0.5%), cellulose (24.69 ± 1.5; 37.39 ± 1.2%), hardness (0.17 ± 0.1; 0.6 ± 0.3 cm/g^{-1}), nitrogen (13.16 ± 1.3; 16.41 ± 1.0 g/kg^{-1}) and phosphorus (0.46 ± 0.05; 0.53 ± 0.07 g/kg^{-1}) in *E. cloeziana* and *I. laurina*, respectively. The breakdown rates for these two leaf types (collected in nets 1 m^2 in area placed 1.5 m from the ground) were measured individually by the loss of weight of 3 g (± 0.1 g dry weight) of leaves, correcting for the initial humidity and transport loss [46], incubated in litter bags (15 × 15 cm, 10 mm mesh size).

In total, 252 litter bags were placed at a depth of 0.3 m in pool areas at the 14 sampling sites in 1st, 2nd, 3rd and 4th order streams [47] in the 4 sub-basins (Gama – SB1, Capetinga – SB2, Taquaras – SB3 and Cabeça de Veado – SB4 sub-basins). The removal of the *E. cloeziana* and *I. laurina* leaves occurred initially after 10 days of incubation, which corresponds to the expected time required for leaching and initiating microbial colonization [46]. After this sampling, the principal leaf breakdown rate (*k*) was used to estimate the next sampling time for each detritus type (at approximately 75% of the remaining mass), which was determined to be 40 days for *E. cloeziana* and 85 days for *I. laurina*. This stage of decomposition occurs when microbial activity is high and the invertebrate community is already established. Subsequently, the *k* value was used a second time to perform corrections and acquire new values to obtain approximately 50% of the remaining mass for each detritus type. This sampling time was determined to be 75 days for *E. cloeziana* and 125 days for *I. laurina*. After this period, the community established during the ecological process of degradative succession is affected by the reduction in the available detritus (additional information about k, see also Chapter 6 of [46]).

The sampling times were calculated by dividing the initial weight (W0) by the estimated value of *k*. This calculation yields the time for the total course of leaf processing (TLP, days). From the

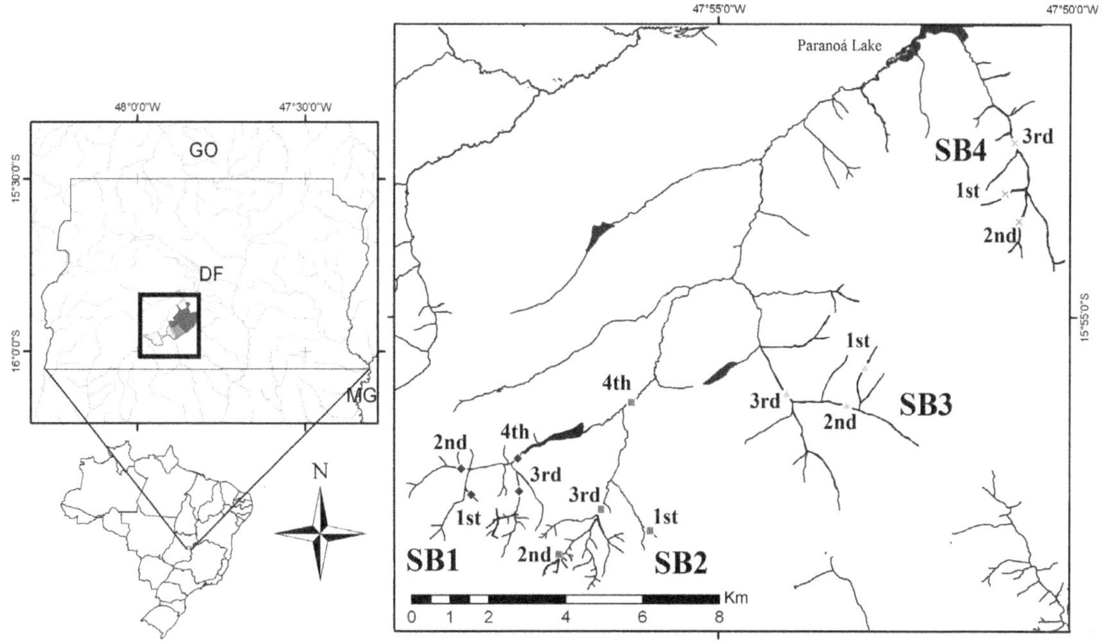

Figure 1. Sampling sites. Geographic location of the sampling sites consisting of streams of the 1st, 2nd, 3rd and 4th orders replicated in 4 sub-basins in the Gama-Cabeça de Veado watershed, Federal District (Brazil).

equation $W0/k = TLP$, we can calculate how many days will be required to reach a desired percentage of the initial weight (Wt). The first sample was collected after 10 days of incubation for both species, so that TLP for 10 days/0.25 = day on which Wt = 75%. The next sample was collected after 40 days for *E. cloeziana* and after 75 days for *I. laurina*, so that TLP 40/75 days/0.5 = day by which Wt = 50%. The above procedure was performed for each sample site (based on the mean value) and type of detritus. However, it was not possible to determine the final value for *I. laurina* because the dry season ended after 120 days, before 50% of the mass had been lost. Measurements after the end of the dry season would not have been meaningful because variations in rainfall and associated variations in other physical and chemical conditions would have influenced the results.

On removal from the streams, the litter bags were placed individually into insulated plastic bags and transported in thermal containers (± 4°C) to the laboratory. Temperature, electrical conductivity, pH, dissolved oxygen and water turbidity were obtained *in situ* with a multi-analyzer measured each time leaf bags were removed. The depth and average speed of the right, left and central portions of the watercourse were measured with a flow-meter, and the instantaneous discharge of water was then calculated. We collected 1 L of water to determine the nitrate [48], ammonia [49] and orthophosphate [50] concentrations. The canopy openings were quantified using hemispherical photographs taken with a digital camera equipped with a fish-eye lens. These photographs were later analyzed using Gap Light Analyzer software (2.0). The leaves were washed with tap water in a 120 µm mesh sieve. The invertebrates retained on the sieve were preserved in 70% alcohol for later identification and counting [51,52]. The numbers of *taxa* and individuals were calculated for the aquatic invertebrate community, and biomass was obtained by desiccation at 60°C for 72 h. The invertebrates were classified into five feeding categories [51–53]: gathering-collectors (G–C), filtering-collectors (F–C), shredders (Sh), scrapers (Sc) and predators (P).

Five leaves from each sample were randomly collected, and three disks (1.2 cm diameter) were extracted from each leaf, resulting in three five-disk sets. One set was used to determine the remaining ash-free dry mass (AFDM; calculated after incineration in a muffle furnace at 550°C for 4 h), and the other sets were used to assess the ergosterol and ATP concentrations. The remaining material was oven-dried at 60°C for 72 h to determine its dry weight. The leaf breakdown rates (k) were calculated using the negative exponential model of percent mass lost over time ($W_t = W_0 e^{-kt}$; Wt = remaining weight; W0 = initial weight; −k = decay rate; t = time). After the leaves had been dried and weighed, they were pulverized for further analysis of the total polyphenol and tannic acid concentration [54], lignin and cellulose contents [55] and the resistance of leaves to rupture (hardness of intact leaves [46]). Values for total nitrogen were obtained using a CHN basic analyzer (Carlo Erba 1500 for WI; Thermo Electron Corp. Milan, Italy), and values for total phosphorus were obtained using the ascorbic acid method after acid digestion. The total micro-organism biomass was measured by quantifying ATP [56]. The biomass of aquatic Hyphomycetes was evaluated by quantifying ergosterol, a lipid exclusive to fungal membranes in this community [57].

Data Analysis

An analysis of variance (function lm, package stats for R version 2.12.1; [58]) was used to analyze the physical and chemical parameters of the water (temperature, electrical conductivity, pH, dissolved oxygen, turbidity, nitrites, nitrates, orthophosphates and mean velocity) and the structure of stream stretches (instantaneous discharge of the stream and canopy openings in riparian vegetation) as dependent variables, using two categorical factors, namely, sub-basins and stream order. Stream order was also used as a co-variate (continuous variable). We also used the leaf mass remaining, invertebrate communities (number of *taxa*, density and biomass), the relative abundance of functional trophic groups of invertebrates (gathering-collectors, filtering-collectors, shredders, scrapers and predators) and microbial biomass (ATP and ergosterol) as dependent variables against the same two categorical factors. Stream order was also used as a co-variate (continuous variable). This procedure was performed similarly for both types of detritus. All models used a Gaussian distribution (link = log; test = F). We used an analysis of contrasts to discriminate among categorical variables. The normality of the data was tested using a Kolmogorov-Smirnov test, the homogeneity of variance was determined with a Levene test, and the data were transformed whenever necessary with the Naperian logarithm (ln) to obtain the best fit [58].

Results

Abiotic Variables

The values of instantaneous discharge, electrical conductivity and nitrates were the highest in the 3rd and 4th order streams. In contrast, the 1st order streams had the highest values for temperature, canopy opening and nitrite concentrations in the water, and the water velocity was the lowest. Dissolved oxygen, pH, turbidity and orthophosphates did not differ among the stream orders (analysis of contrasts, $p < 0.05$; Table S1, Table 1, Figure 2). In sub-basin 2 (SB2), we observed high electrical conductivity, high nitrite and nitrate concentrations and low water temperatures. The values for canopy opening and water velocity were highest in sub-basins 1 (SB1) and 4 (SB4), respectively. Dissolved oxygen and orthophosphates were highest in sub-basin 3 (SB3). Instantaneous discharge, pH and turbidity did not differ among the sub-basins (analysis of contrasts, $p < 0.05$; Table S1, Figure 2). We observed that the higher percentages of sums of squares and variance in instantaneous discharge and water velocity could be explained by differences in the stream order. However, dissolved oxygen, electrical conductivity, temperature, pH, turbidity, canopy opening, nitrates, nitrites and orthophosphates exhibited a high level of variance among the sub-basins (Table 1).

Leaf Breakdown Rates

The leaf breakdown rates (k) were the highest in the 2nd order streams, with values of −0.0083 and −0.0022 for *E. cloeziana* and *I. laurina*, respectively, followed by the 3rd order stream sections, with values of −0.0071 and −0.0022. We also observed higher k values in 1st order streams (−0.0053 and −0.0015) than in 4th order streams (−0.0051 and −0.0018) for *E. cloeziana* and *I. laurina*, respectively. However, the remaining mass did not differ among stream orders for either of the detritus types (Figure S1, Table 2, Figure 3A and 4A). The highest k values were observed in SB4 (−0.0105 and −0.0030), followed by SB3 (−0.0088 and −0.0022 for *E. cloeziana* and *I. laurina*, respectively). For *E. cloeziana*, the values were −0.0062 and −0.0049, whereas *I. laurina* exhibited values of −0.0018 and −0.0016 (for SB1 and SB2, respectively). The remaining mass showed the lowest values in SB4 and SB3 among the sub-basins studied. The variance in the remaining mass was higher (by sums of squares) and also explained the variations in the sub-basins (Figure S1, Table 2, Figure 3A and 4A).

Biotic Community

The density of invertebrates was higher in 4th order streams for both detritus types (means of 27 and 21 ind/g for *E. cloeziana* and *I. laurina*, respectively). Differences for *I. laurina* were observed among sub-basins, with high values in SB3 (mean 24 ind/g) and SB4 (mean 26 ind/g). The number of *taxa* was significantly different among the stream orders and sub-basins, with the highest values in SB3 (mean of 7 and 6 *taxa* for *E. cloeziana* and *I. laurina*, respectively) and SB4 (mean of 6 *taxa* for *E. cloeziana* and *I. laurina*), primarily in 3rd order streams for both detritus types (mean of 7 and 6 *taxa* for *E. cloeziana* and *I. laurina*, respectively). However, the biomass (total mean 0.003 and 0.002 ind/g for *E. cloeziana* and *I. laurina*, respectively) did not differ among the stream orders or sub-basins for either of the detritus types. The high variances in density, richness and biomass (in terms of the percentage of the sums of squares) were explained by differences in the sub-basins for both detritus types, except for the density of invertebrates in *I. laurina* (Figure S2 and S3, Table 2).

The functional trophic groups differed significantly among stream orders only for the filtering-collectors, with the highest values in the 3rd order streams (mean 20% for *E. cloeziana* and *I. laurina*) and the lowest in the 1st order streams (mean of 10% for *E. cloeziana* and *I. laurina*) for both detritus types. The relative abundance of predators was higher in SB3 (mean 24 and 25% for *E. cloeziana* and *I. laurina*, respectively) and SB4 (mean of 32 and 34% for *E. cloeziana* and *I. laurina*, respectively), whereas the values for shredders were higher in SB3 (mean of 27 and 25% for *E. cloeziana* and *I. laurina*, respectively) for both detritus types. However, SB1 exhibited high abundances of gathering-collectors (mean of 41 and 50% for *E. cloeziana* and *I. laurina*, respectively) and scrapers (mean of 18 and 16% for *E. cloeziana* and *I. laurina*, respectively) but a low abundance of filtering-collectors (mean of 6 and 8% for *E. cloeziana* and *I. laurina*, respectively) for both detritus types. The high variance in the relative abundance for all functional trophic groups (by the sums of squares) was also explained by changes in sub-basins for both detritus types (Figure S2 and S3, Table 2, Figure 3B and 4B).

The ATP values differed only among sub-basins for both detritus types, with the highest values in SB3 (mean of 2155.8 nmoles/g AFDM) for *E. cloeziana* (total mean 991.8 nmoles/g AFDM) and in SB3 (mean 633.3 nmoles/g AFDM) and SB4 (mean 2023.9 nmoles/g AFDM) for *I. laurina* (total mean 847.8 nmoles/g AFDM) (Table 2; Fig. 2C and 3C). There were no differences in the ergosterol concentrations among the hydrological stream orders and sub-basins for *E. cloeziana* (total mean 541.2 µg/g). However, we found higher ergosterol concentrations for *I. laurina* (total mean 382.9 µg/g) in SB3 (mean 392.6 µg/g) and SB4 (mean 464.1 µg/g), although they did not differ among the hydrological stream orders. The variances in ATP and ergosterol concentrations were also explained by changes in sub-basins for both detritus types, except for ergosterol in *E. cloeziana*, which showed a high level of variation with stream order (Figure S2 and S3, Table 2, Figure 3D and 4D).

Discussion

Scale Analysis

The instantaneous discharge, water velocity, turbidity and nitrogen series (nitrate and nitrite) were more influenced by changes in stream order (high heterogeneity), with higher values downstream (increasing from 1st to 4th order), as expected according to Vannote et al. [1], except that nitrites exhibited the inverse pattern. The finding of relatively few influences at local scales can be explained by the large discontinuities inherent in smaller geomorphological units (habitat patches create discontinuities in space) that increase the potential influence from the local characteristics of the environment [4,5,14,18]. Flow changes, for example, create hydrological discontinuities along stream corridors and isolate habitats. However, the other abiotic variables were influenced by changes in sub-basins (high homogeneity) that correspond to regional scales [19,21]. These factors worked at the watershed level and may increase its fragility in the face of intense

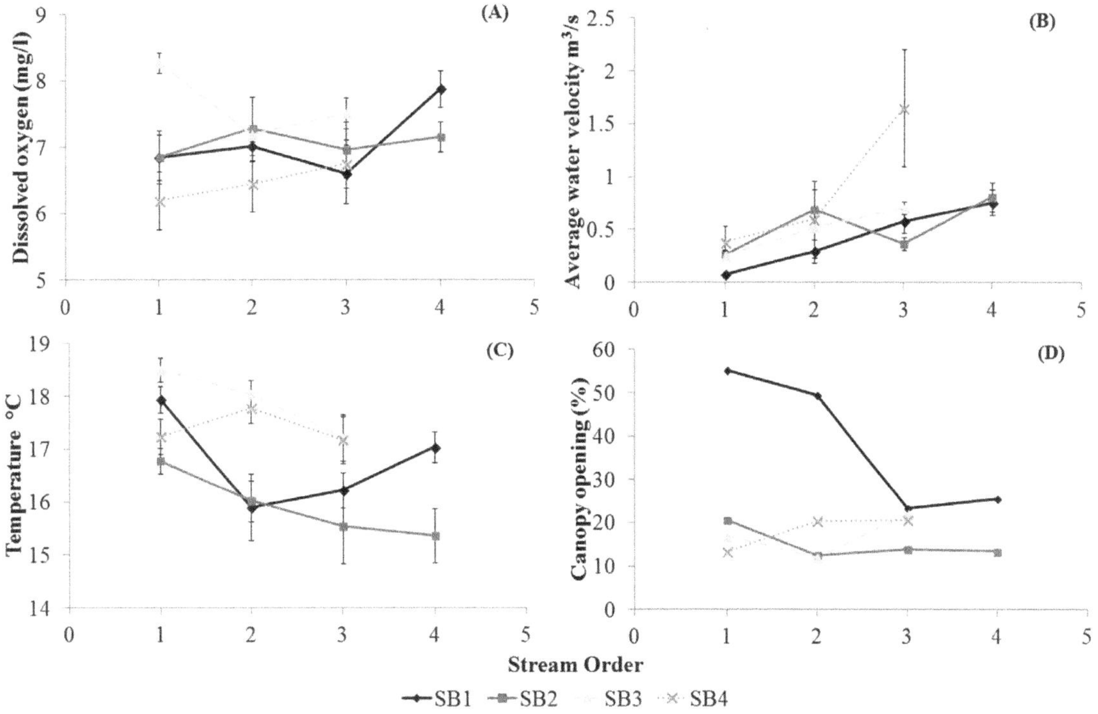

Figure 2. Abiotic variables. Mean values and standard errors for dissolved oxygen (A), water velocity (B), water temperature (C) and canopy opening (D) for the stream orders and among sub-basins.

climatic changes because the climate is the primary controlling factor at large scales [22]. The basins are influenced by environmental factors that systematically change across longitudinal (upstream/downstream), vertical (sediment/water) and lateral (terrestrial/aquatic) gradients, forming different spatial and temporal patterns at regional and local scales [5,17,18].

As expected, the high-quality detritus (*E. cloeziana*) showed more rapid leaf breakdown rates (*k*) than the low-quality (*I. laurina*) detritus, indicating that the rate could be driven by micro-scale processes [14]. These findings represent important evidence that the riparian vegetation could be responsible for determining the ecosystem characteristics [30,31,59], as also proposed by Gonçalves et al. [35] for tropical systems, highlighting the need to study this vegetation. For example, if riparian vegetation is composed of plant species that have a higher stoichiometric ratio (higher quality), we expect more rapid response for organic matter cycling (higher decomposition). This direct relationship between terrestrial and aquatic ecosystems demonstrates that any modification in a riparian ecosystem would affect the function, primarily in areas composed of palatable plants (high quality and decomposition), corroborating the work of Frauendorf et al. [60]. Thus, lower quality vegetation (slower decomposition) will be less sensitive to other factors, and this may explain the resistance of the Brazilian savanna in comparison with other Brazilian tropical systems [61]. Moreover, the leaf breakdown at the macro-scale was more influenced by the regional scale (sub-basin) than by the local scale (stream order) [14], confirming our hypothesis. This result indicates that the patterns observed in studies covering a given time period (timely studies), common in tropical literature (for more see also [13,32]), cannot be generalized from local to regional scales [17,29,62] or to whole watersheds [14–16,18]. In addition, this finding may indicate fragility in the upstream basins due to the slower leaf breakdown rates. However, the upstream area is a source of nutrients and organisms for the downstream basins [1]. The upstream basins can give support productivity and may be responsible for extending the depuration capability of the system (as represented by the microbial pools) and maintaining functionality downstream [60]. Therefore, we believe that the association between detritus quality (important at the micro-scale) and the environmental features of the watershed (important at the macro-scale) is responsible for shaping organic matter cycling in the watershed and should be further investigated in future studies.

Previous studies indicating that the microbial community is the principal decomposer [6,30,35], are confirmed by our results, as we found that the high-quality leaf species was also more susceptible to leaching and microbial action, whereas the low-quality leaf species was influenced primarily by fungal colonization. Both leaf samples were consumed by shredders, but a higher abundance of shredders was observed in the *E. cloeziana* detritus. We also found a higher loss of mass due to the high water velocity (mechanical fragmentation and leaching), dissolved oxygen and temperatures, which accelerate biological metabolism [37,63]. The detritus quality is important only for defining the local rates and their pathways for leaf breakdown [6,35,64]. However, the detritus quality has little influence on the general pattern along the "riverscape" and at any specific scale [15]. Therefore, based on an analysis of the samples after a certain percentage of mass has been lost (25, 75 and 50%) and not simply at predefined time points (e.g., 7, 15, 30 days), it is possible to show a clear colonization effect independent of quality. To be sure, detritus quality is a highly important determinant of the abundance of shredders and explains the importance of shredders for both detritus types [39]. We cannot study the variations associated with spatial scale in terms only of the local context because there are many factors in the ecological levels (community and ecosystem) that are responsible for variability found in the large scale [18,29].

Table 1. Degrees of freedom (DF), residuals, sums of squares (%), F tests and analyses of variance and contrasts (AC) for dissolved oxygen, electrical conductivity, water temperature, pH, turbidity, water velocity, canopy openness and nitrites, nitrates and orthophosphates in sub-basins and along the stream orders in the Gama-Cabeça de Veado Basin.

		Df	Sum Sq (%)	F value	Pr(>F)	AC
Instantaneous discharge	Sub-Basin	3	5.71	1.97	0.127	
	Order	1	31.56	32.71	**<0.001**	1st = 2nd < 3rd = 4th
	Residuals	65	62.72			
Dissolved oxygen	Sub-Basin	3	11.34	2.80	**0.047**	SB4 = SB2 = SB1 < SB3
	Order	1	0.87	0.65	0.425	
	Residuals	65	87.78			
Electrical conductivity	Sub-Basin	3	42.63	27.22	**<0.001**	SB4 = SB1 = SB3 < SB2
	Order	1	23.42	44.86	**<0.001**	1st = 2nd < 3rd = 4th
	Residuals	65	33.93			
Temperature	Sub-Basin	3	22.29	6.66	**<0.001**	SB2 < SB1 = SB4 = SB3
	Order	1	5.15	4.62	**0.035**	4th = 3rd = 2nd < 1st
	Residuals	65	72.55			
pH	Sub-Basin	3	5.44	1.25	0.298	
	Order	1	0.47	0.33	0.567	
	Residuals	65	94.07			
Turbidity	Sub-Basin	3	2.66	0.62	0.605	
	Order	1	4.23	2.96	0.090	
	Residuals	65	93.10			
Average water velocity	Sub-Basin	3	10.44	3.79	**0.014**	SB1 = SB3 = SB2 < SB4
	Order	1	29.77	32.37	**<0.001**	1st < 2nd = 4th = 3rd
	Residuals	65	59.77			
Canopy opening	Sub-Basin	3	59.83	41.13	**<0.001**	SB2 = SB3 = SB4 < SB1
	Order	1	8.64	17.83	**<0.001**	4th = 3rd = 2nd < 1st
	Residuals	65	31.52			
Nitrate	Sub-Basin	3	42.56	32.68	**<0.001**	SB3 = SB4 = SB1 < SB2
	Order	1	29.22	67.32	**<0.001**	1st = 2nd < 3rd < 4th
	Residuals	65	28.21			
Nitrite	Sub-Basin	3	35.50	23.86	**<0.001**	SB4 = SB3 < SB1 < SB2
	Order	1	32.25	65.02	**<0.001**	4th < 2nd = 3rd < 1st
	Residuals	65	32.24			
Orthophosphate	Sub-Basin	3	11.62	2.87	**0.043**	SB3 < SB1 = SB4 = SB2
	Order	1	0.54	0.40	0.527	
	Residuals	65	87.83			

However, the local approach has been used in all previous tropical studies of leaf breakdown [12,13].

Leaf Breakdown Rates

In agreement with the proposal of Gonçalves et al. [13] for tropical systems, the k values of *E. cloeziana* were classified as intermediate ($-0.0173 > k > -0.0041$), and those of *I. laurina* were classified as slow ($k < -0.0041$) for all sampling sites, indicating a strong influence of detritus quality. The high leaching due to the high solubility of polyphenols and tannins (secondary compounds) in *E. cloeziana* can accelerate the decay rate [64,65]. Therefore, the rapid leaching of these secondary compounds, which has an inhibitory effect on detritivores, as well as the lower hardness of *E. cloeziana*, did not limit biotic colonization due to the low residence time in this type of detritus [35,64]. Additionally, *Eucalyptus* sp. (an exotic species) is rapidly colonized and decomposed in the Brazilian savannah. It is possible that this pattern is due to the high quality of *Eucalyptus* relative to native species [35,39,66].

In contrast, we found lower breakdown rates for *I. laurina* which were most likely a consequence of a high content of structural compounds (lignin and cellulose) and relative hardness (cuticle thickness), hindering the release of other chemical compounds (e.g., polyphenols, nitrogen and phosphorus [35,65,66]). Therefore, the chemical characteristics of detritus determine the speed of processing (primarily at local scales), showing that leaf breakdown rates increase with quality and palatability [6,7,35,39]. Detritus quality is of lower importance when if it is observed at different

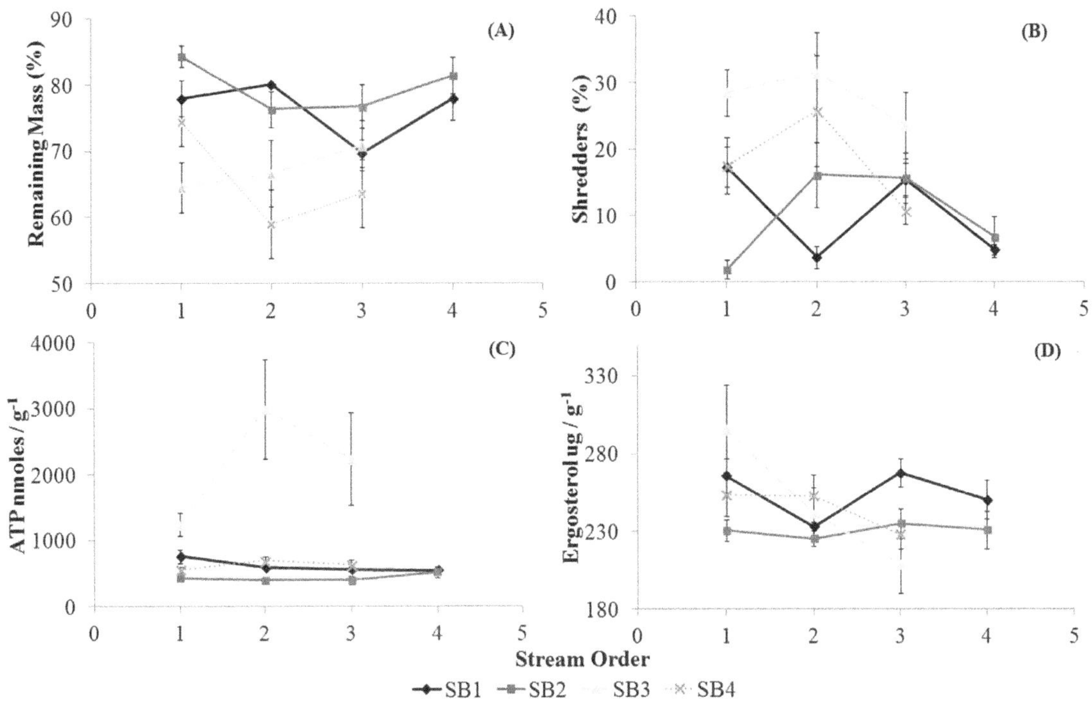

Figure 3. Leaf breakdown process in *E. cloeziana*. Mean values and standard errors for the remaining mass (A), shredder abundance (B), total microbial biomass (ATP; C) and fungal hyphomycetal biomass (Ergosterol; D) for the stream orders and sub-basins for *E. cloeziana*.

scales. In the study area, we observed that the regional scale is decisive for driving the general pattern of this important ecological process along the "riverscapes" [21,22].

Abiotic and Biotic Variables in Leaf Breakdown

The natural environmental changes that occur across stream orders [4] were not sufficient to modify the remaining mass, and

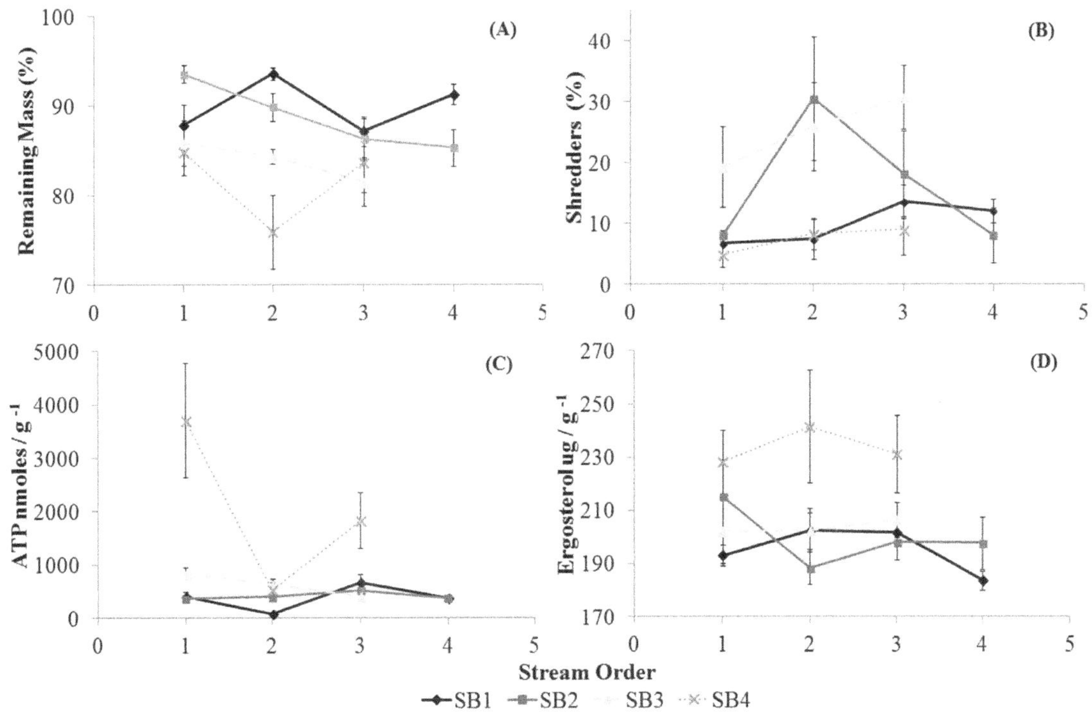

Figure 4. Leaf breakdown process in *I. laurina*. Mean values and standard errors for the remaining mass (A), shredder abundance (B), total microbial biomass (ATP; C) and fungal hyphomycetal biomass (Ergosterol; D) for the stream orders and sub-basins for *I. laurina*.

Table 2. Degrees of freedom (DF), residuals, sums of squares (%), F test and analyses of variance and contrasts (AC) for the remaining mass, density, richness and biomass of invertebrates, functional trophic groups of invertebrates (predators, shredders, gatherer-collectors, filtering-collectors and scrapers), ATP and ergosterol in sub-basins and along the stream orders in the Gama-Cabeça de Veado Basin for *E. cloeziana* and *I. laurina*.

		E. cloeziana					*I. laurina*				
		Df	**Sum Sq (%)**	**F value**	**Pr(>F)**	**AC**	**Df**	**Sum Sq (%)**	**F value**	**Pr(>F)**	**AC**
Remaining mass	Sub-Basin	3	15.05	6.51	**<0.001**	SB4 = SB3 < SB1 = SB2	3	11.01	4.72	**0.004**	SB4 = SB3 < SB2 = SB1
	Order	1	0.18	0.24	0.628		1	1.95	2.51	0.116	
	Residuals	110	84.76				112	87.04			
Invertebrate Density	Sub-Basin	3	0.63	0.25	0.864		3	9.29	4.14	**0.008**	SB2 = SB1 < SB3 = SB4
	Order	1	5.49	6.43	**0.013**	$2^{nd} = 1^{st} = 3^{rd} < 4^{th}$	1	6.88	9.19	**0.003**	$2^{nd} = 1^{st} = 3^{rd} < 4^{th}$
	Residuals	110	93.88				112	83.83			
Invertebrate Richness	Sub-Basin	3	12.42	5.65	**0.001**	SB2 = SB1 < SB4 = SB3	3	25.53	15.24	**<0.001**	SB2 < SB1 = SB3 = SB4
	Order	1	7.00	9.55	**0.003**	$1^{st} = 4^{th} = 2^{nd} < 3^{rd}$	1	11.92	21.33	**<0.001**	$1^{st} = 2^{nd} < 4^{th} = 3^{rd}$
	Residuals	110	80.58				112	62.56			
Invertebrate Biomass	Sub-Basin	3	1.29	0.48	0.697		3	4.10	1.62	0.189	
	Order	1	0.18	0.20	0.652		1	1.47	1.74	0.190	
	Residuals	110	98.53				112	94.43			
Predators	Sub-Basin	3	9.44	3.84	**0.012**	SB1 = SB2 = SB3 < SB4	3	11.87	5.09	**0.002**	SB1 = SB2 < SB3 = SB4
	Order	1	0.45	0.55	0.461		1	1.12	1.45	0.232	
	Residuals	110	90.11				112	87.01			
Shredders	Sub-Basin	3	13.89	5.99	**0.001**	SB2 = SB1 = SB4 < SB3	3	9.76	4.09	**0.008**	SB4 = SB1 = SB2 < SB3
	Order	1	1.12	1.45	0.232		1	1.26	1.58	0.211	
	Residuals	110	84.99				112	88.99			
Gatherer-Collectors	Sub-Basin	3	22.51	10.68	**<0.001**	SB4 = SB3 < SB2 = SB1	3	27.15	14.29	**<0.001**	SB3 = SB2 = SB4 < SB1
	Order	1	0.23	0.33	0.565		1	1.91	3.02	0.085	
	Residuals	110	77.26				112	70.93			
Filtering-Collectors	Sub-Basin	3	11.14	4.89	**0.003**	SB1 < SB3 = SB4 = SB2	3	10.61	4.75	**0.004**	SB1 = SB2 = SB3 < SB4
	Order	1	5.23	6.88	**0.010**	$1^{st} = 2^{nd} = 4^{th} < 3^{rd}$	1	5.98	8.03	**0.005**	$1^{st} < 4^{th} = 2^{nd} = 3^{rd}$
	Residuals	110	83.63				112	83.41			
Scrapers	Sub-Basin	3	21.70	10.21	**<0.001**	SB3 = SB4 = SB2 < SB1	3	11.79	5.07	**0.002**	SB3 = SB4 = SB2 < SB1
	Order	1	0.34	0.47	0.493		1	1.50	1.94	0.166	
	Residuals	110	77.96				112	86.71			
ATP	Sub-Basin	3	26.40	13.19	**<0.001**	SB2 = SB1 = SB4 < SB3	3	17.59	8.07	**<0.001**	SB2 = SB1 = SB3 < SB4
	Order	1	0.20	0.31	0.581		1	1.06	1.46	0.230	
	Residuals	110	73.40				112	81.35			
Ergosterol	Sub-Basin	3	2.25	0.87	0.457		3	9.82	4.07	**0.009**	SB1 = SB2 = SB3 < SB4
	Order	1	3.27	3.81	0.053		1	0.10	0.13	0.720	
	Residuals	110	94.47				112	90.08			

the local scale could not affect the decomposer communities (shredders and microorganisms) for either of the detritus types. Decomposer communities are the driving factors for leaf breakdown, and their absence leads to similar breakdown patterns along the stream order gradient [6]. This finding might indicate that ecological functioning in headwater streams (1st to 3rd order) was similar within the same sub-basin [1]. However, this process might change over a large spatial gradient, as represented by the sub-basin scale [22]. Nevertheless, increases in richness and the density of invertebrates and a decrease in the abundance of filtering-collectors for both detritus types were observed across this large spatial gradient. These results demonstrate that these variables had no effect on leaf breakdown [52,53].

The values of the remaining mass for both detritus types were lower in SB4 and SB3 (high decomposition) than in the other sub-basins. It is probable that the reason for this difference was the higher temperatures, dissolved oxygen concentrations and water velocities resulting from the microclimate of the geographic location (within a valley). Therefore, the higher temperatures [63] and oxygen concentrations [37] observed in SB4 and SB3 may elevate the metabolic activity of the decomposer community [9], especially microorganisms. The higher metabolic activity of the decomposer community, associated with high water velocity (mechanical fragmentation and leaching), which increased the degree of physical abrasion [67], accelerated the leaf breakdown rates. In SB4 and SB3, higher density and richness of invertebrates and higher shredder abundance, with the greatest densities in *E. cloeziana*, were also observed. The shredders directly utilize leaf tissues for feeding, and increasing biological fragmentation [7,42] can also accelerate the leaf breakdown rates [26]. Certain shredders in these locations (genus *Phylloicus*) can build their capsules from leaf tissue, and this use of leaf material also contributes to fragmentation [52,53].

The relative abundance of shredders was influenced by variation, primarily among the sub-basins. A greater relative abundance of shredders in comparison with other tropical systems was observed in SB3 for both detritus types [10,39,42]. A low relative abundance of shredders was found in the other sub-basins. This result is consistent with the findings of previous studies in the Cerrado [13,36,39]. The importance of shredders for leaf breakdown is unclear in the tropics due to their low abundance or absence in these streams [10,39,42], but studies have shown little effect in tropical streams [11,33]. From a global perspective, the strong effects observed in the current study were most likely due to the preference of shredders (primarily Trichoptera and Plecoptera) for high altitudes (due to the lower temperatures) in tropical regions [42]. The preference of this group for *E. cloeziana* indicates that detritus quality is also important [10] and that the composition of the vegetation influences the functioning of aquatic systems. Therefore, higher altitudes (low temperature [42]), high dissolved oxygen, the composition of the flora (ideally including *E. cloeziana* [10]) and moderate values of nutrient concentrations in the water [62] favor a high abundance of shredders. The predominance in the Cerrado of leaves that are low in nutrients [66], is associated with hydric and thermic stress and could be responsible for the absence or low abundance of shredders found in most tropical streams [10,42].

The high-quality detritus (*E. cloeziana*) was shown to be most influenced by the total microbial community and the low-quality detritus (*I. laurina*) by the fungal community in SB4 and SB3 (high decomposition). *E. cloeziana* has elevated amounts of labile compounds, facilitating the activity of bacteria (rapid life cycles) that use compounds derived from the leaching of the leaves of labile detritus as their preferred resource [68]. These bacteria could be important during leaf breakdown and not only, as observed by several authors, at the early stages of the process [6,13,35]. However, this pattern is rarely observed in tropical streams. In lower-quality detritus (*I. laurina*), we observed an interaction between the biomass of the two microbial communities, highlighting fungi as the principal component [63]. Due to its high capacity to metabolize refractory molecules (e.g., cellulose and lignin) and to decompose them, the fungal community is the primary decomposer in tropical streams, and this principle explains the great significance of fungi in *I. laurina* decomposition [6,35,68]. Fungal action can increase the palatability of detritus, as well as its nutritional quality, for other decomposers, and the high biomass of fungi might be another factor responsible for the higher abundance of shredders in these sub-basins [7,35].

In general, we conclude that variations in scale contribute to the variation in the leaf breakdown rate, highlighting the importance of similar studies of this type that determine effects at different scales. The variability of the physical structure of streams (primarily temperature, dissolved oxygen and nutrients) accelerates leaf breakdown from upstream to downstream, but this process was only demonstrated at the sub-basin scale in the location studied, partially corroborating the initial hypothesis. The replacement of shredder invertebrates by microorganisms was observed but was contrary to the prediction of our hypothesis. Shredders were favored by microorganisms (primarily in *E. cloeziana*), with stronger interactions between them than those previously found to drive leaf breakdown rates. Based on our interest in the influence of spatial structure on ecosystem functions, we observed that watersheds with warmer microclimates and streams with higher nutrient levels and oxygen in the water could be accelerating the metabolism of the ecosystem in the watershed, with increased negative effects downstream. For the management of tropical watersheds, we noted that the upstream areas are more fragile and sensitive to environmental impacts but show greater importance in the cycling of nutrients. We performed analyses within and between spatial scales to assess the relative importance of various watershed scales in determining the local breakdown rate for leaves. Local characteristics are responsible for the diversification of this process across the "riverscape", and high heterogeneity underscores the difficulty of making predictions based on local studies.

Supporting Information

Table S1 Abiotic variables in sampling sites. Average values and the standard deviation of outflow, dissolved oxygen in the water (mg l^{-1}), electrical conductivity (μS-cm^2), water temperature (Temp. °C), pH, turbidity (NTU), water velocity (m-s), nitrite, nitrate, orthophosphate (mg l^{-1}) and the percentage of canopy openness (%) in sub-basin and stream order along the Gama-Cabeça de Veado Basin.

Figure S1 Remaining mass over time in sampling sites. Percentages of remaining mass along of the day in *E. cloeziana* (A and C) and *I. laurina* (B and D), between stream order (A e B) and sub-basin (C and D).

Figure S2 Biotic Community over time in *E. cloeziana* detritus. Average values and standard error of density (A and B), richness (C and D), biomass (E and F) of aquatic invertebrates, total microbial biomass (ATP; G and H) and fungal hyphomicetos biomass (I and J) along of the days in *E. cloeziana*, among stream order (A, C, E, G and I) and sub-basin (B, D, F, H and J).

Figure S3 Biotic Community over time in *I. laurina* detritus. Average values and standard error of density (A and B), richness (C and D), biomass (E and F) of aquatic invertebrates, total microbial biomass (ATP; G and H) and fungal hyphomicetos biomass (I and J) along of the days in *I. laurina*, among stream order (A, C, E, G and I) and sub-basin (B, D, F, H and J).

Acknowledgments

The first author also received a scholarship from CAPES during his PhD research in the Graduate Program in Ecology of UFSC. We are also grateful to Cássia Alves Lima for valuable logistical support for the field and laboratory work.

Author Contributions

Conceived and designed the experiments: RSR MMP JFG. Performed the experiments: RSR MMP JFG. Analyzed the data: RSR MMP JFG. Contributed reagents/materials/analysis tools: RSR MMP JFG. Wrote the paper: RSR MMP JFG.

References

1. Vannote RL, Minshall GW, Cummins KW, Sedell JR, Cushing CE (1980) River Continuuum Concept. Canadian Journal of Fisheries and Aquatic Sciences 37: 130–137.
2. Fabre E, Chauvet E (1998) Leaf breakdown along an altitudinal stream gradient. Archiv für Hydrobiologie 141: 167–179.
3. Statzner B, Higler B (1985) Questions and comments on the river continuum concept. Canadian Journal of Fisheries and Aquatic Sciences 42: 1038–1044.
4. Benda L, Poff NL, Miller D, Dunne T, Reeves G, et al. (2004) The network dynamics hypothesis: How channel networks structure riverine habitats. BioScience 54: 413–427.
5. Larned ST, Datry T, Arscott DB, Tockner K (2010) Emerging concepts in temporary-river ecology. Freshwater Biology 55: 717–738.
6. Gessner MO, Chauvet E, Dobson M (1999) A perspective on leaf litter reakdown in streams. Oikos 85: 377–384.
7. Graça MAS, Cressa C, Gessner MO, Feio MJ, Callies KA, et al. (2001) Food quality, feeding preferences, survival and growth of shredders from temperate and tropical streams. Freshwater Biology 46: 947–957.
8. Mathuriau C, Chauvet E (2002) Breakdown of leaf litter in a neotropical stream. Journal of the North American Benthological Society 21: 384–396.
9. Suberkropp K, Chauvet E (1995) Regulation of Leaf Breakdown by Fungi in Streams: Influences of Water Chemistry. Ecology 76: 1433–1445.
10. Wantzen KM, Wagner R (2006) Detritus processing by invertebrate shredders: a neotropical-temperate comparison. Journal of the North American Benthological Society 25: 216–232.
11. Moulton TP, Magalhaes-Fraga SAP, Brito EF, Barbosa FA (2010) Macro-consumers are more important than specialist macroinvertebrate shredders in leaf processing in urban forest streams of Rio de Janeiro, Brazil. Hydrobiologia 638: 55–66.
12. Abelho M (2001) From litterfall to breakdown in streams: a review. Scientific World Journal 1: 656–680.
13. Gonçalves JFJ, Martins RT, Ottoni BMP and Couceiro SRM (2013) Uma visão sobre a decomposição foliar em sistemas aquáticos brasileiros. In: N Hamada, J. L Nessimian and R. B Querino, editors. Insetos aquáticos: biologia, ecologia e taxonomia.
14. Wiens JA (1989) Spatial Scaling in Ecology. Functional Ecology 3: 385–397.
15. Tiegs SD, Akinwole PO, Gessner MO (2009) Litter decomposition across multiple spatial scales in stream networks. Oecologia 161: 343–351.
16. Lowe WH, Likens GE, Power ME (2006) Linking scales in stream ecology. Bioscience 56: 591–597.
17. Schneider DC (2001) The rise of the concept of scale in ecology. BioScience 51: 545–554.
18. Wiens JA (2002) Riverine landscapes: taking landscape ecology into the water. Freshwater Biology 47: 501–515.
19. Allen TFH, Starr TB (1982) Hierarchy: Perspectives for Ecological Complexity. Chicago: University of Chicago Press.
20. Del Arco AI, Ferreira V, Graça MAS (2012) The performance of biological indicators in assessing the ecological state of streams with varying catchment urbanisation levels in Coimbra, Portugal. Limnetica 31: 141–154.
21. Frissell CA, Liss WJ, Warren CE, Hurley MD (1986) A Hierarchical Framework for Stream Habitat Classification: Viewing Streams in a Watershed Context. Environmental Management 10: 199–214.
22. Allan JD (2004) Landscapes and Riverscapes: The Influence of Land Use on Stream Ecosystems. Annual Reviews of Ecology, Evolution, and Systematics 35: 257–284.
23. Royer TV, Minshall GW (2003) Controls on leaf processing in streams from spatial-scaling and hierarchical perspectives. Journal of North American Benthological Society 22: 352–358.
24. Irons JG, Oswood MW, Stout RJ, Pringle CM (1994) Latitudinal patterns in leaf litter breakdown: is temperature really important? Freshwater Biology 32: 401–411.
25. Minshall GW, Petersen RC, Cummins KW, Bott TL, Sedell JR, et al. (1983) Interbiome Comparison of Stream Ecosystem Dynamics. Ecological Monographs 53: 2–25.
26. Sponseller RA, Benfield EF (2001) Influences of Land Use on Leaf Breakdown in Southern Appalachian Headwater Streams: A Multiple-Scale Analysis. Journal of the North American Benthological Society 20: 44–59.
27. Hagen EM, Webster JR, Benfield EF (2006) Are leaf breakdown rates a useful measure of stream integrity along an agricultural landuse gradient? Journal of the North American Benthological Society 25: 330–343.
28. Hieber M, Gessner MO (2002) Contribution of Stream Detrivores, Fungi, and Bacteria to Leaf Breakdown Based on Biomass Estimates. Ecology 83: 1026–1038.
29. Young RG, Collier KJ (2009) Contrasting responses to catchment modification among a range of functional and structural indicators of river ecosystem health. Freshwater Biology 54: 2155–2170.
30. Cummins KW (1974) The importance of different energy sources in freshwater ecosystems. In: K. W Cummins, editor editors. Productivity of World Ecosystems. Washington: National Academy of Sciences. pp. 166.
31. Petersen RC, Cummins KW (1974) Leaf processing in a woodland stream. Freshwater Biology 4: 343–368.
32. Bianchini Jr I (1999) Aspectos do processo de decomposição nos ecossistemas aquáticos continentais. In: M. L. M Pompêo, editor editors. Perspectivas da Limnologia no Brasil. São Luis: Gráfica e Editora União. pp. 21–43.
33. Moretti M, Gonçalves JJF and Callisto M (2007) Leaf breakdown in two tropical streams: Differences between single and mixed species packs. Limnologica 37: 250–258.
34. Bruder A, Schindler MH, Moretti MS, Gessner MO (2013) Litter decomposition in a temperate and a tropical stream: the effects of species mixing, litter quality and shredders. Freshwater Biology: n/a-n/a.
35. Gonçalves JF Jr., Rezende RS, Martins NM, Gregorio RS (2012) Leaf breakdown in an Atlantic Rain Forest stream. Austral Ecology 37: 807–815.
36. Gonçalves JF Jr., Graça MAS, Callisto M (2007) Litter decomposition in a Cerrado savannah stream is retarded by leaf toughness, low dissolved nutrients and a low density of shredders. Freshwater Biology 52: 1440–1451.
37. Medeiros AO, Pascoal C, Graça MAS (2009) Diversity and activity of aquatic fungi under low oxygen conditions. Freshwater Biology 54: 142–149.
38. Wright MS, Covich AP (2005) Relative importance of bacteria and fungi in a tropical headwater stream: Leaf decomposition and invertebrate feeding preference. Microbial Ecology 49: 536–546.
39. Gonçalves JF, Jr., Rezende RS, França J, Callisto M (2012) Invertebrate colonisation during leaf processing of native, exotic and artificial detritus in a tropical stream. Marine and Freshwater Research 63: 428–439.
40. Li AOY, Ng LCY, Dudgeon D (2009) Effects of leaf toughness and nitrogen content on litter breakdown and macroinvertebrates in a tropical stream. Aquatic Sciences 71: 80–93.
41. Ligeiro R, Moretti MS, Goncalves JF Jr, Callisto M (2010) What is more important for invertebrate colonization in a stream with low-quality litter inputs: exposure time or leaf species? Hydrobiologia 654: 125–136.
42. Boyero L, Pearson RG, Dudgeon D, Ferreira V, Graça MAS, et al. (2012) Global patterns of stream detritivore distribution: implications for biodiversity loss in changing climates. Global Ecology and Biogeography 21: 134–141.
43. Larned ST (2000) Dynamics of coarse riparian detritus in a Hawaiian stream ecosystem: a comparison of drought and post-drought conditions. Journal of the North American Benthological Society 19: 215–234.
44. Rueda-Delgado G, Wantzen KM, Tolosa MB (2006) Leaf-Litter Decomposition in an Amazonian Floodplain Stream: Effects of Seasonal Hydrological Changes. Journal of the North American Benthological Society 25: 233–249.
45. Sutherland WJ, Freckleton RP, Godfray HCJ, Beissinger SR, Benton T, et al. (2013) Identification of 100 fundamental ecological questions. Journal of Ecology 101: 58–67.
46. Graça MAS, Barlocher F, Gessner MO (2005) Methods to Study Litter Decomposition. Dordrecht: Springer.
47. Strahler AN (1952) Hypsometric (Area Altitude) Analysis of Erosional Topology. Geological Society of America Bulletin 63: 1117–1142.
48. Golterman HL, Clymo RS, Ohnsta DMAM (1978) Methods for chemical analysis of freshwater. Oxford: Blackwell Scientific Publications.
49. Koroleff F (1976) Determination of nutrients. In: K Grasshoff, editor editors. Methods of sea water analysis. Verlag Chemie Weinhein. pp. 117–181.
50. Strickland JDH, Parsons TR (1960) A manual of seawater analysis. Ottawa: Fisheries Research Board of Canada. 188 p.
51. Pérez GP (1988) Guía para el studio de los macroinvertebrados acuáticos del departamento de Antioquia. Bogotá: Editorial Presencia Ltda.

52. Merrit RW, Cummins KW (1996) An introduction to the aquatic insects of North America. Dubuque: Kendall/Hunt Publishing Company.
53. Cummins K, Merritt R, Andrade P (2005) The use of invertebrate functional groups to characterize ecosystem attributes in selected streams and rivers in south Brazil. Studies on Neotropical Fauna and Environment 40: 69–89.
54. Bärlocher F, Graça MAS (2005) Total Phenolics. In: M. A. S Graça, F Barlocher and M. O Gessner, editors. Methods to Study Litter Decomposition. Dordrecht: Springer. pp. 97–100.
55. Gessner MO (2005) Proximate Lignin and Cellulose. In: M. A. S Graça, F Barlocher and M. O Gessner, editors. Methods to Study Litter Decomposition. Dordrecht: Springer. pp. 115–120.
56. Abelho M (2005) Extraction and quantification of ATP as a measure of microbial biomass. In: M. A. S Graça, F Bärlocher and M. O Gessner, editors. Methods to Study Litter Decomposition. Springer. pp. 223–229.
57. Gessner MO (2005) Ergosterol as a measure of fungal biomass. In: M. A. S Graça, F Barlocher and M. O Gessner, editors. Methods to Study Litter Decomposition. Dordrecht: Springer. pp. 189–195.
58. Crawley MJ (2007) The R Book. England: John Wiley & Sons Ltd.
59. Naiman RJ and Décamps H (1997) The ecology of interfaces: riparian zones. Annu Rev Ecol Syst 28: 621–658.
60. Frauendorf TC, Colón-Gaud C, Whiles MR, Barnum TR, Lips KR, et al. (2013) Energy flow and the trophic basis of macroinvertebrate and amphibian production in a neotropical stream food web. Freshwater Biology 58: 1340–1352.
61. Carvalho FMV, De Marco P, Ferreira LG (2009) The Cerrado into-pieces: Habitat fragmentation as a function of landscape use in the savannas of central Brazil. Biological Conservation 142: 1392–1403.
62. Bae M-J, Kwon Y, Hwang S-J, Chon T-S, Yang H-J, et al. (2011) Relationships between three major stream assemblages and their environmental factors in multiple spatial scales. Annales de Limnologie - International Journal of Limnology 47: S91–S105.
63. Gulis V, Suberkropp K (2003) Interactions between stream fungi and bacteria associated with decomposing leaf litter at different levels of nutrient availability. Aquatic Microbial Ecology 30: 149–157.
64. Ardon M, Pringle CM (2008) Do secondary compounds inhibit microbial- and insect-mediated leaf breakdown in a tropical rainforest stream, Costa Rica? Oecologia 155: 311–323.
65. Ardon M, Stallcup LA, Pringle CM (2006) Does leaf quality mediate the stimulation of leaf breakdown by phosphorus in Neotropical streams? Freshwater Biology 51: 618–633.
66. Oliveira-Filho AT, Ratter JA (1994) A study of the origin of central Brazilian forests by the analysis of plant species distribution patterns. Journal of Botany 52: 141–194.
67. Santos Fonseca AL, Bianchini I, Pimenta CMM, Soares CBP, Mangiavacchi N (2012) The flow velocity as driving force for decomposition of leaves and twigs. Hydrobiologia 703: 59–67.
68. Weyers HS, Suberkropp K (1996) Fungal and Bacterial Production during the Breakdown of Yellow Poplar Leaves in 2 Streams. Journal of the North American Benthological Society 15: 408–420.

13

Early Warning Signals of Ecological Transitions: Methods for Spatial Patterns

Sonia Kéfi[1]*[⁹], Vishwesha Guttal[2⁹], William A. Brock[3,4], Stephen R. Carpenter[4,5], Aaron M. Ellison[6], Valerie N. Livina[7], David A. Seekell[8], Marten Scheffer[9], Egbert H. van Nes[9], Vasilis Dakos[10]

1 Institut des Sciences de l'Evolution, CNRS, Université de Montpellier II, Montpellier, France, **2** Centre for Ecological Sciences, Indian Institute of Science, Bangalore, India, **3** Department of Economics, University of Wisconsin, Madison, Wisconsin, United States of America, **4** Department of Economics, University of Missouri, Columbia, Missouri, United States of America, **5** Center for Limnology, University of Wisconsin, Madison, Wisconsin, United States of America, **6** Harvard Forest, Harvard University, Petersham, Massachusetts, United States of America, **7** National Physical Laboratory, Hampton Road, Teddington, United Kingdom, **8** Department of Environmental Sciences, University of Virginia, Charlottesville, Virginia, United States of America, **9** Department of Aquatic Ecology and Water Quality Management, Wageningen University, Wageningen, The Netherlands, **10** Integrative Ecology Group, Estacion Biologica de Donana, Sevilla, Spain

Abstract

A number of ecosystems can exhibit abrupt shifts between alternative stable states. Because of their important ecological and economic consequences, recent research has focused on devising early warning signals for anticipating such abrupt ecological transitions. In particular, theoretical studies show that changes in spatial characteristics of the system could provide early warnings of approaching transitions. However, the empirical validation of these indicators lag behind their theoretical developments. Here, we summarize a range of currently available spatial early warning signals, suggest potential null models to interpret their trends, and apply them to three simulated spatial data sets of systems undergoing an abrupt transition. In addition to providing a step-by-step methodology for applying these signals to spatial data sets, we propose a statistical toolbox that may be used to help detect approaching transitions in a wide range of spatial data. We hope that our methodology together with the computer codes will stimulate the application and testing of spatial early warning signals on real spatial data.

Editor: Ricard V. Solé, Universitat Pompeu Fabra, Spain

Funding: The authors are grateful to the Santa Fe Institute and the Arizona State University for the financial support. SK acknowledges support from the European Union's Seventh Framework Programme (FP7/2007-2013) under grant agreement no. 283068 (CASCADE). VG acknowledges support from a Ramalingaswami Fellowship, Department of Biotechnology, Government of India, and the Ministry of Environment and Forests, Government of India. SRC's work is supported by NSF. AME is supported by NSF award 11-44056. VNL is supported by NERC (NE/F005474/1) postdoctoral fellowship of the AXA Research Fund and grant of the National Measurement Office (2013–2016). VD acknowledges a RUBICON (NWO funded) grant and an EU Marie Curie fellowship. The funders had no role in study design, data collection and analysis, decision to publish, or preparation of the manuscript.

Competing Interests: The authors have declared that no competing interests exist.

* E-mail: sonia.kefi@univ-montp2.fr

⁹ These authors contributed equally to this work.

Introduction

A range of ecosystems, from lakes and forests to rangelands and coral reefs, can exhibit multiple stable states [1]. In such ecosystems, abrupt shifts between ecological states may lead to ecological and economic losses. This happens when ecosystems reach a 'tipping point', at which they may rapidly reorganize into an alternative state with contrasting features. Such shifts have been documented not only in ecosystems, but also in a wide spectrum of complex systems including physiological systems, financial markets, and human societies [1]. However, the enormous complexity of such systems and the lack of detailed understanding of their underlying processes make it difficult to identify the points at which these systems may experience major changes. To circumvent this problem, recent research has focused on devising early warning signals of imminent transitions [2].

A number of early warning signals for ecological transitions has been proposed based on a phenomenon called 'critical slowing down' that generally occurs prior to a 'bifurcation' [3,4]. The closer a system is to a bifurcation point, the longer time it takes to recover to its stable state upon any disturbance. Theoretical studies of ecological models suggest that either a direct measure of slow recovery rate [3–5] or its manifestations in the temporal and spatial dynamics of the system can potentially act as generic early warning signals of an impending transition [2,6–8]. This phenomenon of slowing down is expected to occur before a broad range of transitions, including, but not limited to, the so-called 'catastrophic shifts' [9]. Catastrophic shifts are a particular case of transitions that are especially relevant because of their possible association with hysteresis and their lack of reversibility [1].

Generic early warning signals evaluated on time series have attracted a lot of attention in the literature [2]. However, recent theoretical studies suggest that for ecosystems that are not well-mixed (such as drylands, boreal wetlands, or heterogeneous habitats which host mobile predators), changes in spatial characteristics of the system could provide early warnings of approaching transitions as well [5,10–13]. More generally, the spatial structure of ecosystems can provide information about the ecosystem degradation level [14–22]. Spatial information allows us to devise additional kinds of indicators thus adding to our arsenal of early warning signals. At the same time, well-resolved spatial data is becoming increasingly available at low cost due to improved technology (such as remote sensing).

Empirical verification of these indicators, however, has not been able to keep pace with the rapid growth in theoretical studies, and a number of recent studies question the ability and practical efficiency of these indicators to anticipate upcoming regime shifts in real systems [23,24]. A few recent laboratory and field experiments [25–29], as well as analysis of climatic paleo-records [30], suggest that generic leading indicators (i.e. variance, skewness and autocorrelation at-lag-1) may indeed be detected in time series of real systems prior to transitions, but the empirical validation of the spatial indicators remains scarce [15,17,20,21,25,31].

The discrepancy between theoretical developments and their empirical validation arises from a number of issues, such as the lack of sufficiently resolved and long term data, as well as the lack of a coherent methodological framework that outlines the steps and statistical tools necessary to detect those signals. For indicators based on time series, these issues have been addressed in a recent paper which provides a methodological guide to practitioners and managers to detect early warnings in time series [32] (see also http://www.early-warning-signals.org).

Here, we complement the previous work on detecting early warnings in time series data by providing a step-by-step methodology for detecting early warning signals in spatial data sets. We gather, for the first time, all the early warning signals proposed in the literature so far in a spatial context. We apply these metrics on model-generated data sets along a degradation gradient, and we discuss their interpretation based on a few potential null models. Our analysis mimicks a situation where an ecosystem would be degrading and where we would have access to several snapshots of an ecosystem's spatial structure taken over a period of time, or at different locations along a degradation gradient. We hope that our methodology together with the computer codes will stimulate testing and applications of spatial early warning signals on spatial data (R-code for the spatial analysis can be found at https://github.com/earlywarningtoolbox/spatial_warnings).

Methods

Spatial indicators

We first give a brief overview of the spatial early warning signals proposed in the literature so far. More details about the indicators and their precise mathematical formulation are provided in Appendix S1. Table 1 summarizes the spatial indicators and their expected trends along a degradation gradient.

Slowing-down based indicators: spatial correlation and spectral properties (DFT). Due to increased recovery time to local equilibrium after a perturbation, neighboring units become more like each other when a system approaches a bifurcation point, i.e. they become increasingly correlated [5]. The increasing spatial coherence can be quantified by the spatial correlation function, or Moran's I, between ecological states separated by a certain distance. The near-neighbor spatial correlation, the analog of autocorrelation at lag 1 for time series, is calculated for the distance between nearest neighboring units of the system.

Spatial spectral properties change as the system approaches a tipping point [13]. To quantify spectral properties, we compute the Discrete Fourier Transform (DFT) that decomposes spatial data into components of sine and cosine waves of different wavenumbers [33]. A wavenumber can be thought of as a 'spatial frequency', or the number of times that a pattern is repeated in a unit of spatial length. In a spatial data set, periodicity is visualized as wavelength, which is inversely related to wavenumber (i.e., a small wavenumber corresponds to a large wavelength and vice-versa). As DFT is generally a complex number, we often plot the power spectrum (also 2D-periodogram) which is the magnitude of the complex DFT matrix (see Appendix S1 for details). Increased memory manifests itself as spectral reddening, i.e. spatial variation becomes increasingly concentrated at low wavenumbers [13]; in other words, long wavelength fluctuations become dominant prior to a transition [34].

Two metrics that help characterize spatial patterns for periodicity and directionality can be evaluated from the power spectrum. First, the radial-spectrum (r-spectrum) is obtained by summing the power spectrum at constant distances from the origin of the power spectrum, i.e. along concentric circles at different distances from the center. It allows evaluating the periodicity of the patterns. Periodic patterns are characterized by a peak in the power spectrum. The wavenumber at which a peak occurs corresponds to the number of times that a pattern reproduces itself within a unit area of the spatial data, and therefore contains information about the scale of the pattern. Second, the angular-spectrum (θ-spectrum) is obtained by summing the values of the power spectrum using angular sectors. It allows evaluating the isotropy (orientation) of the spatial pattern. For an isotropic data set, the θ-spectrum will show uniform amplitude at all angles, whereas for an anisotropic data set, the amplitude of the spectrum will show strong amplitudes for specific orientations [35].

Variability based indicators: Spatial variance and spatial skewness. Increased recovery time enroute to a bifurcation point may lead to stronger fluctuations around the equilibrium state of the system [36]. This can cause spatial variance of the system to increase prior to a transition [10,11]. Spatial variance is formally defined as the second moment around the spatial mean of the state variable. It has also been shown that the fluctuations around the mean can become increasingly asymmetric as the system approaches a bifurcation point. This is because the fluctuations in the direction of the alternative stable state take longer to return back to the equilibrium than those in the opposite direction [11]; this asymmetry can also arise due to local flickering events (i.e. occasional jumps of local units between their current and alternative state) [37]. The spatial asymmetry can be measured by spatial skewness, which is the third central moment scaled by the standard deviation.

Patch based indicators: shapes and sizes of patches. Many ecological systems, such as shrublands in semi-arid ecosystems and mussel beds in the intertidal, exhibit striking spatial self-organized patterns [38]. It has been suggested that the nature of local ecological interactions, such as the relative scales of competition and facilitation, can strongly influence the type of emerging spatial patterns, leading to i) regular, periodic patches with a characteristic patch size [22,38–40], or to ii) no characteristic scale of patchiness [15,39–42]. These different types of spatial structures have been observed in a range of ecosystems, however their use as potential indicators of degradation has mostly been developed in the case of drylands, where both types of spatial structures exist. In drylands, it has been shown that the early warnings depend on the type of patchiness exhibited by the ecological system [12].

In ecosystems exhibiting periodic patterns, as the level of external stress increases, a predictable sequence of self-organized patterns based on 'Turing instability' occurs. In isotropic areas (i.e. no preferred orientation of the pattern) the shape of the patterns shifts from gaps to labyrinths and to spots as the system becomes more degraded. Thus, spotted vegetation patterns have been proposed to be an early warning signal of imminent desertification

Table 1. Early warning signals of transitions in spatial data.

Method/Indicator	Phenomenon			Expected trend	Ref.
	Rising memory	Rising variability	Patchiness		
Spatial correlation	x			increase	[5]
Return time	x			increase	[12]
Discrete Fourier Transform	x			spectral reddening	[13]
Spatial variance		x	x	increase	[10,11]
Spatial skewness		x	x	peaks (see caption)	[11]
Patch-size distributions			x	change in shape of the dist.	[15]
Regular spotted patterns			x	change in patch shape	[22]
Power spectrum			x	spectral reddening	[43]

Leading spatial indicator, the primary underlying phenomenon, the expected trend along a degradation gradient, and the original references in which these were proposed. The trend of spatial skewness depends on the nature of test data: It can show a nonmonotonic behavior (thus, a peaking) for transition from a low density state to higher density state. For discrete data, it typically shows a monotonic behavior (increasing or decreasing depending on whether it is transitioning from fully covered to bare state or the other way transition).

in drylands characterized by periodic patterns [16,22]. In anisotropic areas with band-like patterns, the wavelength increases as the system approaches a transition [43].

In contrast to periodic patterns, there are cases where spatial processes give rise to non-periodic (irregular) patterns. In these cases, we can quantify the size of each patch and calculate the frequency of occurrence of different patch sizes. It is common practice to characterize the patchiness of these systems by a function that best describes the distribution of patch sizes. Irregular patterns may be characterized by a scale-free patch-size distribution, which means that there is no typical patch size in the ecosystem. Such a distribution may be well approximated by a pure power law [44] or by other heavy-tailed functions, such as a log-normal, a stretched exponential or a power law with cutoff [44,45]. Scale-free patch-size distributions have been observed in several ecosystems [15,20,41,42]. Computational models of drylands predict that larger vegetation patches become fragmented into smaller ones as aridity or grazing pressure increases and show an increasing deviation from a theoretical power law as the ecosystem approaches the desertification point [15,46]. Therefore, it has been hypothesized that an increasing deviation from power-law distribution of patch sizes can signal increasing degradation (but see [17,18]).

Statistical significance tests

There are two steps in computing the spatial indicators. The first one is to compute the spatial metrics for a given snapshot. The second is to evaluate the trend of the spatial metrics along a degradation gradient (see Table 1). In doing so, we need to ensure that the spatial metrics for each snapshot and their trends differ from what would be expected by chance. A standard way to produce null models is to generate surrogate data and compare the trends in the indicators obtained from the original data to the trends obtained from the surrogate data [32]. Here, we discuss ways of obtaining null models for spatial early warning signals.

Null models. One way of obtaining a null model is to randomly permute or shuffle the elements of the spatial matrix, and this is also called bootstrapping. This is equivalent to a randomization procedure that removes any spatial structure from the original data but conserves the values of spatial variance and spatial skewness since these moments do not depend on the spatial arrangement of the data points. Therefore, such surrogate data cannot act as a null model for spatial variance or skewness but only for other metrics such as spatial autocorrelation, DFT, and patch-size distribution.

To devise a null model for spatial variance and skewness, Eby, Guttal and others (unpublished data) propose a coarse-graining method which should be applied for both the reshuffled matrix and the real data matrix. In this method, we first divide the full matrix of dimension $n \times n$ into nonoverlapping submatrices of size $s \times s$. We then replace each submatrix by its average to obtain a smaller 'coarse-grained matrix' of size $c_g \times c_g$ (note that $c_g = n/s$). The basic intuition behind the method is as follows: consider any two non-overlapping submatrices of dimension (e.g. 5×5) from the reshuffled matrix. Since the reshuffled matrix is equivalent to a random matrix, the average of the entries of the two sub-matrices chosen would be roughly equal to the average of the full matrix. This exercise of 'coarse-graining' necessarily reduces variability across submatrices in the case of a reshuffled (thus, random) matrix. Now, consider two non-overlapping submatrices in the real data. If we expect that the real data contains a non-random spatial pattern, the average of the entires of the two submatrices need not be of comparable value to each other nor with the average of the full real data matrix. Therefore, in contrast to the reshuffled matrix case, coarse-graining will not necessarily reduce variability in the real data, especially if it contains a spatial pattern. Since variability determines spatial variance and skewness, the coarse-graining applied to the reshuffled matrix provides a null model for spatial variance and spatial skewness.

An alternative method of building null models is to construct a spatial matrix from a continuous stochastic process. This method is applicable when continuous data (such as biomass density) is available at each spatial point as in data set 1 and 3. More specifically, we construct a null model matrix where each entry is a random number (e.g., from a normal distribution) whose mean and variance are equal to the mean and variance of the original data matrix, respectively. This approach provides a null model for spatial skewness, correlation, and DFT, but not for spatial variance since variance is, by construction, fixed to be the same as the one from the original matrix (see figures in Appendix S3). However, it may be claimed, following the same arguments as above, that the coarse-graining method can help estimate statistical significance of spatial variance.

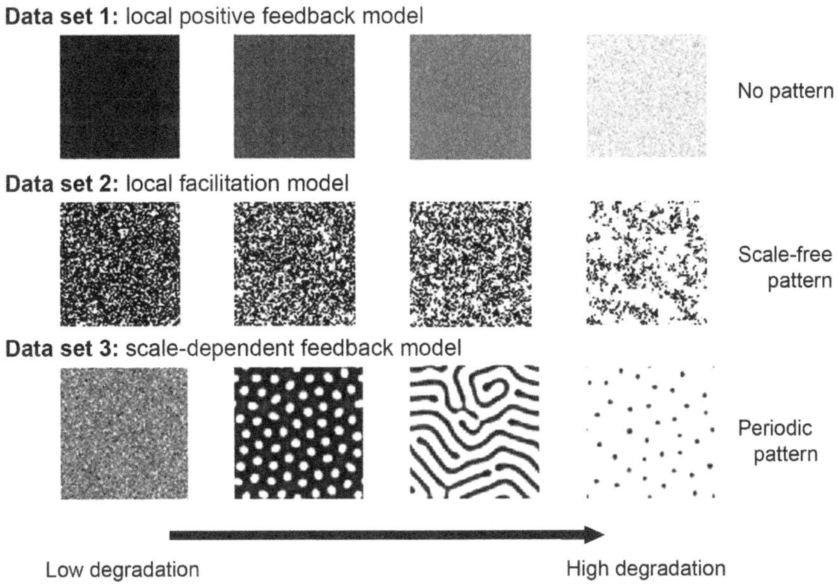

Figure 1. Spatial patterns along a degradation gradient in the three data sets. In each row, the system approaches the bifurcation point from left to right (see Fig. S1 in Appendix S2 to visualize the location of the four snapshots along the degradation gradient.) First row: local positive feedback model (data set 1). Middle row: local facilitation model (data set 2). Bottom row: scale-dependent feedback model (data set 3). In each panel, darker cells correspond to higher vegetation biomass.

The particular case of patchy ecosystems. A system with patchiness may show characteristic scales in the r- and θ-spectra. These r- and θ-spectra of the data can be compared to those obtained from a null model. In the case where there is no spatial structure, it is known analytically that the values of the scaled power spectrum observed in each bin when calculating the r- or θ-spectrum should be distributed as $\frac{\chi^2_{2k}}{2k}$ with k the number of values in the bin (see for instance [35,47]; this gives very similar results as the confidence intervals presented in Appendix S3).

Once the patches are identified and their size evaluated, various heavy tailed (e.g. power law, power law with a cut-off, log-normal, etc) and non-heavy tailed distributions (e.g. exponential) can be fitted to the patch-size distribution. One way of fitting a given distribution to data has been to use ordinary least square regression on the log-log transformed probability distribution function of patch sizes. However, this method is known to have substantial bias in estimating the parameter values, especially for small data sets [44,45]. Maximum likelihood methods or least-square fits of the inverse-cumulative distribution (which quantifies the number of patches whose size is larger than a given value s for different values of s) provide more accurate estimates of the parameters of most heavy-tailed functions [48].

Trends. The above methods inform us about whether the spatial indicators for a given spatial data set are significantly different from those of random patterns. However, to anticipate ecological transitions, we also need to know how these spatial indicators are changing along a degradation gradient. Trend statistics like the Kendall's τ [12,30] or Pearson's correlation coefficient [25,49] can be used to quantify the strength of the trend in the indicator along the gradient.

Simulated spatial data sets

Spatial data in ecology are typically obtained by field studies, data collecting devices placed at various locations of an ecosystem or extracted from spatial imagery. In any of these cases, the nature of data at a given spatial location can be of two types: (a) a discrete occupancy data, such as presence or absence of vegetation (or species) at each pixel of an image, or (b) a continuous variable, such as NDVI (Normalized Difference Vegetation Index) at each pixel or nutrient concentration at each sampling point.

Here, to serve better our method-illustration purposes, we chose to work on model-generated data rather than real data. We thereby circumvent limitations of missing or noisy data, and avoid issues of misinterpretation arising from potential insufficient knowledge about the underlying degradation gradients in real data. We generated three synthetic data sets using three representative models of tipping points and self-organization in ecological systems. The three models treat ecological variables in a spatially-explicit framework with stochasticity. They all describe vegetation dynamics under resource limitation or grazing pressure, but they differ in the nature of ecological interactions and the emerging spatial vegetation structure. A detailed description of the models can be found in Appendix S2 but a brief description follows.

- Data set 1 was obtained from a local positive feedback model resulting in a non-patchy vegetation structure [50,51] (Fig. 1 first row). Space is represented as a two-dimensional lattice [52,53]. Locally, vegetation density grows logistically and is lost due to grazing. Biomass and water are exchanged between neighboring sites at a certain rate, such that a site with high biomass (or water) will have the tendency to diffuse biomass (or water) to its neighboring sites. As rainfall falls below a certain threshold, the ecosystem undergoes an abrupt transition from a globally high vegetation density to a bare state due to the nearly synchronous shifts of each of the sites to a desert state [52].
- Data set 2 is based on a local facilitation model that exhibits spatial patterns characterized by a scale-free patch-size distribution [15]. In this stochastic cellular automaton model, an ecosystem is represented by a grid of cells, each of which

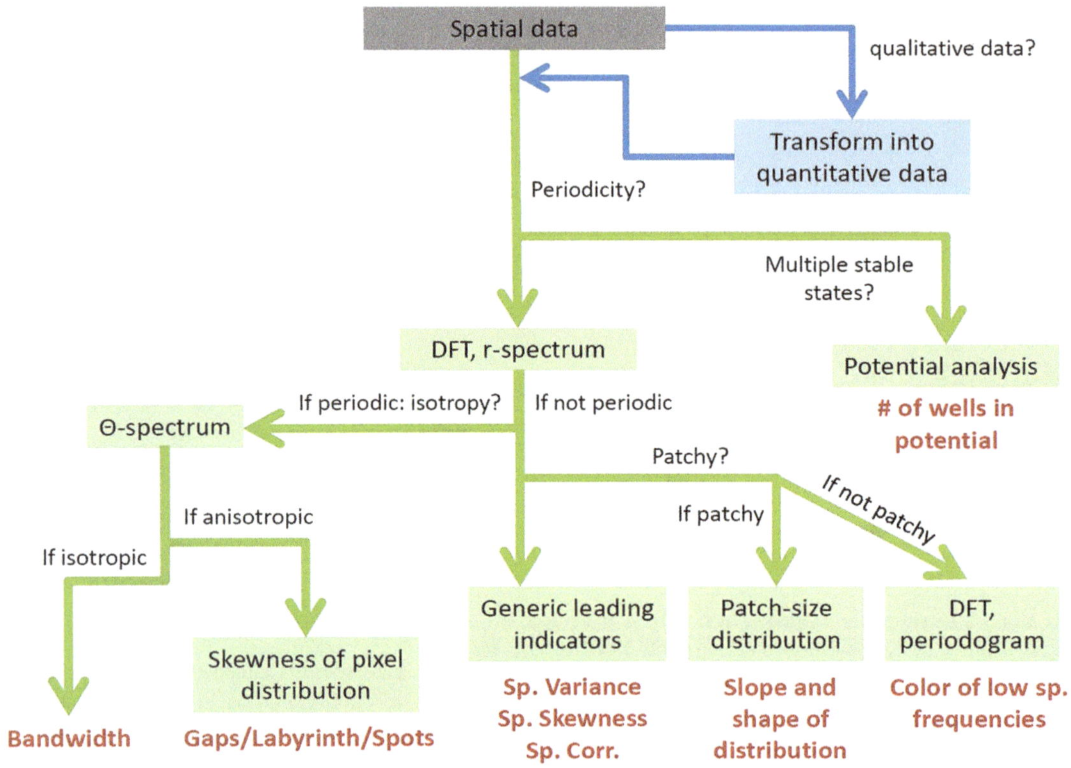

Figure 2. Flow chart of analysis to perform on a spatial data set.

can be in one of three discrete states: vegetated (+), empty (o) or degraded (−). Empty cells represent fertile soil whereas degraded cells represented eroded soil patches unsuitable for recolonization by vegetation. A key ecological mechanism is the positive effect of vegetation on its local neighborhood through increased regeneration of degraded cells. Because of this local facilitation, vegetated cells tend to form clusters (Fig. 1 second row). When the environmental conditions become harsher, there is a point at which the vegetation dies out and the system becomes a desert resembling a saddle-node (or fold) bifurcation.

- Data set 3 is based on a scale-dependent feedback model that results in periodic (Turing-like) spatial vegetation patterns [54]. This model is based on a three partial differential equations model describing the dynamics of vegetation biomass, soil water and surface water. Plants grow due to soil water availability and die due to natural mortality and/or grazing. The infiltration rate of water in the soil is higher in areas with vegetation than in bare soil, leading to the accumulation of water under vegetation and to its depletion further away, resulting in a scale-dependent feedback responsible for the formation of regular vegetation patterns [54] (Fig. 1 last row). When water availability becomes limited, a homogeneous vegetated state becomes unstable leading to self-organized patterns such as gaps, labyrinths and spots. A further reduction in water availability leads to a transition into a desert state, again mimicking a fold-like bifurcation.

The three models exhibit a bifurcation from one state (e.g., vegetated) to an alternative state (e.g., desert) as an external parameter (such as rainfall, grazing, etc) changes. See Appendix S2 for underlying mathematical equations and parameter values of these models. In a previous study [12], it has been shown that the three systems take increasingly longer to recover to their equilibrium after perturbation, thus demonstrating that critical slowing down is a generic feature of the transitions observed in these three ecological models, regardless of their different underlying mechanisms and their different types of spatial structures.

For our analysis, we selected ten snapshots (i.e. two-dimentional space discretized into matrices recording the spatial spread of the vegetation at the end of the simulation) for each of these models at different points along a gradient of degrading conditions prior to the transition. We illustrate our analyses using only ten of these points which are not equally spaced along the gradient (their location is shown on Fig. S1 in Appendix S2).

We are interested in quantifying how the spatial characteristics of these matrices change when approaching a tipping, or bifurcation, point. We are considering cases where the whole ecosystem shifts to an alternative state. In our mathematical representation of the ecosystem, this is equivalent to the entire matrix undergoing a shift. The degradation sequence of the matrices might correspond to snapshots in time (e.g. temperature changing through time) or in space (e.g. herbivory pressure changing in space depending on a distance to a water point). Both types of data are relevant to evaluate and test early warning signals. However, shifts of a given spatial ecosystem in time are more commonly the type of phenomena that we are trying to anticipate.

Results

We suggest a step by step process to decide which spatial indicators should be used (Fig. 2). The spatial statistics that need to be evaluated depend on the type of data set (with discrete or

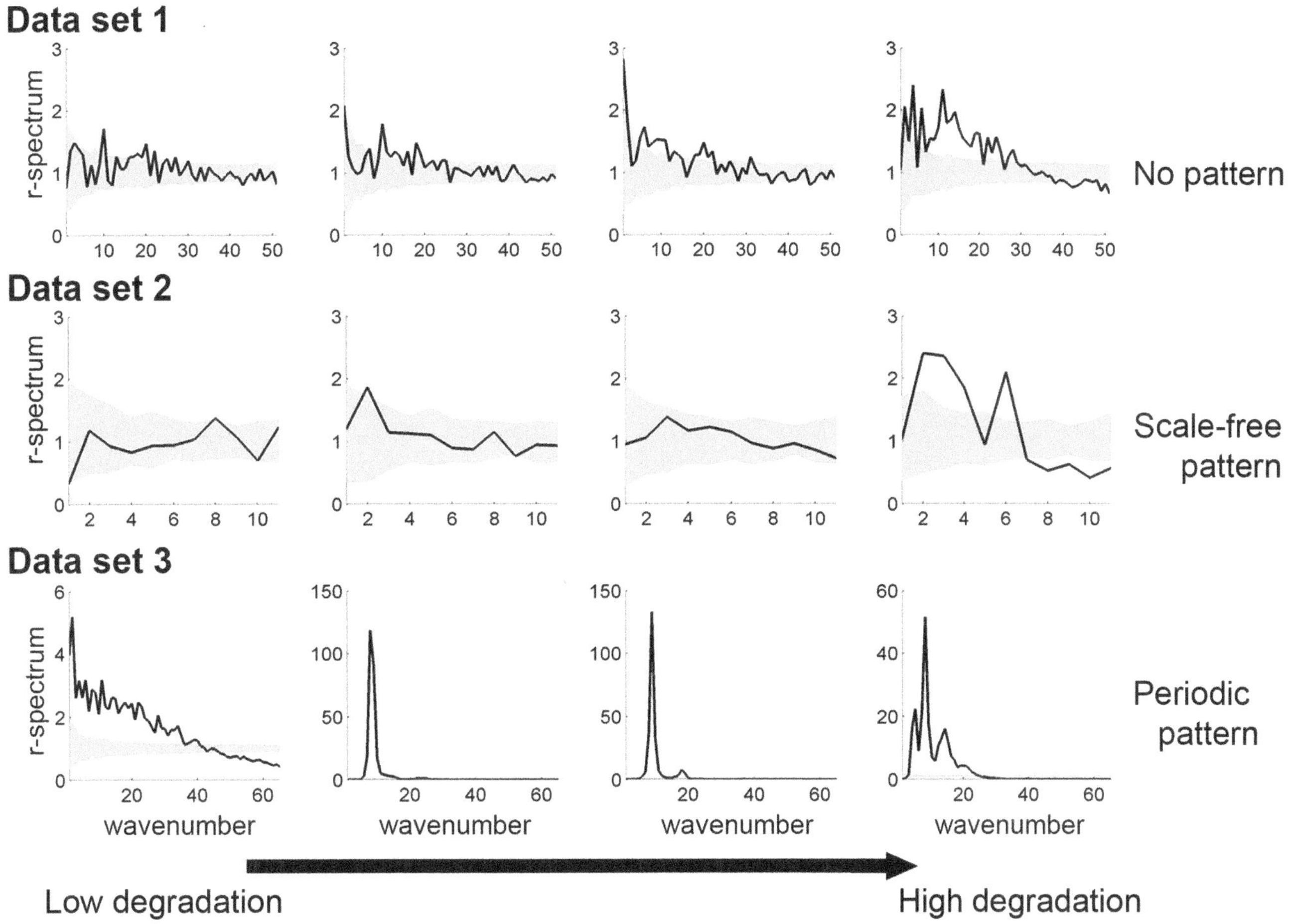

Figure 3. Radial-spectrum along a degradation gradient in the three data sets. In a row, each panel corresponds to the radial-spectrum of the system at a different location along the degradation gradient. The system approaches the bifurcation point from left to right column (as in Fig. 1). First row: local positive feedback model. Middle row: local facilitation model (original data transformed using 5×5 submatrices). Bottom row: scale-dependent feedback model. Gray areas correspond to 95% confidence intervals obtained using 200 simulations of a null model (i.e. data sets of same size generated by reshuffling the original data set).

continuous values) at hand. Two of the three datasets used in this paper provide quantitative, continuous data, i.e. vegetation biomass (data set 1 and 3), whereas data set 2 gives qualitative, discrete data indicating the presence or absence of vegetation in each cell. For this latter data set, we transformed the original matrix by the coarse-graining procedure described in "Stastistical significance tests". More specifically, we used submatrices of 5×5 cells in which we counted the number of cells occupied by vegetation [12]. We obtained matrices that were 25 times smaller than the original ones and where values in each of the cells ranged between 0 and 25, indicating the local abundance of vegetation. The size of the submatrix used to transform the original data may affect the behavior of the indicators.

The first question to ask is whether the patterns observed are periodic or not. The *r*-spectrum obtained from the DFT analysis provides information about whether the patterns are periodic, while the θ-spectrum indicates whether the patterns are isotropic (i.e. with no specific orientation) or anisotropic (i.e. with a specific orientation, e.g. band-like patterns). If the patterns are not periodic, the generic leading indicators (i.e. spatial variance, spatial skewness and spatial correlation between nearby sites) may be used [12] and the power spectrum should be checked for possible reddening [13]. In addition, if the patterns are not only irregular but also patchy (e.g. can be characterized by two phases, one vegetated and one bare), the patch-size distribution may be plotted and estimated [15,46]. If the patterns are periodic and anisotropic, the wavelength of the pattern should be evaluated. The wavelength is equivalent to the dominant length scale of the pattern provided by the *r*-spectrum [43]. For periodic isotropic patterns, the skewness of the distribution of values of the data set (e.g. grey pixels in the case of a greyscale image) indicates the type of patterns (i.e. spots, gaps or labyrinths) [43]. Additionally, it is noteworthy that if the data set includes several replicates at each stress level, potential analysis may be performed [55–57] (see more details in Appendix S1).

Next, we present how this methodology can be applied to our three data sets.

1. Distinguishing periodic from non-periodic patterns

We used DFT analysis to estimate the *r*-spectra as a function of wavenumbers for all the three data sets (Fig. 3). The first and second data sets (Fig. 3, first and second row) show a noisy pattern indicating that contribution to *r*-spectra is not significant for all wavenumbers. However, the *r*-spectrum for the third data set, the scale-dependent feedback model, shows a clear peak (Fig. 3 last row) even far from the transition. The peak indicates that there is dominant wavelength (corresponding to a characteristic patch size) which is a signature of periodic patterns. Note that periodicity can

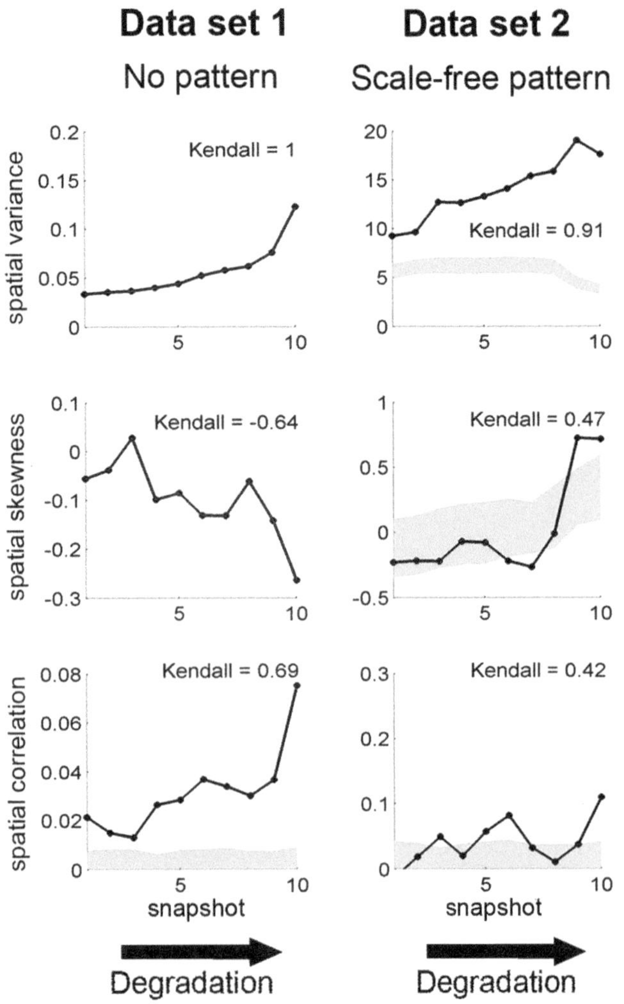

Figure 4. Generic leading indicators in data sets 1 and 2 along a degradation gradient. In each panel, the x-values correspond to the rank of the snapshot of the system along the degradation gradient. This mimicks a scenario where we would not know the exact value of the driver but where we can order the data set along a degradation gradient and see Fig. S1 in Appendix S2 to visualize the location of the four snapshots along the degradation gradient. For each of the three data sets, 10 snapshots were used. Left: local positive feedback model (data set 1). Right: local facilitation model (data set 2; original data transformed using 5×5 sub-matrices). First row: spatial variance. Second row: spatial skewness. Third row: spatial correlation at lag one. In each panel, Kendall's τ, quantifying the trend of the indicator, is indicated. Gray areas correspond to 95% confidence intervals obtained using 200 simulations of a null model (i.e. data sets of same size generated by reshuffling the original data set).

also be seen by plotting the power spectrum, where the periodicity is visible as a ring corresponding to the dominant wavelength of the data set (see upcoming paragraph "DFT and reddening"). Therefore, we conclude that the two first data sets are not periodic whereas the last one shows some spatial periodicity.

2. Probing spatial early warnings for non-periodic patterns

We first focus on the case of the non-periodic patterns, i.e. data set 1 showing no clear spatial structure and data set 2 with vegetation clusters.

Spatial correlation, variance and skewness. In data set 1, spatial correlation at lag 1 and spatial variance increase whereas spatial skewness decreases toward the bifurcation point (Fig. 4), as expected from theory. The behavior of these indicators is very similar for data set 2, except that the spatial variance decreases just before the collapse. This difference in the behaviour of the spatial variance is due to the fact that data set 2 measures only presence or absence of vegetation (i.e qualitative data) whereas data set 1 provides biomass density at each location in space (i.e. quantitative information). We note that the spatial variance and skewness for data set 1 are identical to those of null model which was obtained by a random reshuffling; therefore, we do not see error bars. However, we used the coarse-graining method for the discrete data set 2 which provides a null model for spatial variance and skewness. See a later section on 'Probing statistical significance' for further comments.

DFT and reddening. The spatial power spectrum (or 2D-periodogram, see Eq. 5 in Appendix S1) shows a reddening of the signal, i.e., the amplitude of the power spectra increases at low wavenumbers, as the system approaches the bifurcation point (Fig. 5 first and second rows). The reddening of the power spectra provides advance warning of the transition in all data sets. That trend is even clearer on the r-spectra, which sums the values of the 2D-periodogram for all the wavenumbers and shows that the lower wavenumbers contribute more to the total variance of the data set as the system approaches the bifurcation point (Fig. 3 first and second rows).

Non-periodic and patchy: patch-size distribution. Data set 2 was not only characterized by non-periodic patterns, but our visual examination reveals that it also exhibits distinct patches of vegetation and bare ground. In that case, it makes sense to look at the distribution of patch sizes. A way of plotting such data is to calculate the inverse cumulative distribution, i.e. plotting the number of patches whose size is larger than a given value s as a function of s. The inverse cumulative distribution is nearly scale-free far from the bifurcation point, while its slope decreases and the distribution becomes bent (toward less large patches) as the system approaches the bifurcation point (Fig. 6 top row) [46].

For comparison, we plotted the inverse cumulative patch-size distribution of data set 3 that is also patchy but periodic. Far from the transition, after the onset of pattern formation, the periodic patterns presented a patch-size distribution characterized by a sharp cutoff (Fig. 6 bottom row). As the system approaches the bifurcation point, the value of the cutoff decreases indicating decreasing patch size in the periodic pattern.

3. Probing spatial early warnings for periodic patterns

In contrast to the first two data sets, data set 3 exhibits periodic patterns (Fig. 1 and 3 last rows). The θ-spectra does not indicate a strong and clear signal at any specific angle, suggesting that the patterns do not have a clear orientation, i.e. patterns are isotropic (Fig. 7 first row). When the system approaches the bifurcation point (from left to right panel on Fig. 7 second row), the distribution of values of the data set goes from one peak reflecting the absence of patterns, to a two-peak distribution due to the occurrence of both vegetation and bare soil in the system after the emergence of spatial patterns, and finally to a distribution that is skewed toward small values because of the dominance of bare soil in the system. The last distribution observed before the bifurcation point characterizes spot patterns [43] which has been hypothesized to be an indicator of imminent desertification [22].

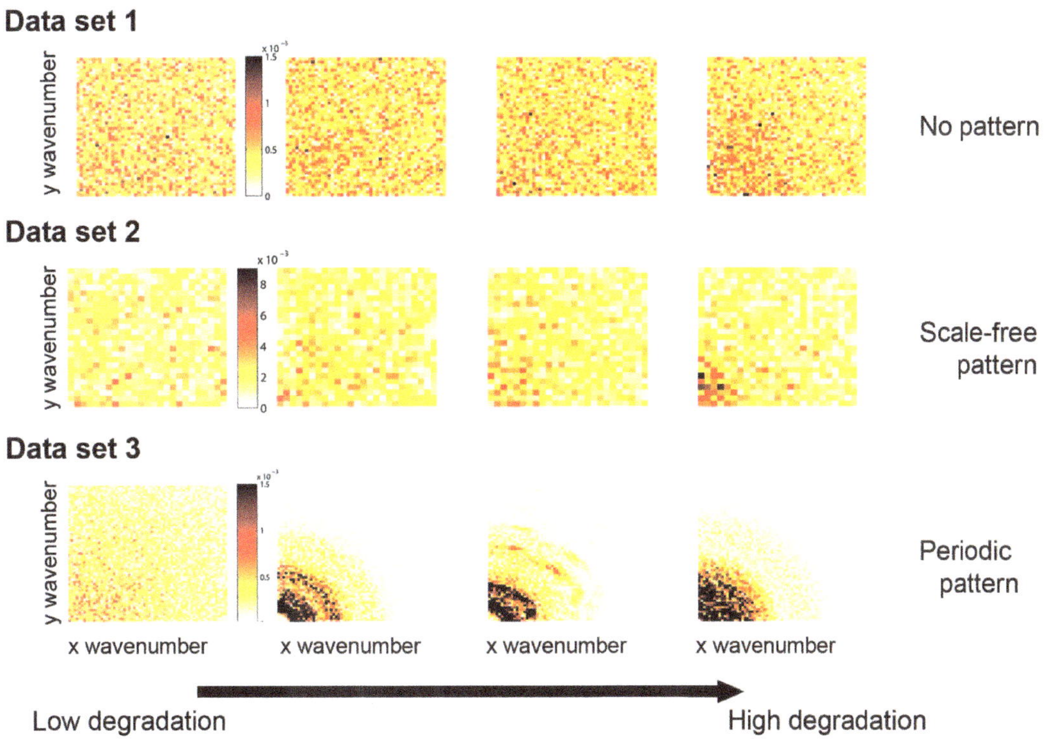

Figure 5. Power spectrum along a degradation gradient in the three data sets. In a row, the system approaches the bifurcation point, from left to right column (as in Fig. 1). For a data set of size $M \times N$, the power spectrum is typically plotted for wavenumbers up to $p = \frac{M}{2}$ and $q = \frac{N}{2}$ [47] and is scaled by the spatial variance σ^2 (i.e. the scaled power spectrum is evaluated as $\frac{I}{\sigma^2}$) [35]. Red color indicates higher values of the scaled power spectrum, $\frac{I}{\sigma^2}$. The x and y-axis correspond to the wavenumbers along these directions. First row: local positive feedback model (data set 1). Middle row: local facilitation model (data set 2; original data transformed using 5×5 sub-matrices). Bottom row: scale-dependent feedback model (data set 3).

4. Probing statistical significance

We make further comments on the statistical significance for indicators of spatial data, especially in the context of null models for spatial variance and spatial skewness. As an illustration of the coarse-graining method proposed in the "Methods" section, we compared the trends in the generic leading indicators on coarse-grained matrix obtained by different dimensions of submatrix, $s \times s$ with $s = 1$ (in this case the coarse-grained matrix is identical to the original matrix), 2 and 5 but starting from the same original matrix of size 100×100 from our data set 2. As shown in the first row of Fig. 8, the values of spatial variance along the degradation gradient differ substantially from the ones of the null model only in the coarse-grained case (second and third column). The same result seems to be true, although the differences are not as pronounced, for spatial skewness (second row). In contrast, the coarse-graining method does not offer a good null model for spatial correlation (last row, Fig. 8). In summary, a random matrix that has the same dimensions and average as the original spatial data can act as a null model for computing spatial correlation only. On the other hand, a comparison between indicators of coarse-grained matrices of both original and random matrix data can act as a null model for spatial variance and spatial skewness.

Discussion

In this manuscript, we presented a systematic methodology for applying spatial early warning signals of abrupt ecological transitions. To demonstrate the methods, we employed data generated from simulations of spatially-explicit ecological models with stochasticity that showed abrupt transitions from one state to an alternative state. It is increasingly being recognized that spatial dynamics pose challenges and provide opportunities for both basic science as well as management [58]. Research on spatial dynamics in ecology is continually uncovering new patterns and mechanisms. Some of these processes are likely related to regime shifts. In this context, the main objective of this manuscript was to provide a methodological guide that can stimulate the application of spatial indicators for ecological transitions on empirical data sets from real case studies.

In recent years, much effort has been devoted to the search for 'generic' indicators based on the idea that there are some common behaviors across a range of complex systems as they approach a bifurcation point [59]. Taking into account the spatial organization of natural systems has revealed that many indicators may not behave in spatially-structured systems as they would in other systems. More specifically, recent theoretical studies have suggested that trends of generic leading indicators (spatial variance, spatial skewness, and spatial correlation between nearby sites) could be different in self organized patterned systems [12]. For such ecosystems, system-specific indicators may be more appropriate. In particular, when spatial patterns are periodic or regular (Turing-like), the shape of the patterns may give an idea of the proximity to the threshold where the system may undergo a regime shift [22]; specifically, spots could warn of approaching

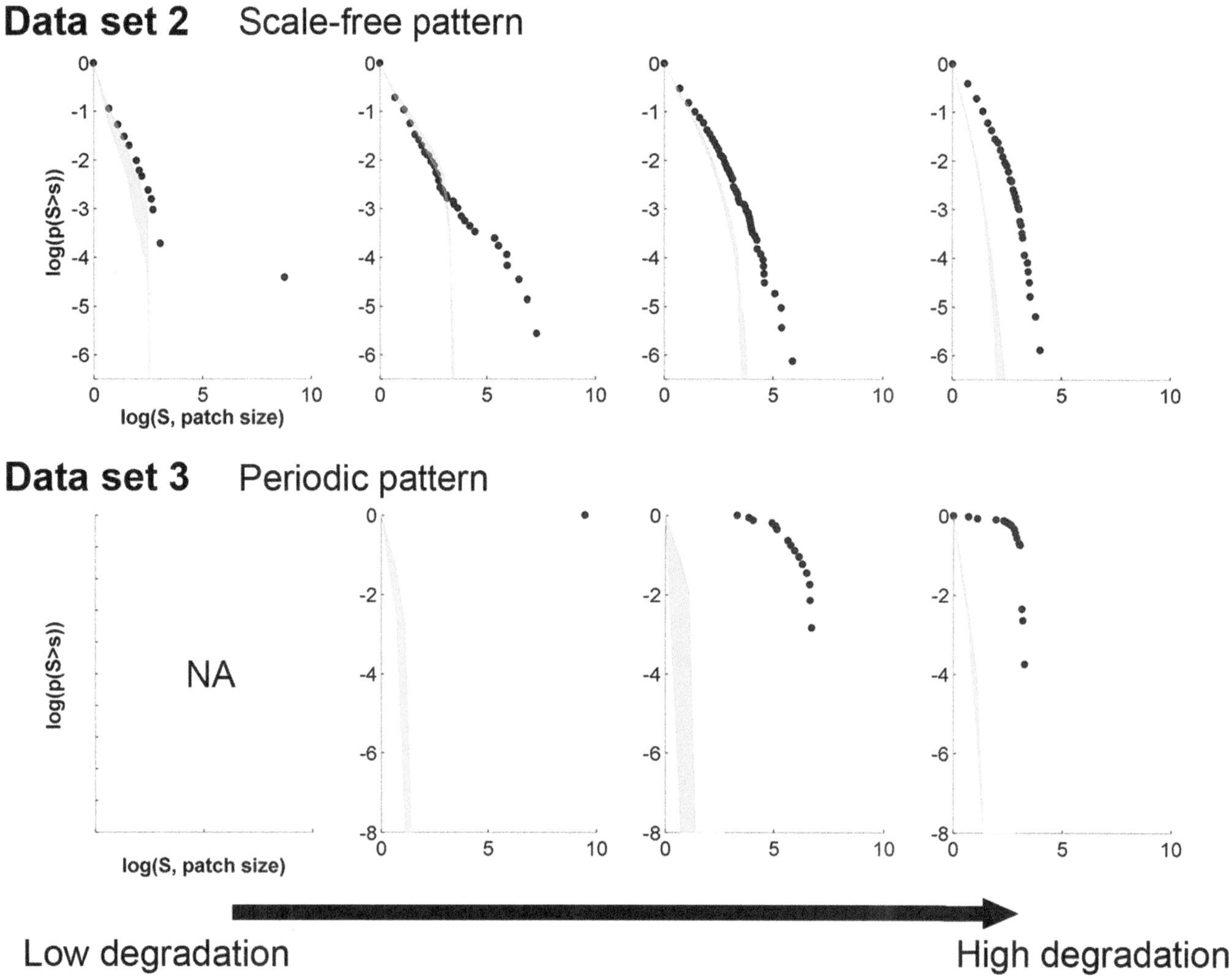

Figure 6. Inverse cumulative patch-size distributions along a degradation gradient in data sets 2 and 3. Along each row, the system approaches the bifurcation point from left to right colum (as in Fig. 1)n. First row: local facilitation model (original data not transformed). Bottom row: scale-dependent feedback model. Gray areas corresponds to 95% confidence interval obtained using 200 simulations of a null model (i.e. data sets of same size generated by reshuffling the original matrix).

desertification [16,22]. If the patterns are non-periodic or irregular (in particular, if the patch-size distribution is described by a heavy-tailed distribution), the patch size distribution may contain information about the degradation level of the ecosystem but more needs to be known about the underlying ecological mechanisms to interpret the changes in the shape of this distribution [15,20,46,60]. In other words, the nature of spatial organization of ecosystems seems to be a key factor in determining what type of indicators may be employed to detect an impending ecological transition in spatially-structured systems. Therefore, a first and essential step when starting to analyze a spatial data set is to get an idea of the type of spatial organization that one is dealing with. A good knowledge of the system and its underlying ecological mechanisms (specifically those responsible for the spatial structure) are required to know which indicator to use and how to interpret the changes. Theoretical studies have started developing methodologies for inferring underlying mechanisms from a limited number of spatial snapshots [61,62]. Such knowledge will facilitate to assess the risks of ecological transitions while accounting for potential false and failed alarms.

Clearly, the usefulness of the indicators presented here depends greatly on the underlying mechanisms driving the change in the ecosystems studied. In all our analyses, we assumed that a single driver is monotonically changing, while the underlying environmental conditions are assumed relatively stable and the environmental stochasticity relatively weak. This may not always be the case. For example, spatial correlation can increase due to changes in the underlying spatial heterogeneity of the environment, or due to alterations in local 'mixing' (or diffusion) levels in the landscape [5]. Effects of increasing spatial variance near a critical point could be confounded by various intrinsic factors such as demographic noise arising from changes in population sizes, state-dependent/multiplicative noise. The nature of the dispersal processes between patches in fragmented landscapes may affect expected trends in the indicators. In addition, extrinsic factors such as environmental fluctuations that vary in space and time can complicate our interpretations of early warning signals. In a similar way, system-specific conditions may affect the behavior and thereby interpretation of the indicators. For example, in Mediterranean drylands, degradation is accompanied by a change in the patch-size distribution toward less large patches, whereas the opposite was

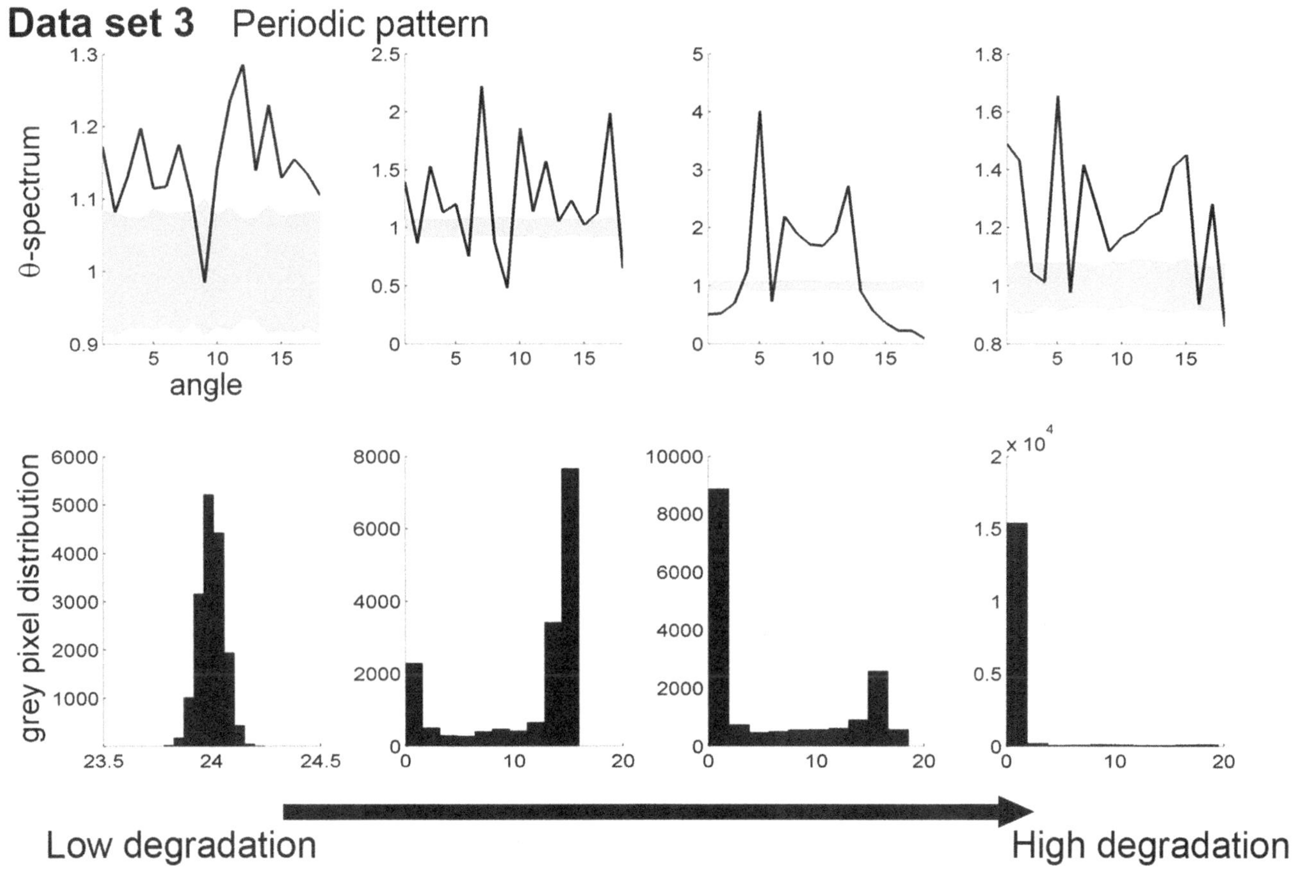

Figure 7. Analysis of the periodic patterns of data set 3. In a row, the system approaches the bifurcation point from left to right column (as in Fig. 1). First row: θ-spectrum. Gray areas corresponds to 95% confidence interval obtained using 200 simulations of a null model (i.e. data sets of same size generated by reshuffling the original matrix). Second row: histogram of the values of the data set (or pixels of the image).

observed in salt marshes because the underlying mechanisms driving the formation of the patterns differ in the two ecosystems [15,46,60]. Therefore, our interpretations of leading indicators are prone to both false positives and false negatives.

To prevent such errors, knowledge of the underlying heterogeneity, repeated observations of the system [60], or a sufficiently well-described system so that it can be modelled are necessary to know what to expect along a degradation trend. As this is not usually the case, general models have been suggested to be fitted to time series data[63,64] to simulate surrogate scenarios for comparing trends obtained from the original data sets. In this case the model provides an expectation of i) whether and when a shift is likely, and ii) what trends should look like as the system is approaching a shift. Such approaches, however, are yet to be developed for spatially-explicit systems. Although we presented a couple of ways to develop null models, these are rather simplistic as they entirely neglect any underlying spatial structuring. Therefore, the design and selection of null models for spatial data, which is an area of active research in ecology, is another important avenue for further research [65]. There are other promising avenues of research to be pursued. Composite metrics that combine spatial patterns with their temporal dynamics could potentially offer new and potentially more reliable indicators of imminent transitions [66]. Ecological transitions and leading indicators in the context of metapopulation dynamics, where factors that stabilize or make species more vulnerable to extinction have been extensively studied, may be a fertile ground for further research.

More importantly, despite a few recent studies, we still lack empirical tests of spatial early warnings. Researchers have studied spatial warnings in laboratory populations of microbial organisms such as Daphnia and yeast [25,31]. In these studies, individual populations are maintained in locally well-mixed small beakers or petri-dishes and they are 'connected' to other populations by 'controlled dispersal' where the researcher transfers a fraction of the local population to its nearest neighbors. In the field, researchers have employed space-for-time substitution; in this approach, it is assumed that spatial patterns at locations with different values of stressors (e.g. grazing or rainfall) are equivalent to the dynamics of spatial patterns where the stressor is changing with time. This is a widely used approach in ecology as an alternative to long term ecological studies, for example to investigate ecological succession or how ecosystems may respond to climate change [67]. In the context of alternative stable states and tipping points, this method has been employed to establish the existence of (multiple) stable states in savanna ecosystems as a function of rainfall [57,68], and to forecast how spatial self-organization of semi-arid vegetation may respond to increasing stressors such as grazing [15]. However, the application of space-for-time method is not without limitations [69]. One needs to be cautious about the possibility of existence of various other sources of heterogeneity, both biotic and abiotic, along a spatial gradient of stressor. As we have argued before, the interpretation of the

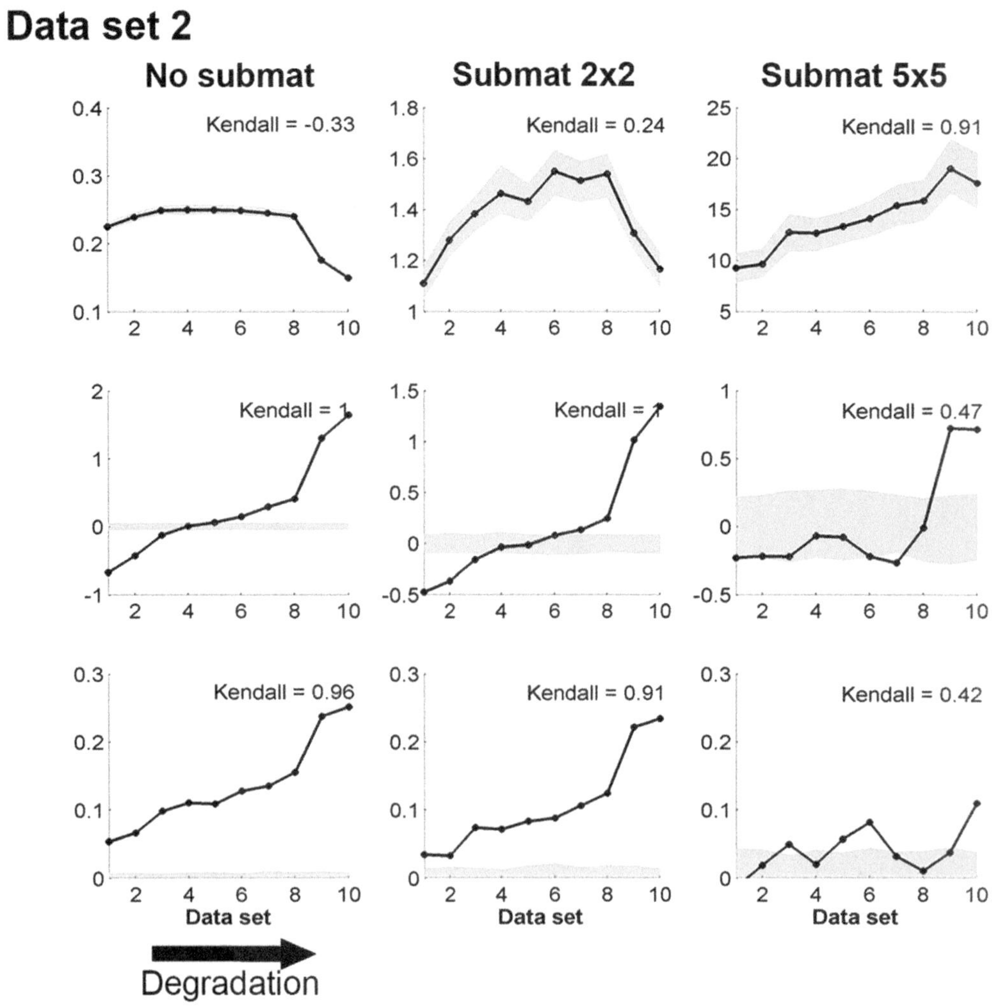

Figure 8. Generic leading indicators in data set 2 along a degradation gradient using different coarsening of the original data. In each panel, the x-value indicates the rank of the snapshot along the degradation gradient. This mimicks a scenario where we would not know the exact value of the driver but where we can order the data set along a gradient and See Fig. S1 in Appendix S2 to visualize the location of the four snapshots along the degradation gradient. Left: original data. Middle: original data transformed using 2×2 sub-matrices. Middle: original data transformed using 5×5 sub-matrices. First row: spatial variance. Second row: spatial skewness. Third row: spatial correlation at lag one. In each panel, Kendall's τ quantifies the strength of the trend of the indicator. Gray areas correspond to 95% confidence intervals obtained using 200 simulations of a null model (i.e. data sets of same size generated by reshuffling the original matrix).

trends of early warnings are crucially dependent on the nature of local ecological interactions and the patterns they produce. Therefore, sources of heterogeneity, especially those that alter the ecological processes generating spatial patterns, may therefore compound the complexity of interpretation of results based on space-for-time substitutions.

In conclusion, both the theory and the application of the spatial indicators lag behind the development of the temporal ones. Spatial patterns may however offer advantages for anticipating or detecting ecological transitions of the types studied here [11,32]. Unlike temporal indicators which require long, unbroken time series of frequent observations, spatial indicators may be evaluated even if measurements are irregular or infrequent over time. While spatial pattern measurements require intensive data collection at each time point, in many cases this may be easier than high-frequency time series sampling [5,11,13]. In both space and time, more empirical validation of the indicators currently proposed in the literature is needed. We do not yet have an example where early warning signals were used to avert an upcoming shift (they have been used in models, experiments or retroactively). We hope that this work will stimulate further development, testing and application of spatial indicators in a broad range of ecosystems.

Acknowledgments

This paper was based on the workshop 'Practical Methods for Analysis of Early Warnings for Regime Shifts' organized by VD, SRC and MS in the Santa Fe Institute, in Santa Fe, New Mexico, USA, from 10 to 12 of October 2011. We thank Vincent Deblauwe for useful comments on previous versions of the manuscript and for sharing his r - and θ - spectrum code.

Author Contributions

Conceived and designed the experiments: SK VG VD SRC DAS VNL. Performed the experiments: SK VG. Analyzed the data: SK VG. Contributed reagents/materials/analysis tools: SK VG VD. Wrote the paper: SK VG WAB SRC AME VNL DAS MS EHvN VD.

References

1. Scheffer M, Carpenter SR, Foley JA, Folke C, Walker B (2001) Catastrophic shifts in ecosystems. Nature 413: 591–596.
2. Scheffer M, Bascompte J, Brock W, Brovkin V, Carpenter S, et al. (2009) Early-warning signals for critical transitions. Nature 461: 53–59.
3. Wissel C (1984) A universal law of the characteristic return time near thresholds. Oecologia 65: 101–107.
4. van Nes EH, Scheffer M (2007) Slow recovery from perturbations as a generic indicator of a nearby catastrophic regime shift. American Naturalist 169: 738–747.
5. Dakos V, van Nes E, Donangelo R, Fort H, Scheffer M (2010) Spatial correlation as leading indicator of catastrophic shifts. Theoretical Ecology 3: 163–174.
6. Held H, Kleinen T (2004) Detection of climate system bifurcations by degenerate fingerprinting. Geophysical Research Letters 31: L020972.
7. Carpenter SR, Brock WA (2006) Rising variance: a leading indicator of ecological transition. Ecology Letters 9: 308–315.
8. Guttal V, Jayaprakash C (2008) Changing skewness: an early warning signal of regime shifts in ecological systems. Ecology Letters 11: 450–460.
9. Kéfi S, Dakos V, Scheffer M, Van Nes E, Rietkerk M (2013) Early warning signals also precede non-catastrophic transitions. Oikos 122: 641–648.
10. Obórny B, Meszéna G, Szabó G (2005) Dynamics of populations on the verge of extinction. Oikos 109: 291–296.
11. Guttal V, Jayaprakash C (2009) Spatial variance and spatial skewness: leading indicators of regime shifts in spatial ecological systems. Theoretical Ecology 2: 3–12.
12. Dakos V, Kéfi S, Rietkerk M, van Nes E, Scheffer M (2011) Slowing down in spatially patterned ecosystems at the brink of collapse. The American Naturalist 177: E153–E166.
13. Carpenter S, Brock W (2010) Early warnings of regime shifts in spatial dynamics using the discrete fourier transform. Ecosphere 1: art10.
14. Barbier N, Couteron P, Lejoly J, Deblauwe V, Lejeune O (2006) Self-organized vegetation patterning as a fingerprint of climate and human impact on semi-arid ecosystems. Journal of Ecology 94: 537–547.
15. Kéfi S, Rietkerk M, Alados CL, Pueyo Y, Papanastasis VP, et al. (2007) Spatial vegetation patterns and imminent desertification in mediterranean arid ecosystems. Nature 449: 213–217.
16. Kéfi S, Eppinga M, de Ruiter P, Rietkerk M (2010) Bistability and regular spatial patterns in arid ecosystems. Theoretical Ecology 3: 257–269.
17. Maestre F, Escudero A (2010) Is the patch size distribution of vegetation a suitable indicator of desertification processes? reply. Ecology 91: 3742–3745.
18. Kéfi S, Alados C, Chaves R, Pueyo Y, Rietkerk M (2010) Is the patch size distribution of vegetation a suitable indicator of desertification processes? comment. Ecology 91: 3739–3742.
19. Pueyo S (2011) Desertification and power laws. Landscape ecology 26: 305–309.
20. Lin Y, Han G, Zhao M, Chang S (2010) Spatial vegetation patterns as early signs of desertification: a case study of a desert steppe in inner mongolia, china. Landscape ecology 25: 1519–1527.
21. Litzow M, Urban J, Laurel B (2008) Increased spatial variance accompanies reorganization of two continental shelf ecosystems. Ecological applications 18: 1331–1337.
22. Rietkerk M, Dekker SC, de Ruiter PC, van de Koppel J (2004) Self-organized patchiness and catastrophic regime shifts in ecosystems. Science 305: 1926–1929.
23. Hastings A, Wysham D (2010) Regime shifts in ecological systems can occur with no warning. Ecology Letters 13: 464–472.
24. Boettiger C, Hastings A (2013) Tipping points: From patterns to predictions. Nature 493: 157–158.
25. Drake J, Griffen B (2010) Early warning signals of extinction in deteriorating environments. Nature 467: 456–459.
26. Carpenter S, Cole J, Pace M, Batt R, Brock W, et al. (2011) Early warnings of regime shifts: A whole-ecosystem experiment. Science 332: 1079–1082.
27. Veerat AJ, Faassen EJ, Dakos V, van Nes EH, Lurling M, et al. (2012) Recovery rates reflect distance to a tipping point in a living system. Nature 481: 357–359.
28. Dai L, Vorselen D, Korolev K, Gore J (2012) Generic indicators for loss of resilience before a tipping point leading to population collapse. Science 336: 1175–1177.
29. Wang R, Dearing J, Langdon P, Zhang E, Yang X, et al. (2012) Flickering gives early warning signals of a critical transition to a eutrophic lake state. Nature 492: 419–422.
30. Dakos V, Scheffer M, Van Nes E, Brovkin V, Petoukhov V, et al. (2008) Slowing down as an early warning signal for abrupt climate change. Proceedings of the National Academy of Sciences 105: 14308–14312.
31. Dai L, Korolev K, Gore J (2013) Slower recovery in space before collapse of connected populations. Nature 496: 355–358.
32. Dakos V, Carpenter SR, Ellison AM, Guttal V, Ives AR, et al. (2012) Early warning signals of critical transitions: Methods for time series. PLOS ONE 7: e41010.
33. Legendre P, Legendre L (1998) Numerical ecology. Amsterdam: Elsevier Science.
34. Guttal V (2008) Applications of nonequilibrium statistical physics to ecological systems (phd thesis). The Ohio State University 3: 39–41.
35. Couteron P (2002) Quantifying change in patterned semi-arid vegetation by fourier analysis of digitized aerial photographs. International Journal of Remote Sensing 23: 3407–3425.
36. Sethna JP (2007) Statistical Mechanics: Entropy, Order Parameters, and Complexity. Oxford: Oxford University Press.
37. Dakos V, van Nes EH, Scheffer M (2013) Flickering as an early warning signal. Theoretical Ecology : DOI:10.1007/s12080-013-0186-4.
38. Rietkerk M, van de Koppel J (2008) Regular pattern formation in real ecosystems. Trends in Ecology & Evolution 23: 169–175.
39. Manor A, Shnerb M (2008) Facilitation, competition, and vegetation patchiness: From scale free distribution to patterns. Journal of Theoretical Biology 253: 838–842.
40. von Hardenberg J, Kletter A, Yizhaq H, Nathan J, Meron E (2010) Periodic versus scale-free patterns in dryland vegetation. Proceedings of the Royal Society B: Biological Sciences 2077: 1771–1776.
41. Scanlon T, Caylor K, Levin S, Rodriguez-Iturbe I (2007) Positive feedbacks promote power-law clustering of Kalahari vegetation. Nature 449: 209–212.
42. Foti R, del Jesus M, Rinaldo A, Rodriguez-Iturbe I (2013) Signs of critical transition in the everglades wetlands in response to climate and anthropogenic changes. Proceedings of the National Academy of Sciences 110: 6296–6300.
43. Deblauwe V, Couteron P, Lejeune O, Bogaert J, Barbier N (2011) Environmental modulation of self-organized periodic vegetation patterns in sudan. Ecography 34: 990–1001.
44. Newman M (2005) Power laws, pareto distributions and zipf's law. Contemporary physics 46: 323–351.
45. White E, Enquist B, Green JL (2008) On estimating the exponent of power-law frequency distributions. Ecology 89: 905–912.
46. Kéfi S, Rietkerk M, Roy M, Franc A, De Ruiter P, et al. (2011) Robust scaling in ecosystems and the meltdown of patch size distributions before extinction. Ecology letters 14: 29–35.
47. Mugglestone MA, Renshaw E (1998) Detection of geological lineations on aerial photographs using two-dimensional spectral analysis. Computers and geosciences 24: 771–784.
48. Clauset A, Shalizi CR, Newman ME (2009) Power-law distributions in empirical data. SIAM review 51: 661–703.
49. Seekell D, Carpenter S, Cline T, Pace M (2012) Conditional heteroskedasticity forecasts regime shift in a whole-ecosystem experiment. Ecosystems 15: 741–747.
50. Shnerb NM, Sarah P, Lavee H, Solomon S (2003) Reactive glass and vegetation patterns. Physical Review Letters 90: 038101.
51. Guttal V, Jayaprakash C (2007) Impact of noise on bistable ecological systems. Ecological Mod-elling 201: 420–428.
52. van Nes E, Scheffer M (2005) Implications of spatial heterogeneity for catastrophic regime shifts in ecosystems. Ecology 86: 1797–1807.
53. Keitt T, Lewis M, Holt R (2001) Allee effects, invasion pinning, and species borders. the American Naturalist 157: 203–216.
54. Rietkerk M, Boerlijst MC, van Langevelde F, HilleRisLambers R, van de Koppel J, et al. (2002) Self-organization of vegetation in arid ecosystems. American Naturalist 160: 524–530.
55. Livina V, Kwasniok F, Lenton T (2010) Potential analysis reveals changing number of climate states during the last 60 kyr. Climate of the Past 6: 77–82.
56. Livina V, Kwasniok F, Lohmann G, Kantelhardt J, Lenton T (2011) Changing climate states and stability: from pliocene to present. Climate dynamics 37: 2437–2453.
57. Hirota M, Holmgren M, Nes EV, Scheffer M (2011) Global resilience of tropical forest and savanna to critical transitions. Science 334: 232–235.
58. Walters C, Pauly D, Christensen V (1999) Ecospace: Prediction of mesoscale spatial patterns in trophic relationships of exploited ecosystems, with emphasis on the impacts of marine protected areas. Ecosystems 2: 539–554.
59. Gilmore R (1981) Catastrophe Theory for Scientists and Engineers. New York: John Wiley and Sons.
60. Weerman E, Van Belzen J, Rietkerk M, Temmerman S, Kéfi S, et al. (2012) Changes in diatom patch-size distribution and degradation in spatially self-organized intertidal mudflat ecosystem. Ecology 93: 608–618.
61. Marcos-Nikolaus P, Martin-González J, Solé R (2002) Spatial forecasting: Detecting determinism from single snapshots. International Journal of Bifurcation and Chaos 12: 369–376.

62. Mocenni C, Facchini A, Vicino A (2010) Identifying the dynamics of complex spatio-temporal systems by spatial recurrence properties. Proceedings of the National Academy of Sciences of the United States of America 107: 8097–8102.
63. Boettiger C, Hastings A (2012) Quantifying limits to detection of early warning for critical transitions. Journal of the Royal Society Interface 9: 1–29.
64. Lade S, Gross T (2012) Early warning signals for critical transitions: a generalized modeling approach. Plos computational biology 8: e1002360.
65. Fortin M, Dale M (2005) Spatial analysis: a guide for ecologists. Cambridge, UK: Cambridge University Press.
66. Corrado R, Cherubini AM, Pennetta C (2013) Signals of critical transitions in ecosystems associated with fluctuations of spatial patterns. In: Noise and Fluctuations (ICNF), 2013 22nd International Conference on. IEEE, pp. 1–4.
67. Blois JL, Williams JW, Fitzpatrick MC, Jackson ST, Ferrier S (2013) Space can substitute for time in predicting climate-change effects on biodiversity. Proceedings of the National Academy of Sciences 110: 9374–9379.
68. Staver AC, Archibald S, Levin SA (2011) The global extent and determinants of savanna and forest as alternative biome states. Science 334: 230–232.
69. Johnson EA, Miyanishi K (2008) Testing the assumptions of chronosequences in succession. Ecology Letters 11: 419–431.

14

Spatial Structures of the Environment and of Dispersal Impact Species Distribution in Competitive Metacommunities

Dexiecuo Ai[1], Dominique Gravel[2], Chengjin Chu[1], Gang Wang[1]*

1 State Key Laboratory of Grassland and Agro-Ecosystems, School of Life Sciences, Lanzhou University, Lanzhou, China, **2** Département de Biologie, Chimie et Géographie, Université du Québec à Rimouski, Rimouski, Canada

Abstract

The correspondence between species distribution and the environment depends on species' ability to track favorable environmental conditions (via dispersal) and to maintain competitive hierarchy against the constant influx of migrants (mass effect) and demographic stochasticity (ecological drift). Here we report a simulation study of the influence of landscape structure on species distribution. We consider lottery competition for space in a spatially heterogeneous environment, where the landscape is represented as a network of localities connected by dispersal. We quantified the contribution of neutrality and species sorting to their spatial distribution. We found that neutrality increases and the strength of species-sorting decreases with the centrality of a community in the landscape when the average dispersal among communities is low, whereas the opposite was found at elevated dispersal. We also found that the strength of species-sorting increases with environmental heterogeneity. Our results illustrate that spatial structure of the environment and of dispersal must be taken into account for understanding species distribution. We stress the importance of spatial geographic structure on the relative importance of niche vs. neutral processes in controlling community dynamics.

Editor: Matteo Convertino, University of Florida, United States of America

Funding: The study was financially supported by the National Natural Science Foundation of China (No. 31270472). The funders had no role in study design, data collection and analysis, decision to publish, or preparation of the manuscript.

Competing Interests: The authors have declared that no competing interests exist.

* E-mail: wgmg36@lzu.edu.cn

Introduction

The relationship between species distribution and the environment is a central topic in ecology [1–5] and is important to various fields in ecology such as biogeography, evolution and, more recently, in conservation biology and climate change research [6–12]. The central aspect of this relationship for modelers is obviously its strength – that is, how frequently species inhabit their most favorable conditions, or the other way around, how frequently a given location is occupied by the most competitive species [4]. It is however unrealistic that species are optimally distributed in their most suitable conditions because dispersal often constrains the potential of species from accessing some habitable areas [13] and biotic interactions, including human impacts, may further prevent establishment in some areas [4,14]. Emergent ecological drift could also blur the relationship between distribution and the environment despite strong niche differentiation [15].

Metacommunity theory has advanced understanding of how spatial dynamics and local interactions shape community structure and biodiversity [16–18]. It tells us that the correspondence between species distribution and the environment depends on species' ability to track favorable environments (via dispersal), maintain competitive hierarchy against the constant influx of migrants (mass effect) and demographic stochasticity (ecological drift) [15,19,20]. We understand from theory there are two important aspects of the landscape spatial structure susceptible to impact the strength of species-sorting: the connectivity matrix (i.e. the spatial arrangement among localities and dispersal rate among them) and the environmental heterogeneity (i.e. variance and range of environmental conditions and their spatial autocorrelation). However, we only have a limited understanding of how these factors interact to affect species distribution.

A metacommunity was simply defined as a set of local communities linked by dispersal [16,21,22]. Dispersal was considered homogeneous throughout the landscape in most original metacommunity models [20,23]. It is however a much more natural and convenient way to represent landscapes as a network of localities connected by dispersal [24]. An advantage of the network framework is the sophisticated set of quantitative tools available for characterizing network structure [25–27]. For instance, metrics such as network centrality are used to measure the contribution of node position to the importance, influence, prominence of an actor in a network. In this context, the object of interest is no longer the effect of average dispersal, but rather its variance. The network approach has been influential for the development of landscape-scale conservation plans [24,25,28,29] and provides a more realistic representation of spatial dynamics [30,31]. The metacommunity of early neutral models is often represented as a large species pool [32] and held constant during local community dynamics [33–35]. Recent studies on neutral dynamics in spatially explicit landscapes however shown that the spatial arrangement of localities has considerable impacts on the structure at both the local scale [36,37] and the regional scale

[38,39]. For example, it was found that local species richness increases with the centrality of the community in the spatial network, while global diversity increases with higher connectivity and dispersal for metacommunity consisted of fewer local communities.

Niche differentiation over a spatially heterogeneous environment is often considered as the main explanation for species coexistence [40–43]. Niche theory predicts that species richness will increase with environmental heterogeneity. The number of species that could be packed along an environmental gradient however depends on the overlap of their niches. May and McLean [44] defined the limiting similarity as the maximal overlap of species' utilization of resources enabling their coexistence. It was later found that metacommunity dynamics could alter significantly the limiting similarity among species [23,45]. Dispersal limitations prevent a species to fight for all its favorable locations, thereby allowing the coexistence with a competitively inferior, but similar species. Alternatively, a competitively superior species could be displaced from a location by the constant influx (a mass effect) of propagules of an inferior species. Even if both dispersal and environmental heterogeneity has been extensively studied, their effect (and interaction) on the strength of species sorting has seldom been studied.

Ecological drift is an emergent phenomenon that could also impact significantly species distribution. It is defined as population changes arising from stochastic population dynamics [32] and can be measured as the variance in community composition (e.g. with metrics of beta-diversity) between replicated time series [45]. Ecological drift is an emergent property of community dynamics affecting species distribution and is the driver underlying species abundance distributions in the neutral theory [32,46,47]. In non-neutral communities, ecological drift is acting along with deterministic processes to determine the species distribution [45,48,49]. Ecological drift can impose a limit to similarity according to Tilman [50].

Our first objective in this study is to investigate how the landscape structure influences the relative importance of species-sorting and ecological drift. Our second objective is to examine how environmental heterogeneity can affect the relative importance of these two processes. The metacommunity is represented as a set of local communities linked by dispersal in a realistic spatial network. Local dynamics follow the lottery competition for space in a heterogeneous environment proposed by Gravel et al. [15]. We use network centrality measures to quantify the position of a node in a metacommunity [37]. We consider the impact of the variance and range of environmental conditions, and the spatial structure of the heterogeneity (variance within versus between localities).

Model Description

We consider an individual-based, spatially explicit stochastic model of metacommunity dynamics. The metacommunity consist of n local communities connected by dispersal. Each local community is represented as a square lattice of size $L \times L$ with a torus shape, where each site is occupied by a single individual. The dynamics follow the zero-sum rule of the lottery model [32]. At each time step, an individual dies with probability d and empty sites are filled by recruits either drawn from the local community or the surrounding ones.

Recruitment Dynamics

Recruitment of an individual at a site follows the lottery model used by Gravel et al. [15]. The model is however expanded to account for dispersal within and between local communities. Hence, the recruitment probability $R_{i,x,t}$ of species i in local community t at the site x is:

$$R_{i,x,t} = (1-m_t)\frac{\sum_{f=1}^{\sigma} \lambda_{i,x,t} N_{i,f,t} K(f)}{\sum_{j=1}^{S}\sum_{f=1}^{\sigma} \lambda_{j,x,t} N_{j,f,t} K(f)} + m_t \sum_{q=1,q\neq t}^{v} P_{i,q}, \tag{1}$$

where the parameter m_t is the probability that a recruit is an immigrant coming from surrounding local communities connected to locality t. $N_{j,f,t}$ is the total number of individuals of species j at distance f from the site x. $K(f)$ is the local dispersal kernel. According to Snyder and Chesson [51], short-range dispersal can facilitate coexistence when environmental variation is permanent. Hence we use the four nearest-neighbor dispersal. $P_{i,q}$ is the relative abundance of species i within the local community q connected to t. The probability of an immigrant coming from a surrounding community, m_t, is defined as [38]:

$$m_t = \frac{v\gamma/2}{1+v\gamma/2} \approx \frac{v\gamma}{2}. \tag{2}$$

Where γ is the weight of dispersal between local communities [52] and v is the number of local communities connected with t.

We consider a Gaussian-shaped function to describe the relationship between the environment and the survival probability $\lambda_{i,x,t}$ of offspring from species i at the site x in local community t:

$$\lambda_{i,x,t} = \exp[\frac{-(E_{x,t}-\mu_i)^2}{2{w_i}^2}] \tag{3}$$

where $E_{x,t}$ is the local environmental condition, μ_i is the niche optima of species i, and w_i is the niche breadth of species i (we assume for simplicity identical niche breadth for all species).

The environment is heterogeneous within and between local communities. We consider for simplicity a linear gradient of environmental conditions within each local community. The environmental condition $E_{x,t}$ is thus a linear function of the position on the lattice:

$$E_{x,t} = E_{\min,t} + \frac{x}{X} E_{range}. \tag{4}$$

We equally divide the lattice into X subsections and each subsection has two dimensions. We label all subsections from 0 to X–1. The value of X thus determines the grain size of the environment. $E_{x,t}$ is the E value of section x, E_{range} is the range of E, and $E_{min,\ t}$ is the minimal value of E in local community t. We use $E_{min,\ t}$ and E_{range} to control the environmental heterogeneity of local community t. The variance of environmental conditions between local communities (regional variance) is determined by E_{range} and $E_{min,t}$. We number the local communities from 1 to n and $E_{min,t}$ is defined as t/n, for example, $E_{min,0} = 0/n$, $E_{min,1} = 1/n$. The variance of the environment within given local community is determined by E_{range} and increases with E_{range}.

The distribution of niche optima is evenly spaced across the range of available environments [43]. We number the species from

1 to M, with niche optima of species i being

$$\mu_i = \frac{(i-1)}{M}[(1-E_{\text{interval}})+E_{range}]+\frac{1}{2M}. \quad (5)$$

where $E_{interval}$ is the interval of minimal environmental values between adjacent local communities, and is $E_{min,\ t}-E_{min,\ t-1}$, that is $1/n$. E_{range} determines the similarity among niche optimums.

Network Construction and Centrality Measures

The metacommunity is represented by a graph where each of the n local communities is considered a node. We use the random geometric graph algorithm [53] to generate the network structure. Unlike simple shapes (such as circle and star) and random graphs, the random geometric graph has a heterogeneous degree distribution and represents landscapes more realistically. A random geometric graph is constructed by dropping n points randomly into a unit square and by connecting any two points within a threshold distance (Euclidean) r from each other. We ensure that the network is fully connected (a single graph). We assume equal weights for all edges.

Typical network studies address issues of connectivity and centrality. Connectance (the fraction of realized links between nodes relative to all potential links) is a network-level property quantifying the average connectivity of the network. Centrality is a node-level property, quantifying its position relative to other nodes in the network. We consider a variety of network centrality measures to measure different aspects of network topology. Betweenness centrality is the number of shortest paths that the focal node lies on [52] in other words the number of 'times' that any node needs a given node to reach any other node by the shortest path. This index is a useful measure of the node's importance to the network [54]. Closeness centrality is the inverse of the average path lengths from a node to each other node and degree centrality is the number of neighbors for each node [29,37,55]. Eigenvector centrality assigns relative scores to all nodes in the network based on the idea that connections to high scoring nodes contribute more to the score of the focal node than equal connections to low scoring nodes [37,56]. Centrality measures are computed using the *igraph* package [57] implemented in the R version 2.13.1 [58].

Neutrality and Strength of Species-sorting

We calculate a neutrality index by dividing the variance in species relative abundance among replicated runs for a scenario with species-sorting by the variance among replicates for a neutral model with the same landscape structure [15]. The index ranges from 0 to 1, with a value of 0 for community dynamics dominated by species-sorting and 1 for neutral dynamics (see details in Gravel *et al.* [15]).

We use a conceptually similar approach to measure the strength of the association between species distribution and the environment. We compare simulated results to the expected distribution with perfect species-sorting (i.e. when the species with highest $\lambda_{j,x,t}$ wins the competition). We record 1 for a site that is occupied by the best competitor and 0 if by a weaker one. We get an index for the strength of species-sorting at one locality by calculating the number of sites occupied by the best competitor divided by the local community size (L^2).

Simulations

We investigate the effects of node centrality in the metacommunity and of environmental heterogeneity with simulations for the following combinations of parameters: (1) species-sorting vs. neutral (2) with or without speciation and (3) three levels of network connectance ($r = 0.15$, 0.25 and 0.35, see Fig. S1) (4) four levels of dispersal probability ($m = 0.0001$, 0.001, 0.01 and 0.1) (5) seven different levels of environmental heterogeneity ($E_{range} = 0.5$, 1.0, 1.5, 2.0, 2.5, 3.0 and 3.5). 20 replicates of 10^6 time steps were run for each simulation scenario. Speciation rate is set at 0.0001 and kept constant in all simulations. The niche optima of new species are assigned from $[\mu_1, \mu_M]$. Metacommunity size is fixed at $n = 50$ and local community size at $L = 50$, $d = 0.1$, $X = 50$ and $\sigma = 1$.

Results

The Influence of Node Position

We find that local species richness increases with all metrics of node centrality and immigration (Fig. S2). The relationship between node centrality and the neutrality index interacts with immigration (Fig. 1- for simplicity, we only present the results of $r = 0.35$ in the main text, simulations with other values of r are attached as Supplementary Information). Neutrality (the variance among replicated runs) increases with node centrality at low immigration, while the opposite is found at high immigration. Species-sorting is strongest for peripheral nodes and lowest for central ones at low immigration, while it is systematically low, independently of node centrality, at high immigration (Fig. 2). The relationship between species-sorting and centrality however quickly saturates at high immigration, as there is no more difference between $m = 0.01$ and $m = 0.1$.

The Influence of Environmental Heterogeneity

We find that neutrality decreases with environmental heterogeneity (E_{range}) at all values of immigration (Fig. 3). There is thus a better match between environmental conditions and species distribution with increasing environmental heterogeneity. Meanwhile, the strength of species-sorting increases with environmental heterogeneity at all values of immigration (Fig. 4).

Discussion

Our results show that the spatial structure of the metacommunity (i.e. the way localities are connected by dispersal and the environment heterogeneity) has considerable impacts on the strength of species-sorting, ecological drift, and ultimately, species distribution. Spatial contingencies such as the position of a local community in the landscape and the distribution environmental conditions influence the ability of a species to track its favorable environment. Species-sorting is typically considered a process [59] whereby here it is quantified as an emergent pattern – the realized correspondence between environment and species distributions. We emphasize two conclusions from our analysis. First, neutrality increases and the strength of species-sorting decreases with the centrality of a local community when the average dispersal is relatively low, whereas the opposite relationship is found at higher dispersal. Second, neutrality decreases and the strength of species-sorting increases with the variance of environmental conditions.

Both the average dispersal rate and the position of local communities contribute to the connectivity matrix [60]. We consequently find that species richness increases with both the immigration probability and node centrality (Fig. S2; [37]), as observed in previous studies on the mass effect [61]. High immigration and centrality promote the mass effect because they increase the influx of inferior competitors and therefore the likelihood of their recruitment. The influence of the position of a

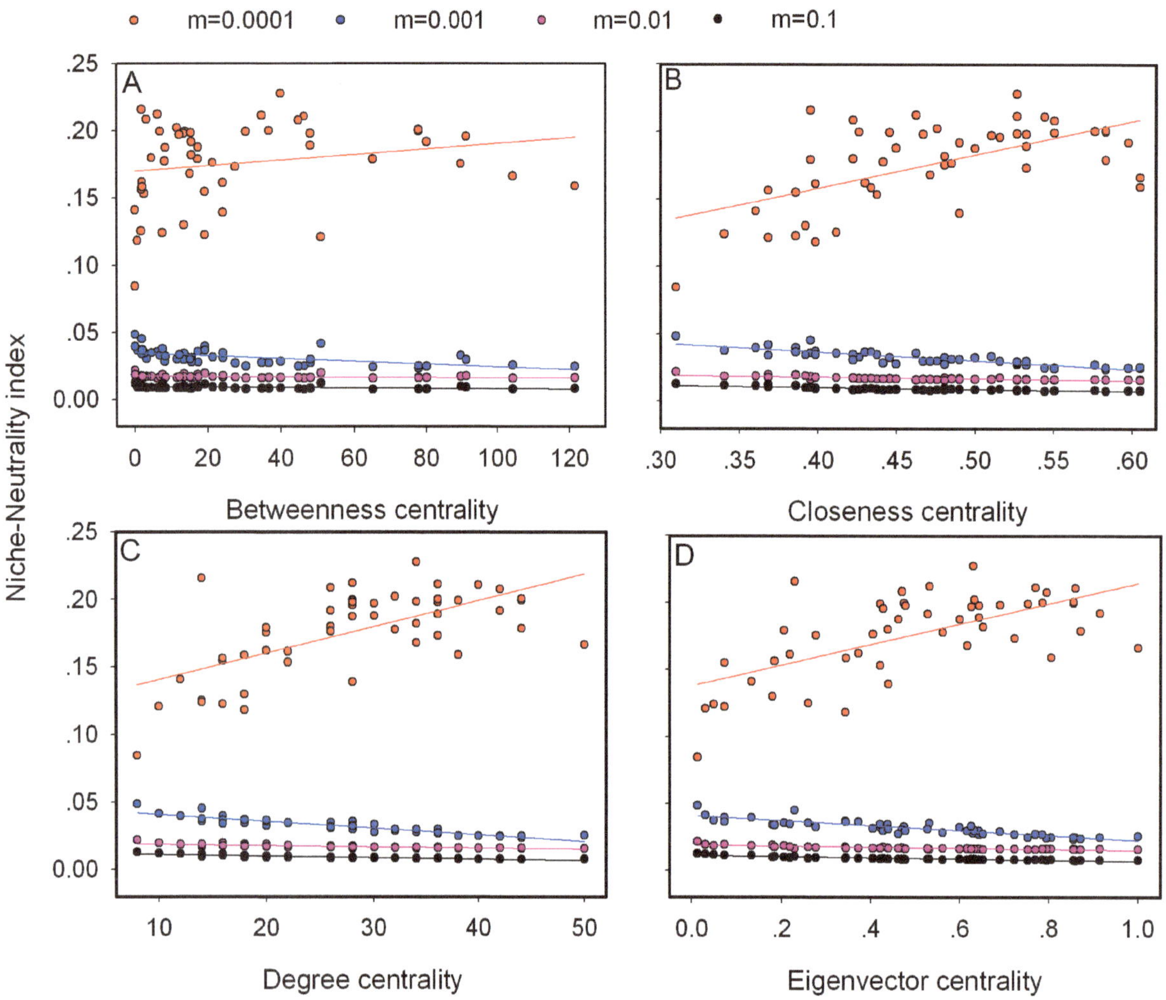

Figure 1. Effect of four centrality metrics in network metacommunity on the niche-neutrality index. There is different migration rates and network connectance $r = 0.35$. Each data point is averaged for 20 replicates.

local community in the landscape on the strength of species-sorting and ecological drift however differs with the average immigration (Fig. 1, Fig. 2, Fig. S3 and Fig. S4).

Under low immigration and in peripheral nodes, there is no mass effect and therefore the best competitor systematically (deterministically) wins the competition. Species-sorting is therefore strong and neutrality low. At central nodes there is a significant mass effect occurring, preventing the top competitor to win. Species identity could however be deterministic as well (and therefore both neutrality and species-sorting low), for instance in a sink node constantly receiving the same influx of migrants from the source node. If there are enough spatial refuges (peripheral nodes where a species could deterministically outcompete other ones), then the neutrality at the regional level will remain low because they will prevent drift to extinction at the regional scale. The situation changes at high immigration because species are moving rapidly throughout the landscape and often the best competitors are outcompeted by chance alone. Regional diversity reduces [20] and the amount of drift increases because there is a much higher niche overlap [15]. Peripheral nodes are subject to a lot of stochasticity because the drift will be maximal (no immigration to prevent random extinction). On the other hand, central nodes will be dominated by the most abundant species at the regional level and the composition might thus appear more stably. The species-sorting is consequently much lower at high immigration.

Our results show that despite species-sorting and ecological drift being conceptually opposite, this relationship does not always hold. The match between species distribution and the environment could obviously only be low under neutral dynamics. We have however found that in the presence of a strong mass effect, there will be low species-sorting by definition (top competitors are replaced by inferior ones). Neutrality might however remain low provided there are locations where some species deterministically win competition and therefore act as constant sources of immigrants. The emergent mixture of coexistence mechanisms found across the landscape is perhaps the most important contribution from our study. These results could have only been found with spatial heterogeneity in the environment and in the connectivity matrix, promoting a variable distribution of sources (refuges) and sinks.

We also find that increasing the environmental heterogeneity strengthens species-sorting (Fig. 3, Fig. 4, Fig. S5 and Fig. S6). The average niche overlap decreases with larger difference between adjacent patches, thereby making it easier for each species to find and fight for their favorable habitat. Snyder and Chesson [51] investigated the joint impacts of dispersal and environmental

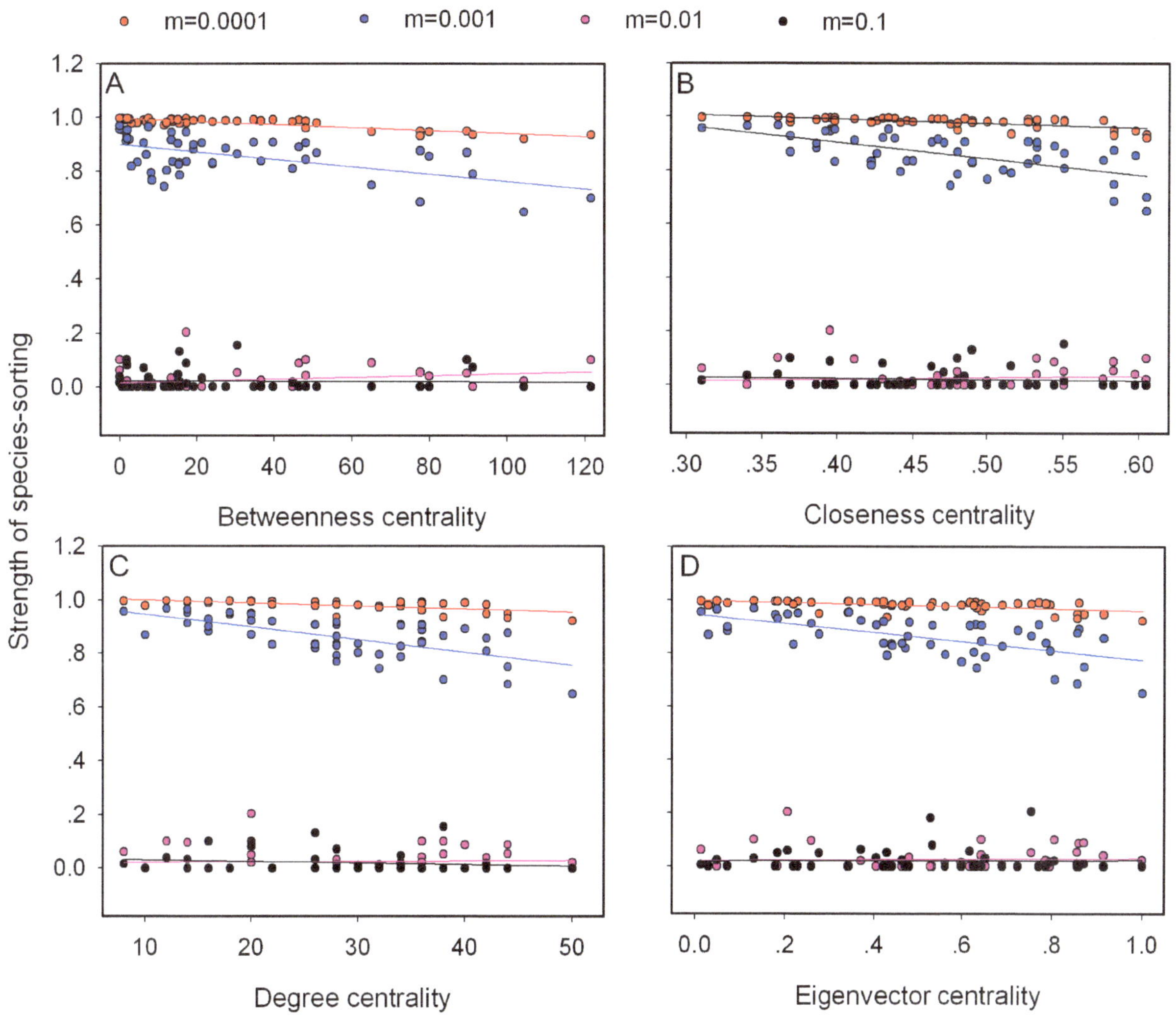

Figure 2. Effect of four centrality metrics on the strength of species sorting. Each data point is averaged for 20 replicates.

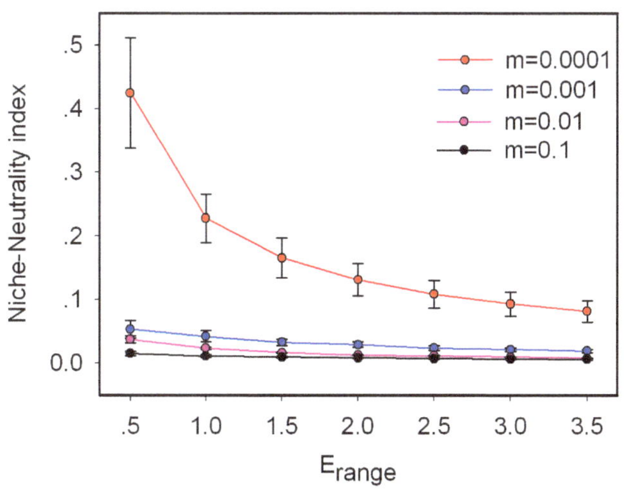

Figure 3. Effect of environmental heterogeneity on the niche-neutrality index. Error bars represent standard deviation over 20 replicate runs.

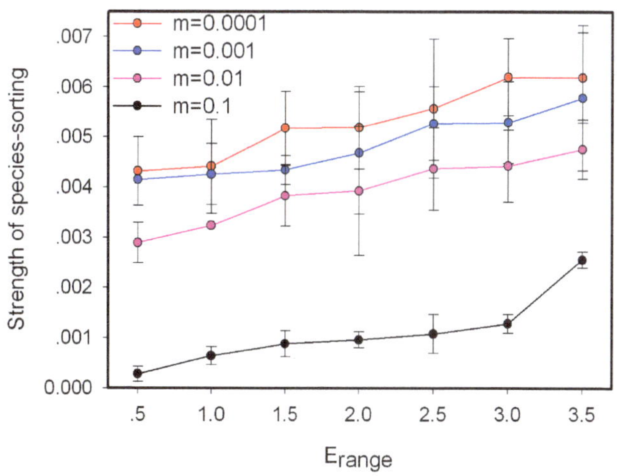

Figure 4. Effect of environmental heterogeneity on the strength of species sorting. Each data point is means for 20 replications.

heterogeneity on coexistence and found that short-range dispersal provide a significant advantage when environmental structure is permanent. The range of environmental variation determines how many niches there will be for a given niche breadth and the effect of dispersal to coexistence depends strongly on environmental variation [51]. Diversity is consequently impacted by environmental heterogeneity (Fig. S7), as studied by Schwilk and Ackerly [43].

Our study adds to current metacommunity theory by considering that beyond the effect of average dispersal among local communities, its variance might also influence biodiversity. We do not vary immigration probability over space in our simulations, thus spatial variance in dispersal nonetheless arises in our simulations from two independent drivers. First, the spatial network structure (the connectivity matrix) determines the relative importance of dispersal among local communities and the structure of the metacommunity [37–39]. Second, spatial heterogeneity of the environment determines species distribution and thereby the composition of the flow of migrants at a given location [62]. With increasing variance in dispersal, some local communities could be neutral or subject to a mass effect, while others might experience a strong species-sorting. There is consequently spatial variance in the coexistence mechanisms at play. For instance, diversity in local communities with low centrality is maintained by species sorting, while diversity in central communities is maintained by neutral drift and the mass effect. This spatial variance of coexistence mechanisms should promote the regional diversity by reducing the regional similarity constraint [20]. If there is variance in dispersal, metacommunity could experience much higher average dispersal without having the homogenizing effect of dispersal [20,63,64].

Species distribution models are important tools to study the distribution of species and forecast their response to environmental changes. Recently, Boulangeat et al. [65] proposed a new framework integrating abiotic constraints, dispersal limitations and biotic interactions to predict the species distribution and abundance. The increased performance of their integrated model is indeed attributable to explicit accounting of dispersal and environmental heterogeneity. In our model, we consider both the connectivity matrix and the environmental heterogeneity, and they interact to affect the species distribution model. Without environmental heterogeneity, our model converge to the neutral model with network structured metacommunity [37]; while species will be sorted out according to the environmental condition in absence of connectivity matrix [66].

Supporting Information

Figure S1 The structure of metacommunity under different connectance $r = 0.15$ (top), $r = 0.25$ (middle) and $r = 0.35$ (bottom).

Figure S2 Effect of four centrality metric in network metacommunity on the number of species.

Figure S3 Effect of four centrality metrics on the niche neutrality index with $r = 0.15$ and $r = 0.25$.

Figure S4 Effect of four centrality metrics on the strength of species sorting. The network connectance $r = 0.15$ and $r = 0.25$, and migration rate is different.

Figure S5 Effect of environmental heterogeneity on the niche-neutrality index with $r = 0.15$ and $r = 0.25$.

Figure S6 Effect of environmental heterogeneity on the strength of species sorting with $r = 0.15$ and $r = 0.25$.

Figure S7 Relationship between α-, β- and γ-diversity and environmental heterogeneity. By setting E_{range} to 0, species niche optima become identical and the model converges to a neutral model. Each data point is the mean of 20 replications for each local community. Error bars represent the standard deviation.

Acknowledgments

We thank two anonymous reviewers and editor for the helpful comments on the article.

Author Contributions

Conceived and designed the experiments: DA DG GW. Performed the experiments: DA. Analyzed the data: DA DG. Contributed reagents/materials/analysis tools: DA GW. Wrote the paper: DA DG CC GW.

References

1. Araújo MB, Pearson RG (2005) Equilibrium of species'distributions with climate. Ecography 28: 693–695.
2. Guisan A, Thuiller W (2005) Predicting species distribution: offering more than simple habitat models. Ecology Letters 8: 993–1009.
3. Araújo MB, Rahbek C (2006) How does climate change affect biodiversity? Science 313: 1396–1397.
4. Araújo MB, Peterson AT (2012) Uses and misuses of bioclimatic envelope modelling. Ecology 93: 1527–1539.
5. Beck J, Ballesteros-Mejia L, Buchmann CM, Dengler J, Fritz SA, et al. (2012) What's on the horizon for macroecology? Ecography 35: 673–683.
6. Austin MP, Nicholls AO, Margules CR (1990) Measurement of the realized qualitative niche: environmental niches of five *Eucalyptus* species. Ecological Monographs 60: 161–177.
7. Leathwick JR (1998) Are New Zealand's Nothofagus species in equilibrium with their environment? Journal of Vegetation Science 9: 719–732.
8. Pearce J, Lindenmayer D (1998) Bioclimatic analysis to enhance reintroduction biology of the endangered helmeted honeyeater (*Lichenostomus melanops cassidix*) in southeastern Australia. Restoration Ecology 6: 238–243.
9. Anderson RP, Peterson AT, Gomez-Laverde M (2002) Using niche-based GIS modeling to test geographic predictions of competitive exclusion and competitive release in South American pocket mice. Oikos 98: 3–16.
10. Araújo MB, Cabeza M, Thuiller W, Hannah L, Williams PH (2004) Would climate change drive species out of reserves? An assessment of existing reserve-selection methods. Global Change Biology 10: 1618–1626.
11. Thomas CD, Cameron A, Green RE, Bakkenes M, Beaumout LJ, et al. (2004) Extinction risk from climate change. Nature 427: 145–147.
12. Thuiller W (2004) Patterns and uncertainties of species' range shifts under climate change. Global Change Biology 10: 2020–2027.
13. Pulliam H R (2000) On the relationship between niche and distribution. Ecology Letters 3: 349–361.
14. Caylor KK, Manfreda S, Rodriguez-Iturbe I (2005) On the coupled geomorphological and ecohydrological organization of river basins. Advances in Water Resources 28: 69–86.
15. Gravel D, Canham CD, Beaudet M, Messier C (2006) Reconciling niche and neutrality:the continuum hypothesis. Ecology Letters 9: 399–409.
16. Leibold MA, Holyoak M, Mouquet N, Amarasekare P, Chase JM, et al. (2004) The metacommunity concept:a framework for multi-scale community ecology. Ecology Letters 7: 601–613.
17. Hugueny B, Cornell,HV, Harrison S (2007) Metacommunity models predict the local-regional species richness relationship in a natural system. Ecology 88: 1696–1706.

18. Logue JB, Mouquet N, Peter H, Hillebrand H, Group TMW (2011) Empirical approaches to metacommunities:a review and comparison with theory. Trends in Ecology & Evolution 0: 1–10.
19. Hurtt GC, Pacala SW (1995) The consequences of recruitment limitation:reconciling chance, history and competitive differences among plants. Journal of Theoretical Biology 176: 1–12.
20. Mouquet N, Loreau M (2002) Coexistence in metacommunities:the regional similarity hypothesis. American Naturalist 159: 420–426.
21. Gilpin ME, Hanski IA (1991) Metapopulation dynamics:empirical and theoretical investigations. London: Academic Press.
22. Wilson DS (1992) Complex interactions in metacommunities, with implications for biodiversity and higher levels of selection. Ecology 73: 1984–2000.
23. Mouquet N, Loreau M (2003) Community patterns in source-sink metacommunities. American Naturalist 162: 544–557.
24. Dale MRT, Fortin MJ (2010) From graphs to spatial graphs. Annual Review of Ecology, Evolution and Systematics 41: 21–38.
25. Urban D, Keitt T (2001) Landscape connectivity:a graph-theoretic perspective. Ecology 82: 1205–1218.
26. Newman ME (2003) The structure and function of complex networks. SIAM Review 45: 167–256.
27. Ai D, Chu C, Ellwood MDF, Hou,R, Wang G (2013) Migration and niche partitioning simultaneously increase species richness and rarity. Ecological Modelling 258: 33–39.
28. Minor ES, Urban DL (2008) A graph-theory framework for evaluating landscape connectivity and conservation planning. Conservation Biology 22: 297–307.
29. Urban D, Minor ES, Treml EA, Schick RS (2009) Graph models of habitat mosaics. Ecology Letters 12: 260–273.
30. Muneepeerakul R, Levin SA, Rinaldo A, Rodriguez-Iturbe I (2007) On biodiversity in river networks: A trade-off metapopulation model and comparative analysis. Water Resources Research 43.
31. Muneepeerakul R, Bertuzzo E, Lynch HJ, Fagan WF, Rinaldo A, et al. (2008) Neutral metacommunity models predict fish diversity patterns in Mississippi-Missouri basin. Nature 453: 220–223.
32. Hubbell SP (2001) The unified neutral theory of biodiversity and biogeography. Princeton: Princeton Univ. Press.
33. McKane AJ, Alonso D, Sole RV (2004) Analytic solution of Hubbells model of local community dynamics. Theoretical Population Biology 65: 67–73.
34. Vallade M, Houchmandzadeh B (2006) Species abundance distribution and population dynamics in a two-community model of neutral ecology. Phys Rev E 74: 051914-051911: 051919.
35. Allouche O, Kadmon R (2009) A general framework for neutral models of community dynamics. Ecology Letters 12: 1–10.
36. Economo EP, Keitt TH (2008) Species diversity in neutral metacommunities:a network approach. Ecology Letters 11: 52–62.
37. Economo EP, Keitt TH (2010) Network isolation and local diversity in neutral metacommunities. Oikos 119: 1355–1363.
38. Desjardins-Proulx P, Gravel D (2012) How likely is speciation in neutral ecology? American Naturalist 179: 137–144.
39. Desjardins-Proulx P, Gravel D (2012) A complex speciation-richness relationship in a simple neutral model. Ecology and Evolution 2: 1781–1790.
40. Grubb P (1977) The maintenance of species richness in plant communities:the importance of the regeneration niche. Biological Reviews 52: 107–145.
41. Chesson P, Warner R (1981) Environmental variability promotes coexistence in lottery competitive systems. The American Naturalist 117: 923–943.
42. Comins H, Noble I (1985) Dispersal, variability, and transient niches: species coexistence in a uniformly variable environment. The American Naturalist 126: 706–723.
43. Schwilk DW, Ackerly DD (2005) Limiting similarity and functional diversity along environmental gradients. Ecology Letters 8: 272–281.
44. May RM, McLean AR (2007) Theoretical ecology principles and applications: OXFORD Univ. Press.
45. Gravel D, Guichard F, Hochberg ME (2011) Species coexistence in a variable world. Ecol Lett 14: 828–839.
46. Hubbell SP (2003) Modes of speciation and the lifespans of species under neutrality:a response to the comment of Robert E.Ricklefs. Oikos 100: 193–199.
47. Maurer BA, McGill BJ (2004) Neutral and non-neutral macroecology. Basic and Applied Ecology 5: 413–422.
48. Purves DW, Pacala SW (2005) Ecological drift in niche-structured communities: neutral pattern does not imply neutral process. In: Burslem D, Pinard M, Hartley S (Eds.), Biotic interactions in the tropics. Cambridge, UK: Cambridge University Press. 107–138 p.
49. Ruokolainen L, Ranta E, Kaitala V, Fowler MS (2009) When can we distinguish between neutral and non-neutral processes in community dynamics under ecological drift? Ecology Letters 12: 1–11.
50. Tilman D (2004) Niche tradeoffs,neutrality,and community structure:a stochastic theory of resource competition,invasion,and community assembly. PNAS 101: 10854–10861.
51. Snyder RE, Chesson P (2003) Local dispersal can facilitate coexistence in the presence of permanent spatial heterogeneity. Ecology Letters 6: 301–309.
52. Proulx SR, Promislow DEL, Phillips PC (2005) Network thinking in ecology and evolution. Trends in Ecology & Evolution 20: 345–353.
53. Penrose M (2003) Random geometric graphs (Oxford studies in probability).
54. Newman MEJ (2010) Networks: an introduction. UK: Oxford University Press.
55. Freeman L (1978) Centrality in social networks conceptual clarification. Social Networks 1: 215–239.
56. Bonacich P (1972) Factoring and weighting approaches to status scores and clique identification. Journal of Mathematical Sociology 2: 113–120.
57. Csardi G (2011) Network analysis and visualization.
58. Team RDC (2009) R: a language and environment for statical computing. R Foundation for Statical Computing, Vienna. http://wwwr-projectorg.
59. Langenheder S, Szekely AJ (2011) Species sorting and neutral processes are both important during the initial assembly of bacterial communities. The ISME Journal 5: 1086–1094.
60. Muneepeerakul R, Azaele S, Levin SA, Rinaldo A, Rodriguez-Iturbe I (2011) Evolution of dispersal in explicitly spatial metacommunities. Journal of Theoretical Biology 269: 256–265.
61. Loreau M, Mouquet N (1999) Immigration and the maintenance of local species diversity. American Naturalist 154: 427–440.
62. Mouquet N, Gravel D, Massol F, Calcagno V (2013) Extending the concept of keystone species to communities and ecosystems. Ecology Letters 16: 1–8.
63. Caswell H (1978) Predator-mediated coexistence: a non-equilibrium model. The American Naturalist 112: 127–154.
64. Forbes AE, Chase JM (2002) The role of habitat connectivity and landscape geometry in experimental zooplankton metacommunities. Oikos 96: 433–440.
65. Boulangeat L, Gravel D, Thuiller W (2012) Accounting for dispersal and biotic interactions to disentangle the drivers of species distributions and their abundances. Ecology Letters 15: 584–593.
66. Tilman D (1982) Resource competition and community structure, Monographs in population biology. Princeton: Princeton Univ. Press.

15

Commonly Rare and Rarely Common: Comparing Population Abundance of Invasive and Native Aquatic Species

Gretchen J. A. Hansen[1*¤a], M. Jake Vander Zanden[1], Michael J. Blum[2], Murray K. Clayton[3], Ernie F. Hain[4], Jennifer Hauxwell[5], Marit Izzo[1¤b], Matthew S. Kornis[1¤c], Peter B. McIntyre[1], Alison Mikulyuk[1,5], Erika Nilsson[1¤d], Julian D. Olden[6], Monica Papeş[1¤e], Sapna Sharma[1¤f]

1 Center for Limnology, University of Wisconsin-Madison, Madison, Wisconsin, United States of America, **2** Department of Ecology & Evolutionary Biology, Tulane University, New Orleans, Louisiana, United States of America, **3** Departments of Plant Pathology and Statistics, University of Wisconsin-Madison, Madison, Wisconsin, United States of America, **4** Department of Biology, North Carolina State University, Raleigh, North Carolina, United States of America, **5** Science Services, Wisconsin Department of Natural Resources, Madison, Wisconsin, United States of America, **6** School of Aquatic and Fishery Sciences, University of Washington, Seattle, Washington, United States of America

Abstract

Invasive species are leading drivers of environmental change. Their impacts are often linked to their population size, but surprisingly little is known about how frequently they achieve high abundances. A nearly universal pattern in ecology is that species are rare in most locations and abundant in a few, generating right-skewed abundance distributions. Here, we use abundance data from over 24,000 populations of 17 invasive and 104 native aquatic species to test whether invasive species differ from native counterparts in statistical patterns of abundance across multiple sites. Invasive species on average reached significantly higher densities than native species and exhibited significantly higher variance. However, invasive and native species did not differ in terms of coefficient of variation, skewness, or kurtosis. Abundance distributions of all species were highly right skewed (skewness>0), meaning both invasive and native species occurred at low densities in most locations where they were present. The average abundance of invasive and native species was 6% and 2%, respectively, of the maximum abundance observed within a taxonomic group. The biological significance of the differences between invasive and native species depends on species-specific relationships between abundance and impact. Recognition of cross-site heterogeneity in population densities brings a new dimension to invasive species management, and may help to refine optimal prevention, containment, control, and eradication strategies.

Editor: Michael Somers, University of Pretoria, South Africa

Funding: GJAH and MJVZ were supported by the NSF-funded North Temperate Lakes Long Term Ecological Research Program (DEB-9632853). JH, AM, and MJVZ were supported by the Wisconsin Department of Natural Resources. MJB, EFH, and PBM were funded by the Department of Defense Strategic Environmental Research and Development Program under project RC-1646. MSK was funded by a grant from the University of Wisconsin Sea Grant Institute, federal grant number NA06OAR4170011 project number R/AI-3, to MJVZ. EN was funded with a post-doctoral scholarship from the Swedish Research Council for Environment, Agricultural Sciences and Spatial Planning. SS was supported by the Natural Sciences and Engineering Research Council of Canada Post-Doctoral Fellowship (NSERC PDF). The funders had no role in study design, data collection and analysis, decision to publish, or preparation of the manuscript.

Competing Interests: One author (M.Izzo) is currently employed by a private company, Triventus Consulting. This does not alter the authors' adherence to all the PLOS ONE policies on sharing data and materials.

* E-mail: gretchen.hansen@wisconsin.gov

¤a Current address: Science Services, Wisconsin Department of Natural Resources, Madison, Wisconsin, United States of America
¤b Current address: Triventus Consulting, Falkenberg, Sweden
¤c Current address: Smithsonian Environmental Research Center, Edgewater, Maryland, United States of America
¤d Current address: County Administrative Board, Kalmar, Sweden
¤e Current address: Department of Zoology, Oklahoma State University, Stillwater, Oklahoma, United States of America
¤f Current address: Department of Biology, York University, Toronto, Ontario, Canada

Introduction

Invasive species are recognized as leading drivers of environmental change, and can produce significant economic and ecological impacts [1]. Recently, Davis et al. [2] argued that conservationists, scientists, and the general public hold a pervasive bias against non-native species, and that species should be judged based on their impact rather than their place of origin. The debate that has followed (e.g., [3],[4]) highlights the importance of understanding variation in the impacts of invasive species for both science and resource management, and serves as a reminder of the need to question and empirically test assumptions about invasive species.

Invasion biology research has aimed to elucidate general patterns, sometimes at the cost of overlooking important sources of heterogeneity [5]. It has long been recognized that only a small fraction of introduced species will establish, spread, and cause impacts [6]. As a result, identifying species likely to cause ecological or economic impacts and predicting locations that are likely to be most vulnerable are important goals of invasive species research and management [7–9]. Although in principle it is recognized that abundance (and therefore impact) of invasive species varies among sites, few attempts to quantify relative impacts of invasive species incorporate spatial variability in abundance [8]. Instead, species invasions are largely seen as binary phenomena, where an introduced species is either invasive or not (e.g., [7]); and a location is either invaded (or invasible) or not (e.g., [9]; but see [10]). Consequently, most invasive species monitoring and databases emphasize occurrence rather than abundance (Table S1 in File S1). Admittedly, documenting occurrence is logistically more feasible than documenting abundance, particularly across large spatial scales. However, we may miss important insights by ignoring variability among sites in invasive species abundance.

In contrast to the largely dichotomous paradigm of invasive species research noted above, ecologists have long documented that the abundance of many species tends to be low at most locations and high only in a few; that is, frequency distributions of abundance are right skewed [11], [12]. This empirical pattern of right-skewed abundance distributions is observed across a wide range of taxonomic groups and spatial scales, and is considered central to ecological theory [13], [14]. Despite the importance of invasive species abundance in determining impacts [15], [16], virtually no studies have examined whether invasive species follow this same pattern of right-skewed frequency distribution of abundance (but see [17]).

Understanding invasive species abundance distributions and how they compare to those of native species is important for analysis and management of species invasions. Competing predictions are possible based on different lines of reasoning. On one hand, invasive species are often considered inherently different from native species [18], in terms of both biological traits (e.g., high reproductive potential [7]), and community interactions (e.g., lack of natural enemies [19]), leading some to define invasive species as non-indigenous species that become abundant or 'dominant' [7], [20]. This view implies that invasive species commonly (or always) reach high densities where they establish populations, which would lead to higher mean abundances with less right-skewed distributions. On the other hand, proposed ecological mechanisms behind right-skewed abundance distributions such as differences in niche suitability among sites [11], [21] likely apply equally to species in both their native and introduced ranges. Recent attempts to quantify differences between invasive and native species abundance have found subtle or no differences between the groups (e.g.,[22], [23]), indicating that perhaps invasive species play by the same ecological rules, following right-skewed abundance frequency distributions similar to those of native counterparts.

In this analysis we compare patterns of population abundance for native and invasive aquatic species, collected simultaneously using the same methods. For each species included in our study, we obtained abundance estimates across many sites. We characterized the abundance distribution of each species in terms of its statistical moments (i.e., mean, variance, skewness, and kurtosis), with the objective of determining whether abundance distributions of invasive and native species differed in terms of these statistical moments. We conclude by discussing the implications of invasive species abundance distributions for ecological theory and the management of invasive species.

Methods

Ethics statement

All data existed prior to the initiation of this study and include our own data collected for other purposes, public data collected by management agencies, and published literature sources. Data sources and sampling methods are described in detail in Supplementary Methods S1 in File S1. For data collected by the authors, all animals were captured following protocols approved by the relevant Institutional Animal Care and Use Committee (IACUC), and permits for sampling were obtained from the appropriate authorities (noted where appropriate in Supplementary Methods S1 in File S1).

Data collation and analysis

To examine variability in abundance across sites, we required abundance records for native and invasive species in multiple locations. We compared cross-site abundance distributions of 17 invasive and 104 native species of aquatic plants, invertebrates (crayfish, mussels, prawns, snails), and fishes from three distinct geographic regions (Hawai'i, North America, and Europe; Figure 1). We grouped species into eight taxonomic/geographical categories (hereafter taxonomic groups) that were sampled using comparable methods: Crayfish, Hawaiian fish, North American fish, Swedish fish, Mussels, Plants, Prawns, and Snails. The invasive species included in this analysis are defined as "species that are non-native[...]to the ecosystem under consideration and whose introduction causes or is likely to cause economic or environmental harm or harm to human health" [24] and are classified as "invasive" or nuisance species by authorities in each study region (Table S2 in File S1). All invasive species included here have been present for at least 19 years in the invaded study region (Table S2 in File S1) suggesting that their populations should no longer reflect any transient expansion dynamics that could prevail during the early stages of invasion. Our analysis included 24,033 non-zero density records, with 20 to 1,252 site-level abundance records per species (Table S3 in File S1). To facilitate comparisons among species abundance reported in different units, abundances were standardized to range from zero to one by dividing each raw abundance value by the maximum observed value within each taxonomic group.

To characterize abundance distributions, we calculated the first four statistical moments (i.e., mean, variance, skewness, and kurtosis) of the abundance distribution of each species. Preliminary analysis showed that mean and variance were highly correlated; we also calculated coefficient of variation (CV) to compare variation independent of this correlation. Moments were natural log transformed and differences in moments of invasive and native species abundances were estimated using a multi-level modeling approach [25] with origin (invasive or native) as a fixed effect, and taxonomic group as a random effect. Effects are reported as restricted maximum likelihood (REML) estimates of fixed effects from the lme4 package [26] in R v2.15.1 [27]. Confidence intervals were calculated as Bayesian highest posterior density (HPD) intervals of parameter estimates generated from 10,000 Markov Chain Monte Carlo (MCMC) simulations using the languageR package, which incorporates variation from random effects [28]. Differences in moments based on species origin (i.e., differences between invasive and native species) were considered statistically significant if the 95% HPD intervals of estimated difference between native and invasive species did not include zero [25], [29]. Adjusted R^2 values for mixed effects models were calculated as the likelihood ratio test R^2 [30], [31].

We generated empirical cumulative distributions (ECD's) for invasive and native species abundance overall and within each taxonomic group. For these distributions we disregarded species identity and constructed ECD's using abundances of all invasive and all native species in each taxonomic group. We then identified the median abundance (ECD = 0.5) of invasive species within each

Figure 1. Abundance distributions for each species used in this analysis. Labels are coded as follows: taxonomic group abbreviation. Origin. Species ID, where taxonomic group codes are Cr = Crayfish, FHI = Hawaiian fishes, FNA = North American fishes, FSw = Swedish fishes, M = Mussel,

Pl = Plant, Pr = Prawn, and S = Snail; origin codes are I = Invasive and N = Native; see Table S3 in File S1 for species identities. Colors correspond to taxonomic groups and in every group the darker shade corresponds to invasive species in that group. The x-axis scale shows standardized abundance (proportion of taxonomic group-level maximum abundance) and ranges from 0 to 1; the y-axis scale shows the number of sites falling into each abundance class and varies by species to accommodate different numbers of observations (sites). Note that all abundance values are greater than zero.

group, and compared the ECD value of native species for the same abundance value. This allowed us to compare the proportion of sites containing native species at abundances equal to or above the median invasive species abundance.

Results

Abundance distributions of all species were highly right-skewed (skewness > 0; Table S3 in File S1), regardless of invasive status; all species occurred at low densities in the vast majority of sites where they were documented (Figure 1). Invasive species on average reached significantly higher densities than native species and exhibited significantly higher variance (Table 1; Figure 2). However, invasive and native species did not differ in terms of CV, skewness, or kurtosis (Table 1, Figure 2). The absolute difference in mean abundance of invasive and native species was small (Table 1). Mean standardized abundance of invasive species was 0.06 and that of native species was 0.02 on the standardized abundance scale. In other words, the average abundance of invasive and native species was 6% and 2%, respectively, of the maximum abundance observed within a taxonomic group.

We were able to compare abundance distributions of three species in their native and invaded ranges: brook trout (*Salvelinus fontinalis*), brown trout (*Salmo trutta*), and signal crayfish (*Pacifastacus leniusculus*). Although small sample size precludes formal statistical inference, no consistent patterns in statistical moments of species in their invaded compared to native ranges were evident (Table S3 in File S1). Two of three species (brown trout and signal crayfish) exhibited patterns similar to those observed in the full dataset: higher mean abundance and variance in their invaded range. However, two of three species (brook trout and brown trout) exhibited patterns distinct from those observed overall; these species were more right-skewed and exhibited higher kurtosis in their invaded range.

Empirical cumulative frequency distributions of native and invasive species abundance for each taxonomic group were similar, but not identical (Figure 3). The maximum observed abundance value within a taxonomic group was that of a native species in four taxonomic groups - North American fish, Swedish fish, aquatic plants, and snails - but this pattern did not lead to higher mean abundance of native species (Figure 2). In six taxonomic groups, native species were present at densities greater than the median density of invasive species at fewer than 50% of sites (Figure 3). This trend was most prominent for crayfish and prawns; native species were present at or above invasive species median abundances at only 8% and 10% of sites, respectively. Snails showed the opposite trend, with native species present at or above invasive species median abundance at 74% of sites.

Discussion

We identified subtle differences in the abundance distributions of invasive and native aquatic species from a wide range of locations and taxonomic groups. On average, invasive species reached higher abundances than native counterparts in the same region; mean abundance of invasive species was three times greater than the mean abundance of native species. The higher observed abundance of invasive species in this study may in part be driven by biased data collection. Invasive species that typically establish small populations and have low ecological or environmental impact are not well represented in existing data sets and/or the literature. These species are generally considered lower priority for study by management and regulatory authorities. Despite this potential bias, absolute differences between mean abundance of invasive and native species were small (a difference of 0.04 on the standardized abundance scale). Indeed, variability in distributional parameters was high, and for most taxonomic groups (4 of 7) a native species was responsible for the maximum abundance observed within the group (Figure 3). The subtle

Table 1. Model results for hierarchical models of statistical moments.

Moment	R^2	Fixed effect	Estimate	Upper HPD	Lower HPD	Random effect	Variance
Mean	0.47	Intercept	−2.81	−2.15	−3.46	Taxa	1.05
		Origin	−0.98	−0.51	−1.52	Residual	0.76
Variance	0.33	Intercept	−4.74	−3.52	−6.09	Taxa	2.22
		Origin	−2.18	−1.16	−3.28	Residual	3.65
CV	0.16	Intercept	0.45	0.68	0.24	Taxa	0.08
		Origin	−0.08	0.12	−0.24	Residual	0.10
Skewness	0.26	Intercept	0.94	1.27	0.61	Taxa	0.20
		Origin	0.06	0.34	−0.19	Residual	0.22
Kurtosis	0.26	Intercept	2.33	2.91	1.79	Taxa	0.56
		Origin	0.18	0.66	−0.26	Residual	0.66

The intercept represents the restricted maximum likelihood (REML) estimate of the value for invasive species, and the origin effect is the difference between invasive and native species values on the natural log scale. Upper and lower highest probability density (HPD) intervals are the 95% confidence intervals of the fixed effects estimates generated from Markov-chain Monte Carlo resampling. Random effects and their explained variance are also presented for each model, where the taxa effect is the variance attributable to differences among taxonomic groups in statistical moments.

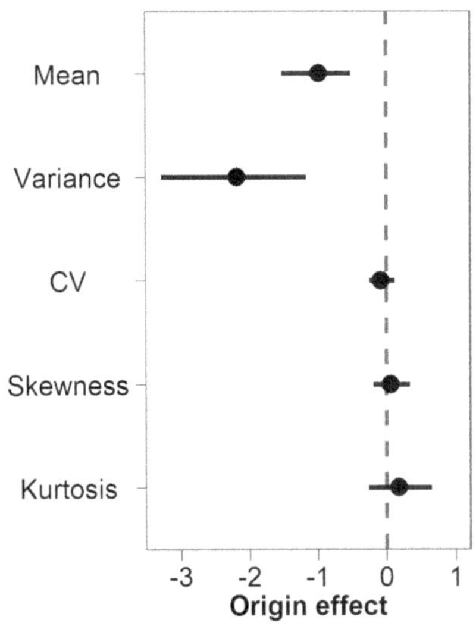

Figure 2. Effect size of origin (invasive vs. native status) on distributional parameters. Effects are presented as the restricted maximum likelihood (REML) estimate of the difference on the natural log scale between invasive and native species (or the natural log of the ratio of invasive:native species values). Bars are 95% highest probability density interval from Markov-chain Monte Carlo (MCMC) resampling; bars that do not overlap zero (dashed line) represent significant differences between invasive and native species.

difference between invasive and native species abundance distributions can be visualized by plotting standardized abundance distributions for all native and all invasive species combined (Figure 4). Although native species are more likely than invasive species to occur in the lowest abundance class (0–0.05 on standardized scale), the vast majority of all species abundances fall within this range regardless of invasive status.

The biological significance of differences between invasive and native species abundance distributions depend on species-specific relationships between density and ecological impact. Unfortunately, knowledge of such relationships is lacking for even the most notorious invaders [32]. Although some species may exert ecological effects disproportional to their abundance (so-called "keystone species"; [33]), the impact of an invasive species is generally positively correlated with its abundance [15], [16], [34]. Our results therefore suggest that the impact of most invasive species will be high in a small number of locations, and relatively low in the majority of invaded sites. However, in the absence of data relating density to impacts, it is impossible to know how impacts will scale with relative abundance– that is, whether a site in which a species reaches 2% of the maximum observed density within a taxonomic group will experience 2% of the maximum observed impact. Similarly, the differences between invasive and native species are likely more important for certain taxonomic groups than others, particularly when invader density is related to impact via a non-linear threshold response [35]. For example, negative effects of one invasive crayfish species occur at densities $\geq 9 \bullet \text{trap}^{-1}$ [36], or 0.28 on the standardized abundance scale. Our analysis shows that invasive crayfish exceed this threshold in ~17% of sampled locations, while native crayfish do so in less than 1% of locations (Figure 3). Thus, the differences in the abundance distributions of invasive and native crayfish could translate into very real differences in ecological impacts on the landscape scale if the per capita effects of crayfish species are similar. In contrast, the impacts of one invasive snail species are highest at densities above 2 m^{-2} [37], or 0.01 on the standardized abundance scale. Native snails exceed this threshold in approximately 35% of sites, while invasive snails do so in only ~15% of sites, suggesting that the ecological effects of invasive snails may be less than those of natives if per capita impacts are similar. Quantifying per capita effects of invasive species (e.g., [38]), as well as the relationship between invasive species abundance and impact (e.g., [10],[35]), is a requisite to interpreting the biological significance of the statistical patterns identified here. Unfortunately, such thresholds have not been identified in any systematic way for most invasive species, and it remains unclear whether generalized thresholds for impact could be identified across diverse groups.

The observed differences in mean abundance of invasive and native species do not necessarily reflect fundamental ecological differences between them. Widespread species tend to be on average more abundant than species restricted to small ranges, driven by higher maximum abundances [39]. Invasive species by definition have spread outside their native range and most are increasing their range size; higher abundances are therefore expected as their range size increases. Indeed, the relationship between range size and abundance does not differ for invasive and native British bird species, but similar to our findings, invasive species reach higher maximum densities than native counterparts [17]. Other studies have demonstrated that invasive plants rarely reach high densities [40] and the majority of invasive species do not reach higher abundances in invaded compared to native ranges [22], [23], supporting the idea that invasive and native species do not follow fundamentally different distributional patterns. However, the results of our study and that of Labra and colleagues [17] suggest invasive species tend to fall toward the high end of the observed range of abundance-distribution relationships. Identifying the mechanisms explaining such patterns is a fruitful area of current research (e.g., [41]), and the existence of similarities as well as differences in invasive and native species could provide insight into the forces behind these widespread ecological patterns [42].

Every invasive species in this analysis occurred at low population densities in the majority of invaded sites (Figure 1), a finding that is notable given the perceived pervasiveness of these species. In fact, invasive and native species abundance distributions were similarly right-skewed. Low abundances are expected at the edges of species ranges, in areas where a species' niche requirements are not fully met or where it has not yet dispersed [12], [43]. Although the mechanisms driving right-skewed species abundance distributions are the subject of debate [13], [42] and outside the scope of this analysis, the similarly right-skewed distributions of invasive and native species abundance observed here support the idea that invasive and native species are governed by similar ecological processes. The highly right-skewed abundance distributions observed for invasive species could reflect management actions rather than natural patterns in cases where control measures effectively reduce invasive species abundances at most sites. However, only two invasive species in our dataset have been subject to widespread management (sea lamprey, *Petromyzon marinus*; and Eurasian water milfoil; *Myriophyllum spicatum*), and abundance distributions of both managed species are less right-skewed than the majority of invasive species ($skewness_{sea\ lamprey} = 2.01$; $skewness_{milfoil} = 1.23$; median skewness of all invasive species = 2.73). Thus, there is no evidence from our analysis that the propensity of invasive species to be present at low abundance arises from control measures.

Another potential explanation for the right-skewed abundance patterns of invasive species is differences among invaded sites in time since species introduction. Temporal variation in species

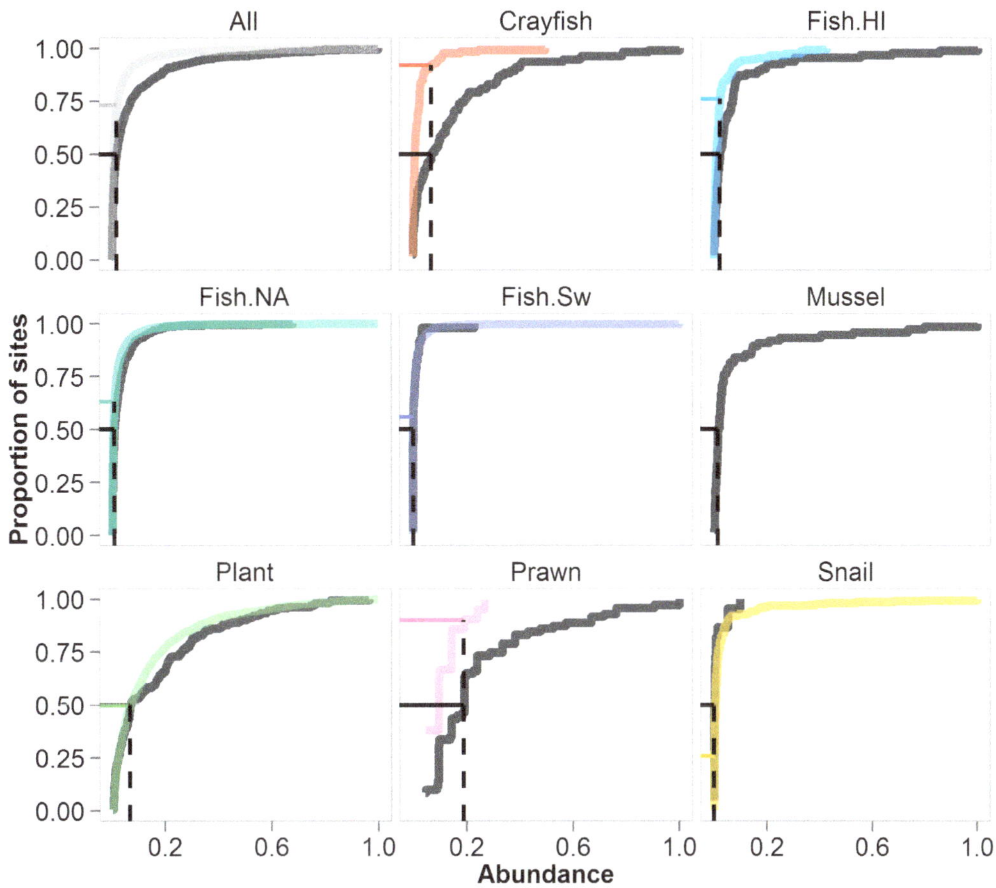

Figure 3. Empirical cumulative distributions (ECD) of invasive and native species. ECDs for invasive (dark grey) and native (colors) species abundance for each taxonomic group. The x-axis is standardized abundance, calculated as the proportion of the maximum abundance observed within a taxonomic group. The probability of finding a species from a given taxonomic group at or below the corresponding x-axis value is plotted on the y-axis. Median abundance for invasive species of each taxonomic group (ECD = 0.5) occurs where the cumulative distribution crosses the black horizontal line, and this median abundance value for invasive species is shown by the vertical black dashed line. The ECD value for native species corresponding to median invader abundance is indicated with a colored horizontal line, and represents the probability of finding native species of each taxonomic group at or below the median invasive value.

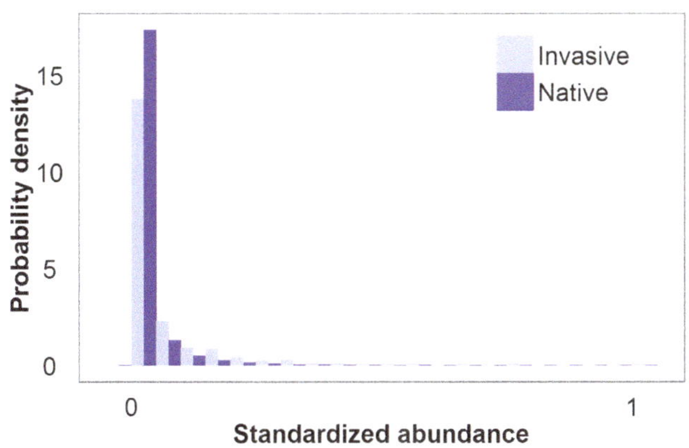

Figure 4. Abundance distributions of all invasive and all native species combined. Probability density of standardized abundance (proportion maximum abundance observed within a taxonomic group) for invasive (light purple) and native (dark purple) species, with all species combined. Abundance values are grouped into 0.05 bins.

abundance has been shown to generate right-skewed abundance patterns similar to those observed across large spatial scales [44], and many invasive populations exhibit some sort of time lag, or extended period of low abundance [45]. It is impossible to ascertain the time since invasion for each of the over 24,000 individual sites included in this analysis. However, we attempted to decrease the influence of lagged temporal patterns in abundance by restricting our analysis to species that have been present in the study region for at least 19 years (Table S2 in File S1). Several invasive species in this study established nearly a century or more ago, and exhibit similar abundance distributions to those that established more recently, suggesting that right-skewed abundance distributions of invasive species are not an artifact of time lags in population increase. Although this does not rule out the possibility that temporal variability in the abundance of invasive populations is responsible for observed abundance distributions, our results demonstrate that on the temporal scales most relevant to management (years to decades), invasive species exhibit highly right-skewed distributions.

The fact that high density populations of invasive species exist in a small number of "hot spots" has important management implications. Sophisticated modeling techniques are used in risk assessments that aim to identify which species are likely to become invasive (e.g., [46], [47]) and which locations are likely to be invaded (e.g., [48–50]), but rarely make predictions about variability in abundance of invasive species. Our results suggest that invasive species control and prevention would benefit from a more nuanced approach that also considers variation in species abundance among invaded sites. For instance, transmissions of infectious diseases follow highly right-skewed distributions similar to species abundances, and the effectiveness of disease control is improved when this heterogeneity is accounted for by predicting the identity of the most potentially infectious individuals and focusing prevention and containment efforts on them prior to a disease outbreak [51]. By analogy, landscape-scale invasive species containment efforts might be enhanced by focusing on the small percentage of "hot spots" containing the majority of individuals of a given invasive species, which would effectively reduce the likelihood of spread to additional sites in the long term. Moreover, because prevention is the most effective tool for combating the negative effects of invasive species [52], the ability to predict which locations are likely to support high densities of invasive species would be useful for targeting prevention so as to minimize impacts. The optimal strategy for invasive species management will be context-dependent, but explicitly considering heterogeneity in invasive species abundance may help guide strategies for invasive species management.

Conclusion

Heterogeneity in the abundance of invasive species known to cause ecological and economic impacts has received surprisingly little attention. Our finding that aquatic invasive species exist at low densities in most locations where they occur runs counter to the perception of invasive species as those that are abundant or dominant wherever they establish [7], [20]. Both invasive and native species are present in low densities in most locations, supporting the notion that applying general ecological and analytical principles to invasive species will advance understanding more so than treating invasions as idiosyncratic occurrences [53]. At the same time, our finding that invasive species are capable of reaching higher average densities than native species has important implications for invasive species impacts and their management, assuming that ecological impact scales with the population density of invasive species (see also [4]). By recognizing the patchiness of invasive species abundance, our results highlight opportunities to improve prevention, control, and eradication strategies.

Supporting Information

File S1 Supplementary information, including Supplementary Methods S1 describing in detail the sampling methodology for each group; Table S1 providing examples of invasive species databases that provide presence-absence data only; Table S2 listing the invasive species included in this analysis, their impacts, and year of invasion; and Table S3 listing all species included in this analysis with their associated statistical moments.

Acknowledgments

We thank the following individuals and organizations for providing data: Evan Childress, Roderick Gagne, James Gilliam, Scott Higgins, J. Derek Hogan, Brad Lamphere, Eric Larson, Gabrielle Lehrer-Brey, Daniel Lindstrom, Rahmat Naddafi, Dan Oele, Chris Solomon, Ryan Walter, The Department of Fisheries and Oceans- Canada, the North Temperate Lakes Long-Term Ecological Research program, the Swedish Electrofishing Register, the United States Fish and Wildlife Service, and the Wisconsin Department of Natural Resources. We are grateful to Steve Carpenter, Tony Ives, and two anonymous reviewers for helpful comments on earlier versions of this manuscript.

Author Contributions

Conceived and designed the experiments: GJAH MJVZ MSK MI EN MP SS. Performed the experiments: GH MJVZ MSK MI AM EN MP. Analyzed the data: GJAH. Contributed reagents/materials/analysis tools: GJAH SS MKC. Wrote the paper: GJAH MJVZ. Provided statistical guidance: MKC JDO SS MJVZ. Provided data: MJB EFH JH MI MSK PBM AM EN JDO MP. Revised manuscript: MJB MKC EFH JH MI MSK PBM AM EN JDO MP SS.

References

1. Sala OE, Chapin III FS, Armesto JJ, Berlow E, Bloomfield J, et al. (2000) Global biodiversity scenarios for the year 2100. Science 287: 1770–1774.
2. Davis MA, Chew MK, Hobbs RJ, Lugo AE, Ewel JJ, et al. (2011) Don't judge species on their origins. Nature 474: 153–154.
3. Simberloff D (2011) Non-natives: 141 scientists object. Nature 475: 36.
4. Simberloff D, Souza L, Nunez MA, Barrios-Garcia MN, Bunn W (2012) The natives are restless, but not often and mostly when disturbed. Ecology 93: 598–607.
5. Melbourne BA, Cornell HV, Davies KF, Dugaw CJ, Elmendorf S, et al. (2007) Invasion in a heterogeneous world: Resistance, coexistence or hostile takeover? Ecol Lett 10: 77–94.
6. Williamson M, Fitter A (1996) The varying success of invaders. Ecology 77: 1661–1666.
7. Kolar CS, Lodge DM (2001) Progress in invasion biology: Predicting invaders. Trends Ecol Evol 16: 199–204.
8. Leung B, Roura-Pascual N, Bacher S, Heikkilä J, Brotons L, et al. (2012) Teasing apart alien species risk assessments: A framework for best practices. Ecol Lett 15: 1475–1493.
9. Peterson AT (2003) Predicting the geography of species' invasions via ecological niche modeling. Q Rev Biol 78: 419–433.
10. Kulhanek SA, Leung B, Ricciardi A (2011) Using ecological niche models to predict the abundance and impact of invasive species: Application to the common carp. Ecol App 21: 203–213.
11. Brown JH, Mehlman DW, Stevens GC (1995) Spatial variation in abundance. Ecology 76: 2028–2043.
12. Brown JH (1984) On the relationship between abundance and distribution of species. The American Naturalist 124: 255–279.
13. McGill BJ (2010) Towards a unification of unified theories of biodiversity. Ecol Lett 13: 627–642.
14. Brown JH (1995) Macroecology. Chicago, IL: University of Chicago Press. 269 p.
15. Parker IM, Simberloff D, Lonsdale WM, Goodell K, Wonham M, et al. (1999) Impact: Toward a framework for understanding the ecological effects of invaders. Biol Invasions 1: 3–19.
16. Thiele J, Kollmann J, Markussen B, Otte A (2009) Impact assessment revisited: Improving the theoretical basis for management of invasive alien species. Biol Invasions 12: 2025–2035.
17. Labra FA, Abades S, Marquet PA (2005) Distribution and abundance: Scaling patterns in exotic and native bird species. In: Sax DF, Stachowicz JJ, Gaines SD, Species invasions insights into ecology, evolution and biogeography. Sunderland, MA: Sinauer Associates. pp. 421–446.
18. Van Kleunen M, Weber E, Fischer M (2010) A meta-analysis of trait differences between invasive and non-invasive plant species. Ecol Lett 13: 235–245.
19. Mack RN, Simberloff D, Lonsdale WM, Evans H, Clout M, et al. (2000) Biotic invasions: Causes, epidemiology, global consequences, and control. Ecol App 10: 689–710.
20. Valery L, Fritz H, Lefeuvre J-C, Simberloff D (2008) In search of a real definition of the biological invasion phenomenon itself. Biol Invasions 10: 1345–1351.
21. Sugihara G (1980) Minimal community structure: An explanation of species abundance patterns. The American Naturalist 116: 770–787.
22. Firn J, Moore JL, MacDougall AS, Borer ET, Seabloom EW, et al. (2011) Abundance of introduced species at home predicts abundance away in herbaceous communities. Ecol Lett 14: 274–281.
23. Parker JD, Torchin ME, Hufbauer RA, Lemoine NP, Alba C, et al. (2013) Do invasive species perform better in their new ranges? Ecology 94: 985–994.
24. Davis MA, Thompson K (2000) Eight ways to be a colonizer; two ways to be an invader: A proposed nomenclature scheme for invasion ecology. Bull Ecol Soc Am 81: 226–230.

25. Gelman A, Hill J (2006) Data analysis using regression and multilevel/hierarchical models. New York, NY: Cambridge University Press.
26. Bates DM, Sarkar D (2007) Lme4: Linear mixed-effects models using s4 classes. R package version 0.99875-6 ed.
27. R Development Core Team (2012) R: A language and environment for statistical computing. 2.15.1 ed. Vienna, Austria. R Foundation for Statistical Computing.
28. Baayen RH, Davidson DJ, Bates DM (2008) Mixed-effects modeling with crossed random effects for subjects and items. J Mem Lang 59: 390–412.
29. Qian SS, Shen Z (2007) Ecological applications of multilevel analysis of variance. Ecology 88: 2489–2495.
30. Magee L (1990) R^2 measures based on wald and likelihood ratio joint significance tests. Am Stat 44: 250–253.
31. Kramer M. R^2 statistics for mixed models (2005) Proceedings of the Conference on Applied Statistics in Agriculture 17. Manhattan, Kansas. Kansas University Press. pp. 148–160.
32. Kulhanek SA, Ricciardi A, Leung B (2011) Is invasion history a useful tool for predicting the impacts of the world's worst aquatic invasive species? Ecol App 21: 189–202.
33. Paine RT (1969) A note on trophic complexity and community stability. Am Nat 103: 91–93.
34. Ricciardi A (2003) Predicting the impacts of an introduced species from its invasion history: An empirical approach applied to zebra mussel invasions. Freshwater Biol 48: 972–981.
35. Yokomizo H, Possingham HP, Thomas MB, Buckley YM (2009) Managing the impact of invasive species: The value of knowing the density–impact curve. Ecol App 19: 376–386.
36. Wilson KA, Magnuson JJ, Lodge DM, Hill AM, Kratz TK, et al. (2004) A long-term rusty crayfish (*Orconectes rusticus*) invasion: Dispersal patterns and community change in a north temperate lake. Can J Fish Aquat Sci 61: 2255–2266.
37. Solomon CT, Olden JD, Johnson PTJ, Dillon RT, Vander Zanden MJ (2010) Distribution and community-level effects of the Chinese mystery snail (*Bellamya chinensis*) in northern Wisconsin lakes. Biol Invasions 12: 1591–1605.
38. Dick JTA, Gallagher K, Avlijas S, Clarke HC, Lewis SE, et al. (2013) Ecological impacts of an invasive predator explained and predicted by comparative functional responses. Biol Invasions 15: 837–846.
39. Gaston KJ, Blackburn TM, Gregory RD, Greenwood JJD (1998) The anatomy of the interspecific abundance-range size relationship for the British avifauna: I. Spatial patterns. Ecol Lett 1: 38–46.
40. Kueffer C, Daehler CC, Torres-Santana CW, Lavergne C, Meyer J-Y, et al. (2010) A global comparison of plant invasions on oceanic islands. Perspect Plant Ecol 12: 145–161.
41. McGill BJ, Nekola JC (2010) Mechanisms in macroecology: AWOL or purloined letter? Towards a pragmatic view of mechanism. Oikos 119: 591–603.
42. McGill BJ, Etienne RS, Gray JS, Alonso D, Anderson MJ, et al. (2007) Species abundance distributions: Moving beyond single prediction theories to integration within an ecological framework. Ecol Lett 10: 995–1015.
43. Lawton JH (1993) Range, population abundance and conservation. Trends Ecol Evol 8: 409–413.
44. Ives AR, Klopfer ED (1997) Spatial variation in abundance created by stochastic temporal variation. Ecology 78: 1907–1913.
45. Strayer DL, Eviner VT, Jeschke JM, Pace ML (2006) Understanding the long-term effects of species invasions. Trends Ecol Evol 21: 645–651.
46. Kolar CS, Lodge DM (2002) Ecological predictions and risk assessment for alien fishes in North America. Science 298: 1233–1236.
47. Goodwin BJ, McAllister AJ, Fahrig L (1999) Predicting invasiveness of plant species based on biological information. Conserv Biol 13: 422.
48. Thuiller W, Richardson DM, Pysek P, Midgley GF, Hughes GO, et al. (2005) Niche-based modelling as a tool for predicting the risk of alien plant invasions at a global scale. Global Change Biology 11: 2234–2250.
49. Mercado-Silva N, Olden JD, Maxted JT, Hrabik TR, Vander Zanden MJ (2006) Forecasting the spread of invasive rainbow smelt in the Laurentian Great Lakes region of North America. Conserv Biol 20: 1740–1749.
50. DeVaney SC, McNyset KM, Williams JB, Peterson AT, Wiley EO (2009) A tale of four "carp": Invasion potential and ecological niche modeling. Plos One 4: E5451; doi 5410.1371/Journal.Pone.0005451.
51. Paull SH, Song S, McClure KM, Sackett LC, Kilpatrick AM, et al. (2012) From superspreaders to disease hotspots: Linking transmission across hosts and space. Front Ecol Environ 10: 75–82.
52. Lodge DM, Williams S, MacIsaac HJ, Hayes KR, Leung B, et al. (2006) Biological invasions: Recommendations for U.S. Policy and management. Ecol App 16: 2035–2054.
53. Sax DF, Brown JH (2000) The paradox of invasion. Global Ecology and Biogeography 9: 363–371.

16

Spatial Variation as a Tool for Inferring Temporal Variation and Diagnosing Types of Mechanisms in Ecosystems

Matthew P. Hammond*, **Jurek Kolasa**

Department of Biology, McMaster University, Hamilton, Ontario, Canada

Abstract

Ecological processes, like the rise and fall of populations, leave an imprint of their dynamics as a pattern in space. Mining this spatial record for insight into temporal change underlies many applications, including using spatial snapshots to infer trends in communities, rates of species spread across boundaries, likelihood of chaotic dynamics, and proximity to regime shifts. However, these approaches rely on an inherent but undefined link between spatial and temporal variation. We present a quantitative link between a variable's spatial and temporal variation based on established variance-partitioning techniques, and test it for predictive and diagnostic applications. A strong link existed between spatial and regional temporal variation (estimated as Coefficients of Variation or CV's) in 136 variables from three aquatic ecosystems. This association suggests a basis for substituting one for the other, either quantitatively or qualitatively, when long time series are lacking. We further show that weak substitution of temporal for spatial CV results from distortion by specific spatiotemporal patterns (e.g., inter-patch synchrony). Where spatial and temporal CV's do not match, we pinpoint the spatiotemporal causes of deviation in the dynamics of variables and suggest ways that may control for them. In turn, we demonstrate the use of this framework for describing spatiotemporal patterns in multiple ecosystem variables and attributing them to types of mechanisms. Linking spatial and temporal variability makes quantitative the hitherto inexact practice of space-for-time substitution and may thus point to new opportunities for navigating the complex variation of ecosystems.

Editor: Bazartseren Boldgiv, National University of Mongolia, Mongolia

Funding: This study was supported by Natural Sciences and Engineering Research Council of Canada (NSERC - http://www.nserc-crsng.gc.ca) grant 5-31314-6100. The funders had no role in study design, data collection and analysis, decision to publish or preparation of the manuscript.

Competing Interests: The authors have declared that no competing interests exist.

* E-mail: mhammond@gmx.com

Introduction

The spatial texture of a landscape is a fundamental reflection of the ecological processes underpinning it. Thus, spatial distributions of population [1], geological [2] and climatological variables [3] can impart key details about the forces, operating over time, that forged them. Spatial patterns are *diagnostic* when they are used to uncover hidden mechanisms in the landscape, and *predictive* when they indicate the likely future behavior of a process. Ecology is full of examples of the former, diagnostic approach where spatial patterns are mined for evidence of mechanisms like dispersal, competition or environmental structuring [4–7]. But the latter, predictive approach is also commonplace. Because obtaining long time series is difficult, inferring temporal patterns from spatial data is used in such varied contexts as: (i) chronosequences, where gradients of different-aged sites are used to track how a process (e.g., succession) changes from one state to another over time [8–11], (ii) boundary dynamics, where spatial snapshots can reveal the rate of species spread [12], (iii) complex dynamics, where spatial data helps detect chaos [13], and (iv) regime or phase shifts, where changes in spatial variation can expose the incipient reorganization of an ecosystem [14–16].

Using spatial patterns to infer temporal patterns ("space-for-time substitution"; *sensu* [8]) or mechanisms quickly encounters the hard problem of interpreting dynamics [17,18]. We, as others [18,19], note that progress will require a deep understanding of what spatial patterns reveal about temporal patterns, and how these come together in the spatiotemporal patterns of landscapes. We further suggest that such insight will help both *predictive* and *diagnostic* efforts.

Here we focus on the variability of values over time (i.e., the inverse of stability [20] or constancy [21]) as opposed to properties like resistance or resilience [21]. We focus, therefore, on the dissimilarity of values and whether this variability (e.g., fluctuations in density) can be predicted from a snapshot of spatial variability. This application, in particular, would be useful given the rarity of long time series in ecology [9], the widespread nature of anthropogenic impacts [22], the increasing attention to ecological stability [20,23–26], and the need for clearer links between spatial and temporal concepts [27].

To our knowledge, no links explicitly tie the temporal and spatial variation of a variable. However, a theoretical foundation for doing so is available through ANOVA variance partitioning [28–30] because overall spatiotemporal variation can be broken down into its spatial and temporal components. Crucially, these components can be re-expressed in terms of temporal variance at the regional scale (i.e., spatially-aggregated at time *k*; given as

Var(Y) where $Y_k = \sum_{i=1}^{n} X_{ik}$). This scale reflects the net sum of what occurs in all patches and thus reflects variation in resources and ecological functions at the landscape level. Our derivation makes regional temporal variance itself decomposable (Fig. 1A). Traditional variance partitioning methods, in contrast, only decompose total spatiotemporal variance. We show in Fig. 1A that regional temporal variance (Var(Y)) obeys a simple relationship with a spatial quantity - the sum of spatial variances measured at time *k* ($\sum$var(X_k)). This relationship, in turn, is modified by two spatiotemporal patterns [28–31], *inter-patch synchrony* and *persistence of spatial variation* (Figs. 1 & 2; see File S1 for derivation and details).

Inter-patch synchrony (summed inter-patch covariances; $\sum$cov(X_i,X_j)) refers to temporal changes that happen simultaneously in patches *i* and *j*. It is well known to boost temporal variation at the regional scale [32] (e.g., widespread population decline during drought). On the other hand, *persistence of spatial variation* (summed inter-time covariances; $\sum$cov(X_k,X_l)) - *persistence* for short - describes differences or gradients between patches *i* and *j* that are retained from time *k* to *l* (e.g., fixed or permanent differences between locations). Opposite in sign to synchrony, persistence is associated with lower temporal variance. This is because a pattern of spatial variation is retained over time only if most patches are relatively stable. Temporal variance thus depends critically on the balance of synchrony and persistence.

Implications of Analytical Framework

The relationship captured in Fig. 1A points out the basis for *predictive* applications like space-for-time substitution. It does so by showing that spatial and temporal variance will scale exactly (by a factor of n_i/n_k) for stochastic processes. Stochastic processes enable

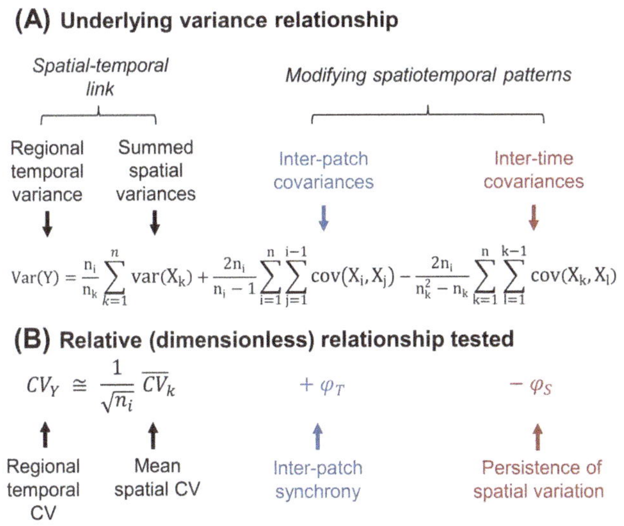

Figure 1. The spatial-temporal variability link. (A) We derived an analytical relationship linking regional temporal variance of a process (Var(Y)) to summed spatial variances at time *k* ($\sum$var(X_k)). Inter-patch synchrony ($\sum$cov(X_i,X_j)) and persistence ($\sum$cov(X_k,X_l)) modify this link and lead temporal and spatial variance to scale as a function of number of patches (n_i) and time points (n_k) when these terms are zero. **(B)** We evaluate the usefulness, for prediction and description, of the corresponding (relative) relationship that uses dimensionless coefficients: Regional temporal CV (CV_Y), mean spatial CV ($\overline{CV_k}$), and indices of synchrony (φ_T) and persistence (φ_S). While an exact solution exists (Eq. S28, File S1), we use a more useful approximation, $CV_Y \cong (1/n_i^{1/2})\overline{CV_k}$, that gives an expected temporal CV for a given spatial CV when synchrony and persistence are negligible (Eq. S31,File S1).

this because their values are uncorrelated between patches *i* and *j*, as well as between times *k* and *l*, and this sets synchrony and persistence terms to zero ($\sum$cov(X_i,X_j) = 0, $\sum$cov(X_k,X_l) = 0). This is a form of ergodicity [33] that can be illustrated by an analogy. Imagine a seascape in which wave peaks are independent of each other: In this null case, wave amplitudes from trough to peak would be equally large or small whether waves were measured from a fixed point (i.e., waves passing over time) or from a transect across the seascape (i.e., a snapshot of waves across space). Our formulation merely adds that this match between temporal and spatial variability applies at the regional (seascape) scale as well as at the patch (wave) scale. Fig. 2 summarizes this mechanism, showing how temporal fluctuations are recorded as spatial variability.

Diagnosis, where inferences are made about how patterns came about, may also be made possible by the analytical solution. This is because components of temporal variance from Fig. 1A also describe and summarize spatiotemporal patterns that are the net result of ecological mechanisms. Moreover, because these terms are linked to temporal variability, they may provide a new view of dynamics and their consequences for stability.

Because they are commonly used in ecology, we extended our analytical framework to include common indices (Fig. 1B) like the Coefficient of Variation (CV), and indices of synchrony (φ_T) and persistence (φ_S). We test the validity of these formulations and turn them to answering three questions: (i) Is the spatial CV of a landscape variable a meaningful proxy for its (regional) temporal CV? (ii) Under what conditions is it predictive? And (iii) what do departures from an exact match between spatial and temporal CV tell us about the forces shaping dynamics of variables? We apply our approach to 136 biotic and physicochemical variables from three landscape types: Laboratory arrays of connected aquatic microcosms (measured for 20 weeks), a natural array of Jamaican coastal rock pool ecosystems (13 years), and a set of seven lakes from the North Temperate Lakes LTER site (30 years). Results shed light on what real world inferences can be drawn when the relationship between spatial and temporal variation is known.

Materials and Methods

Ethics Statement

Invertebrate species were sampled with permission on land owned by University of West Indies (Discovery Bay Marine Lab) and are not protected by law. Laboratory experiments used invertebrate species that do not require permits or procedural approvals.

Analytical Relationship: Linking Spatial and Regional Temporal Variance

Values in a landscape vary over time (*k…n*), and across patches (*i…n*). These dimensions of variation both contribute to regional temporal variance, which is the variance of the spatially-aggregated time series (i.e., Var(Y) where $Y_k = \sum_{i=1}^{n} X_{ik}$). Spatial and temporal variation can be precisely linked through two mathematical truisms: (i) Spatial and temporal variances, estimated from the same site × time data matrix, are related by rules that underlie ANOVA variance partitioning and (ii) these variances, which capture variation at the aggregate scale for both time (i.e., Var(Y)) and for space (i.e., temporally-aggregated; Var(Z) where $Z_i = \sum_{k=1}^{n} X_{ik}$), can be further decomposed into variances and covariances of patches *i* and *j* or time points *k* and *l* [34]. See File S1 for derivation. Var(Y) can thus be re-expressed as in Fig. 1A where; var(X_k) is the spatial variance at time *k*, cov(X_i,X_j) is the covariance of patch *i* with *j* (synchrony), and cov(X_k,X_l) is the

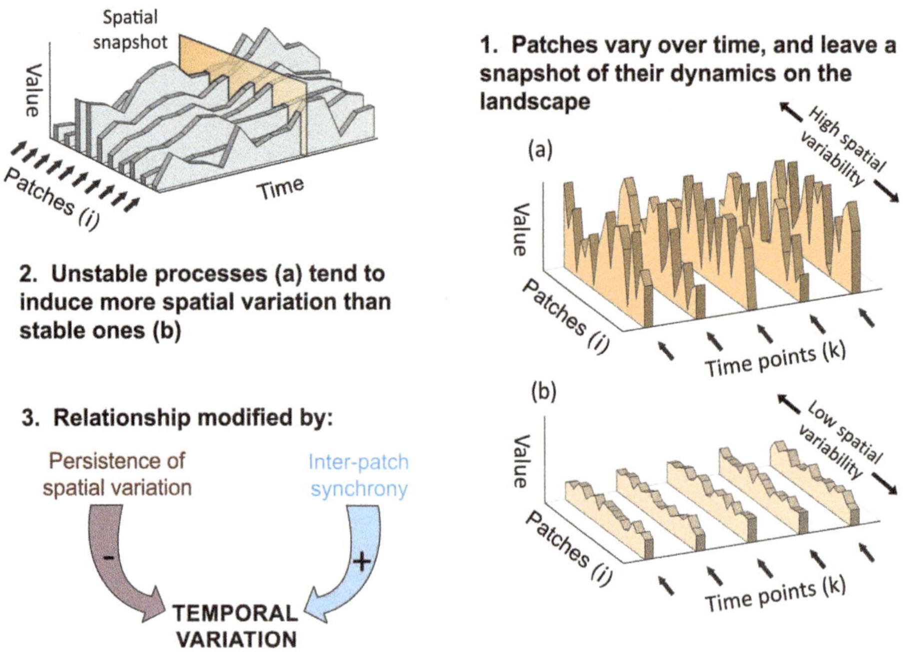

Figure 2. Spatial imprinting of ecosystem processes. Theorized mechanism by which temporal fluctuations of patches create spatial variability in the landscape, which may in turn be a proxy for temporal variability. Spatiotemporal patterns (inter-patch synchrony and persistence) modify the correspondence of spatial and temporal variability (Fig. 1), so it is unknown whether the link is strong enough for predictive (e.g., space-for-time substitution) applications and whether modifying terms have diagnostic/descriptive value.

covariance of time k with l (persistence of spatial variation). These three components of spatiotemporal pattern are consistent with prior theory and statistical concepts [28–31].

We converted the above analytical relationship into dimensionless quantities (Eq. S28; File S1) – regional temporal CV (CV_Y), spatial CV (CV_k), and indices of synchrony (φ_T) and persistence (φ_S). While exact, this relationship does not yield a clear null relationship between CV_Y and spatial CV. We therefore used an approximation of it (Eq. S31; File S1) that gives the expected value of temporal CV from spatial CV in the absence of synchrony or persistence (Fig. 1B). Temporal CV values calculated using this approximation were 94–98% correlated (1:1) with values from random number simulations where synchrony and persistence were close to zero. These null values, in turn, were used to plot the lines of "independent dynamics" shown in spatial-temporal CV plots (Figs. 3–5).

Data Analysis

Biotic variables included population densities of invertebrate and fish species and ecosystem-level quantities like NPP, while physicochemical variables ranged from temperature and pH to ion concentrations (Table S1, S2). For each variable, we estimated all indices in Fig. 1B. Regional temporal CV was estimated as the quotient of the time series standard deviation and mean. Mean spatial CV of a variable was defined as the average of spatial CV's measured at time k. These were calculated either across the three microcosms of each experimental replicate, across 49 Jamaican rock pools, or across the seven LTER lakes. If a species was absent from all water bodies spatial CV could not be calculated for that time point. In these cases, mean spatial CV was calculated as an average of the time points in which it was present (mean frequency of occurrence = 65% of years).

We estimated inter-patch synchrony using a variance ratio φ_T [35] which is, roughly speaking, a ratio of aggregate (regional) to component variances:

$$\varphi_T = \frac{\mathrm{Var}(Y)}{(\sum \mathrm{SD}(X_i))^2} \tag{1}$$

where $\mathrm{SD}(X_i)$ is the standard deviation of a patch. As patches synchronize, the value of φ_T grows from zero to one. The counterpart of inter-patch synchrony is persistence where, instead of temporal changes being similar from patch i to j, spatial differences are similar from time k to time l. We therefore estimated persistence with the spatial counterpart of φ_T, which we call φ_S. φ_S was calculated by replacing Var(Y) and $\mathrm{SD}(X_i)$ in Eq. 1 with their spatial equivalents; the variance of the temporally-aggregated series (i.e., Var(Z)) and spatial standard deviation at time k ($\mathrm{SD}(X_k)$). Analogous to φ_T, φ_S values increase from zero to one as differences among patches persist more through time.

We used General Linear Models (GLM) and multiple regression in Statistica v. 8.0 (StatSoft Inc., 2007) to predict regional temporal CV from mean spatial CV, φ_T, and φ_S (i.e., ~Fig. 1B). Temporal and spatial CV values were log transformed for analysis because, when plotted, they tended to form fan-shaped data clouds that were best described by power functions. We tested residuals of all analyses for normal distribution using Kolmogorov-Smirnov tests. Surfaces (Fig. 5) were fitted by distance-weighted least squares.

Microcosm Connectivity Experiment

We assembled replicate arrays of three ×700 mL aquatic microcosms. Each array contained community types that were relatively stable under laboratory conditions: (i) impoverished, containing ubiquitous microbes initially surviving in distilled water, (ii) phytoplankton and microbes, and (iii) 10 invertebrate species (cladocerans, ostracods) and phytoplankton and microbes. We arranged microcosms such that each initially contained a

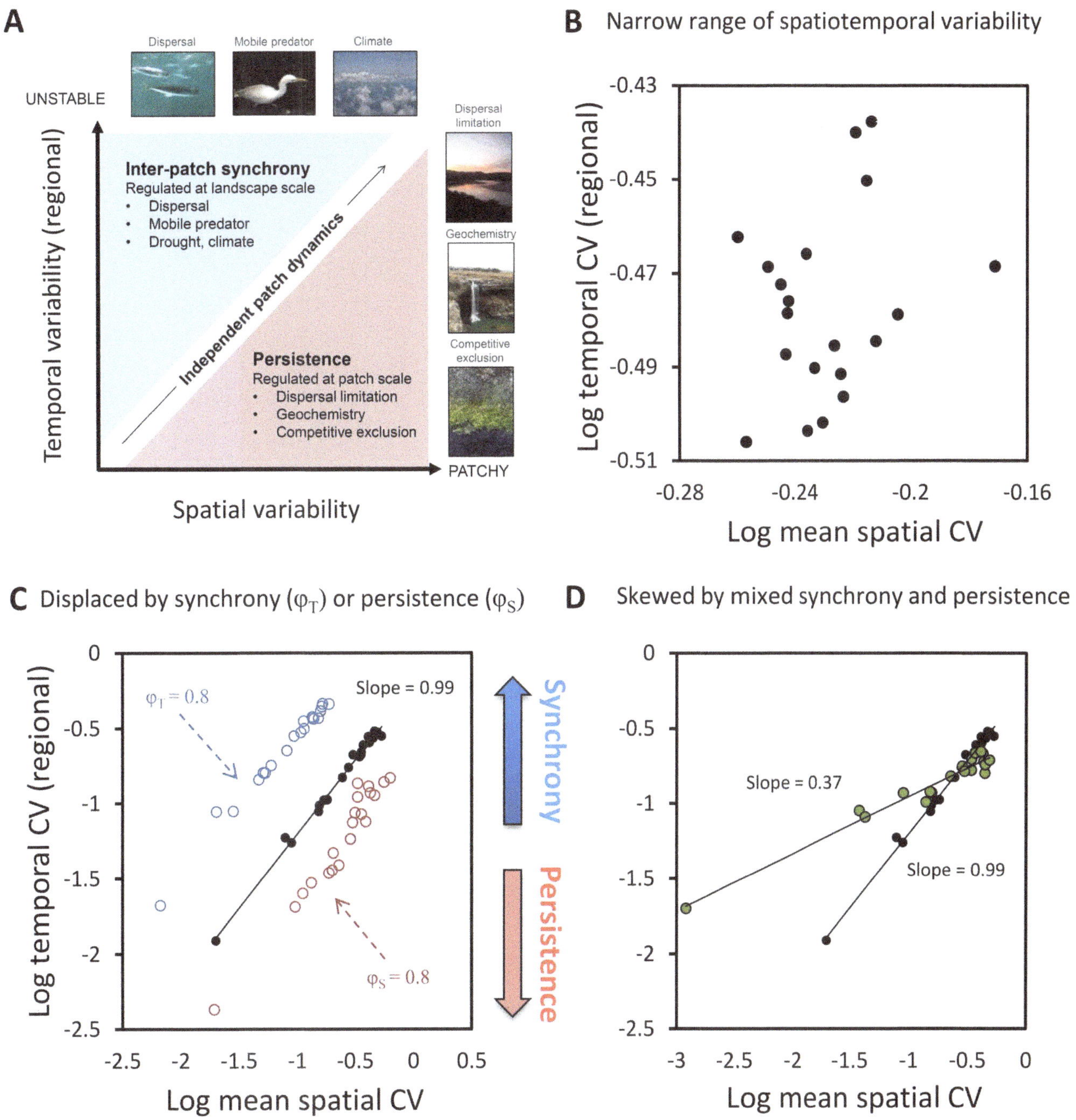

Figure 3. Anatomy of a plot between spatial and regional temporal variability. A stochastic null model of a three patch mosaic illustrated several features of a plot between log mean spatial CV and log regional temporal CV. (**A**) Three regions exist in which a variable (point) can fall - an "independent dynamics region" when values are independent between patches *i* and *j* and time points *k* and *l*, a "synchrony region" when inter-patch synchrony boosts temporal CV, and a "persistence region" when spatial gradients are retained over time; (**B**) Weak linear relationship when variables share similar spatiotemporal variability, leading to scatter from small variations in synchrony or persistence; (**C**) Strong linear relationships when variables differ in spatiotemporal variabililty and occupy the "independent dynamics region" (black circles), but also when all variables are equally dispaced by synchrony (blue circles) or by persistence (red circles); (**D**) Deviation of regression slope from ~1 (black circles) when variables change in synchrony or persistence as a function of variability. Here, a gradient exists from variables with low variability and synchrony to variables with high variability and persistence. Spatial CV values are means of spatial CV measured at time point *k*. Each point represents a variable and is a mean of ten replicates.

distinct community type, with all three types represented in an array.

Spatial exchange was manipulated by connecting component microcosms with clear Tygon® tubes. Treatments were: No connection among microcosms and bi-directional connection among all three microcosms. Connector tube diameters were increased by ~70% at week 10 of the 20-week experiment. Seven ecosystem-level variables were measured weekly in each micro-

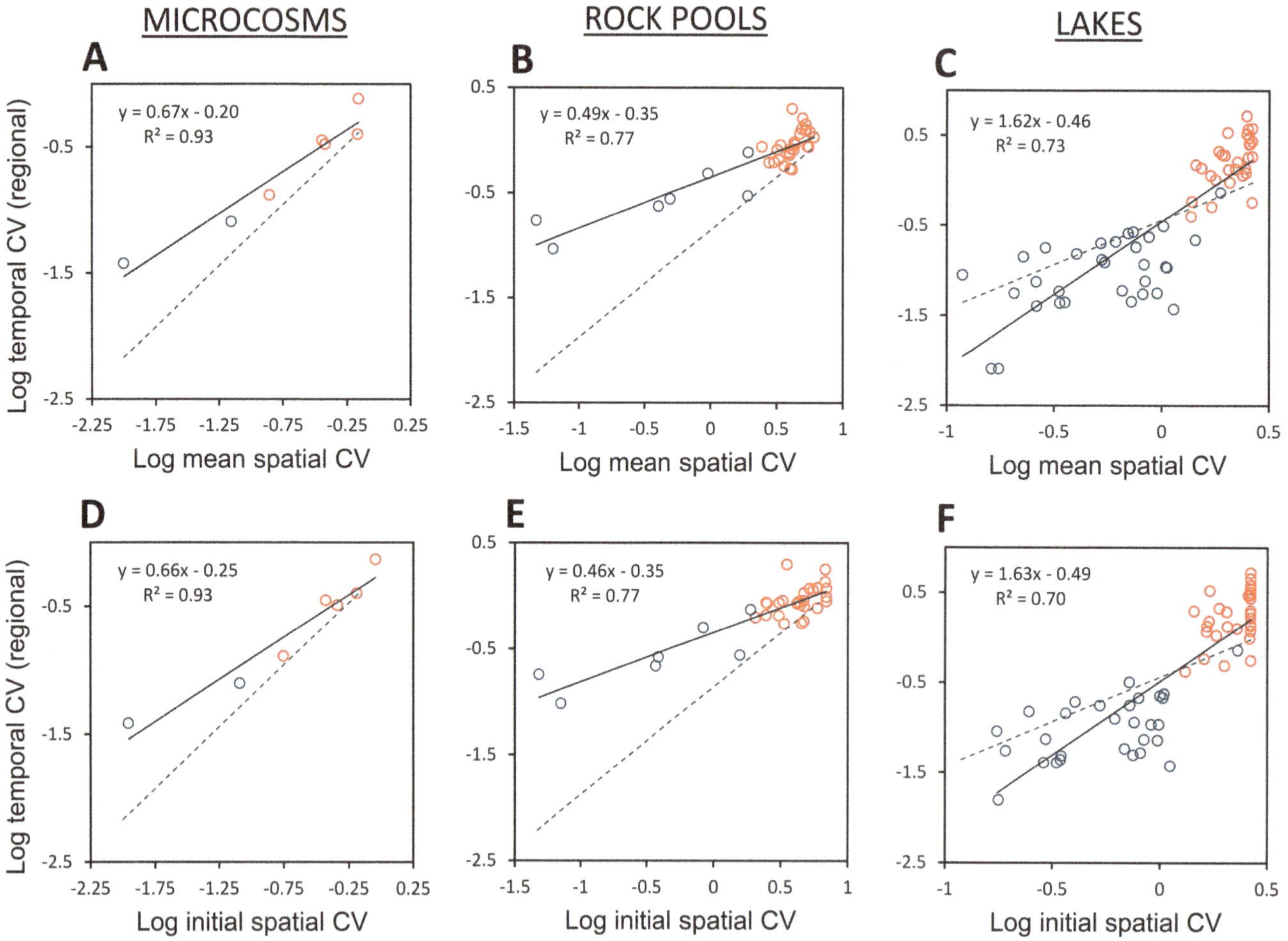

Figure 4. Empirical CV plots illustrating an underlying spatial-temporal link (Fig. 1B). The regional temporal CV of an ecosystem variable (data point) was predictable from its spatial CV in microcosm (n = 7) (**A, D**), rock pool (n = 33) (**B, E**), and lake systems (n = 60) (**C, F**). The predictive value of spatial variability was consistent in that linear associations emerged whether spatial variability was estimated as the mean of spatial CV's at time *k* (**A–C**), or whether a spatial CV from an initial time point (*k*) was used to predict temporal CV of the remaining (*k+1…n*) time series (**D–F**). Dashed lines denote the relationship expected for stochastic processes i.e., when values are independent across space and time. These were obtained by simulating random numbers with the same data structure as empirical data sets. Abiotic variables (blue circles) were consistently more stable and less spatially patchy than biotic variables (red circles).

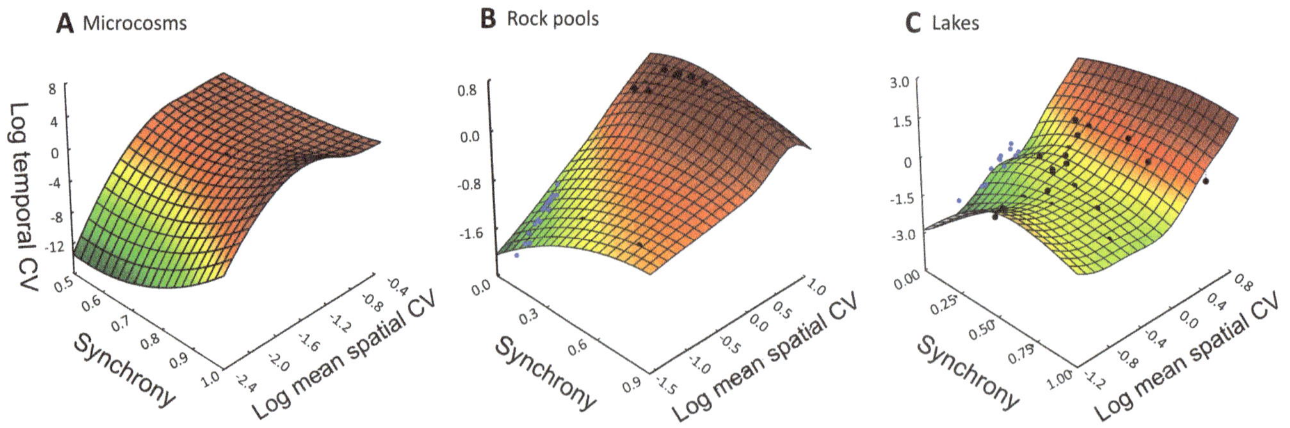

Figure 5. The modifying role of inter-patch synchrony. Relationship between spatial CV and regional temporal CV as modified by the degree of inter-patch synchrony in (**A**) microcosm, (**B**) rock pools and (**C**) lakes. Synchrony increased regional temporal CV relatively little over that explained by spatial CV. Black points = empirical variables, blue points = simulated, randomly-generated variables (n = 20; see File S1) to represent "independent dynamics region."

cosm of the array for 20 weeks using light-dark bottle methods, chlorophyll extractions and environmental sensor probes (Table S1). Microcosm NPP data were rescaled, bringing the lowest value to zero to correct spatial and temporal CV's for negative values.

Natural Rock Pool Ecosystem

We collected data over thirteen annual surveys (1989–2002) in a Jamaican rock-pool system of 78 invertebrate species, dispersing among 49 rock pools. The system lies near Discovery Bay Marine Laboratory, University of the West Indies, on the northern coast of Jamaica (18°28′ N, 77°25′ W). Pools create a mosaic 25 m in radius on a fossil reef no further than 10 m from the ocean and have volumes ranging from 0.5 to 78.4 L. Pools are, on average, within 1 m of the nearest neighbor and never more than 5 m away. Ocean tides occasionally flood a few of the most seaward pools. But most are refilled only by precipitation or, on some occasions, ocean spray. We treated the 49 pools as a single system linked by material fluxes and organism dispersal.

The 70+ invertebrate species in rock pools disperse predominantly by propagules transported via wind, ocean spray, animal vectors and, very occasionally, by overflow from neighboring pools after heavy rainfall [36]. Invertebrate species include: Ostracods (20 species), copepods (five species), cladocerans (five species), worms (15 species), aquatic insects (18 species) and other crustaceans (six species). Most species occurred rarely, some only once (for more details, see [37]). We therefore confined all analyses, except for contributions of variance components (Fig. 6) to 26 common species and temperature, pH, salinity, dissolved oxygen, oxygen saturation, and chlorophyll-a.

Invertebrate densities were estimated for each pool as the number of animals in a 0.5 L sample of water, which was withdrawn after stirring the pool to dislodge organisms from rock walls and to homogenize contents. Each sample was filtered through 63 μm mesh to isolate invertebrates, which were immediately preserved in 50% ethanol. Community samples were sorted, identified to highest possible taxonomic resolution and counted by microscope.

Environmental variables like salinity and pH (Table S2) were measured in each pool using multiprobe sondes (DataSonde, Yellow Springs Instruments, Yellow Springs, Ohio, USA or Hydrolab, Austin, Texas, USA) during biotic surveys for 6–11 of the survey years, depending on the variable.

Small rock pools occasionally dried up, preventing community sampling. These events were recorded as blank data entries, and were <10% of total observations. For our main analyses (Figs. 4–5), we replaced blank entries with zeroes, assuming that a desiccated pool harbored no living, adult invertebrates. To check if this assumption introduced bias, analyses were repeated using two alternative procedures; (i) leaving blank cells unchanged or (ii) interpolating by replacing blanks with the pool mean. All procedures produced similarly significant results indicating no major effect of our assumption. For abiotic variables, blank entries had no logical association with zero (e.g., desiccation does not suggest a 0°C temperature), so these cells were left blank.

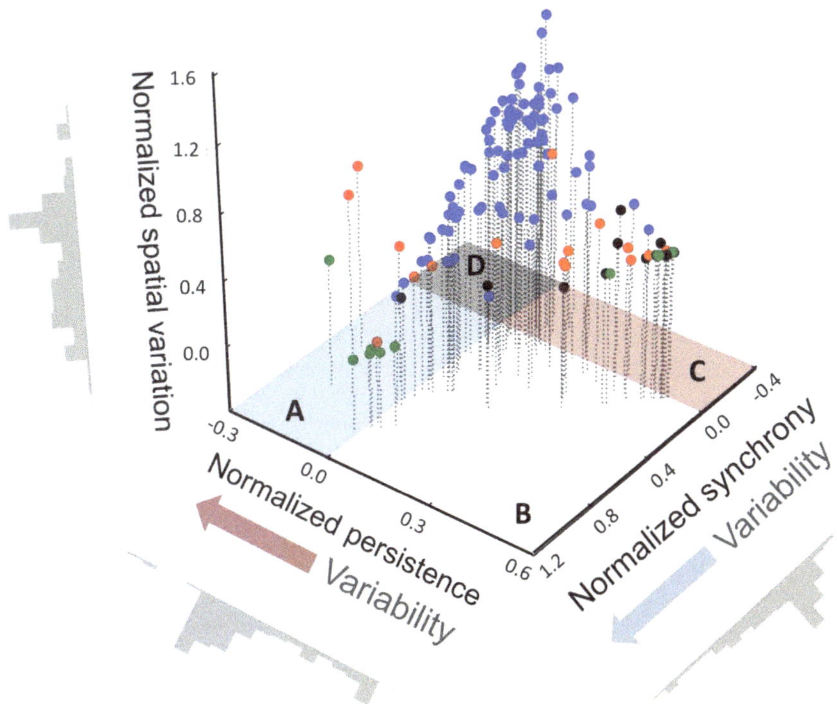

Figure 6. Patterns of spatiotemporal variation underlying the temporal variability of ecosystem variables. The interplay of the three components of temporal variance - spatial variation, synchrony and persistence - was captured by plotting the normalized values of each term in Fig. 1A against each other. Values of each term were standardized to the sum of all three terms such that the resulting proportions summed to one. Variables were assigned to *a priori* groupings based on their likely genesis and mode of regulation, where blue points = species populations, green = atmospheric, red = non-population biotic, black = watershed. n = 136, and includes an additional 36 rare rock pool species that were excluded from earlier analyses due to sparseness of data. Points scatter across theorized modes of dynamics described in Table 1: A = destabilized by synchrony, B = stabilized by persistence, destabilized by synchrony, C = stabilized by persistence, D = stabilized by compensatory dynamics, Intersection of A-D = stabilized by asynchrony. Gray histograms show frequency distributions for each component of temporal variance.

Table 1. Theorized modes of dynamics in landscape variables, their effect on regional temporal variation, and ecological examples.

Graph region	Synchrony/persistence	Dominant effect on temporal variation	Ecological scenario
A	High/Low	Destabilized by synchrony	Local factors less important, landscape-scale factors synchronize dynamics. E.g., synchrony of isolated mammal populations via weather [63]
B	High/High	Stabilized by persistence and destabilized by synchrony	Local factors establish permanent spatial gradients, landscape-scale factors synchronize dynamics. E.g., synchrony of source-sink fish populations via dispersal [64]
C	Low/High	Stabilized by persistence	Local factors establish permanent spatial gradients, dynamics differ among sites. E.g., stable spatial distributions of organisms across habitats [65]
D	Low/Low	Stabilized by compensatory dynamics	Dynamics negatively correlated from time to time, site to site. E.g., spatiotemporal refugia of competing soil nematodes [60]
Intersection of A,B,C,D	Zero/Zero	Stabilized by asynchrony	Dynamics appear stochastic, independent from time to time, site to site. E.g., settlement of broadcast oceanic larvae [66]

Modes reflect different mixtures of inter-patch synchrony and persistence, and correspond to regions of Fig. 6, where variables from three natural ecosystems are plotted by their spatiotemporal patterns.

North Temperate Lakes Long-Term Ecosystem Research Program

We used data from seven Wisconsin lakes (Allequash, Big Muskellunge, Crystal, Sparkling, Trout, Crystal Bog and Trout Bog), collected by the North Temperate Lakes Long-Term Ecological Research program. Lake data were obtained from a public database hosted by the North Temperate Lakes LTER, NSF, Center for Limnology, University of Wisconsin-Madison, available at http://lter.limnology.wisc.edu. We included up to 30 years of data from 60 biotic and abiotic variables across five datasets (Table S2). The following datasets, collected and maintained by LTER associates, were used:

- Chemical limnology of primary study lakes: Major ions
- Chemical limnology of primary study lakes: Nutrients, pH and carbon
- Physical limnology of primary study lakes
- Pelagic macroinvertebrate abundance
- Fish abundance.

Collection methods corresponding to datasets can be found as metadata on the online database (http://lter.limnology.wisc.edu). For fish data, only fyke net catches were used and were standardized by effort (i.e., catch per unit effort) to facilitate comparison. Values for a given lake were annual, obtained by averaging organism densities or physicochemical values across depths, across sampling dates and across stations. Density data were used to equalize the contribution from each lake because Trout Lake is up to 3200 times larger than other lakes, and therefore dominates the landscape spatiotemporal pattern. Results therefore emphasize patterns owing to ecological differences among lakes, rather than to lake size.

Data are available from the LTER database or upon request from the authors.

Results

Three quantities jointly accounted for 87–100% of a variable's regional temporal CV (Table S3) - mean spatial CV, inter-patch synchrony φ_T, and persistence φ_S. This result verifies that spatial and temporal CV's are related and substitutable to the degree that synchrony or persistence does not interfere. Perfect substitution occurs when values are uncorrelated between patches *i* and *j* and between times *k* and *l*. In this case, regional temporal CV is roughly $1/n_i^{1/2}$ times the mean spatial CV (Fig. 1B; Eq. S31). A null, stochastic model (see File S1) showed that such a variable (data point) lies within an "independent dynamics region" on a plot of spatial and temporal CV (Fig. 3A). It falls on a line of slope $\sim 1/n_i^{1/2}$ or ~ 1 on a log-scale (Fig. 3C). Inter-patch synchrony (e.g., from climatic forcing) increases regional temporal variability, shifting a variable to an upper "synchrony region" of the plot (Fig. 3A). But the data point shifts to a lower, "persistence region" (Fig. 3A) when spatial gradients are retained over time (e.g., from patch-specific factors).

A range of plot patterns can emerge depending on the variables sampled and their spatiotemporal dynamics. First, weak or no linear association exists when variables' spatial and temporal CV's span only a narrow range and stochastic differences in synchrony and persistence create scatter (Fig. 3B). Second, strong linear association emerges if all landscape variables have stochastic behavior (black circles, Fig. 3C), similar degrees of inter-patch synchrony (blue circles, Fig. 3C) or similar degrees of persistence (red circles, Fig. 3C). The latter two cases occur because all points are equally displaced up or down from the independent dynamics region. Finally, regression slopes deviate from the expected slope of 1 when some variables display synchrony and others persistence to produce skew (Fig. 3D).

Significant linear regressions existed between spatial CV and regional temporal CV for real ecosystem variables (Fig. 4). In all three data sets, the most stable variables (e.g., hydrological and environmental variables) had low spatial variability, while unstable variables (e.g., species populations) were spatially patchy in the landscape (Fig. 4A–C). Some statistical dependence exists in these plots because mean spatial CV and regional temporal CV are calculated from the same data matrix. However, results were almost perfectly conserved when a single sampling event (*k*) served to estimate spatial CV and predict subsequent (*k+1…n*) temporal CV (Fig. 4D–F). Results are also unlikely to have arisen from biased estimators, since trends were confirmed using several alternative indices (Table S3).

Not all variables lay close to the independent dynamics region. This was reflected in regression slopes and intercepts (solid lines, Fig. 4) which departed from the independent dynamics case (dashed lines, Fig. 4). For instance, regression slopes for microcosms and rock pools were reduced because physicochemical variables (blue circles) at one end of the data cloud exhibited more inter-patch synchrony (Fig. 4A,B,D,E). Meanwhile, lake physicochemical variables showed considerable persistence, lowering temporal variability and steepening the slope (Fig. 4C,F). Though slopes and intercepts deviated from the independent dynamics region, r^2 values of 0.70–0.93 suggest that the rank order of variable's CV's was preserved. We do note some outliers, however, such as pH, Mg and Ca in the persistence region of Fig. 4C, F. Also, some variable types (e.g., species populations; red circles) had greater scatter. Overall, however, synchrony and persistence interfered relatively little with the scaling of spatial and temporal CV. Multiple-regression beta coefficients, for instance, revealed that temporal CV increased 2.9–6.2 times more with a variable's spatial CV than with its synchrony (Fig. 5).

Variables differed in the relative importance of spatial variability, synchrony or persistence to their landscape dynamics. We generated a fingerprint of these dynamics by normalizing the right hand terms of the variance equation (Fig. 1A) for each variable and then plotting them (Fig. 6). A division existed between biotic and physicochemical dynamics, and variables spread across several regions of the plot (quadrants A-D) corresponding to different spatiotemporal behaviors leading to temporal variability. Table 1 synthesizes results by describing these modes of behavior and how they may arise in nature.

Species populations (blue points) clumped together, their dynamics dominated by spatial variability with little synchrony or persistence (Fig. 6; intersection of quadrants A-D). Meanwhile, most atmospherically-driven processes (green points e.g., temperature, dissolved oxygen) were set apart by little persistence but were destabilized by synchrony (quadrant A). In contrast, many watershed-associated variables (black points e.g., ion concentrations) were characterized by persistence but little synchrony (quadrant B), a combination leading to lower temporal variability at the regional scale.

Discussion

Spatial Signatures of Temporal Variability

Variables from three aquatic ecosystems showed a striking and tight correspondence between their regional temporal CV and mean spatial CV. This trend may be considered predictive because it held even when the spatial CV was known from only one time point. Moreover, trends emerged in ecosystems ranging from large to small, and from tropical to temperate, suggesting a potentially general and widespread phenomenon. Applying a space-time correspondence follows more than a century of studies involving substitution [8–10,38], but our formulation extends usefulness in two ways; (i) it is quantitative in the form of equations in Fig. 1 rather than qualitative (e.g., chronosequence studies; see [33] for critical review) and (ii) the logic applies equally when substituting CV's of a single variable or when plotting many ecosystem variables for a multivariate view of landscape variation.

Tight linear dependence between spatial and temporal CV's likely owes to two reasons: First, when dynamics are stochastic or independent, variation in space roughly matches that in time as in ergodicity. Thus small fluctuations in time render equally small fluctuations across space. Second is the empirical observation that the factors which theoretically interfere with this correspondence - synchrony and persistence - do so little, at least when using the CV. While variables can lie anywhere on the plot (shifted up the y-axis by synchrony, down by persistence; Fig. 3C), they adhered more to the "independent dynamics" region than being shifted (Fig. 5). This makes sense in that a variable (e.g., a population) with low temporal CV in each patch will still be relatively stable regionally even when patches partially synchronize. This, in turn, registers as a low spatial CV because of how small temporal fluctuations beget small spatial variation (Fig. 2). There are exciting hints that this type of correspondence also applies to other temporal properties, such as recovery time or deterministic chaos, that leave a telling trace of their temporal dynamics in space [13,16].

Analytical solutions and simulations show that spatial CV has value as a signature of temporal variability under certain conditions. When values of a variable are relatively uncorrelated in time and space (e.g., Fig. 4A), the temporal CV can be recovered with an analytical approximation (Fig. 1B; Eq. S31). Accuracy wanes when ecological forces synchronize patches and shift the variable into the synchrony region of the plot (Fig. 3) where spatial CV underestimates temporal CV. Here, synchrony simultaneously boosts temporal variability and lowers spatial variability by aligning the peaks and troughs of fluctuations. Accuracy is also lower when ecological forces cause spatial gradients to persist through time and shift a variable into the persistence region of the plot (Fig. 3). Here, spatial variability exists, but it is created by patches that are stable over time.

Dynamics must be reasonably assumed to be stochastic to use spatial CV as a quantitative proxy for temporal CV. This assumption will often not hold in nature (e.g., when climate swings induce synchrony). And whether it does hold will likely depend on the types of variables chosen (fast or slow, broad-scale or fine-scale) as well as the spatial and temporal sampling scales (Table 2). Even when the assumption does not hold, however, regressions suggest that the rank order of temporal CV's (e.g., highest to lowest) might still be recovered from spatial CV's for qualitative substitution. This should be possible when: (i) All variables are thought to be synchronized or persistent to the same degree (Fig. 3C), (ii) they smoothly intergrade from synchrony to persistence (Figs. 3D; 4C,F), and/or (iii) the distorting effect of synchrony and persistence can be estimated and corrected for (Table 2). Thus, the link between spatial and temporal variation may be valuable for understanding when CV's are interchangeable, and how to interpret them when they are not.

Some variable types may be better suited to substitution than others. Interestingly, species populations (Fig. 4; red circles) illustrated a tradeoff between precision and accuracy. Precision to distinguish species with high versus low temporal CV's using spatial CV's was weak. This was for the sampling reasons that species had a limited range of variability (i.e., range of CV's< scatter) or possibly greater measurement error; or for the ecological reason that they differed in synchrony or persistence which created scatter. Such differences can occur when species respond even slightly differently to the spatial over the temporal environment [39]. But on the other hand, accuracy was probably higher for species because these variables lay closer to the independent dynamics line than physicochemical variables.

Results captured strong associations between spatial and temporal CV when data were separated by one time point (Fig. 4C–F) or were overlapping (Fig. 4A–C). Future work must address how far into the future a spatial CV can be a proxy for temporal CV. Additional error will certainly accrue for substitution if unexpected events alter a variable's temporal variability in ways not indicated by its initial spatial variability. The exact impact of this non-stationarity, however, is a matter of scale and of research question asked. For instance, high resolution prediction of single-species dynamics following a stochastic disturbance (e.g., forest fire) may be untenable if the disturbance drastically alters variability, synchrony or spatial persistence. But spatial CV should

Table 2. Conditions under which the spatial variability of an ecological process is a precise or accurate substitute for its regional temporal variability.

Condition	Description	Substitution	Requirements	Ecological scenario
Independent dynamics	Values must be reasonably assumed independent from place to place, time to time	Spatial and temporal CV values roughly interchangeable (see Fig. 1B)	Scale of process ≤ scale of measurement in both space and time	Isolated communities with fast dynamics; intercontinental, long-term comparisons
Constant levels of synchrony or persistence across variables	Spatiotemporal behavior (e.g., synchrony) that is shared by all variables shifts all points equally on plot (Fig. 3C). Values shifted by a constant	Order (rank) of temporal CV's is conserved among variables	Approximately equivalent spatiotemporal behavior among variables	Variables similarly shaped by spatial constraints (e.g., microclimates) or by landscape-scale temporal shocks (e.g., weather)
Mixed synchrony and persistence of known magnitude	Variables differ in their synchrony and persistence (Fig. 3D), but degree of divergence is known and corrected for	Temporal CV value or order (rank) can be back-calculated	Estimates of synchrony and/or persistence; Fairly stationary dynamics	Variables responding differentially to spatiotemporal variation in the landscape with synchrony or persistence

be a rough proxy – either quantitative or qualitative – for regional temporal CV under the ecological and sampling conditions in Table 2.

Diagnostic Signatures of Complex Dynamics

The terms of Fig. 1 also allowed a window into diagnosing landscape dynamics. Ecosystems conceal ecological information in an eclectic range of spatiotemporal patterns [40,41]. Populations of exploited species [42], biodiversity hotspots [43], harmful algal blooms [44], pest outbreaks [45] and wildfires [46] all display complex patterning in space and time. A range of analyses explore these patterns by examining underlying frequencies (e.g., spectral [47] and wavelet analysis [48]) and patterns of correlation (e.g., correlograms [49]), or by fitting predictive models (e.g., autoregressive [50]). But these approaches are not designed for linking spatiotemporal pattern to a variable's temporal variation.

We used plots of spatial versus temporal CV's as convenient and unique summaries of landscape dynamics (Fig. 4). Some variables occupied the synchrony region (above dashed line), others the persistence region (below dashed line). Such a mixture may be typical when a wide array of ecosystem processes is sampled. These mixtures left their mark on regression slopes. Slopes <1 in Fig. 4A–B show that abiotic variables in microcosms and rock pools were more stable, but also more prone to synchrony, than populations. Y-intercepts also changed to the degree that multiple variables displayed synchrony or persistence. These therefore offer multivariate indices of the degree of synchrony ($\beta_0 >$ expected) or persistence ($\beta_0 <$ expected) experienced by a landscape (Fig. 3C). Such plots may be fruitful ground for streamlined comparisons of landscapes containing diverse variables and dynamics, like pre-and post-disturbance ecosystems.

The components of temporal variance themselves – spatial variance, synchrony and persistence (Fig. 1A) – may also prove useful for describing patterns and mechanisms driving temporal variation. As we have seen, spatial variance can signal instability at the local, patch scale (e.g., from demographic [51], community [52], natural enemies [53], spatial [26] or local environmental causes; Table S4). Synchrony, in turn, indicates landscape-scale causes of variation like dispersal [53,54] or weather [55,56]. Finally, persistence points to the existence of long-term differences in mean value or state among patches, such as regulation by local communities or physical conditions. Combined, these components of variation gave an alternative view of dynamics.

Fine distinctions emerged when all elements of Fig. 1A were normalized to create a fingerprint or signature of dynamics. The grouping of variables controlled by different parts of the biosphere, like the atmosphere (e.g., temperature, dissolved oxygen) and the watershed (e.g., pH, ion concentrations), suggests that unique signatures may exist for types of ecological processes. Potential may thus exist for predicting the likely dynamics of a variable based on its type (e.g., atmospheric). Meanwhile, the breadth of spatiotemporal behaviors seen suggests a range of spatial variation, synchrony and persistence combinations leading to temporal variation in nature. Our framework may be useful for cataloguing these types of spatiotemporal dynamics in ecosystems (e.g., Table 1), and for making broad-stroke inferences about spatiotemporal mechanisms (e.g., population rescue effects [26,57], predator-prey cycles [58], species coexistence [59,60], and invasive species spread [61]).

Conclusions

Unexplained variation is common in nature, both across heterogeneous landscapes and over timespans of interest. By illustrating the link between spatial and temporal variation, we bring more clarity to the problems of *prediction* and *diagnosis* from spatial or spatiotemporal patterns. More work is needed to test the limits of prediction across scales, variables and ecosystem types. Yet, indications here suggest usefulness in; substituting spatial for temporal variability (either quantitatively or qualitatively), judging when substitution will not work, and interpreting the manifold changes of multivariate landscapes. Such efforts are hoped to add momentum towards the Rosetta Stone of landscape and ecosystem ecology, in which process and mechanism may be deeply and easily discerned from landscape pattern [62].

Supporting Information

File S1.

Table S1 Community and ecosystem variables measured over the 20-week microcosm connectivity experiment.

Table S2 Variables from lake, rock pool and microcosm data sets used in analyses. An additional 36 rock pool species (not shown), known from fewer occurrences, were included for calculating spatiotemporal signatures (Fig. 6).

Table S3 Comparison of CV-based results with alternative indices of variability. General Linear Models were fit between indices of aggregate temporal variability and three spatiotemporal descriptors: Spatial variability, inter-patch synchrony and persistence (see Fig. 1B). Indices of variability included CV and four others. Asterisks denote statistical significance and p-value. *p<0.05, **p<0.01, ***p<0.001. R^2 values in parentheses.

Table S4 Mechanisms that may dampen regional variability and reduce spatial variability by stabilizing local patches.

Acknowledgments

K. Abbott, P. Abrams, M. Anand, S. Carpenter, L. Dai, C. Jones, K. McCann, T. Parker, J. Quinn, D. Rollo, D. Vasseur, R. Vogt, D. Waring, J. Wu and several anonymous reviewers provided comments on the manuscript. We also thank the North Temperate Lakes LTER and Discovery Bay Marine Lab as data sources.

Author Contributions

Conceived and designed the experiments: MPH JK. Performed the experiments: MPH JK. Analyzed the data: MPH. Contributed reagents/materials/analysis tools: MPH JK. Wrote the paper: MH JK.

References

1. Reed DC, Raimondi PT, Carr MH, Goldwasser L (2000) The role of dispersal and disturbance in determining spatial heterogeneity in sedentary organisms. Ecology 81: 2011–2026.
2. Barnes JB, Ehlers TA, Insel N, Mcquarrie N, Poulsen CJ (2012) Linking orography, climate, and exhumation across the central Andes. Geology 40: 1135–1138.
3. Leavitt PR, Fritz SC, Anderson NJ, Baker PA, Blenckner T, et al. (2009) Paleolimnological evidence of the effects on lakes of energy and mass transfer from climate and humans. Limnology and Oceanography 54: 2330–2348.
4. Seabloom EW, Bjørnstad ON, Bolker BM, Reichman OJ (2005) Spatial signature of environmental heterogeneity, dispersal, and competition in successional grasslands. Ecological Monographs 75: 199–214.
5. Fangliang H, Legendre P, Lafrankie J V (2008) Distribution patterns of tree species in a Malaysian tropical rain forest. Journal of Vegetation Science 8: 105–114.
6. Burrows MT, Harvey R, Robb L, Poloczanska ES, Mieszkowska N, et al. (2009) Spatial scales of variance in abundance of intertidal species: Effects of region, dispersal mode, and trophic level. Ecology 90: 1242–1254.
7. Anderson KE, Hilker FM, Nisbet RM (2012) Directional biases and resource-dependence in dispersal generate spatial patterning in a consumer-producer model. Ecology Letters 15: 209–217.
8. Pickett STA (1989) Space-for-time substitution as an alternative to long-term studies. Long-term studies in ecology: Approaches and alternatives. 110–135.
9. Fukami T, Wardle DA (2005) Long-term ecological dynamics: Reciprocal insights from natural and anthropogenic gradients. Proceedings of the Royal Society B 272: 2105–2115.
10. Walker LR, Wardle DA, Bardgett RD, Clarkson BD (2010) The use of chronosequences in studies of ecological succession and soil development. Journal of Ecology 98: 725–736.
11. Blois JL, Williams JW, Fitzpatrick MC, Jackson ST, Ferrier S (2013) Space can substitute for time in predicting climate-change effects on biodiversity. Proceedings of the National Academy of Science 110: 9374–9379.
12. Eppinga MB, Pucko CA., Baudena M, Beckage B, Molofsky J (2013) A new method to infer vegetation boundary movement from "snapshot" data. Ecography 36: 622–635.
13. Sole R V, Bascompte J (1995) Measuring chaos from spatial information. Journal of Theoretical Biology 175: 139–147.
14. Litzow MA, Urban JD, Laurel BJ (2008) Increased spatial variance accompanies reorganization. Ecological Applications 18: 1331–1337.
15. Guttal V, Jayaprakash C (2009) Spatial variance and spatial skewness: Leading indicators of regime shifts in spatial ecological systems. Theoretical Ecology 2: 3–12.
16. Dai L, Korolev KS, Gore J (2012) Slower recovery in space before collapse of connected populations. Nature 496: 355–358.
17. Marcos-Nikolaus P, Martin-Gonzalez JM (2002) Spatial forecasting: Detecting determinism from single snapshots. International Journal of Bifurcation and Chaos 12: 369–376.
18. McIntire EJB, Fajardo A (2009) Beyond description: The active and effective way to infer processes from spatial patterns. Ecology 90: 46–56.
19. Levin SA (1992) The problem of pattern and scale in ecology: The Robert H. MacArthur Award lecture. Ecology 73: 1943–1967.
20. Lehman CL, Tilman D (2000) Biodiversity, stability, and productivity in competitive communities. The American Naturalist 156: 534–552.
21. Grimm V, Wissel C (1997) Babel, or the ecological stability discussions: An inventory and analysis of terminology and a guide for avoiding confusion. Oecologia 109: 323–334.
22. Turner MG (2010) Disturbance and landscape dynamics in a changing world. Ecology 91: 2833–2849.
23. Pimm SL (1984) The complexity and stability of ecosystems. Nature 307: 321–326.
24. Micheli F, Cottingham KL, Bascompte J, Bjørnstad ON, Eckert GL, et al. (1999) The dual nature of community variability. Oikos 85: 161–169.
25. Cottingham KL, Brown BL, Lennon JT (2001) Biodiversity may regulate the temporal variability of ecological systems. Ecology Letters 4: 72–85.
26. Abbott K (2011) A dispersal-induced paradox: Synchrony and stability in stochastic metapopulations. Ecology Letters 14: 1158–1169.
27. Bolliger J, Wagner HH, Turner MG (2007) Identifying and quantifying landscape patterns in space and time. In: Kienast F, Wildi O, Ghosh S, editors. A changing world: Challenges for landscape research. Springer. 177–194.
28. Lewis WM (1978) Comparison of temporal and spatial variation in the zooplankton of a lake by means of variance components. Ecology 59: 666–671.
29. Larsen DP, Kincaid TM, Jacobs SE, Urquhart NS (2001) Designs for evaluating local and regional scale trends. BioScience 51: 1069–1078.
30. Legendre P, De Caceres M, Borcard D (2010) Community surveys through space and time: Testing the space-time interaction in the absence of replication. Ecology 91: 262–272.
31. Chesson PL (1985) Coexistence of competitors in spatially and temporally varying environments: A look at the combined effects of different sorts of variability. Theoretical Population Biology 287: 263–287.
32. Earn DJD, Levin SA, Rohani P (2000) Coherence and conservation. Science 290: 1360–1364.
33. Paine ADM (1985) "Ergodic" reasoning in geomorphology: Time for a review of the term? Progress in Physical Geography 9: 1–15.
34. Feller W (1950) An introduction to probability theory and its applications. New York: Wiley.
35. Loreau M, De Mazancourt C (2008) Species synchrony and its drivers: Neutral and nonneutral community dynamics in fluctuating environments. The American Naturalist 172: E48–E66.
36. Sciullo L, Kolasa J (2012) Linking local community structure to the dispersal of aquatic invertebrate species in a rock pool metacommunity. Community Ecology 13: 203–212.
37. Pandit S, Kolasa J, Cottenie K (2009) Contrasts between habitat generalists and specialists: An empirical extension to the basic metacommunity framework. Ecology 90: 2253–2262.
38. Cowles HC (1899) The ecological relations of vegetation on the sand dunes of Lake Michigan. Botanical Gazette 27: 167–202.
39. Jongejans E, De Kroon H (2005) Space versus time variation in the population dynamics of three co-occurring perennial herbs. Journal of Ecology 93: 681–692.
40. Kareiva P, Wennergren U (1995) Connecting landscape patterns to ecosystem and population processes. Nature 373: 299–302.
41. Bascompte J (1998) Spatiotemporal patterns in nature. Trends in Ecology and Evolution 13: 173–174.
42. Erisman B, Aburto-oropeza O, Gonzalez-abraham C, Mascarenas-Osorio I, Moreno-Baez M, et al. (2012) Spatio-temporal dynamics of a fish spawning aggregation and its fishery in the Gulf of California. Nature Scientific Reports: 1–11.
43. Rutledge DT, Lepczyk CA, Xie J, Liu J (2001) Spatiotemporal dynamics of endangered species hotspots in the United States. Conservation Biology 15: 475–487.
44. Edwards M, Johns DG, Leterme SC, Svendsen E, Richardson AJ (2012) Regional climate change and harmful algal blooms in the northeast Atlantic. Limnology and Oceanography 51: 820–829.
45. Aukema BH, Carroll AL, Zheng Y, Zhu J, Raffa KF, et al. (2008) Movement of outbreak populations of mountain pine beetle: Influences of spatiotemporal patterns and climate. Ecography 31: 348–358.
46. Senici D, Chen HYH, Bergeron Y, Cyr D (2010) Spatiotemporal variations of fire frequency in central boreal forest. Ecosystems 13: 1227–1238.
47. Michele C De, Bernardara P (2005) Spectral analysis and modeling of space-time rainfall fields. Atmospheric Research 77: 124–136.
48. Grenfell BT, Bjornstad ON, Kappey J (2001) Travelling waves and spatial hierarchies in measles epidemics. Nature 414: 716–723.
49. Ranta E, Kaitala V, Lundberg P (1997) The spatial dimension in population fluctuations. Science 278: 1621–1623.
50. Crabtree R, Potter C, Mullen R, Sheldon J, Huang S (2009) A modeling and spatio-temporal analysis framework for monitoring environmental change using NPP as an ecosystem indicator. Remote Sensing of the Environment 113: 1486–1496.

51. Watson A, Moss R, Rothery P, Parr R (1984) Demographic causes and predictive models of population fluctuations in Red Grouse. Journal of Animal Ecology 53: 639–662.
52. Jiang L, Pu Z (2009) Different effects of species diversity on temporal stability in single-trophic and multitrophic communities. The American Naturalist 174: 651–659.
53. Vogwill T, Fenton A, Brockhurst MA (2009) Dispersal and natural enemies interact to drive spatial synchrony and decrease stability in patchy populations. Ecology Letters 12: 1194–1200.
54. Dey S, Joshi A (2006) Stability via asynchrony in Drosophila metapopulations with low migration rates. Science 312: 434–436.
55. Liebhold A, Koenig WD, Bjørnstad ON (2004) Spatial synchrony in population dynamics. Annual Review of Ecology, Evolution and Systematics 35: 467–490.
56. Paradis E, Baillie R, Sutherland WJ, Gregory RD (1999) Dispersal and spatial scale affect synchrony in spatial population dynamics. Ecology Letters 2: 114–120.
57. Blasius B, Huppert A, Stone L (1999) Complex dynamics and phase synchronization in spatially extended ecological systems. Nature 399: 354–359.
58. Tobin PC, Bjørnstad ON (2003) Spatial dynamics and cross-correlation in a transient predator–prey system. Journal of Animal Ecology 72: 460–467.
59. Hassell MP, Comins HN, May RM (1994) Species coexistence and self-organizing spatial dynamics. Nature 370: 290–292.
60. Ettema CH, Rathbun SL, Coleman DC (2000) On spatiotemporal patchiness and the coexistence of five species of Chronogaster (Nematoda: Chronogasteridae) in a riparian wetland. Oecologia 125: 444–452.
61. Pysek P, Hulme PE (2005) Spatio-temporal dynamics of plant invasions: Linking pattern to process 1. Ecoscience 12: 302–315.
62. Keitt TH, Urban DL (2005) Scale-specific inference using wavelets. Ecology 86: 2497–2504.
63. Post E, Forchhammer MC (2002) Synchronization of animal population dynamics by large-scale climate. Nature 420: 168–171.
64. Isaak DJ, Thurow RF (2006) Network-scale spatial and temporal variation in Chinook salmon (Oncorhynchus tshawytscha) redd distributions: Patterns inferred from spatially continuous replicate surveys. Canadian Journal of Fisheries and Aquatic Sciences 296: 285–.
65. Morfin M, Fromentin J-M, Jadaud A, Bez N (2012) Spatio-temporal patterns of key exploited marine species in the Northwestern Mediterranean Sea. PLoS One 7: 1–14.
66. Siegel DA, Mitarai S, Costello CJ, Gaines SD, Kendall BE, et al. (2008) The stochastic nature of larval connectivity among nearshore marine populations. Proceedings of the National Academy of Sciences 105: 8974–8979.

17

Quantifying Water Flow within Aquatic Ecosystems Using Load Cell Sensors: A Profile of Currents Experienced by Coral Reef Organisms around Lizard Island, Great Barrier Reef, Australia

Jacob L. Johansen*

ARC Centre of Excellence for Coral Reef Studies, and School of Marine and Tropical Biology, James Cook University, Townsville, Queensland, Australia

Abstract

Current velocity in aquatic environments has major implications for the diversity, abundance and ecology of aquatic organisms, but quantifying these currents has proven difficult. This study utilises a simple and inexpensive instrument (<$150) to provide a detailed current velocity profile of the coral-reef system around Lizard Island (Great Barrier Reef, Australia) at a spatial and temporal scale relevant to the ecology of individual benthos and fish. The instrument uses load-cell sensors to provide a correlation between sensor output and ambient current velocity of 99%. Each instrument is able to continuously record current velocities to >500 cms^{-1} and wave frequency to >100 Hz over several weeks. Sensor data are registered and processed at 16 MHz and 10 bit resolution, with a measuring precision of 0.06±0.04%, and accuracy of 0.51±0.65% (mean ±S.D.). Each instrument is also pressure rated to 120 m and shear stresses ≤20 kNm^{-2} allowing deployment in harsh environments. The instrument was deployed across 27 coral reef sites covering the crest (3 m), mid-slope (6 m) and deep-slope (9 m depth) of habitats directly exposed, oblique or sheltered from prevailing winds. Measurements demonstrate that currents over the reef slope and crest varies immensely depending on depth and exposure: Currents differ up to 9-fold within habitats only separated by 3 m depth and 15-fold between exposed, oblique and sheltered habitats. Comparisons to ambient weather conditions reveal that currents around Lizard Island are largely wind driven. Zero to 22.5 knot winds correspond directly to currents of 0 to >82 cms^{-1}, while tidal currents rarely exceed 5.5 cms^{-1}. Rather, current velocity increases exponentially as a function of wave height (0 to 1.6 m) and frequency (0.54 to 0.20 Hz), emphasizing the enormous effect of wind and waves on organisms in these shallow coral reef habitats.

Editor: Joshua S. Madin, Macquarie University, Australia

Funding: This study was funded by the Lizard Island Research Station Doctoral Fellowship. http://australianmuseum.net.au/Lizard-Island-Research-Station. The funders had no role in study design, data collection and analysis, decision to publish, or preparation of the manuscript.

Competing Interests: The author has declared that no competing interests exist.

* E-mail: jacob.johansen@my.jcu.edu.au

Introduction

Current is a fundamental feature of aquatic ecosystems and has implications for all aspects of aquatic life [1–4]. Variation in current velocity can shape the relative distribution and abundance of numerous species and families of aquatic organisms [5–6], and directly affect settlement, growth and survival of individuals [7]. While many important ecological patterns have been attributed to the level of currents within marine, brackish and freshwater ecosystems [5,7–10], few studies have directly quantified these currents over the spatial and temporal scales that affect individuals and populations.

A plethora of current sensors is currently being used by oceanographers worldwide including Electromagnetic Current Meters and Acoustic Doppler Current Profilers (ADCP) [11]. These instruments allow long term monitoring for months to years and some are capable of high resolution profiling of the entire water column around each instrument including near bottom velocity measures [12–14]. However, instruments such as the ADCP's are very expensive ($10,000–25,000 per instrument) and in most cases not economically viable for profiling currents in multiple locations simultaneously. The high cost of the commercially available sensors have so far impede current related studies in many fields other than physical oceanography and forced researchers to choose between long-term measurements at few points and high spatial resolution mapping of the current regime [15–16].

Ecological studies that require measures of current velocity have often resorted to low cost plaster dissolution methods or spring-type dynamometers [5,8,17–18]. Plaster dissolution correlate ambient current velocity with the rate of dissolution under different current regimes. While relatively inexpensive and can be deployed in multiple locations to increase the spatial scale of measurement, this method only provides estimates of bulk volume flux over the entire period of deployment. Ecologically, however, maximum current velocities and variations in velocity are of greater importance than average velocities since habitat use by many aquatic organisms is limited by the strongest commonly occurring currents [5,19–20]. These currents impose the greatest physical and physiological demands on resident species, and may directly dislodge sedentary organisms and reduce foraging ability

Table 1. Load cell performance characteristics.

Load cell terminology	*Recommended range*	*Definition*
Creep	±<0.01%/hr	Creep is the change in output occurring over time while under load with all environmental conditions and other variables remaining constant. Creep is due to thermoelastic effects, i.e. the adiabatic heating and cooling of elastic load supporting elements within a load cell as they undergo deflection in response to changes in the applied force [39].
Creep recovery	0.1% of R.O./20 min	The change in rated no-load output (R.O.) occurring with time after removal of a load which had been applied for a specific period of time.
Drift	±<0.02% hr^{-1}	A random change in output under constant load conditions.
Error	±0.01%	The algebraic difference between the indicated and true value of the load being measured.
Hysteresis	±0.01%	The maximum difference between load cell output readings for the same applied load.
Nonlinearity	±0.01%	The maximum deviation of the calibration curve from a straight line drawn between the no-load and rated outputs.
Repeatability	±0.01%	The maximum difference between load cell output readings for repeated loadings under identical loading and environmental conditions.
Resolution	>1024 divisions	The smallest change in mechanical input which produces a change in the output signal.
Response time	<5 ms	The time it takes the load cell to stop oscillating and settle at a reading. A quick return-to-zero is needed for areas of highly oscillating currents (e.g. high wave exposure)
Temperature effect on rated output	±<0.002%/°C	The change in rated output due to a change in ambient temperature.

Load cells must be chosen to fit the conditions where the instrument will be used. Although creep is unavoidable in most high precisions sensors, the effect can be accounted for by choosing a low creep load cell and including zero measures in the logged data for continuous zero verification and adjustment. Zero creep affects bias only and has no effect on the scale factor ensuring measurement stability during long term deployment.

[8–9,21]. Consequently, ecologically meaningful measures of current must include maximum velocities encountered within the spatial scales of the individual as well as the duration and frequency of these maximum velocities.

Spring-type dynamometers can produce low resolution measures of maximum current velocities in an environment [17]. However, each instrument is only capable of recording a single value during deployment. To tease out the duration and frequency of these maximum forces, continuous measures are necessary. Tilt sensors embedded within floats tethered to the bottom have been used with some success for continuous measures [22–23], however these designs often suffer from low resolution particularly at higher current velocities. Donelan and Motycka [24] proposed to use a strain gauge instrument to provide high resolution and continuous measures of currents based on drag forces exerted on a sphere. Although the concept was proven to work [25], this instrument had high power requirements and no internal power or data storage capacity. Lack of instrument autonomy limited deployment to areas and conditions where power source and data could be maintained at a surface station, and severely reduced the number of instruments which could be deployed simultaneously [26]. A recent study by Mach et al [27] demonstrated that autonomy can be achieved in this type of instrument, although build cost remained high. To elucidate the importance of currents for organism ecology and the conditions under which individual aquatic organisms survive, there is a great need for an automated but also practical, inexpensive and easily applicable method that will allow a greater number of researchers to conduct continuous and detailed profiling of currents in environments such as rivers, estuaries and coral reefs.

This study uses of a low power and inexpensive strain gauge instrument to produce a detailed current velocity profile across depth and exposure gradients of a tropical coral reef ecosystem, and thereby help cover a significant gap in our knowledge of the currents encountered by individual benthos and fishes on coral reefs. Each instrument costs <$150 in materials and is capable of obtaining continuous, accurate and precise measures of maximum current velocities and variations in velocities over time scales ranging from seconds to months

Materials and Procedures

Current meter construction

An instrument for continuously measuring and logging current velocity was developed in June 2009 using load cell technology commonly used by industry as a simple and reliable measure of force. A load cell sensor is simply a metal alloy force transducer which produces an analog output signal proportional to the applied weight or force. As force increases on the cell, the internal resistance increases causing a change in output voltage. These sensors require little power, provide linearly increasing output voltage with increasing force and are capable of retaining high precision (<0.01%) and high accuracy (<0.01%) for extended periods (see Table 1 for load cell performance characteristics).

A single point load cell sensor (model 628A from www.hzloadcell.com) capable of measuring ≤3 kg loads with 0.015% precision was chosen for this experiment. This sensor has a natural resonance frequency of ~500 Hz, allowing signals to be recorded to >100 Hz without incurring significant error due to resonance. The sensor was coated with a thin layer of silicon (i.e. submersible) and securely fastened to a submersible instrument housing (PVC pipe: diameter: 11 cm; length: 20 cm; pressure rated to 120 m water depth) with only the force measuring end of the cell protruding horizontally out of the housing (see schematics in Fig. 1 as well as component list and assembly instructions in Fig. S1 and Table S1). Single point load cell sensors measure force in one end

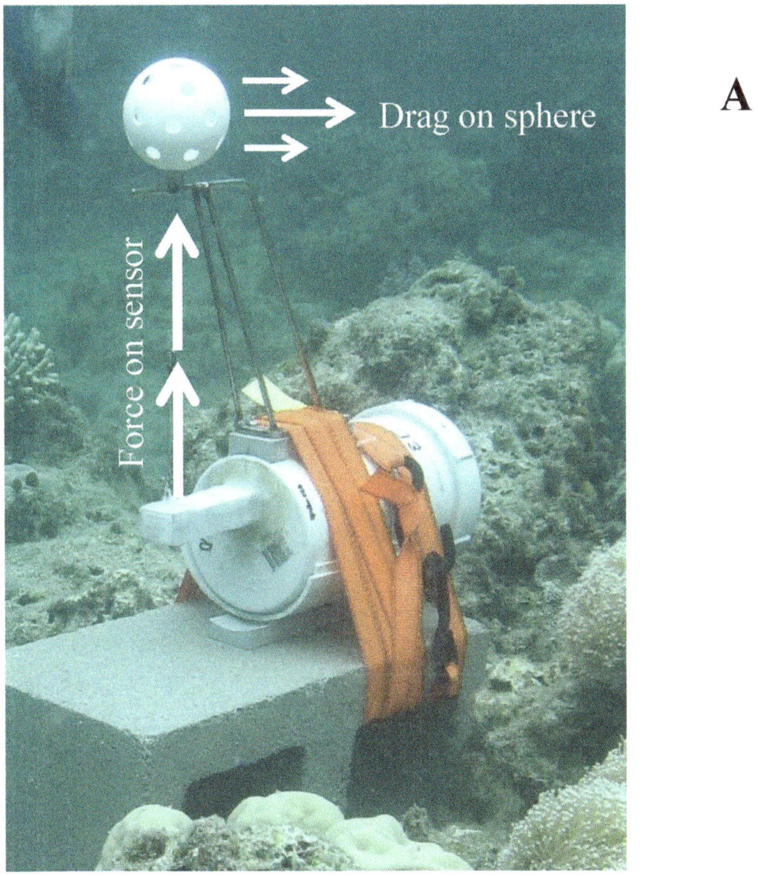

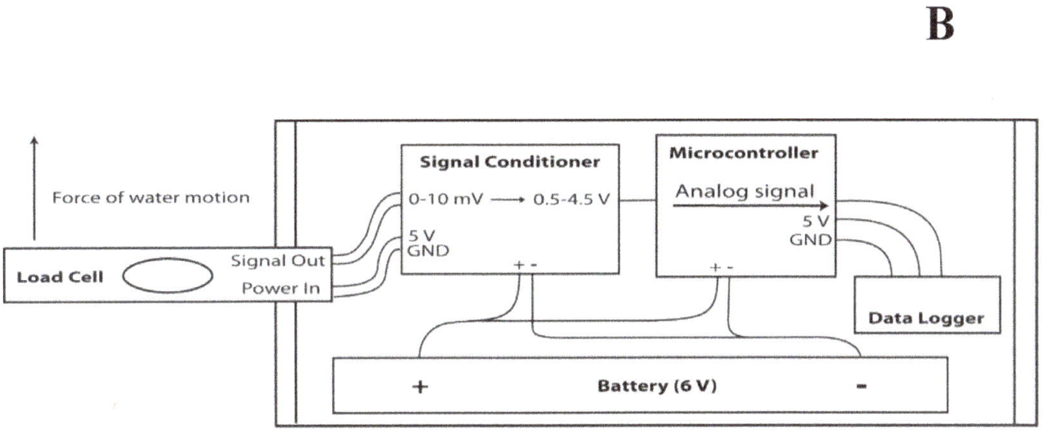

Figure 1. A picture and a diagram of current meter components. Top **A** shows an instrument deployed on a sheltered reef location with clear depictions of the guide rod construction and drag-sphere placement. Bottom **B** illustrates the internal components of the current meter and wiring connections. Notice that the height above the substratum where current measures are conducted can be adjusted by increasing or decreasing the guide rod length.

only whilst the other end acts as an anchoring point. These sensors require all exertion forces to be perpendicular to the surface of the sensor. A stainless steel frame was therefore attached to the outside of the instrument housing with a 10 mm stainless steel guide rod positioned 30 cm directly above the sensor (see photo in Fig. 1). This guide rod can be placed at any desired height above the sensor, allowing the instrument to be tailored to suit a variety of organisms by measuring current closer or farther from the substratum. A 0.5 mm zero stretch line was then attached to the sensor measuring point, passed through the guide rod, and attached to a neutrally buoyant 7 cm diameter drag-sphere (whiffle ball) approximately 0.5 cm above the guide rod (Fig. 1). As a result, currents that produced a drag on the sphere were instantaneously registered by the sensor below (see picture in Fig. 1).

The instrument housing was then fitted with a signal conditioner, a data logger and a battery pack. The signal conditioner was connected between the battery and the sensor and had two functions: 1) to ensure any change in voltage output from the sensor was due to the forces acting on the sensor rather

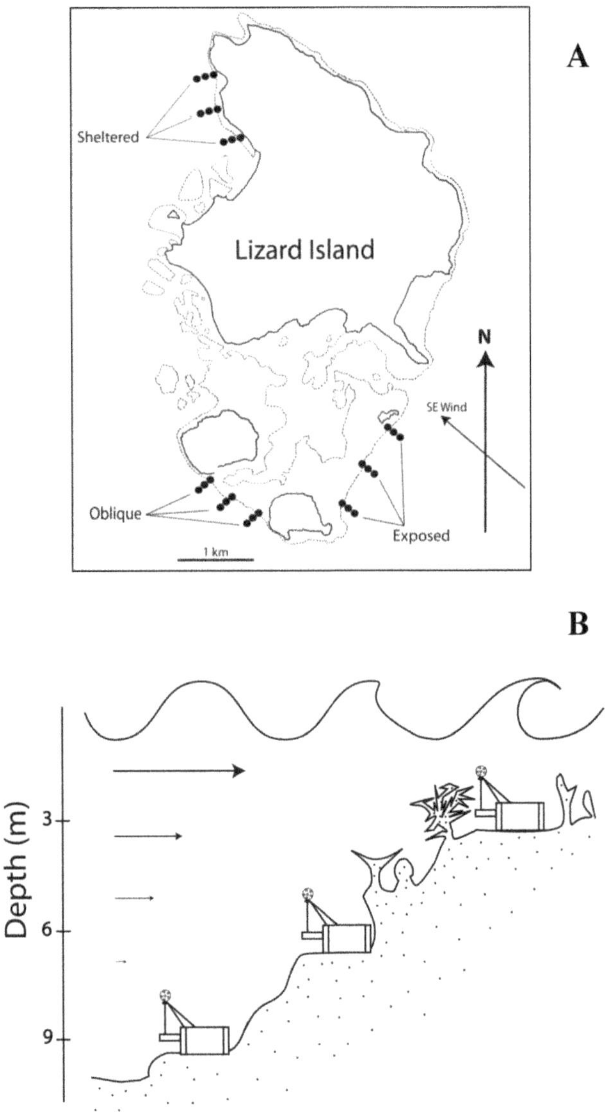

Figure 2. An illustration of measuring locations (**A**) and current meter placement (**B**) on each reef slope around Lizard Island, Northern Great Barrier Reef, Australia (14°40S, 145°28E). Current meters were placed at 3, 6, and 9 m depths at mid-tide in exposed, oblique and sheltered habitats relative to the prevailing south-easterly trade winds. These placements followed the approximate depths of the crest, mid and deep-slope of each habitat.

than slight variations in the voltage supplied by the battery pack; and 2) to amplify the output voltage from the sensor to a signal recognized by the data logger. Numerous commercial signal conditioners exist: This instrument used an INA122P instrumentation amplifier from Texas Instruments (electrical circuits can be seen at http://www.ti.com/lit/ds/symlink/ina122.pdf, accessed May 2013)

The data logger (consisting of a microcontroller and logger) was connected directly to the signal conditioner, and allowed the analog signal coming from the load cell sensor to be registered and recorded. Because analog sensors send out continuous signals, measurements of current velocity can be conducted at any frequency required, depending on the level of detail needed and the type of signal conditioner and data logger used. This instrument used a programmable Arduino Pro 328 5 V/16 MHz microcontroller and a Sparkfun Openlog data logger (for schematics see www.sparkfun.com). This microcontroller registers analog signals from 0–5 V at 10 bit and relays signals at user determined frequencies ≤16 MHz to the data logger. The Openlog data logger then automatically logs every signal sent from the microcontroller to a micro SD-card at a rate of ≤1 MHz. The load cell sensor and data logger used for this setup had resolutions of 60,000 and 1,024, respectively, providing a total measuring resolution of 1,024 divisions between 0–100% drag on the sphere. All components were run by a 6 V 7.2 Ah sealed Acid-lead rechargeable battery. These batteries have little heat build-up and do not release gaseous fumes, making them suitable for usage in hermetically sealed instruments and under elevated atmospheric pressures. All components and the battery were fitted within the watertight PVC housing, which had an O-ring fitted screw cap access point (Fig. 1).

Current meter calibration and measurement validation

Current meter instruments were calibrated in two ways before deployment. First, the recorded signal from each sensor was calibrated to ensure equal accuracy and precision between instruments under equal drag force: The output signal from the load-cell sensor was 0–10 mV and the data logger recorded signals from 0.5–4.5 V, which required the signal conditioner to amplify the load-cell signal by (4.5−0.5 V)/(0.01−0.00 V) = 400 times. Although the data logger could record signals from 0–5 V, this study controlled for potential drift in the signal over time by specifying 0–100% drag on the sensor as 0.5–4.5 V. The exact amplification of the signal may be slightly different between instruments due to minute manufacturing differences (e.g. wiring resistance), but provide accurate signals to the data logger at any load between 0–100% when adjusted correctly. Signal amplification of each instrument was therefore adjusted using a variable resistor by placing 100% load on the sensor (i.e. 3 kg for this load-cell). Additionally, although the sensor had a precision of 0.015%, additional error is expected in electrical circuits due to internal component voltages causing imprecise output signals. The instrument amplifier used here(INA 122P) created a maximum electrical noise in the amplified output signal of 0.28 mV. This precision error was corrected in each instrument using another variable resistor by placing 0% load on the sensor and offsetting the output signal to 0.5 V. Calibrations of amplification levels and offset, therefore, allowed equally accurate and precise sensor signals from 0–100% load on all instruments.

Following these calibrations, instrument performance was validated by testing the precision and accuracy of each instrument. Precision was calculated from the standard deviation of three repeated measures obtained by hanging precision weights off the sensor of 13, 27, 40, 53, 89 and 100% of max load (i.e. 3 kg). Accuracy was calculated as ((measured load − applied load)/applied load) × 100.

Second, ambient water velocities were correlated with measures of drag on the sensor sphere: To increase accuracy of calibrations at low current velocities, the instruments were initially placed in a flume (29 × 40 × 360 cm: width × depth × length) and exposed to incrementally increasing current velocities between 0 and 50 cms^{-1} (resolution of 0.25 cms^{-1}). The flume was calibrated using video recordings of neutrally buoyant particles drifting past a grid. For higher current velocities up to 400 cms^{-1} the instruments were placed in a unidirectional current immediately next to a Vernier LabQuest Flow rate sensor with a resolution of 0.5 cms^{-1} following the boat-tow methodology of Utter and Denny [28].

Table 2. Statistical comparison of current measures from 27 different sites on a coral reef spanning 3 levels of wind exposure (exposed, oblique and sheltered) and 3 different depths (3, 6 and 9 m).

Kruskall-Wallis ANOVA (by rank)															
A) Comparison of individual locations						**B) Multiple comparisons among pooled habitats**									
Habitat	**Depth (m)**	**H**	**df**	**P**	**N**	**Habitat**	**Exp 3 m**	**Exp 6 m**	**Exp 9 m**	**Obl 3 m**	**Obl 6 m**	**Obl 9 m**	**Shl 3 m**	**Shl 6 m**	**Shl 9 m**
Exposued	3	4.924	2	0.09	1659	Exp 3 m		<0.01	<0.01	<0.01	<0.01	<0.01	<0.01	<0.01	<0.01
	6	4.522	2	0.10	1523	Exp 6 m	27.86		<0.01	<0.01	<0.01	<0.01	<0.01	<0.01	<0.01
	9	2.686	2	0.26	717	Exp 9 m	32.37	10.11		<0.01	**1.00**	<0.01	**1.00**	<0.01	<0.01
Oblique	3	0.394	2	0.82	299	Obl 3 m	9.15	6.54	12.67		<0.01	<0.01	<0.01	<0.01	<0.01
	6	0.992	2	0.61	362	Obl 6 m	24.28	7.18	**0.59**	10.67		<0.01	**1.00**	<0.01	<0.01
	9	4.178	2	0.12	252	Obl 9 m	14.95	8.33	5.95	5.1	4.84		<0.01	<0.01	<0.01
Sheltered	3	5.175	2	0.08	1474	Shl 3 m	38.95	11.1	**1.15**	12.92	**0.25**	5.62		<0.01	<0.01
	6	5.453	2	0.07	619	Shl 6 m	45.43	24.25	12.72	22.29	11.12	15.17	15.67		0.02
	9	1.296	2	0.52	285	Shl 9 m	37.3	21.74	13.5	21.95	12.42	15.97	15.42	3.46	

A) No significant differences in drag forces were found among locations of equal depth and exposure. **B)** When sites were pooled by habitat and depth (exposed, oblique and sheltered habitats each at 3, 6 and 9 m depth), multiple comparisons revealed significant differences among all habitats, except the exposed 9 m, oblique 6 m and sheltered 3 m habitats (z-values on lower section, P-values on upper section in **bold**). FDR corrections were used to avoid Type I errors ($\alpha = 0.025$) (Benjamini and Hochberg 1995).

Field application on a coral reef

Instruments were then utilised to create a profile of wave and tidal-driven currents that was adequately detailed for ecological studies of individual benthos and fishes. The profile was created across a section of coral reef surrounding Lizard Island, Northern Great Barrier Reef (GBR), Qld, Australia (14°40S, 145°28E), with permission from the GBR Marine Parks Authority (permit G09/32235.1), during March, June and Feb 2010–2011, but could equally have been conducted in estuarine or river environments. This location is a shallow water mid-shelf coral reef directly exposed to the surrounding water basin but largely protected from open ocean swell (see also studies by Madin *et al.*, 2006 [8] and Fulton & Bellwood 2005 [5] for further details on waves and flow conditions across this reef environment). Twenty-seven sites were chosen for measures of current velocity and wave movement: three exposure habitats ("exposed", "oblique" or "sheltered" reef relative to the prevailing south-easterly trade winds) were chosen and three locations within each habitat were sampled at three different depths. Locations within each habitat were separated by ~300 m, and situated at 3, 6 and 9 m depth at mid tide following the crest, mid-slope and deep-slope of these reefs (Fig. 2). At every site, a current meter was secured to anchoring points in the substratum floor using ratchet straps. Each instrument was set to register changes in drag forces at 10 Hz, and recorded the time, date and the single highest point measure of current velocity every 10 seconds. Each instrument also recorded a series of 80 point measures at 10 Hz once every 60 seconds to allow calculations of wave frequency based on the wave driven pulses of drag on the drag-sphere. This equated to 8.800 drag measures per megabyte used, theoretically allowing the instruments to be deployed for over 35 days with a 500 mb SD-card. Current and wave data were recorded at wind velocities from 0 to 22.5 knots in each habitat, which corresponds to the range of weather conditions commonly encountered in this location (Australian Bureau of Meteorology and Australian Institute of Marine Science). A total of 2,258 hrs of current measures were recorded in the field, with an average of 251±47 hrs (mean ±S.E., range 161–528 hrs) in each habitat.

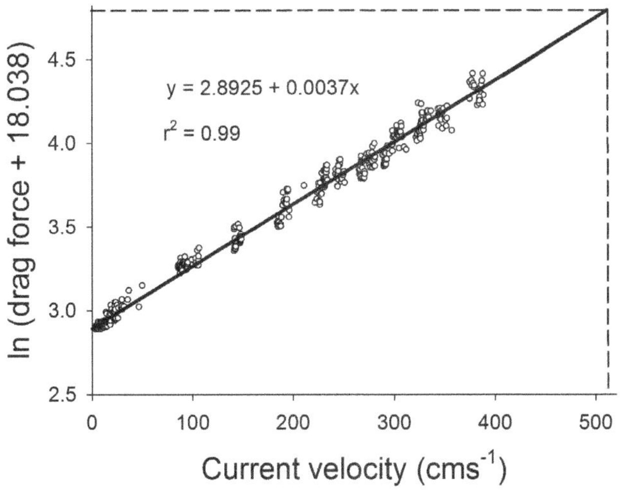

Figure 3. Correlation between drag forces (%) on the current meter sensor and water velocity (cms^{-1}). Calibrations for low velocity (0–50 cms^{-1}) were made by placing the current meter in a unidirectional current within a flume, whilst high velocity (>50–400 cms^{-1}) was calibrated next to a Vernier LabQuest Flow rate sensor following the boat-tow methodology of Utter and Denny [28]. The log-linear correlation is highly significant ($F = 12{,}090.7$, $p < 0.0001$).

Data analysis

Precision and accuracy was compared across instruments using a 2-way ANOVA with applied load and instrument as fixed factors. Data were square-root transformed to meet assumptions for ANOVA.

To account for the effect of winds and waves during field deployment, velocity measures were compared directly to concurrent weather conditions. Prior to September 2010, the most detailed weather data available for this location was hourly recordings of wind velocity, direction and wave height from the Australian Bureau of Meteorology [29]. Post September 2010, wind velocity and direction were available at 10 min intervals from the Australian Institute of Marine Science [30]. In both cases

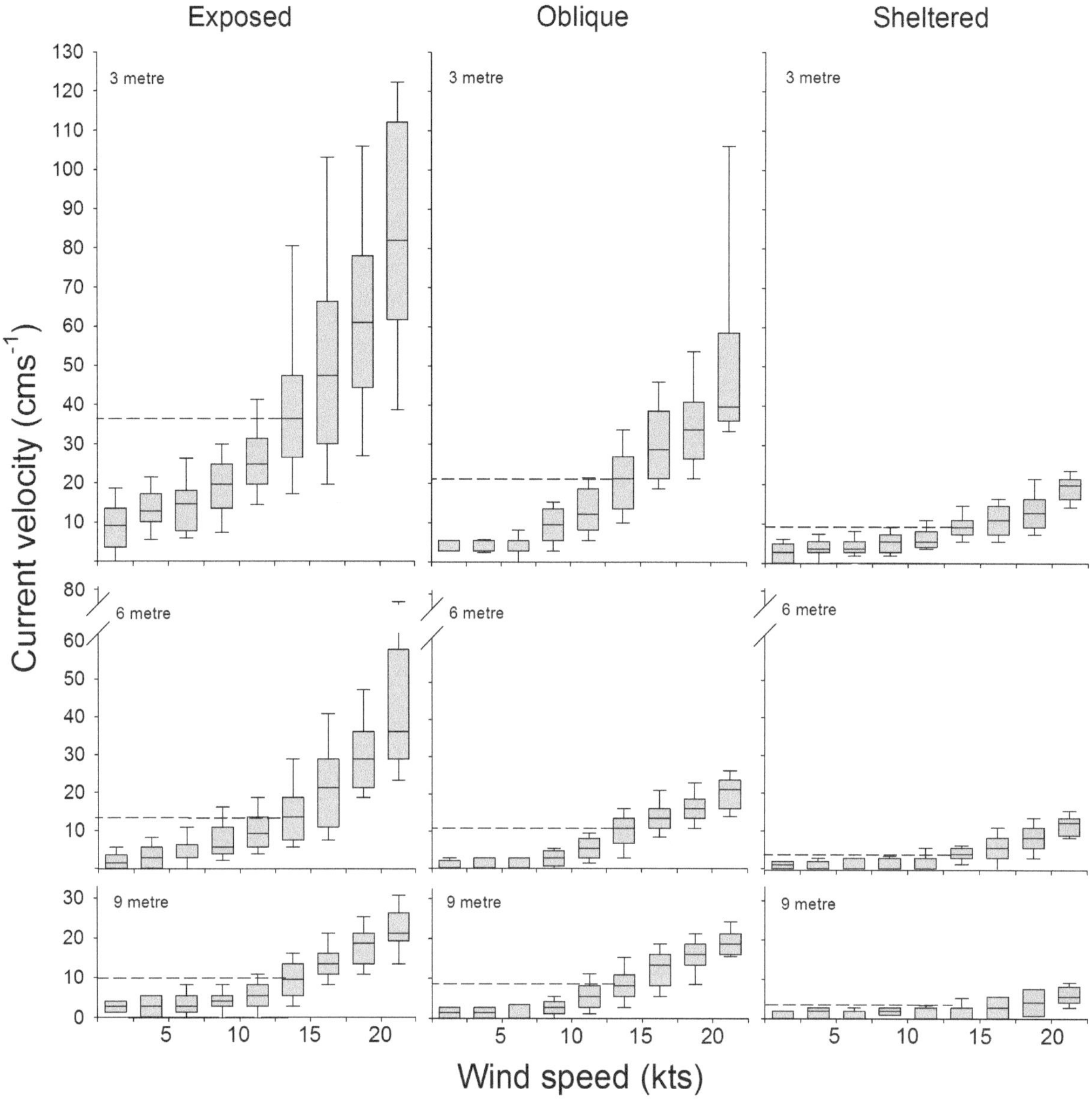

Figure 4. Current velocity profile of the coral reefs surrounding Lizard Island, Northern Great Barrier Reef, Australia. Maximum water velocities (cms^{-1}) in exposed, oblique and sheltered habitats at 3, 6 and 9 m depth relative to a wind exposure of 0–22.5 knots hr^{-1} are shown. Boundary lines indicate the 25th and 75th percentiles, lines within the boxes represent median velocity values and the error bars indicate the 90th and 10th percentiles. Stippled lines indicate the most commonly encountered current velocity in each habitat relative to average daily wind velocities.

wind recordings reflected the average of the 10 min period prior to recording time (see Weather Station No 31213, www.bom.gov.au and Lizard Island Weather Station, http://data.aims.gov.au).

To avoid pseudo-replicating multiple current measures onto each individual weather recording, the analysis only used the single highest measure of current velocity and associated wave frequency of each concurrent weather recording. These highest measures were chosen because wave driven current velocities are highly variable and the ability of aquatic organisms to occupy a habitat is limited by the highest commonly occurring current velocities (e.g. [15–16,20,31]). In total, the 2,258 hrs of current measures provided 7,190 independent measures of current velocities concurrent with available weather data.

Due to high variation in current velocity within and among habitats, the distributions of drag measures were non-normal and could not be transformed to comply with assumptions of normality and homogeneity of variances necessary for linear models. Consequently, differences among locations of equal depth and exposure were examined using a series of Kruskall-Wallis ANOVA's (location as fixed factor). Where no difference in drag force was found among locations, data were pooled to represent nine distinct habitats: exposed, oblique and sheltered habitats at each of the three depths (3, 6 and 9 m, see Table 2). Differences among these nine habitats were then compared using Kruskall-Wallis ANOVA Multiple Comparisons test, followed by False Detection Rate (FDR) correction to avoid Type I errors [32]. Finally, field measures of drag forces in Newton (N) were converted to current velocities in cms^{-1} and plotted relative to wind velocity, whilst wave frequency was plotted relative to concurrent wave height estimates [30]. To account for time delay

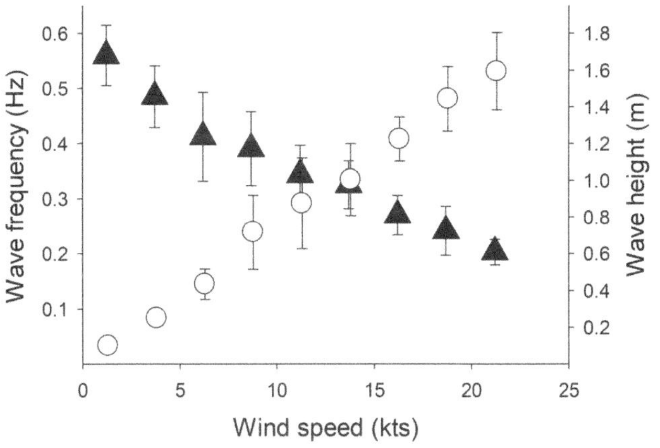

Figure 5. Wind induced wave frequency and amplitude in an exposed coral reef habitat (mean ±S.E.). Wave frequency (black markers) was measured on the exposed slope in front of Lizard Island, and wave amplitude (white markers) was estimated from marine forecasts for this location (www.bom.gov.au).

in wind driven wave propagation and current velocity [33–34] each velocity measure was correlated with the average wind over the preceding 3 hrs.

Results

Current meter performance

Instrument calibration revealed that measures from 0–100% load had a precision of 0.055±0.044% and an accuracy of 0.513±0.653% (mean ±S.D.) with no detectable difference among individual instruments (Precision: $F_{8,53} = 1.23$, $p = 0.31$; Accuracy: $F_{8,53} = 1.13$, $p = 0.36$, Observed Power 0.80). Precision was also independent of applied load ($F_{5,53} = 1.00$, $p = 0.43$, Observed power 0.99) whilst accuracy increased from 1.285% to 0.046% as the applied load increased to 100% ($F_{5,53} = 8.91$, $p<0.01$). This increase in accuracy was caused by identical measuring increments from 0 to 100% load (10 bit resolution).

Recorded values of current velocities (cms^{-1}) displayed a significant log-linear correlation with drag forces (%) on the load cell sensors (Best fit: $y = 2.8925 + 0.0037x$, $r^2 = 0.99$, See Fig. 3), and at current velocities between 0 and 400 cms^{-1} the instruments showed a measuring resolution of ±0.26 cms^{-1} (Fig. 3).

Current velocity profile of a coral reef ecosystem on the Great Barrier Reef, Australia

Current velocities across all examined habitats ranged between 0.8 and 82.0 cms^{-1} depending on habitat exposure, depth and wind velocity (Fig. 4). In general, habitats of different exposures and depths showed distinct and highly significant differences in current velocities (Kruskall-Wallis ANOVA, $H_8 = 3499.93$, $p<0.01$, $N = 7190$). Only the exposed 9 m, oblique 6 m and sheltered 3 m habitats showed no significant differences in recorded velocities (Table 2). The highest current velocities were found at the crest of the exposed habitats (3 m depth, median: 9.2–82.0 cms^{-1}) while the lowest velocity was found at 9 m depth in sheltered habitats (0.8–5.9 cms^{-1}, Fig. 4). Across all habitats, current velocity increased exponentially with wind velocity and became increasingly more variable at the highest wind velocities (see Fig. 4, 5). Currents that coincided with periods of negligible wind (≤1 knots) revealed that tidal driven currents explained only 3.3±0.4 cms^{-1} (mean ±S.E.) of total water movement within both exposed and sheltered habitats (range 0–5.5 cms^{-1}). Wave frequency reduced from 0.54±0.03 Hz to 0.20±0.02 Hz (mean ±S.E.) as wind velocity increased from 0 to 22.5knots, and estimated wave height increased to 1.6 m (Fig. 5).

Discussion

This study uses load cell technology to measure water currents in aquatic ecosystems, at a resolution necessary for ecological studies on benthos and fish, and results notably illustrate the many advantages of this method in terms of automation, costs, data detail and acquisition flexibility in field environments.

The current meter instruments described here can register and process current velocity and wave movement at a frequency of 16 MHz and record the measures at >1 MHz, which covers the 0.6–1 MHz sampling frequencies often used for wave studies [35–36]. Low energy consumption allows data collection to run for several weeks continuously and the sensor can maintain a signal precision and accuracy of ≥99% for all current velocities. The instruments also offer 10 bit measuring resolution and a 98.1% correlation between sensor signal and current velocity to >400 cms^{-1}. In particular, the maximum current velocities encountered in this study was equivalent to just 57.0% of the maximum drag force on the sensors, suggesting that these instruments are capable of measuring water velocity to >500 cms^{-1}. Furthermore, the instrument housing is pressure rated to >194 psi theoretically allowing deployment to depths of 120 m, whilst the 0.5 mm zero stretch line leading to the load cell is capable of withstanding current or wave driven shear stresses on the drag sphere of >25 kNm^{-2}. At less than $150 in component costs, these durable instruments provide affordable access to highly reliable velocity measures and allow simultaneous recordings of minimum, average and maximum current velocity and wave movement in multiple locations over time.

The field measures presented here also cover a significant gap in the current literature [15–16] by presenting a detailed profile of current velocities across depth and exposure gradients for a large section of a tropical coral reef system. Greater than 15-fold differences in maximum velocities were recorded among individual coral reef habitats (5.5 to >82 cms^{-1}) and up to 7-fold differences among habitat sites that differ by only a few meters in depth. Increasing wind velocities from >0 to 22.5 knots caused up to 9-fold increases in current velocities within individual exposed habitats. Based on daily wind velocity data from this location between 2003–2011 [30], average yearly wind velocities are 12.6±0.3 knots with slightly stronger winds in winter (14.5±0.3 knots) and weaker winds in summer (10.8±0.3 knots, mean ±S.E.). Consequently, the most commonly occurring current velocities for these reefs range from 1.3 to 36.4 cms^{-1} depending on habitat exposure and depth (see Fig. 4).

Given the ease of instrument deployment and detail of data acquired, these instruments clearly provide a real-world alternative to existing methods for measuring currents in aquatic ecosystems (see e.g. [27]). These instruments may be useful for a wide variety of habitats and ecological questions. In rivers and streams with planar beds, for instance, these instruments could be used to measure the water velocity along different sections of the bed over prolonged timescales and demonstrate variations in current to different zones of the river and associated benthic community. Equally, continuous measures of currents in wave exposed coastal ecosystems and large lakes could clarify the physical forces impacting on resident species and help explain species compositions and community structures in individual habitats.

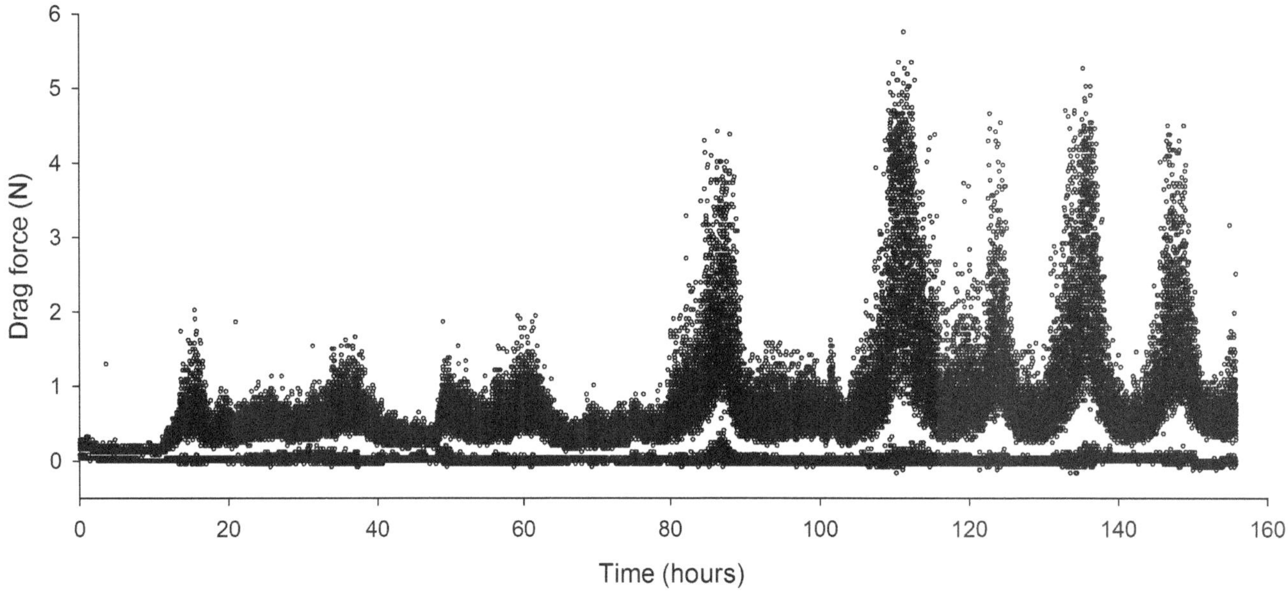

Figure 6. A raw data stream from a 7 day continuous measure of current velocity. Water velocity is measured as drag on the current meter drag-sphere and is here reported in Newton (N). In this data stream, drag forces were continuously measured at 16 MHz and for every 10 sec interval the single highest and single lowest measure were recorded. Notice how variation in drag forces increases with increasing current velocity. Notice also the stable minimum measures over time highlighting that no ascendible zero-drift occurred. The high measure (~1.4 N) at the beginning of the data stream is a test pull made by the diver immediately after securing the current meter to the substratum, and indicates the start of data recording in the field location.

Results of this study allow performance measures of individual species to be compared to their particular habitat use and distributions patters. For instance, the current velocities $\geq$82 cms^{-1} recorded here had shear stresses exceeding 100 Nm^{-2} (calculated following τ = Force (N)/Area (m2)). Such forces arguably require benthic organisms to have tremendous structural strength, grip on the substrate and/or significant swimming abilities in order to survive. Given these conditions are measured on a mid-shelf reef which is relatively protected from direct open ocean swell, even harsher conditions may be expected in more exposed locations such as the outer Great Barrier Reef. Consequently, these results clearly highlight why the distribution and abundance of many reef species such as corals and teleosts has been linked to current velocities within individual habitats [19,31,37].

The presented instrument specifically measured currents over a 7 cm diameter drag-sphere and at a distance of 35 cm above the substratum befitting the size and location of many reef fishes swimming above the substratum [38]. However the scale of the instrument could easily be change to accommodate species that are closer or further from the substratum by extending or reducing the guide rod length (Fig. 1). Equally, a smaller combination of drag-sphere and load cell could be used for measuring current over smaller spatial scales, or the instrument could be inverted on a stand or buried to place the drag-sphere much closer to the substratum and thereby measure near bed velocities experienced by benthic organisms.

Using a programmable microcontroller allows for a high degree of freedom in deployment time and in the specific type and amount of data recorded. Once programmed, the Arduino Pro 328 used here can monitor time and date and specifically record data such as peak, minimum and average values (see example raw data trace in Fig. 6) which can be used to calculate water displacement and acceleration in addition to velocity. Importantly, this microcontroller can also be programmed to power down between measurements, thereby vastly reducing power consumption and significantly extending field deployment time to several months. Reduction in sampling frequency may equally lower power consumption and significantly extend deployment time.

Although current direction was not recorded in the present study, these data could be included by using a bi-directional or multi-directional load cell rather than the single point type described here. These cells measure forces in the x, y and z direction and would enable this instrument to measure the precise direction of water motion, whilst keeping power consumption at a minimum. Velocity time series during periods of steady conditions in e.g. stream beds could equally provide info on turbulence by calculating the standard deviation or root mean square of force measures. Importantly, measures of current direction, turbulence, or wave frequency significantly above the $<$1 Hz recorded here, may require an instrument housing and load cell sensor with greater structural stiffness to minimize instrument flex and maintain frequency response. However, such inclusions may substantially increase versatility for some users.

This instrument was deployed in an environment where shear stresses $>$100 Nm^{-2}, emphasizing its durability under real-world harsh conditions. Whilst currents have previously been described for some tropical coral reefs (e.g. [5]), this is the first time a current velocity profile has been recorded for such a wide combination of habitats and weather conditions. Consequently, this instrument may significantly enhance our ability to link organism ecology with ambient currents in aquatic environments.

Supporting Information

Figure S1 A detailed schematic and assembly diagram of the instrument housing and sensor array. Component descriptions and manufacturer can be seen in Table S1.

Table S1 Flow meter component list.

Programming information S1 An example of the programming used to control the flow meter instrument.

Acknowledgments

I thank the LIRS staff for logistical support, Dominique Roche, Sandra Binning, Patrick Vespa and Matt Mitchell for field assistance, Andy Hoey for proof reading, and Chris Fulton and John Steffensen for calibration assistance.

Author Contributions

Conceived and designed the experiments: JLJ. Performed the experiments: JLJ. Analyzed the data: JLJ. Contributed reagents/materials/analysis tools: JLJ. Wrote the paper: JLJ. Conceived the idea, designed and built the instrument, conducted the field experiment, analysed the data, and wrote the manuscript: JLJ.

References

1. Vogel S (1996) Life in Moving Fluids: The Physical Biology of Flow, 2nd ed. 467 pp. Princeton University Press, Princeton UK
2. Lancaster J, Downes BJ (2010) Linking the hydraulic world of individual organisms to ecological processes: Putting ecology into ecohydraulics. River Research and Applications 26: 385–403.
3. Nikora V (2010) Hydrodynamics of aquatic ecosystems: An interface between ecology, biomechanics and environmental fluid mechanics. River Research and Applications 26: 367–387.
4. Nilsson C, Brown RL, Jansson R, Merritt DM (2010) The role of hydrochory in structuring riparian and wetland vegetation. Biological Reviews 85: 837–858.
5. Fulton CJ, Bellwood DR (2005) Wave-induced water motion and the functional implications for coral reef fish assemblages. Limnology and Oceanography 50: 255–264.
6. Kennen JG, Riva-Murray K, Beaulieu KM (2010) Determining hydrologic factors that influence stream macroinvertebrate assemblages in the northeastern US. Ecohydrology 3: 88–106.
7. Poff NL, Allan JD, Bain MB, Karr JR, Prestegaard KL, et al. (1997) The natural flow regime. BioScience 47(11): 769–784.
8. Madin JS, Connolly SR (2006) Ecological consequences of major hydrodynamic disturbances on coral reefs. Nature 444: 477–480.
9. Johansen JL, Fulton CJ, Bellwood DR (2007) Avoiding the flow; refuges expand the swimming potential of coral reef fishes. Coral Reefs 26: 577–583.
10. Shutter M, Crocker J, Pajimans A, Janse M, Osinga R, et al. (2010) The effect of different flow regimes on the growth and metabolic rates of the scleractinian coral *Galaxea fascicularis*. Coral Reefs 29:737–748.
11. MacVicar BJ, Beaulieu E, Champagne V, Roy G (2007) Measuring water velocity in highly turbulent flows: field tests of an electromagnetic current meter (ECM) and an acoustic Doppler velocimeter (ADV). Earth Surface Processes and Landforms 32: 1412–1432.
12. Collar P, Griffiths G (2001) Single point current meters. *In* J.H. Steele, S.A. Thorpe and K.K. Turekian [eds.], Encyclopedia of Ocean Sciences, pp. 2796–2803. San Diego, USA, Academic Press.
13. Stone MC, Hotchkiss RH (2007) Evaluating velocity measurement techniques in shallow streams. Journal of Hydraulic Research 45(6): 752–762.
14. McIntyre N, Marshall M (2008) Field verification of bed-mounted ADV meters. Proceedings of the Institution of Civil Engineers. Water Management 161: 199–206.
15. Monismith SG (2007) Hydrodynamics of coral reefs. Annu Rev Fluid Mech 39: 37–55
16. Hearn CJ (2011) Perspectives of coral reef hydrodynamics. Coral Reefs 30:1–9
17. Bell CE, Denny MW (1994) Quantifying "wave exposure": a simple device for recording maximum velocity and results of its use at several field sites. Journal of Experimental Marine Biology and Ecology 181: 9–29.
18. Porter ET, Sanford LP, Suttles SE (2000) Gypsum dissolution is not a universal integrator of 'water motion'. Limnology and Oceanography 45: 145–158.
19. Grigg RW (1998) Holocene coral reef accretion in Hawaii: a function of wave exposure and sea level history. Coral Reefs 17: 263–272
20. Storlazzi CD, Brown EK, Field ME, Roders K, Jokiel PL (2005) A model for wave control on coral breakage and species distribution in the Hawaiian Islands. Coral Reefs 24: 43–55
21. Finelli CM, Clarke RD, Robinson HE, Buskey EJ (2009) Water flow controls distribution and feeding behavior of two co-occuring coral reef fishes: I. Field measurements. Coral Reefs 28: 461–473.
22. Figurski JD, Malone D, Lacy JR, Denny MW (2011) An inexpensive instrument for measuring wave exposure and water velocity. Limnol Oceanogr Methods 9: 204–214.
23. Marchant R, Stevens T, Choukroun S, Coombes G, Santarossa M, et al. (accepted for future issue) A Buoyant Tethered Sphere for Marine Current Estimation. IEEE Journal of Oceanic Engineering 10.1109/JOE.2012.2236151
24. Donelan MA, Motycka J (1978) Miniature drag sphere velocity probe. Review of Scientific Instruments 49: 298–304.
25. Gaylord B (1999) Detailing agents of physical disturbance: wave-induced velocities and accelerations on a rocky shore. Journal of Experimental Marine Biology and Ecology 239: 85–124.
26. Gaylord B, Denny MW, Koehl MAR (2008) Flow forces on seaweeds: Field evidence for roles of wave impingement and organism inertia. Biol Bull 215: 295–308.
27. Mach KJ, Tepler SK, Staaf AV, Bohnhoff JC, Denny MW (2011) Failure by fatigue in the field: a model of fatigue breakage for the macroalga *Mazzaella*, with validation. The Journal of Experimental Biology 214: 1571–1585
28. Utter B, Denny MW (1996) Wave-induced forces on the giant kelp Macrocystis pyrifera (Agardh): field test of a computational model. Journal of Experimental Biology 199: 2645–2654.
29. Australian Institute of Marine Science, Australia. Available: http://data.aims.gov.au. Accessed 12 March 2012.
30. Beureau of meterology, Australia. Available: www.bom.gov.au. Accessed 12 March 2012.
31. Fulton CJ, Bellwood DR, Wainwright PC (2005) Wave energy and swimming performance shape coral reef fish assemblages. Proceedings of the Royal Society of London: B Biological Science. 272: 827–832.
32. Benjamin Y, Hochberg Y (1995) Controlling the False Discovery Rate: a Practical and Powerful Approach to Multiple Testing. Journal of the Royal Statistical Society Series B-Statistical Methodology 57(1): 289–300.
33. Swain J, Baba M (1992) Parametric wave prediction model based on time delay concept-An evaluation. Proceedings of the Indian Academy of Science 101(3): 215–226.
34. Kinsman B (2002) Wind Waves: Their Generation and Propagation on the Ocean Surface. 676 pp, Dover Phoenix Editions, Dover Publications, UK.
35. Lowe RJ, Falter JL, Bandet MD, Pawlak G, Atkinson MJ, et al. (2005) Spectral wave dissipation over a barrier reef. J Geophys Res 110: C04001
36. Herbers THC, Lentz SJ (2009) Observing directional properties of ocean swell with an Acoustic Doppler Current Profiler. J Atmosp Ocean Techn27: 210–225
37. Veron JEN, Stafford-Smith M (2000) Corals of the world. 490pp. Australian Institute of Marine Science, Nature press, Australia.
38. Randall JE, Allen GR, Steene RC (1997) Fishes of the Great Barrier Reef and Coral Sea. University of Hawaii Press, Honolulu, HI.
39. Pontius PE, Mitchell RA (1982) Inherent problems in force measurement. Some results in the study of force sensor-machine interaction, hysteresis, thermoelastic effect and creep are discussed. Experimental Mechanics 22(3): 81–88.

18

Combining Aesthetic with Ecological Values for Landscape Sustainability

Dewei Yang[1,2], Tao Luo[1,2]*, Tao Lin[1,2], Quanyi Qiu[1,2], Yunjian Luo[1,2]

1 Key Lab of Urban Environment and Health, Institute of Urban Environment, Chinese Academy of Sciences, Xiamen, China, **2** Xiamen Key Lab of Urban Metabolism, Xiamen, China

Abstract

Humans receive multiple benefits from various landscapes that foster ecological services and aesthetic attractiveness. In this study, a hybrid framework was proposed to evaluate ecological and aesthetic values of five landscape types in Houguanhu Region of central China. Data from the public aesthetic survey and professional ecological assessment were converted into a two-dimensional coordinate system and distribution maps of landscape values. Results showed that natural landscapes (i.e. water body and forest) contributed positively more to both aesthetic and ecological values than semi-natural and human-dominated landscapes (i.e. farmland and non-ecological land). The distribution maps of landscape values indicated that the aesthetic, ecological and integrated landscape values were significantly associated with landscape attributes and human activity intensity. To combine aesthetic preferences with ecological services, the methods (i.e. field survey, landscape value coefficients, normalized method, a two-dimensional coordinate system, and landscape value distribution maps) were employed in landscape assessment. Our results could facilitate to identify the underlying structure-function-value chain, and also improve the understanding of multiple functions in landscape planning. The situation context could also be emphasized to bring ecological and aesthetic goals into better alignment.

Editor: Ben Bond-Lamberty, DOE Pacific Northwest National Laboratory, United States of America

Funding: This study was supported by the National Natural Science Foundation of China (41371535, 41001098, 41371192 and 40971111) and the Fujian Provincial S&T Programme (2013Y0083 and 2011J01280). The funders had no role in study design, data collection and analysis, decision to publish, or preparation of the manuscript.

Competing Interests: The authors have declared that no competing interests exist.

* Email: tluo@iue.ac.cn

Introduction

Various landscapes continually provide human society with multiple benefits, which are derived from ecological services and aesthetic attractiveness [1–5]. However, few researches gave consideration to aesthetic preferences in ecological assessment of landscapes [6,7], and thus efforts towards landscape sustainability encounter the dilemma of making eco-aesthetically appealing landscapes [8–12].

Studies on landscape aesthetics have revealed two contrasting paradigms, which were either from an objective perspective (intrinsic attribute of the landscape) or a subjective point of view (human preferences) [1,13,14]. The employed methods varied from formalized indicators to aesthetic surveys [13,15]. In the urban and regional planning, it still remains challengeable to construct eco-aesthetically appealing landscapes that meet multiple human needs [7,10,11,16–19].

In attempt to combine aesthetic with ecological values, a number of models and/or indicators have been employed in landscape assessment [7,20,21]. Some studies have found that people appreciate aesthetically appealing landscapes that tend to be natural in ecological assessment [1–3], while some studies have indicated that ecologically sound landscapes may not be aesthetically pleasing [2,22,23]. This contradiction poses a challenge for efforts to combine aesthetic preferences and ecological service using a reliable assessment framework [7,24,25]. Moreover, the scale and spatial patterns of various landscape types need to be taken into account for both ecologically vital and aesthetically attractive.

Land use/cover change is a key socio-economic footprint of human activity and potentially affects aesthetic and ecological functions of landscapes [24,26,27]. The ecological services vary with different land-use types [28,29], while the spatial patterns of land-use types give rise to various aesthetic experiences. Thus, to what extent do landscape types meet human needs in terms of ecological services and aesthetic attractiveness [3,17,23,30,31]? This would be useful to determine what measures of both ecological and aesthetic landscapes are potentially positive [20,25].

Therefore, the aims of the present study included that: (1) a hybrid framework, which combine aesthetic and ecological landscape values, were proposed to assess integrated functional performances of different landscape types; (2) the framework was used to examine integrated functional performances and interrelationships of different landscape types in the Houguanhu Region of Wuhan City, central China; and (3) potential strategies for landscape planning were discussed. Our results could encourage an attempt of interdisciplinary research in a shared vision of landscape sustainability.

Methodology

Case study

The Houguanhu Region (113°41′–114°13′E, 30°15′–30°41′N) lies within the southwestern Wuhan city of Hubei Province in central China. Houguanhu Region covers an area of approximately 148.18 km^2, and is mainly comprised of lakes and flatlands with a height of less than 100 m. The average annual temperature is 16.5°C and the total annual precipitation is 1,100–1,450 mm. Houguanhu Region has a population of approximately 63,300 people living in 78 settlements, and 79% of residents are classified as rural population. Regional planning for Houguanhu currently classifies it as an emerging "Ecological Livable Area" and "Tourist Resort", and this region is experiencing rapid changes in landscape structures [5].

The Houguanhu Region has a variety of diverse landscape resources and a rich historical culture. Five general landscape types, consisting of forest, farmland, grassland, water body and non-ecological land (i.e., desert in [28]), are shown in Figure 1, where the classification criteria follow Costanza et al. [28]. Non-ecological land includes rural settlement, built-up areas, roads, mines, industry zones and historic areas. The water body covers 30.56% of the total area of the Houguanhu Region, where the Zhiyin Lake and the Houguan Lake are the two largest lakes in this region. Farmland and scattered non-ecological land cover 46.75% and 15.98% of total area, respectively. Forest mainly in hilly areas and scattered grasslands occupy 0.21% of the total area. Forest and grasslands are mainly distributed in the south region, while farmland and non-ecological land are scattered across the entire area. The water body is mainly located in the central region.

In recent years, diverse landscapes have attracted a large number of tourists from outside the region. At the same time, however, rapid population growth and urban expansion have exerted intensive pressures on the natural environment and landscape. In response, since 2009 the local government has developed a series of regional landscape planning for sustainable development [5,32]. Since cultural functions of landscapes usually receive less attention in landscape planning in China, this study may also benefit ongoing landscape planning in other similar regions.

Field survey

The perception-based aesthetic landscape value evaluation method [1,33] was employed in this study. A revised survey instrument of Von Haaren [33] was used to evaluate the landscape aesthetic value. The survey instrument consists of basic information, an aesthetic landscape value score and a viewshed map. Basic information (i.e. survey date, survey site coordinate, and site description) provided a landscape background for the 146 sample sites. The aesthetic landscape value score ($Vt = V1 - V2 + V3$) for each site included a positive score ($V1$), negative score ($V2$), and positive impact score of surrounding landscapes ($V3$). In each site, the biophysical landscape components (e.g., lakes, grassland, forest, hedgerows, electricity pylons, mines, and unsightly isolated buildings) were considered in the evaluation process. Landscape features (e.g., naturalness, openness, uniqueness, diversity, accessibility and visual comfort) also contributed to evaluation scores. The aesthetic landscape value was coded from 1 (lowest) to 5 (highest) on a 5-point rating scale. The viewshed map was sketched on each site for further revision using ArcGIS 9.2. Each viewshed map would be assigned an aesthetic landscape value score.

A total of 146 grid samples of 1 km^2 were established within the study area using the uniform-grid-square sampling method. Basic maps (e.g., topographical map, landscape type map, administrative map) and socio-economic data (e.g., regional planning documents, and regional eco-economic statistic materials) were collected for spatial analysis using ArcGIS 9.2 tools.

The site survey of aesthetic landscape value was conducted from June 9th to 15th, 2010. Six surveyors from Institute of Urban Environment participated in a 3-hour field training program, and then conducted the formal survey as three two-person groups. During the survey, surveyors compared each other's score and discussed until an agreement on a final score. After a 7-day survey, 141 valid viewshed maps with corresponding aesthetic scores were obtained. Blind viewshed areas without field aesthetic scores and five invalid survey samples were given aesthetic scores using spatial interpolation and joint method in ArcGIS 9.2. Consequently, a

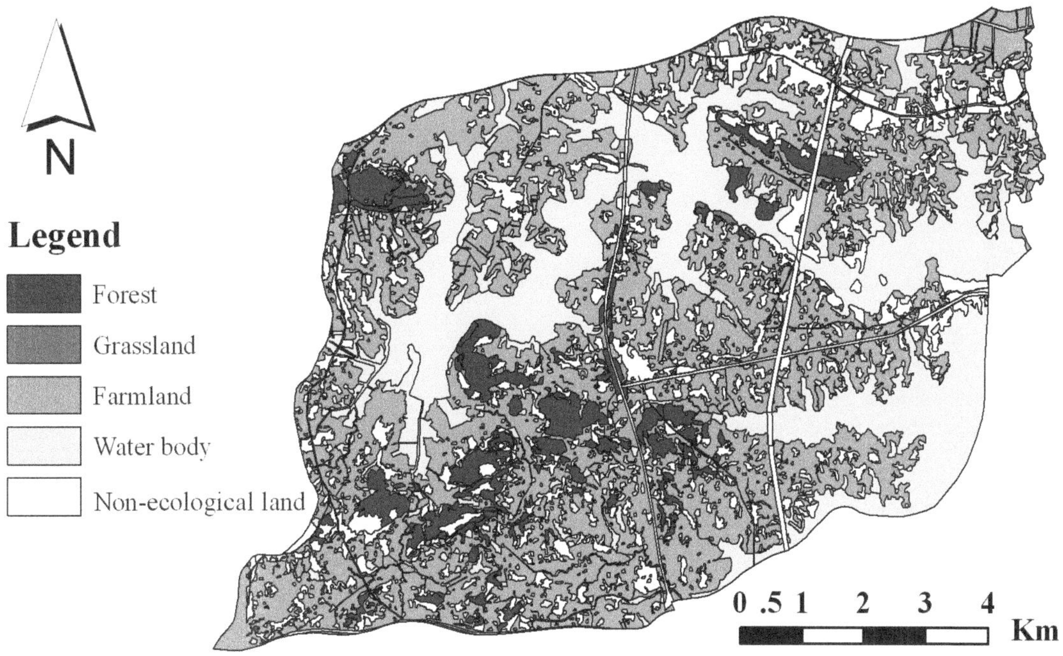

Figure 1. Landscape type map in Houguanhu Region. This Figure and following Figure 3a are redrawn according to Luo et al. [5] and Yang et al. [32].

distribution map of aesthetic landscape value with 146 assigned samples was established. In this map, the study area was divided into graded aesthetic landscape value units based on a 5-point rating scale, with 5 being the highest value. The higher value will be more aesthetic attractive. Yang et al. [32] discussed in detail the survey instrument and data processes.

Aesthetic landscape value coefficients

The landscape type map and the aesthetic landscape value distribution map were overlaid in ArcGIS 9.2 to identify the aesthetic value of different landscape types. Then the area of each landscape type was extracted for each graded aesthetic landscape value region. The aesthetic landscape value coefficient, which refers to the aesthetic landscape value of unit area per landscape type, was calculated as:

$$Q_i = \frac{\sum P_i}{A_i} = \frac{\sum M_i \bullet N_i}{A_i} \tag{1}$$

Where Q_i is the aesthetic landscape value coefficient (point/hm^2) of the ith land type (forest, grassland, farmland, water body, and non-ecological land); P_i is the aesthetic landscape value of the ith landscape types; M_i is the aesthetic landscape value score of the ith landscape type; N_i is the area of the ith landscape type in 5-grade aesthetic landscape value units; A_i is total area of the ith landscape type.

Ecological landscape value coefficients

According to the ecological service valuation framework of Costanza et al. [28], ecological service values of Chinese terrestrial ecosystems were evaluated by export judgment, which were developed by Xie et al. [29]. Ecological landscape value coefficients were calculated for five landscape types based on the sum of the eight ecosystem service values (see Table 1). The ecological landscape value coefficients were added up, where the recreation and cultural service were excluded, because they were inappropriate to simply add and subtract when comparing with monetary value. Based on the ecological value coefficients and the areas of the five landscape types, an ecological landscape value distribution map was calculated using Equation (2):

$$H_i = U_i \cdot V_i \tag{2}$$

Where H_i is the ecological landscape value (\$/hm^2) of the ith landscape type (forest, grassland, farmland, water body and non-ecological land); U_i is the ecological landscape value of the ith landscape type on a scale of 1–5; V_i is the area of the ith landscape type.

Integrated evaluation of landscape values

The coefficients of ecological and aesthetic landscape value were normalized to evaluate integrated landscape values. The ecological and aesthetic landscape value coefficients of five landscape types were converted into values from −2 to 2 using Equation (3):

$$K = \frac{X - X_{min}}{4(X_{max} - X_{min})} - 2 \tag{3}$$

Where K is the normalized aesthetic or normalized ecological landscape value coefficient of each landscape type; X is the aesthetic or ecological landscape value coefficient of each landscape type; X_{max} and X_{min} are the maximum and minimum value of the aesthetic or ecological landscape value coefficient.

We assumed that the aesthetic and ecological landscape values contributed equally to the integrated landscape value. In order to quantify the functional roles of different landscape types in the integrated landscape values, the two normalized coefficients were visually displayed in a two-dimension coordinate system. The distribution maps of the aesthetic and ecological landscape value were overlapped into an integrated one of landscape value in ArcGIS 9.2, which were used to identify spatial characteristics of landscapes.

Results

Landscape value coefficients

Functional performances and value contribution of the five landscape types were quantified using a 5-point Likert Scale with 5 being the highest value. Aesthetic landscape value per unit area was in decreasing order: grassland, water body, forest, farmland, and non-ecological land (Table 2). The natural landscapes performed better than the artificial landscapes in terms of aesthetic functional performance.

The normalized aesthetic and ecological value coefficients were obtained using Equation (3) (Table 3). The aesthetic landscape value per unit area of water bodies was highest, followed by forest, grassland, farmland, and non-ecological land. Similarly to the aesthetic landscape values, natural landscapes had higher ecological values than cultivated or man-made landscapes.

Combination of Landscape values

We identified aesthetic, ecological and integrated functional performances of the five landscape types using a two-dimension coordinate system (Figure 2). In the two-dimension coordinate system, the five landscape types contributed unevenly, or even in opposite directions, to both landscape values. Water body was the only type where both landscape values were positive. Grassland was aesthetically attractive, but it was not ecologically valuable. The other three landscape types (i.e. forest, farmland, and non-ecological land) showed a positive relationship on aesthetic and ecological landscape value, respectively, but made a negative contribution to aesthetic and ecological landscape value.

Distribution of landscape values

We drew the aesthetic landscape value distribution map after the aesthetic landscape survey (Figure 3a), and the ecological landscape value distribution map based on Equation (2) (Figure 3b). The integrated landscape value distribution map was obtained from the overlay of the aesthetic and ecological landscape value distribution maps in ArcGIS 9.2 (Figure 4). The landscape value scores ranged from 1 (lowest) to 5 (highest).

The aesthetic, ecological and integrated landscape values varied with the human activity intensity, and gradually decreased from the external water body to the transitional farmland and central non-ecological land. Central water body and adjacent areas received the highest values, the lowest was in northeastern built-up areas, and the moderate scores were in most other areas.

The distribution of aesthetic landscape value showed a large-scale spatial continuity, with the exception of the northeastern region. The central water body with visual characteristics of broad sight, good order and naturalness obtained higher aesthetic landscape value scores, while scattered non-ecological land and farmland with disorderliness, high-intensity buildings and fragmentation received lower evaluation scores. In contrast, the distribution of ecological landscape values exhibited a mosaic distribution with an unclear spatial trend. However, water body and their adjacent areas received higher ecological value scores.

Table 1. Ecosystem service values of Chinese terrestrial ecosystems (based on Xie et al. [29] and ecological landscape value coefficients of the five landscape types.

Ecosystem services	Forest ($/hm^2)	Grassland ($/hm^2)	Farmland ($/hm^2)	Water body ($/hm^2)	Non-ecological land ($/hm^2)
Atmospheric regulation	463.62	105.97	66.23	0.00	0.00
Climate regulation	357.65	117.12	115.81	59.85	0.00
Water conservation	423.88	104.10	78.07	2651.94	3.90
Soil formation and protection	516.60	253.75	189.99	1.30	2.60
Waste treatment	173.53	170.47	213.41	2365.68	1.30
Biodiversity conservation	431.83	141.84	92.38	324.01	44.24
Food production	13.25	39.04	130.13	13.01	1.29
Raw materials	344.40	6.50	13.01	1.29	0.00
Ecological landscape value coefficients	2724.76	938.80	899.04	5417.10	53.33

Discussion

Metrics for evaluating landscape values

The functional performances and interrelationships between the five landscape types were depicted as the integrated landscape value in a two-dimension coordinate system and displayed in landscape value distribution maps. In contrast to objective aesthetic approaches [3,13,24,34], the subjective landscape aesthetics in this study reflects the intrinsic perception of the forms and spatial configuration of landscape elements. Compared with conceptual models [7,35,36], our metrics (i.e., survey, coefficients, the two-dimension coordinate system and scopes) may quantitatively capture the functional performance of different landscape types.

Landscape functions/values and landscape attributes

Our results showed that landscape values were strongly related to landscape attributes and the human activity intensity. Unlike semi-natural and human-induced landscapes, such as farmland and non-ecological land, the natural landscapes (i.e., water bodies and forest) contributed positively to both aesthetic and ecological functions. The controversial relationships (i.e., coincidence or disjuncture) between aesthetic performances and ecological quality have raised wide concerns [2,7,8,20,37]. These relationships can be quantified from the two-dimension coordinate system and distribution maps in this study, e.g., the negative or positive contribution level of landscape types in landscape values.

Human activities can result in the decrease of landscape values by changing natural landscapes into semi-natural and artificial landscapes. It was found that the landscape values increase from the human-dominated and semi-natural landscapes to more natural landscapes. The underlying structure-function-value chain can be induced from the modified landscape gradient influenced by human activities, which has been discussed in some landscape assessments [6,36,38,39].

Implications for landscape planning

The functional performances and distribution of landscape values were identified in present study, which may help optimize landscape planning in terms of developing priorities and intensity. The two-dimension coordinate system indicated that replacing one landscape type with another may result in a positive or negative change in the comprehensive landscape value. For example, more value would decrease if the water body was replaced by the non-ecological land than by the farmland. Thus, in order to bring aesthetic needs and ecological goals into better alignment, a

Table 2. Aesthetic landscape value scores (P, point$\cdot$hm^2) of five landscape types.

Aesthetic landscape value units	Forest	Grassland	Farmland	Water body	Non-ecological land
1	0.00	0.00	34.32	4.80	20.82
2	0.00	0.00	313.94	112.48	160.72
3	144.76	0.82	2239.91	670.59	1350.75
4	2672.88	47.30	16046.76	8326.75	5069.18
5	1238.10	98.48	9884.85	10807.28	2743.78
Total	4055.74	146.60	28519.77	19921.89	9345.24
Aesthetic landscape value coefficients	4.21	4.61	4.12	4.40	3.95

Table 3. Normalized aesthetic and ecological landscape value coefficients.

	Forest	Grassland	Farmland	Water body	Non-ecological land
Ecological landscape value coefficients ($/hm²)	2724.76	938.80	899.04	5417.10	53.33
Normalized ecological value coefficients	−0.01	−1.34	−1.37	2.00	−2.00
Aesthetic landscape value coefficients (point/hm²)	4.21	4.61	4.12	4.40	3.95
Normalized aesthetic value coefficients	−0.44	2.00	−0.97	0.73	−2.00

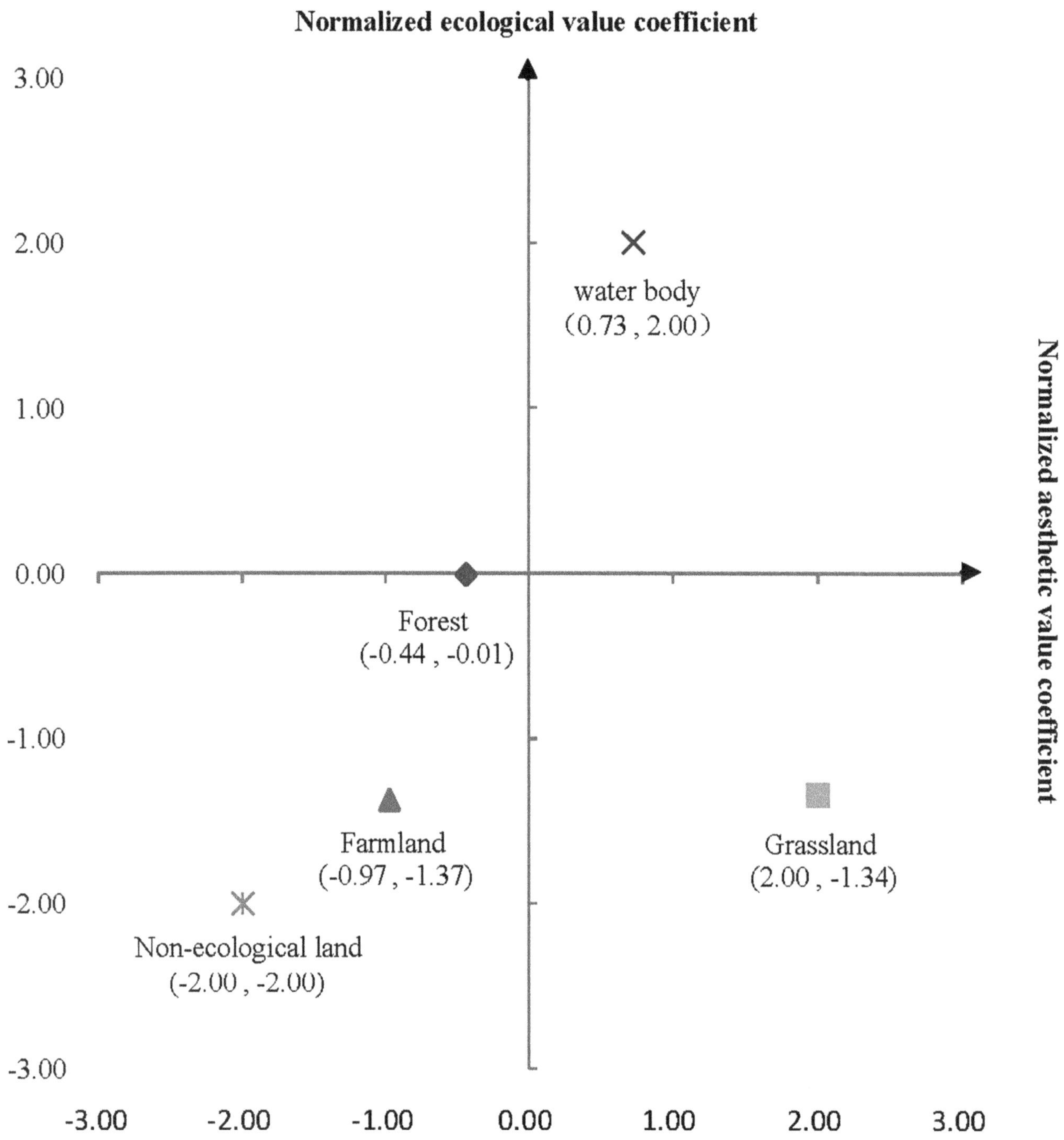

Figure 2. Two-dimensional coordinate system for landscape values.

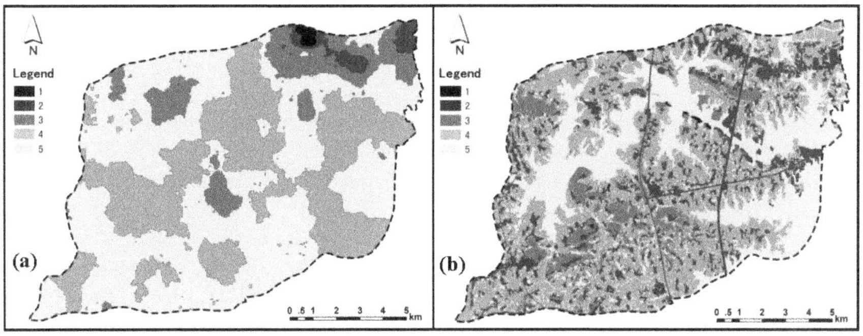

Figure 3. (a) Aesthetic landscape value distribution map and (b) ecological landscape value distribution map in Houguanhu Region.

functional compromise may encounter in landscape planning. In the landscape value distribution maps, the gradients of inherent landscape values and the concerns of underlying landscape value can be used to identify appropriate strategies, such as protection, restoration, and reconstruction. The protection and restoration measures may be taken to expand natural landscape spaces, while human-influenced landscapes should be reconstructed in an orderly way. The reconstruction measures (e.g., building hedgerow buffer zones, reducing the cutting effect of a road network on the overall landscape, or increasing accessibility between landscapes) may be involve in landscape planning.

The situational context (e.g. the social phase, indigenous culture and stakeholder demands) should also be incorporated into landscape planning in response to various concerns regarding the landscape functions [2,24,40]. Tourists are likely to care more about farmland aesthetics, while farmers put a higher value on

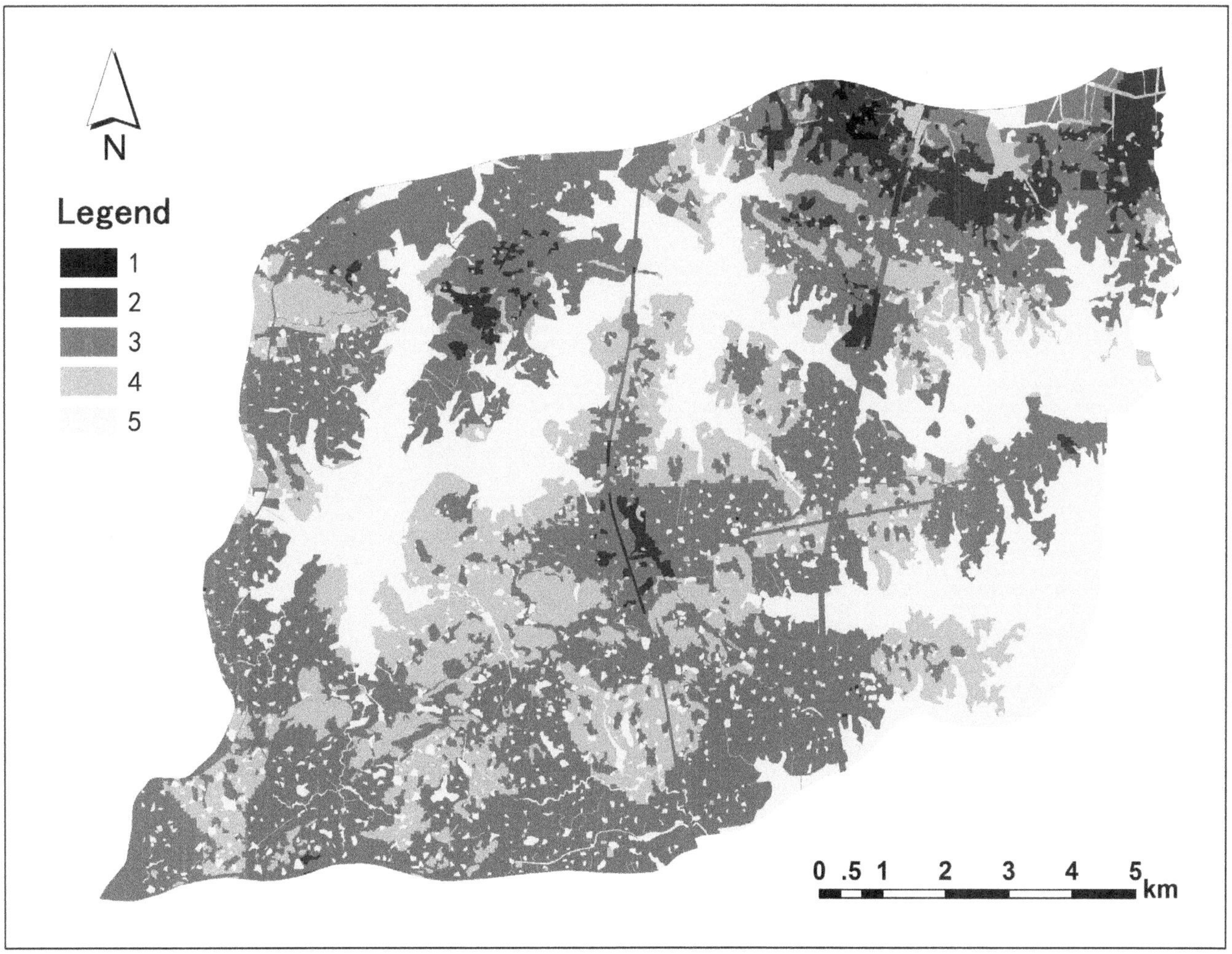

Figure 4. Integrated landscape value distribution map.

agricultural productivity. However, both farmland attractiveness enjoyed by tourists and production features valued by farmers are likely to undermine ecological functions. Therefore, it was recommended for policy makers to align aesthetic features to better support ecological health. It was also recommended that the disorderly non-scenic regions which were important to the overall landscape (e.g. rural settlements, roads, and historic sites) should be carefully enhanced using green corridors and an ordered appearance to foster a more positive aesthetic experience.

Limitations and future improvements

Some limitations and possible improvements of this study included:

(1) In order to make the survey easier, site aesthetic evaluation was kept simple, and only five major landscape types were identified. It is recommended that the subdivisions of landscapes should be adopted for future research to convey more detailed information. In addition, surveyors from a variety of cultural backgrounds could be involved to incorporate a greater diversity of perspectives into landscape planning.

(2) More comprehensive aspects of cultural landscape values (e.g. knowledge systems, social relations, and aesthetic values) could be incorporated into future studies to address the overall performance of landscape resources in the decision-makings.

(3) Subjective aesthetic landscape coefficients can only convey information related to the relative importance of landscape values of the five landscape types. This may limit the use of aesthetic landscape coefficients when comparing with ecological evaluations.

Conclusions

Due to the difficulties in integrating different data types, quantitative models and interdisciplinary studies, it was a far-reaching challenge to connect subjective (aesthetic) and objective (ecological) aspects in an integrated landscape assessment. Nevertheless, the landscape sustainability requires a careful investigation of the relationship between ecology and aesthetics. Compared to conventional conceptual frameworks, the metrics (i.e. field survey, normalized landscape value coefficients, and a two-dimensional coordinate system) were employed in this study in order to quantify the functional performances of landscape types and to present an integrated landscape value. The results bring aesthetic and ecological goals into a better alignment in landscape planning. The interdisciplinary approach may lead to a revolution in landscape designing and planning, because it balances ecological and aesthetic functions for landscape sustainability.

Acknowledgments

We would like to express our gratitude to the people and institutions in Wuhan who assisted us during the field survey. We sincerely thank the editor and anonymous reviewers for their useful comments and suggestions on our paper. We are grateful to Yan Zhu and Jonathan Vause for improving the written quality of the paper.

Author Contributions

Conceived and designed the experiments: DY T. Luo. Performed the experiments: DY T. Luo. Analyzed the data: DY. Contributed reagents/materials/analysis tools: DY T. Luo T. Lin QQ YL. Wrote the paper: DY.

References

1. Daniel TC (2001) Whither scenic beauty? Visual landscape quality assessment in the 21st century. Landscape Urban Plan 54: 267–281.
2. Gobster PH, Nassauer JI, Daniel TC, Fry G (2007) The shared landscape: What does aesthetics have to do with ecology? Landscape Ecol 22: 959–972.
3. Tyrväinen L, Silvennoinen H, Kolehmainen O (2003) Ecological and aesthetic values in urban forest management. Urban Forestry Urban Green 1: 135–149.
4. Klein LR (2013) Quantifying relationships between ecology and aesthetics in agricultural landscapes. Washington State University. 70–98.
5. Luo T, Wang H, Qiu QY, Liu J, Yang ZY (2010) Assessment of non-ecological landscape functions in environmental impact assessment for plan. Res Environ Sci 23(10): 1320–1327. (in Chinese).
6. Wu J (2010) Landscape of culture and culture of landscape: does landscape ecology need culture? Landscape Ecol 25: 1147–1150.
7. Musacchio LR (2009) The ecology and culture of landscape sustainability: emerging knowledge and innovation in landscape research and practice. Landscape Ecol 24: 989–992.
8. Parsons R (1995) Conflict between ecological sustainability and environmental aesthetics: conundrum, canärd or curiosity. Landscape Urban Plan 32: 227–244.
9. Tress B, Tress G, Décamps H, D'Hauteserre AM (2001) Bridging human and natural sciences in landscape research. Landscape Urban Plan 57: 137–141.
10. Naveh Z (2007) Landscape ecology and sustainability. Landscape Ecol 22: 1437–1440.
11. Wu J (2013) Landscape sustainability science: ecosystem services and human well-being in changing landscapes. Landscape Ecol 28: 999–1023.
12. Porto M, Correia O, Beja P (2014) Optimization of landscape services under uncoordinated management by multiple landowners. PLoS ONE 9(1): 1–16.
13. Frank S, Fürst C, Koschke L, Witt A, Makeschin F (2013) Assessment of landscape aesthetics–Validation of a landscape metrics-based assessment by visual estimation of the scenic beauty. Ecol Indic 32: 222–231.
14. Lothian A (1999) Landscape and the philosophy of aesthetics: is landscape quality inherent in the landscape or in the eye of the beholder? Landsc Urban Plan 44: 177–198.
15. Roth M, Gruehn D (2012) Visual landscape assessment for large areas–using GIS, internet surveys and statistical methodologies in participatory landscape planning for the Federal State of Mecklenburg-Western Pomerania, Germany. Latvian Acad Sci A: Humanit Soc Sci 129–142.
16. Forman RTT (2008) The urban region: natural systems in our place, our nourishment, our home range, our future. Landscape Ecol 23: 251–253.
17. Mozingo LA (1997) The aesthetics of ecological design: seeing science as culture. Landscape J 16: 46–59.
18. Nassauer JI (1997) Cultural sustainability: aligning aesthetics and ecology. In: Nassauer JI (Eds.) Placing Nature: Culture and Landscape Ecology. Washington, DC: Island Press.
19. Mikusiński G, Blicharska M, Antonson H, Henningsson M, Göransson G, et al. (2013) Integrating ecological, social and cultural dimensions in the implementation of the landscape convention. Landscape Res 38(3): 384–393.
20. Junker B, Buchecker M (2008) Aesthetic preferences versus ecological objectives in river restorations. Landscape Urban Plan 85: 141–154.
21. Norton LR, Inwood H, Crowe A, Baker A (2012) Trialing a method to quantify the 'cultural services' of the English landscape using Countryside Survey data. Land Use Policy 29: 449–455.
22. Hands DE, Brown RD (2002) Enhancing visual preference of ecological rehabilitation sites. Landscape Urban Plan 58: 57–70.
23. Williams KJH, Cary J (2002) Landscape preferences, ecological quality, and biodiversity protection. Environ Behav 34: 257–274.
24. Fry G, Tveit MS, Ode Å, Velarde MD (2009) The ecology of visual landscapes: Exploring the conceptual common ground of visual and ecological landscape indicators. Ecol Indic 9: 933–947.
25. Hale BW, Steen-Adams MM, Predick K, Fisher N (2005) Ecological conservation through aesthetic landscape planning: A case study of the Lower Wisconsin State Riverway. Environ Manage 35: 381–395.
26. Bastian O, Lütz M (2006) Landscape functions as indicators for the development of local agri-environmental measures. Ecol Indic 6: 215–227.
27. Zhao JW, Luo PJ, Wang RH, Cai YL (2013) Correlations between aesthetic preferences of river and landscape characters. J Environ Eng Landscape Manag 21(2): 123–132.
28. Costanza R, d'Arge R, Groot DR, Farber S, Monica G, et al. (1997) The values of the world's ecosystem services and natural capital. Nature 387: 253–260.
29. Xie GD, Lu CX, Leng YF, Zheng DU, Li SC (2003) Ecological assets valuation of the Tibetan Plateau. J Nat Resour 18: 189–196. (in Chinese).
30. Nassauer JI (2004) Monitoring the success of metropolitan wetland restorations: cultural sustainability and ecological function. Wetlands 24: 756–765.
31. Purcell AT, Lamb RJ (1998) Preference and naturalness: an ecological approach. Landscape Urban Plan 42: 57–66.
32. Yang ZY, Luo T, Zhu ZC, Wang H, Song Y, et al. (2012) Landscape aesthetic assessment based on experiential paradigm assessment technology. Environ Sci Technol 35(3): 187–205. (in Chinese).
33. Von Haaren C (2004) Landschaftsplanung. Stattgart: Eugen GmbH & Co. 245–272. (in German).

34. Dandy N, Van Der Wal R (2011) Shared appreciation of woodland landscapes by land management professionals and lay people: An exploration through field-based interactive photo-elicitation. Landscape Urban Plan 102: 43–53.
35. Gómez-Sal A, Belmontes JA, Nicolau JM (2003) Assessing landscape values: a proposal for a multidimensional conceptual model. Ecol Model 168: 319–341.
36. Termorshuizen JW, Opdam P (2009) Landscape services as a bridge between landscape ecology and sustainable development. Landscape Ecol 24: 1037–1052.
37. Yang B, Brown TC (1992) A cross-cultural comparison of preferences for landscape styles and landscape elements. Environ Behav 24: 471–507.
38. Yang DW, Kao WTM, Zhang GQ, Zhang NY (2014) Evaluating spatiotemporal differences and sustainability of Xiamen urban metabolism using emergy synthesis. Ecol Model, 272: 40–48.
39. Yang DW, Qiu QY, Liu WJ (2012) Complementary zoning frameworks for sustainable management of eco-economic capital: a case study of Lijiang, China. J MT Sci, 9: 760–769.
40. Aretano R, Petrosillo I, Zaccarelli N, Semeraro T, Zurlini G (2013) People perception of landscape change effects on ecosystem services in small Mediterranean islands: A combination of subjective and objective assessments. Landscape Urban Plan 112: 63–73.

Spatial Structure of Seagrass Suggests That Size-Dependent Plant Traits Have a Strong Influence on the Distribution and Maintenance of Tropical Multispecies Meadows

Jillian L. S. Ooi[1,2,3,4*], Kimberly P. Van Niel[2,3], Gary A. Kendrick[1,3], Karen W. Holmes[2,5]

1 School of Plant Biology, Faculty of Natural and Agricultural Sciences, The University of Western Australia, Perth, Western Australia, Australia, **2** School of Earth and Environment, Faculty of Natural and Agricultural Sciences, The University of Western Australia, Perth, Western Australia, Australia, **3** The UWA Oceans Institute, The University of Western Australia, Perth, Western Australia, Australia, **4** Department of Geography, Faculty of Arts and Social Sciences, Universiti Malaya, Kuala Lumpur, Malaysia, **5** Centre for Ecohydrology, School of Environmental Engineering, The University of Western Australia, Perth, Western Australia, Australia

Abstract

Background: Seagrass species in the tropics occur in multispecies meadows. How these meadows are maintained through species co-existence and what their ecological drivers may be has been an overarching question in seagrass biogeography. In this study, we quantify the spatial structure of four co-existing species and infer potential ecological processes from these structures.

Methods and Results: Species presence/absence data were collected using underwater towed and dropped video cameras in Pulau Tinggi, Malaysia. The geostatistical method, utilizing semivariograms, was used to describe the spatial structure of *Halophila* spp, *Halodule uninervis, Syringodium isoetifolium* and *Cymodocea serrulata*. Species had spatial patterns that were oriented in the along-shore and across-shore directions, nested with larger species in meadow interiors, and consisted of multiple structures that indicate the influence of 2–3 underlying processes. The Linear Model of Coregionalization (LMC) was used to estimate the amount of variance contributing to the presence of a species at specific spatial scales. These distances were <2.5 m (micro-scale), 2.5–50 m (fine-scale) and >50 m (broad-scale) in the along-shore; and <2.5 m (micro-scale), 2.5–140 m (fine-scale) and >140 m (broad-scale) in the across-shore. The LMC suggests that smaller species (*Halophila* spp and *H. uninervis*) were most influenced by broad-scale processes such as hydrodynamics and water depth whereas large, localised species (*S. isoetifolium* and *C. serrulata*) were more influenced by finer-scale processes such as sediment burial, seagrass colonization and growth, and physical disturbance.

Conclusion: In this study, we provide evidence that spatial structure is distinct even when species occur in well-mixed multispecies meadows, and we suggest that size-dependent plant traits have a strong influence on the distribution and maintenance of tropical marine plant communities. This study offers a contrast from previous spatial models of seagrasses which have largely focused on monospecific temperate meadows.

Editor: Ruth H. Carmichael, Dauphin Island Sea Lab; University of South Alabama, United States of America

Funding: Fieldtrips were partly funded by the MOSTI E-Science grant (04-01-03-SF0177) and the Grant for Research Student Training (GRST, UWA) awarded to JLSO. JLSO was supported by the Endeavour International Postgraduate Research Scholarship, the University Postgraduate Award, the PhD Completion Scholarship (University of Western Australia) and the Hadiah Cuti Belajar program (Universiti Malaya). The funders had no role in study design, data collection and analysis, decision to publish, or preparation of the manuscript.

Competing Interests: The authors have declared that no competing interests exist.

* E-mail: jillian_03@um.edu.my

Introduction

In tropical marine ecosystems, seagrass species occur in dense, well-mixed stands where habitat partitioning is subtle and less defined than in temperate waters. How these multispecies meadows are maintained through species co-existence and what their ecological drivers may be has been an overarching question in seagrass biogeography (see [1]). The three broad categories of ecological drivers in plant communities – life strategy and morphology, interspecific interactions within a community, and response to environmental gradients [2] – operate together to produce an overall combined effect, resulting in the non-random development of species gradients and patches across the habitat range (i.e. *spatial structure*). Thus, descriptions of spatial structure provide insight into the relationships between biological patterns and processes [3] and is a useful starting point for understanding what drives species distributions.

Seagrass species distributions have been linked to differential rates and strategies of growth, reproduction and dispersal [4,5,6,7], interspecific competition [8,9] and environmental gradients such as water depth, light, temperature and hydrodynamics [10,11,12,13,14]. Following hierarchy theory, Legendre

[15] suggested that physical processes create spatial structure in biological systems at broad scales, while contagious biotic processes (e.g. reproduction, biological interactions, food availability) cause further structuring at finer scales. In the case of seagrasses, Vermaat [1] argued for light and hydrodynamics as the prime physical processes that determine meadow extent by governing depth penetration and dispersal (i.e. broad-scale drivers), while implying that clonal growth strategies and physical disturbance by herbivory and storms maintain the characteristic patchiness of meadows (i.e. finer-scale drivers). However, the link between these underlying processes and species patterns remains inconclusive because very few studies have been spatially explicit. Studies have more commonly focused on the seagrass shoot or ramet at distances of centimetres to metres. This creates a problem in scaling-up from local process to broader-scale patterns [13,16,17], leading to the observation that spatial scales of field measurement have rarely been aligned with specific ecological processes in seagrass research [18].

Seagrass studies that have explicitly linked processes to spatial structure do so either through cross-scale studies that identify scale-dependent differences in species distribution and abundance, or by using techniques which explicitly incorporate spatial structure, e.g. variogram modelling. Both approaches have demonstrated spatial structures in the distribution of seagrasses across multiple spatial scales, but what we know is drawn mostly from studies in meadows where species differentiation in space is strong and species form individual monospecific meadows. Examples include *Amphibolis griffithii*, *Posidonia coriacea*, and *P. sinuosa* occurrence in southwest Australia [16,19]; *Zostera tasmanica* morphology and bed architecture in southeast Australia [20]; *Zostera marina* and *Halodule wrightii* occurrence in Beaufort, southeast USA [13]; *H. decipiens* occurrence in Florida, southeast USA [21]; *P. oceanica* plant and meadow architecture in the Mediterranean [22,23,24,25]; the within-meadow genotypic distribution of *P. oceanica* [26], and the kinship structure of *Zostera marina* [27].

In contrast, spatially-explicit patterns and processes in multispecies meadows have received less attention. Our study addresses this gap by using geostatistics to model the spatial structure of four co-existing tropical species in a forereef system in Pulau Tinggi, Malaysia. Geostatistical techniques were selected because they have the potential to capture small differences in spatial structure, and in doing so provide insight into the spatial relations of species in their environment [28]. We ask two questions (1) what is the spatial structure of *Halophila* spp., *Halodule uninervis*, *Syringodium isoetifolium* and *Cymodocea serrulata* distribution in Pulau Tinggi?; and (2) what are the critical spatial scales at which the spatial structures of taxa are most evident? Based on these spatial scales, we make suggestions about what some of the potential ecosystem drivers may be for this study area.

Methods

Ethics Statement

This project was conducted under the Economic Planning Unit of Malaysia Research Permit no. UPE: 40/200/19/2351 and the Marine Parks Department of Malaysia Research Permit no. JTLM (S) 620-2/1/1 Jld 2(2).

Study Area

The field survey was conducted from 15 April to 15 May 2009 in Pulau Tinggi (pulau = island), a continental island located 12 km off the southeast coast of Peninsular Malaysia. It is ringed by fringing coral reefs, with subtidal seagrass meadows in the forereef zone, i.e. on the seaward side of the coral reefs, extending down to a depth of approximately 10 m. The multispecies meadows consisted of *Halophila ovalis* (R. Br.) Hooker f., *Halodule uninervis* (Forsskal) Ascherson, *Cymodocea serrulata* (R. Brown) Ascherson, *Syringodium isoetifolium* (Ascherson) Dandy, *Halophila minor* (Zollinger) den Hartog, *Halophila decipiens* Ostenfeld, and *Halophila spinulosa* (R. Brown) Ascherson [29].

Data Collection

Towed video was used to characterize seagrass distribution along transects covering the range of sedimentary environments, water depth, and bathymetric features in the study area (Fig. 1). Using an underwater camera mounted on a glider frame and towed from a boat at a speed of 0.5–2.0 knots, visual recordings were made of the substrate. The boat position, heading, and boat speed were logged from a GPS and linked to the video frames at 2 second intervals along each transect. The camera was maintained at a height of <0.5 m from the seabed, with an approximate angle of 45 degrees from the substrate to enable species identification. The towed video footage was analysed by recording seagrass species composition at intervals of 2.5 m within the seagrass beds and 10 m outside seagrass beds. Identification was possible for all species except for *Halophila* spp. What is identified as *Halophila* spp. in this study comprises a complex predominantly of *H. ovalis* but also of *H. decipiens* and *H. minor*.

Drop video was deployed to provide additional data in areas between transects, thus reducing the clustering effect typically caused by a transect-sampling approach. The drop video system consisted of an underwater video in water-tight housing attached to a quadrat measuring 0.5×0.5 m. At each drop video point, the video camera was lowered onto the seabed randomly, ten times. This resulted in ten samples at each point, at an average distance of 1–2 meters apart. Seagrass species composition within the quadrat was recorded for each sample and added to the full data set comprising sampling points from both towed and drop video.

The main data set was divided into a training dataset (n = 8036) and a validation dataset (n = 2164). The former was used for variogram construction to obtain the variogram parameters (nugget, sill, range). The latter was used in cross validation to select the best model type and number of nested structures, based on (1) the predictive power of the models and on (2) the goodness-of-fit and parsimony of the models The final models presented here are those where the mean, median, standard deviation, mean squared and total errors of the predicted values are 0 or closest to 0 (best predictive power), and which have the lowest *AIC* (best goodness-of-fit and parsimony). All data are available on request from the first author.

Indicator Semivariograms

The spatial structure of seagrass species was quantified by the indicator semivariogram [30] separately for each species to depict how the occurrence of a species varies as a function of distance. The indicator variogram is defined as the average squared semi-difference between pairs of observations [30]:

$$\gamma_I(h; z_k) = \frac{1}{2N(h)} \sum_{\alpha=1}^{N(h)} [i(u_\alpha; z_k) - i(u_\alpha + h; z_k)]^2 \qquad (1)$$

Where γ_I is the dissimilarity between data pairs separated by a distance of h; $N(h)$ is the number of pairs of observations at distance h; z_k is the indicator threshold of species presence or

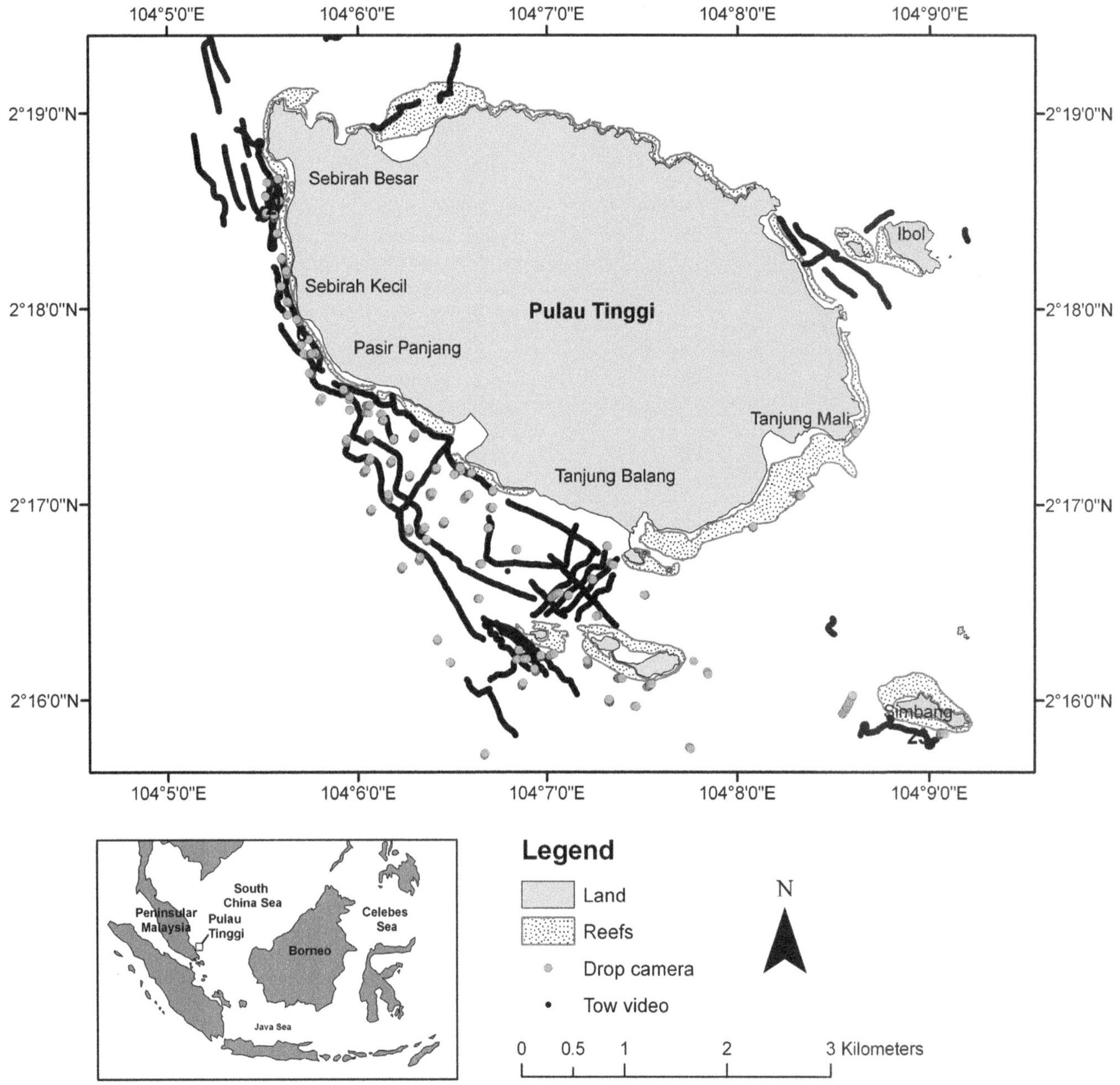

Figure 1. Location of tow video, drop video, and seagrass sampling points around Pulau Tinggi, southeast Peninsular Malaysia. Tow video points have converged to form tracks.

absence at coordinates μ_α. A low variance ($\gamma_I(h; z_k)$) is closer to 0, corresponding to high correlation or continuity.

Variograms were interpreted and fitted using one or a combination of the accepted models (spherical, exponential, Gaussian, nugget-effect) [31]. The variogram model parameters are (1) the *sill* (heterogeneity of species occurrence), i.e. the semivariance on the y-axis where the variogram reaches its asymptote, which is approximately the global variance of the data; (2) the *range* (maximum diameter of species patches or gaps), i.e. the distance on the x-axis where the semivariogram reaches its asymptote and beyond which all points are spatially uncorrelated; and (3) the *nugget*; i.e. the semivariance on the y-axis at the ordinate of the variogram due to measurement error and fine spatial-scale fluctuations in the property of interest [32] (Fig. 2). Geostatistical analyses were implemented in WinGslib 1.5.6.

Anisotropy is the property of having different patterns in different directions. To capture anisotropy, directional variograms were constructed for different azimuths, which were north-south (0°), northeast-southwest (45°), east-west (90°) and southeast-northwest (135°). Calculations were constrained to pairs of points oriented within 45 degrees of a particular azimuth (or 22.5° on either side of a point). Because variogram values are less reliable at large distances [33], variograms were computed up to half the maximum distance between sampling points. These distances were 1800 m (0°), 700 m (45°), 1300 m (90°) and 1700 m (135°).

Out of these four directional variograms, the two with the longest and shortest ranges were used to characterise spatial

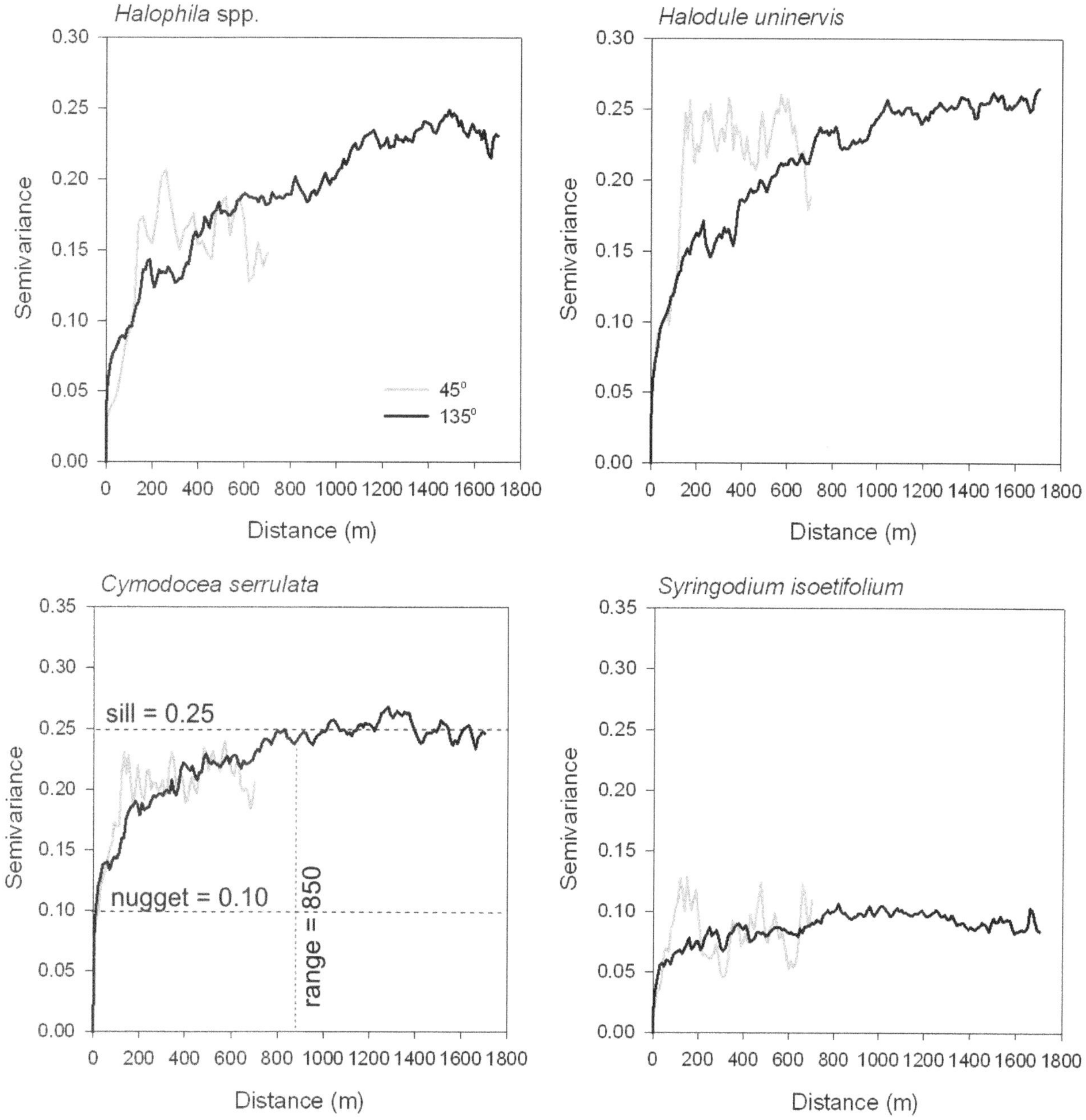

Figure 2. Directional variograms for presence/absence of four seagrass species in the along-shore (135°) and across-shore (45°) of Pulau Tinggi, Peninsular Malaysia. Other directions (0° = north-south and 90° = east-west) were modeled but are not shown. Variograms were plotted for half the maximum distance between points in each direction, lag 10 m, tolerance 5 m. The theoretical variogram structures (nugget, range and sill) are shown for *Cymodocea serrulata.*

structure (question 1) and identify critical scales of study in the Linear Model of Coregionalization (question 2). These variograms were also used to produce kriged maps of species.

Kriged Maps of Species

Ordinary indicator kriging was performed on the variogram model to provide a probability estimate of occurrence for each species in the study area. Indicator kriging predicts the odds of species presence in geographic space by using the parameters of the chosen variogram model to describe the spatial correlation structure with the following equation:

$$I(u; z_k) = \sum_{\alpha=1}^{n(u)} \lambda_\alpha(\mu; z_k)[I(\mu; z_k)] \qquad (2)$$

Where $\lambda_\alpha(\mu; z_k)$ is the weight from the variogram modelling assigned to the indicator data $I(\mu; z_k)$ In this way, predictions of species occurrence can be made for unsampled locations based on occurrence data from samples in a local neighbourhood. If anisotropy is present in the dataset, kriging is performed using variograms of the axes of maximum and minimum continuity combined into a single model. Kriged maps were produced in grids measuring 5×5 m. Kriged output was exported to ArcGIS 9.3 for visualization and map production.

Linear Model of Coregionalization (LMC)

The Linear Model of Coregionalization (LMC) was used to determine critical spatial scales in species distributions (question 2). The basic assumption of the LMC is that the variables of a model are generated by the superimposition of independent physical processes operating at different spatial scales [34]. In the case of co-occurring seagrass species, the separate species variograms may be fitted with variogram functions which allow inference of how strongly a corresponding scale-specific process influences the distribution of each species.

In this study, the LMC procedure was performed separately for the two directions of spatial continuity based on the assumption that these will each have different dominant spatial scales. The LMC procedure was applied by (1) computing individual species variograms, (2) choosing a suitable combination of basic variogram functions common to all species variograms, (3) fitting the selected model to the species variograms and (4) estimating the proportional contribution of the basic functions to each model, based on the rationale that comparison of the scale-specific sills reveals the proportion of variance accounted for at each scale [35]. The total variance at each spatial scale is formally expressed as [36]:

$$\gamma(h) = \sum_{1=1}^{Ns} b^1 g^1(h) \quad (3)$$

Where Ns is the number of common variogram structures $g(h)$, b is the sill, and 1 indicates a particular spatial scale.

Results

A total of 37 towed video transects amounting to 25 kilometres in length were collected. Out of the sampled 10,200 data points, 66% had seagrass (Table 1). *Halophila* spp. was the most frequently identified seagrass (65%), followed by *Halodule uninervis* (50%), *Cymodocea serrulata* (45%), and *Syringodium isoetifolium* (11%).

Table 1. Total number of frames obtained from tow and drop video, counts (*n*) and percentages (%) of frames with and without seagrass, and breakdown by species.

Video data	*n*	% of total frames
Total number of sampled frames	**10200**	
Tow video (~2.5 m intervals)	9586	
Drop video (~1.0 m intervals)	614	
Seagrass present	**6695**	**66**
Seagrass species		
Halophila spp.	6641	65
Halodule uninervis	5065	50
Cymodocea serrulata	4607	45
Syringodium isoetifolium	1096	11
Seagrass absent	**3505**	**34**

Structure of Directional Variograms

All variograms showed strong spatial dependence with obvious sills (Fig. 2), indicating that these species were distributed in clearly defined patches or aggregations, but were structured differently in different directions. For all species, the axes of maximum and minimum continuity were 135° (along-shore direction) and 45° (across-shore direction). In the along-shore, range distances were shortest for *C. serrulata* and *S. isoetifolium*, while *Halophila* spp and *H. uninervis* extended over longer distances (Table 2). Because these species co-occur, the anisotropy ratios indicate that species distribution is spatially nested. *Halophila* spp. (*H. ovalis*, *H. decipiens* and *H. minor*) and *H. uninervis* were present in all patches from edge to edge, with the widest average patch size measuring around 1,000–1,400 m in the along-shore and 250–280 m in the across-shore. In patch interiors are *C. serrulata* and *S. isoetifolium* with average patch sizes of 800–850 m along-shore and 150–200 m across-shore.

The directional variograms were well-structured in the along-shore direction, in that the semivariances increased gradually to reach a sill (Fig. 2). In the across-shore direction, variograms also reached their sills, but undulations were evident, indicating patchiness and repetition of patches in species distribution across nested spatial scales. All variograms required a nugget component plus an additional spherical or exponential component (Table 2), suggesting that taxa occurrence was influenced by at least 2–3 dominant processes.

The kriged maps show a nested pattern of patch sizes and location, with *Halophila* spp. and *H. uninervis* having large extents, and *C. serrulata* and *S. isoetifolium* with smaller extents (Fig. 3). These species co-occur with the highest variability mostly at meadow edges, suggesting strong influence of environmental gradients on species distribution. Striped patterns in the kriged maps are an artefact of the sampling protocol, where strong currents made it difficult to orient transects in the across-shore direction.

Spatial Structure of Species at Different Scales

The Linear Model of Coregionalization (LMC) was constructed for the four taxa as a group based on model parameters of the individual species variograms in Table 2. The main contributing spatial scales in the along-shore direction were <2.5 m (micro-scale), 2.5–50 m (fine-scale) and >50 m (broad-scale). A nugget component plus nested exponential and spherical components provided the best fit (Fig. 4, Table 3). Occurrence of *Halophila* spp and *H. uninervis* were dominated by broad-scale processes, with variance contributions between 60–65% (Table 4a). *C. serrulata* had equal contributions from micro-scale (40%) and broad-scale processes (42%). *S. isoetifolium* had the largest contributions from fine scale (42%) and broad scale (35%) processes.

In the across-shore direction, the main contributing spatial scales were <2.5 m (micro-scale), 2.5 m –140 m (fine-scale) and >140 m (broad-scale). The variograms were best fitted by a nugget component and a double spherical component (Fig. 5, Table 3). *Halophila* spp and *H. uninervis* had large variance contributions of 83–86% by broad-scale processes (Table 4b). The broad-scale had a large influence (45%) on *C. serrulata*, as did the micro-scale (33%). *S. isoetifolium* was dominated by processes at the fine-scale (81%), whereas the micro-scale and broad-scale had little to no influence on its occurrence.

Table 2. Parameters of models fitted independently to variograms of the four seagrass species in Pulau Tinggi, Malaysia.

Species	C_o	C_1	C_2	*Sill*	a_1	a_2	NSR	AR
Along-shore (135°)								
Halophila spp	0.060	0.170_{Exp}	–	0.230	1400	–	26	5.0
H. uninervis	0.060	0.190_{Exp}	–	0.250	1000	–	24	4.0
C. serrulata	0.100	0.150_{Exp}	–	0.250	850	–	40	4.3
S. isoetifolium	0.020	0.030_{Sph}	0.045_{Exp}	0.095	50	800	20	5.3
Across-shore (45°)								
Halophila spp	0.023	0.045_{Sph}	0.092_{Sph}	0.160	220	280	14	–
H. uninervis	0.040	0.030_{Sph}	0.160_{Sph}	0.230	30	250	17	–
C. serrulata	0.070	0.025_{Sph}	0.115_{Sph}	0.210	50	200	23	–
S. isoetifolium	0.015	0.030_{Sph}	0.035_{Hole}	0.080	130	150	19	–

All variograms required a nugget component plus an additional exponential, spherical or sine wave component. These are indicated as *Exp, Sph* and *Hole,* respectively. C_o = nugget, C_1 = variance contribution for the 1st structure, C_2 = variance contribution for the 2nd structure, a_1 = range for the 1st structure (m), a_2 = range for the 2nd structure (m), NSR (nugget-to-sill ratio) = $(C_o/Sill)$*100, AR = anisotropy ratio (a_1 in 135°: a_2 in 45°).

Discussion

1. Spatial Patterns were Directional, Nested and had Multiple Structures

In the forereefs of Pulau Tinggi, seagrass species had spatial patterns that were directional, nested, and consisted of multiple structures. The largest difference in directionality was between the along-shore and across-shore, pointing to the presence of gradients oriented predominantly along those axes. Species distributions were characterized as "nested" because *C. serrulata* and *S. isoetifolium* were present mainly in meadow interiors while *Halophila* spp and *H. uninervis* occupied the larger portion of the gradient. Nestedness implies spatial organization in response to a common environmental gradient interacting with biological mechanisms such as differential colonization abilities and physiological limits of species. Also, the variograms of each of the species were fitted best with more than one model structure, suggesting that species were influenced by multiple additive processes acting across multiple scales.

2. Critical Spatial Scales are Linked to Species Size

By quantifying the spatial structure of the co-existing species, we detected critical scales for each of them. The LMC showed each species to likely be influenced by 2–3 processes across a range of scales. The dominant nested spatial scales in the along-shore were <2.5 m (micro-scale), 2.5–50 m (fine-scale), and >50 m (broad-scale); and in the across-shore were <2.5 m (micro-scale), 2.5–140 m (fine-scale), and >140 m (broad-scale).

Size-based rules were evident. In ascending order of species size and descending order of structure based on the spatial scale of environmental processes and drivers are *H. ovalis, H. uninervis, S. isoetifolium* and *C. serrulata* [37]. *Halophila* spp and *H. uninervis* were likely to be influenced predominantly by broad-scale processes; *S. isoetifolium* by fine-scale processes; and *C. serrulata* equally by micro and broad-scale processes (Table 4). Thus, this spatially-explicit approach suggests that large and localised species were influenced by finer-scale processes across a wider range of spatial scales, whereas small, widespread species appeared to be driven almost entirely by broad-scale processes (Fig. 6).

Other studies have also implied this pattern. In a study linking species diversity and shoot density to a siltation gradient at the broad spatial scale of hundreds of kilometres, *H. ovalis* was found to be the least tolerant of variation in siltation, whereas *S. isoetifolium* and *C. serrulata* had weaker responses [38]. In contrast, interspecific competition at the scale of centimetres to metres did not result in significant variation in *H. ovalis,* whereas it did for larger species such as *Thalassia hemprichii* and *S. isoetifolium* [8,39]. The idea here is that there is a size-dependent relationship between species and the spatial scale of their environmental drivers. Thus, intrinsic plant traits, which in seagrasses are size-dependent [37,40], interact with external drivers to produce structure in the spatial distribution of species.

3. Potential Scale-specific Drivers

With critical spatial scales identified by the LMC, we propose hypotheses about the dominant drivers on species distribution in the study area, on the basis of matching spatial scales.

Micro-scale (<2.5 m). Variation in species occurrence at <2.5 m could be the result of biotic disturbances that cause a reorganisation of seagrass species and the placement of their shoots. These include grazing by dugongs and turtles and bioturbation by infauna such as crustaceans, echinoderms, molluscs, oligochaetes and polychaetes, all of which are common and abundant in this region [41]. Wherever they graze, dugongs impact seagrasses by removing whole plants and causing seagrasses to change their growth patterns and morphology [42], and by selecting for fast-growing species such as *H. ovalis* [43]. Similarly, infauna have the capacity to cause localised substrate disturbance that could have large cumulative effects on the structure of marine ecosystems [44], including seagrass.

However, burrowing activity by infaunal shrimp was the most appreciable biotic driver in our site, evident from the widespread presence of shrimp sediment mounds. These mounds occurred at densities of 1–2 m^{-2} in Pulau Tinggi, and elsewhere in Southeast Asia have been reported in densities of 2–3 m^{-2} on backreef seagrass meadows [45,46]. Sediment mounds result not only in the burial of seagrasses, but possibly also in the germination of seeds exposed by the process of bioturbation [21]. Thus, when coupled with species-specific responses of seagrasses to burial, shrimp mounds may account for much of the random, unpredictable patterns of seagrass distribution observed as a nugget effect in this study.

The variance contribution to *C. serrulata* was double that of the other species at the micro-scale suggesting greater variability in its

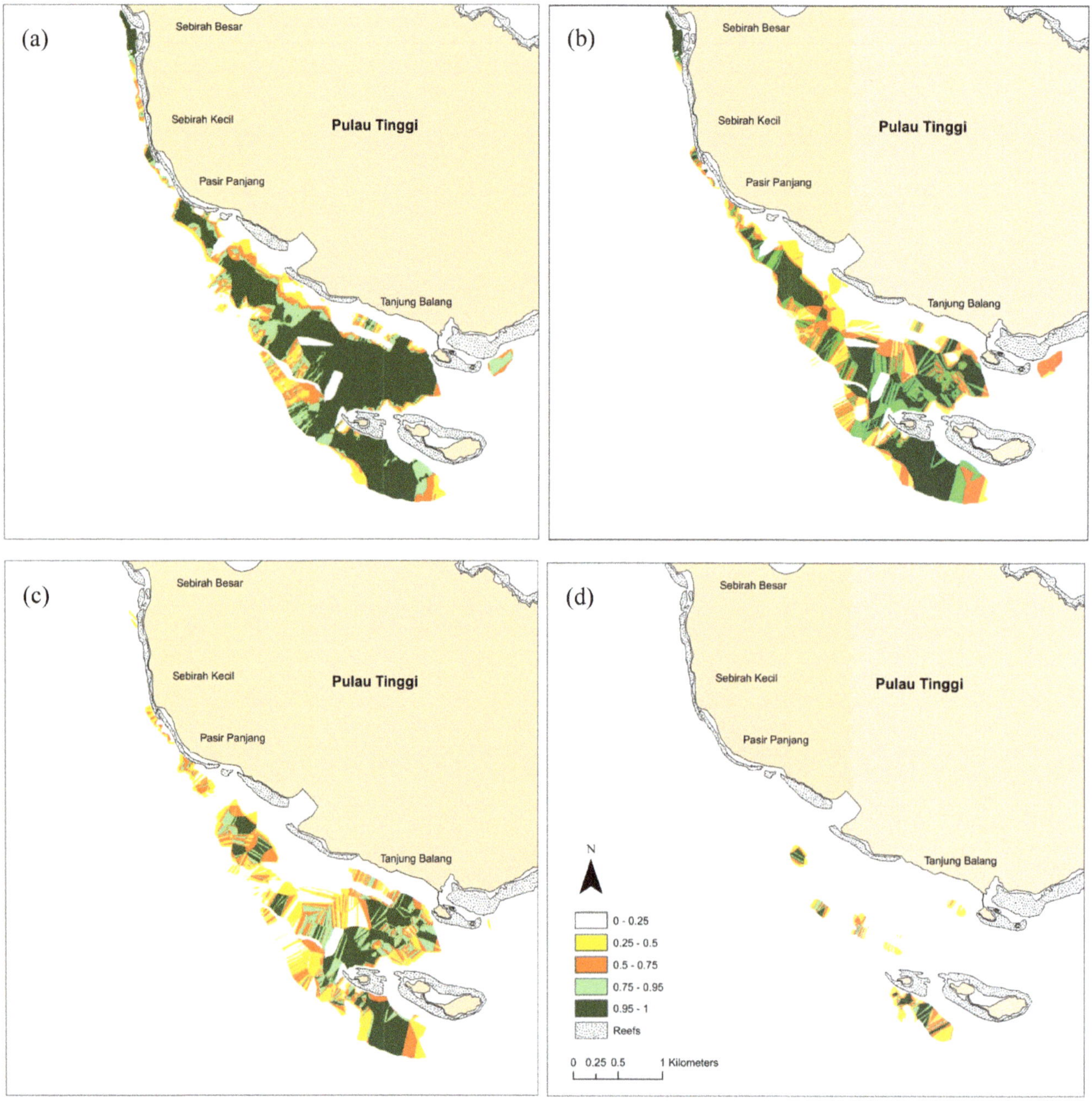

Figure 3. Probability estimates for the occurrence of (a) *Halophila* spp. (b) *Halodule uninervis* (c) *Cymodocea serrulata*, and (d) *Syringodium isoetifolium* in the southwest of Pulau Tinggi, Malaysia, produced by ordinary kriging in 5 m cells.

occurrence as a result of micro-scale processes such as sediment burial. However, past experimental evidence suggested quite the opposite – that *C. serrulata* should be the least variable of all species at the micro-scale because it is the one most likely to survive burial and to remain present in the landscape [37]. Its internal growth program may provide an explanation: as the largest species in this study, *C. serrulata* elongates less rapidly, has larger spacing between successive shoots, and branches less frequently and at more shallow angles than the others [4,47]. Thus, *C. serrulata* may be good at spreading across substrate but does not fill up gaps caused by sediment mounds as efficiently and quickly as *H. ovalis*, *H. uninervis* and *S. isoetifolium*. This has also been demonstrated in a cross-species removal experiment [6]. Such a growth program could result in the random occurrence of *C. serrulata* at the micro-scale despite its ability to out-survive the others when buried under sediment.

Fine-scale (2.5–50 m along-shore, 2.5–140 m across-shore). Fine-scale drivers were considerably more difficult to identify because of the wide range in distances. However, the type of variogram models used is informative about patch characteristics. The exponential model in the along-shore (Table 3) indicates that patches have an irregular extent usually attributed to stochastic processes [48]. Irregular gap formation at this scale could have been caused by the grazing of mega fauna, sediment erosion and deposition both natural and caused by boat movement and anchoring, and surface runoff outlets from land which cause

Table 3. Parameters of the linear coregionalization for four seagrass species in Pulau Tinggi, Malaysia in the two principal directions of continuity.

Spatial scales	Model	Sill			
		Halophila spp.	*Halodule uninervis*	*Cymodocea serrulata*	*Syringodium isoetifolium*
Along-shore (135°)					
Micro-scale (<2.5 m)	Nug	0.060	0.060	0.100	0.020
Fine-scale (2.5–50 m)	Exp	0.020	0.040	0.045	0.042
Broad-scale (>50 m)	Sph	0.150	0.150	0.105	0.033
Across-shore (45°)					
Micro-scale (<2.5 m)	Nug	0.023	0.040	0.070	0.015
Fine-scale (2.5–140 m)	Sph	0.000	0.000	0.045	0.065
Broad-scale (>140 m)	Sph	0.137	0.190	0.095	0.000

Nug = nugget, Exp = exponential model, Sph = spherical model.

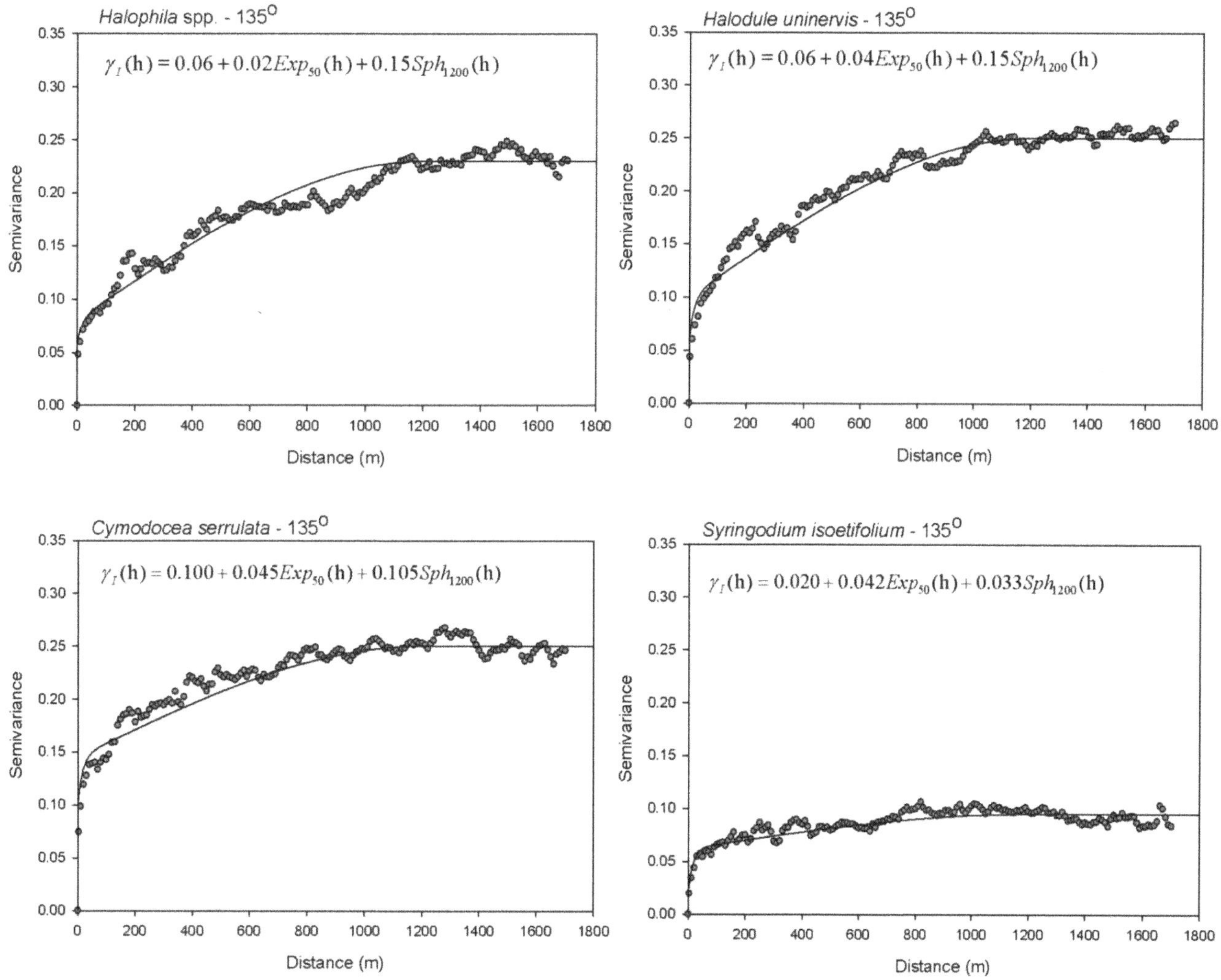

Figure 4. Sample variograms (dots) and the fitted Linear Model of Coregionalization (solid lines) of seagrass species presence/absence in the along-shore direction (135°) in Pulau Tinggi, Malaysia. Sample variograms are displayed with a lag of 10 m and a tolerance of 5 m.

changes in salinity and nutrient input. In the across-shore direction, the spherical model provided the best fit. Thus, drivers in this direction resulted in species patches that were fairly regular.

Broad-scale (>50 m along-shore, >140 m across-shore). Variation in species occurrence at the broad-scale is potentially driven by hydrodynamics in the along-shore. Pulau Tinggi has diurnal tides, with a daily range of 0.5 to 1.5 m (Malaysian Meteorological Department, 2010). During the twice-daily change of tides, the flow of water current was in the along-shore direction at velocities of more than 50 cm s^{-1}. In Pulau Tinggi, hydrodynamics may increase plant dispersal between beds and cause substrate disturbance, resulting in between-meadow differences in seagrass populations, especially for *Halophila* spp. and *H. uninervis*.

In the across-shore direction, the depth gradient (light gradient) is a possible broad-scale driver. At the landward meadow edge where water depth is around 3 m, photosynthetically active radiation (PAR) averages 37% of surface irradiance. The meadow slopes down gradually to around 10 m depth where PAR averages 15% of surface irradiance, beyond which no seagrasses occur [29]. Because *Halophila* spp and *H. uninervis* possess small rhizomes, they have small respiratory demands [49] and they do not integrate

Table 4. Percent of total variance for each species explained at each spatial scale, fitted to the constraints of the Linear Model of Coregionalization in the (a) along-shore, and (b) across-shore.

Species	Micro-scale	Fine-scale	Broad-scale
(a) Along-shore direction	<2.5 m (%)	2.5–50 m (%)	>50 m (%)
Halophila spp.	26	9	65
Halodule uninervis	24	16	60
Cymodocea serrulata	40	18	42
Syringodium isoetifolium	21	42	35
(b) Across-shore direction	<2.5 m (%)	50–140 m (%)	>140 m (%)
Halophila spp.	14	0	86
Halodule uninervis	17	0	83
Cymodocea serrulata	33	21	45
Syringodium isoetifolium	19	81	0

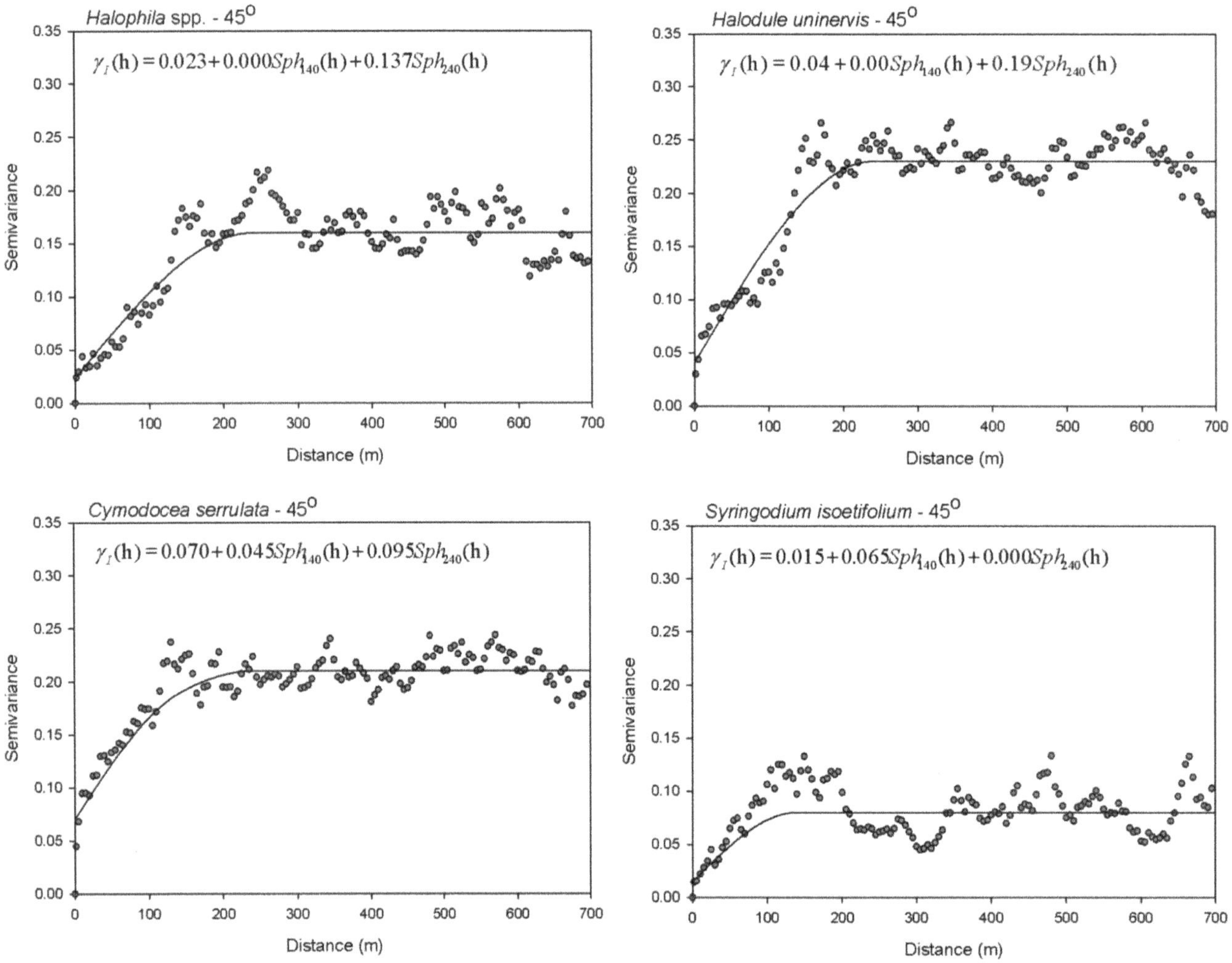

Figure 5. Sample variograms (dots) and the fitted Linear Model of Coregionalization (solid lines) of seagrass species presence/absence in the across-shore direction (45°) in Pulau Tinggi, Malaysia. Sample variograms are displayed with a lag of 5 m and a tolerance of 2.5 m.

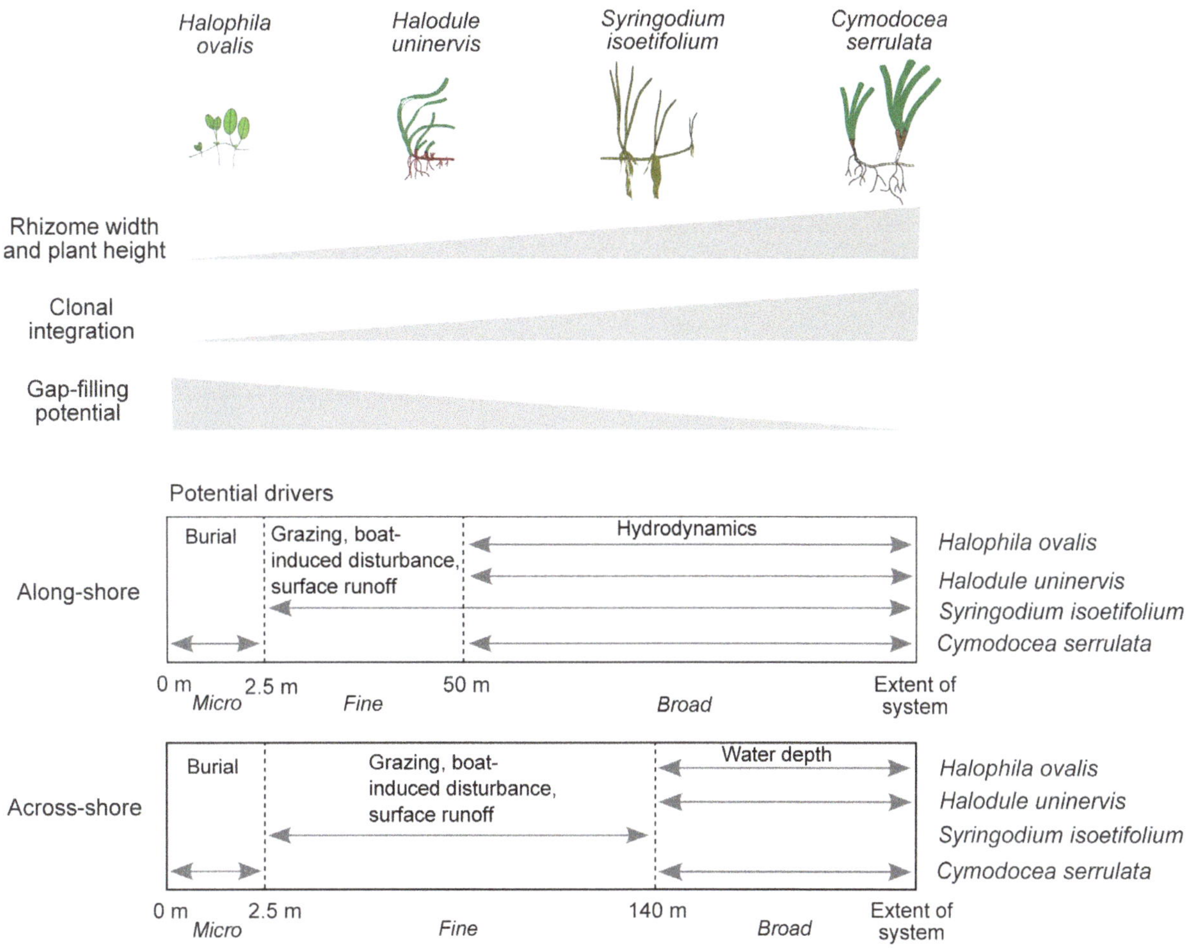

Figure 6. Relative characteristics of the four seagrass species in Pulau Tinggi, Malaysia, including rhizome width and plant height (species size), degree of clonal integration and gap-filling potential. Potential drivers in the along-shore and across-shore are shown. Arrows denote the distances at which a species is most strongly influenced (>30% variance in the Linear Model of Coregionalization).

resources as well as larger species such as *S. isoetifolium* and *C. serrulata* [40,50]. These are size-specific traits which result in small species having greater sensitivity to environmental changes that occur over the broad spatial scale, such as a reduction in light along a gradient of water depth. Light and seasonal environmental variability have previously been shown to affect resource availability for small species [51].

The conceptual model proposed for seagrass species distribution in Pulau Tinggi is based on their spatial structure (Fig. 6). However, it would be simplistic to assume close associations between pattern and process on the basis of matching scales and variogram structure alone, i.e. different ecological processes may present similar variogram structures or one single ecological process may present many different structures. Therefore, these models are meant merely to provide a framework within which competing hypotheses about underlying drivers may be constructed and further tested.

Conclusions

The forereef seagrass system in Pulau Tinggi exhibits clear spatial structure, in that patterns of species distribution were directional, nested and had multiple structures. We found evidence for size-based rules in the distribution of marine plant communities in response to their underlying drivers, where small, widespread species such as *Halophila* spp and *H. uninervis* were most influenced by broad-scale processes whereas large, localised species such as *S. isoetifolium* and *C. serrulata* were most influenced by finer-scale processes over a wider range of distances.

This study demonstrates the potential of using spatial structure to provide further clues about the underlying drivers of species distribution. Based on directionality and the scales of influence, we proposed a range of potential drivers of seagrass structure in the study area. Hydrodynamics and water depth are potentially the broad-scale drivers that operate at distances of >50 m in the along-shore and >140 m in the across-shore, providing support for the view that these are the prime drivers of meadow extent [1]. Sediment burial under bioturbation mounds is a likely driver at distances of <2.5 m. Gap formation caused by grazing, disturbance from boating activities and surface runoff potentially influences the spatial structure of seagrass species at distances of 2.5–50 m in the along-shore and 2.5–140 m in the across-shore.

Acknowledgments

We thank Ben Piek, Renae Hovey, Affendi Yang Amri and the SEABUDS team for assisting in the field; Rosmadi Fauzi and Wong Ching Lee for administrative support.

Author Contributions

Conceived and designed the experiments: JLSO KPVN GAK. Performed the experiments: JLSO KPVN GAK. Analyzed the data: JLSO KWH. Contributed reagents/materials/analysis tools: JLSO GAK KPVN. Wrote the paper: JLSO KPVN GAK.

References

1. Vermaat JE (2009) Linking clonal growth patterns and ecophysiology allows the prediction of meadow-scale dynamics of seagrass beds. Perspectives in Plant Ecology, Evolution and Systematics 11: 137–155.
2. Wagner HH (2003) Spatial covariance in plant communities: Integrating ordination, geostatistics, and variance testing. Ecology 84: 1045–1057.
3. Borcard D, Legendre P, Drapeau P (1992) Partialling out the spatial component of ecological variation. Ecology 73: 1045–1055.
4. Vermaat JE, Agawin NSR, Duarte CM, Fortes MD, Marba N, et al. (1995) Meadow maintenance, growth and productivity of a mixed Philippine seagrass bed. Marine Ecology-Progress Series 124: 215–225.
5. Duarte CM (1991) Allometric scaling of seagrass form and productivity. Marine Ecology Progress Series 77: 289–300.
6. Rasheed MA (2004) Recovery and succession in a multi-species tropical seagrass meadow following experimental disturbance: the role of sexual and asexual reproduction. Journal of Experimental Marine Biology and Ecology 310: 13–45.
7. Lacap CDA, Vermaat JE, Rollon RN, Nacorda HM (2002) Propagule dispersal of the SE Asian seagrasses Enhalus acoroides and Thalassia hemprichii. Marine Ecology-Progress Series 235: 75–80.
8. Duarte CM, Terrados J, Agawin NSR, Fortes MD (2000) An experimental test of the occurrence of competitive interactions among SE Asian seagrasses. Marine Ecology-Progress Series 197: 231–240.
9. Birch WR, Birch M (1984) Succession and pattern of tropical intertidal seagrasses in Cockle Bay, Queensland, Australia: A decade of observations. Aquatic Botany 19: 343–367.
10. Duarte CM (1991) Seagrass depth limits. Aquatic Botany 40: 363–377.
11. Ralph PJ, Durako MJ, Enríquez S, Collier CJ, Doblin MA (2007) Impact of light limitation on seagrasses. Journal of Experimental Marine Biology and Ecology 350: 176–193.
12. Fonseca MS, Kenworthy WJ (1987) Effects of current on photosynthesis and distribution of seagrasses. Aquatic Botany 27: 59–78.
13. Fonseca M, Whitfield PE, Kelly NM, Bell SS (2002) Modeling seagrass landscape pattern and associated ecological attributes. Ecological Applications 12: 218–237.
14. Fonseca MS, Bell S (1998) Influence of physical setting on seagrass landscapes near Beaufort, North Carolina, USA. Marine Ecology-Progress Series 171: 109–121.
15. Legendre P (1993) Spatial autocorrelation - trouble or new paradigm? Ecology 74: 1659–1673.
16. Kendrick GA, Holmes KW, Van Niel KP (2008) Multi-scale spatial patterns of three seagrass species with different growth dynamics. Ecography 31: 191–200.
17. Kendrick GA, Duarte CM, Marbà N (2005) Clonality in seagrasses, emergent properties and seagrass landscapes. Marine Ecology-Progress Series 290: 291–296.
18. Bostrom C, Pittman SJ, Simenstad C, Kneib RT (2011) Seascape ecology of coastal biogenic habitats: advances, gaps, and challenges. Marine Ecology-Progress Series 427: 191–217.
19. Holmes KW, Van Niel KP, Kendrick GA, Radford B (2007) Probabilistic large-area mapping of seagrass species distributions. Aquat Conserv: Mar Fresh-water Ecosyst 17: 385–407.
20. Miller CJ, Campbell SJ, Scudds S (2005) Spatial variation of Zostera tasmanica morphology and structure across an environmental gradient. Marine Ecology-Progress Series 304: 45–53.
21. Fonseca MS, Kenworthy WJ, Griffith E, Hall MO, Finkbeiner M, et al. (2008) Factors influencing landscape pattern of the seagrass Halophila decipiens in an oceanic setting. Estuarine, Coastal and Shelf Science 76: 163–174.
22. Alcoverro T, Duarte CM, Romero J (1995) Annual growth dynamics of *Posidonia oceanica*: contribution of large scale versus local factors to seasonality. Marine Ecology Progress Series 120: 203–210.
23. Balestri E, Cinelli F, Lardicci C (2003) Spatial variation in *Posidonia oceanica* structural, morphological and dynamic features in a northwestern Mediterranean coastal area: a multi-scale analysis. Marine Ecology Progress Series 250: 51–60.
24. Borg JJA, Attrill MMJ, Rowden AAA, Schembri PPJ, Jones MMB (2005) Architectural characteristics of two bed types of the seagrass Posidonia oceanica over different spatial scales. Estuarine, Coastal and Shelf Science 62: 667.
25. Gobert S, Kyramarios M, Lepoint G, Pergent-Martini C, Bouquegneau J-M (2003) Variations at different spatial scales of *Posidonia oceanica* (L.) Delile beds; effects on the physico-chemical parameters of the sediment. Oceanologica Acta 26: 199–207.
26. Migliaccio M, De Martino F, Silvestre F, Procaccini G (2005) Meadow-scale genetic structure in *Posidonia oceanica*. Marine Ecology Progress Series 304: 55–65.
27. Hämmerli A, Reusch TBH (2003) Genetic neighbourhood of clone structures in eelgrass meadows quantified by spatial autocorrelation of microsatellite markers. Heredity 91: 448–455.
28. Rossi RE, Mulla DJ, Journel AG, Eldon HF (1992) Geostatistical tools for modeling and interpreting ecological spatial dependence. Ecological Monographs 62: 277–314.
29. Ooi JLS, Kendrick GA, Van Niel KP, Affendi YA (2011) Knowledge gaps in tropical Southeast Asian seagrass systems. Estuarine, Coastal and Shelf Science 92: 118–131.
30. Goovaerts P (1998) Geostatistical tools for characterizing the spatial variability of microbiological and physico-chemical soil properties. Biology and Fertility of Soils 27: 315–334.
31. Oliver M, Webster R, Gerrard J (1989) Geostatistics in Physical Geography. Part I: Theory. Transactions of the Institute of British Geographers 14: 259–269.
32. Burgess TM, Webster R (1980) Optimal interpolation and isarithmic mapping of soil properties. 1. The semi-variogram and punctual kriging. Journal of Soil Science 31: 315–331.
33. Chiles JP (1999) Geostatistics: Modeling Spatial Uncertainty. New York: Wiley Inter-Science. 695 p.
34. Goovaerts P (1992) Factorial kriging analysis - a useful tool for exploring the structure of multivariate spatial soil information. Journal of Soil Science 43: 597–619.
35. Holmes KW, Kyriakidis PC, Chadwick OA, Soares JV, Roberts DA (2005) Multi-scale variability in tropical soil nutrients following land-cover change. Biogeochemistry 74: 173–203.
36. Dobermann A, Goovaerts P, George T (1995) Sources of soil variation in an acid ultisol of the Philippines. Geoderma 68: 173–191.
37. Ooi JLS, Kendrick GA, Van Niel KP (2011) Effects of sediment burial on tropical ruderal seagrasses are moderated by clonal integration. Continental Shelf Research 31: 1945–1954.
38. Terrados J, Duarte CM, Fortes MD, Borum J, Agawin NSR, et al. (1998) Changes in community structure and biomass of seagrass communities along gradients of siltation in SE Asia. Estuarine Coastal and Shelf Science 46: 757–768.
39. Nakaoka M, Iizumi H (2000) Magnitude of within-patch variation in seagrass Halophila ovalis growth affected by adjacent Thalassia hemprichii vegetation. Ecological Research 15: 415–424.
40. Marbá N, Hemminga MA, Duarte CM (2006) Resource translocation within seagrass clones: allometric scaling to plant size and productivity. Oecologia 150: 362–372.
41. Japar Sidik B, Muta Harah Z, Aziz A (2006) Distribution and significance of seagrass ecosystems in Malaysia. Aquatic Ecosystem Health & Management 9: 1–12.
42. Nakaoka M (2005) Plant-animal interactions in seagrass beds: ongoing and future challenges for understanding population and community dynamics. Population Ecology 47: 167–177.
43. Preen A (1995) Impacts of dugong foraging on seagrass habitats - observational and experimental evidence for cultivation grazing Marine Ecology-Progress Series 124: 201–213.
44. Meadows PS, Meadows A, Murray JMH (2012) Biological modifiers of marine benthic seascapes: Their role as ecosystem engineers. Geomorphology 157: 31–48.
45. Duarte CM, Terrados J, Agawin NSR, Fortes MD, Bach S, et al. (1997) Response of a mixed Philippine seagrass meadow to experimental burial. Marine Ecology-Progress Series 147: 285–294.
46. Nacorda HM (2008) Burrowing shrimps and seagrass dynamics in shallow-water meadows off Bolinao (NW Philippines). Leiden, the Netherlands: CRC Press/ Balkema. 100 p.
47. Marbà N, Duarte CM (1998) Rhizome elongation and seagrass clonal growth. Marine Ecology-Progress Series 174: 269–280.
48. McBratney AB, Webster R (1986) Choosing functions for semi-variograms of soil properties and fitting them to sampling estimates. Journal of Soil Science 37: 617–639.
49. Dennison WC (1987) Effects of light on seagrass photosynthesis, growth and depth distribution. Aquatic Botany 27: 15–26.
50. Marbá N, Hemminga MA, Mateo MA, Duarte CM, Mass YEM, et al. (2002) Carbon and nitrogen translocation between seagrass ramets. Marine Ecology-Progress Series 226: 287–300.
51. Marbá N, Cebrián J, Enríquez S, Duarte CM (1996) Growth patterns of western Mediterranean seagrasses: species-specific responses to seasonal forcing. Marine Ecology Progress Series 133: 203–215.

20

Diving Behavior of the Reef Manta Ray Links Coral Reefs with Adjacent Deep Pelagic Habitats

Camrin D. Braun[1,3]*, Gregory B. Skomal[2], Simon R. Thorrold[3], Michael L. Berumen[1]

1 Red Sea Research Center, King Abdullah University of Science and Technology Thuwal, Kingdom of Saudi Arabia, **2** Massachusetts Division of Marine Fisheries, New Bedford, Massachusetts, United States of America, **3** Biology Department, Woods Hole Oceanographic Institution, Woods Hole, Massachusetts, United States of America

Abstract

Recent successful efforts to increase protection for manta rays has highlighted the lack of basic ecological information, including vertical and horizontal movement patterns, available for these species. We deployed pop-up satellite archival transmitting tags on nine reef manta rays, *Manta alfredi*, to determine diving behaviors and vertical habitat use. Transmitted and archived data were obtained from seven tagged mantas over deployment periods of 102–188 days, including three recovered tags containing 2.6 million depth, temperature, and light level data points collected every 10 or 15 seconds. Mantas frequented the upper 10 m during daylight hours and tended to occupy deeper water throughout the night. Six of the seven individuals performed a cumulative 76 deep dives (>150 m) with one individual reaching 432 m, extending the known depth range of this coastal, reef-oriented species and confirming its role as an ecological link between epipelagic and mesopelagic habitats. Mean vertical velocities calculated from high-resolution dive data (62 dives >150 m) from three individuals suggested that mantas may use gliding behavior during travel and that this behavior may prove more efficient than continuous horizontal swimming. The behaviors in this study indicate manta rays provide a previously unknown link between the epi- and mesopelagic layers of an extremely oligotrophic marine environment and provide evidence of a third marine species that utilizes gliding to maximize movement efficiency.

Editor: Yan Ropert-Coudert, Institut Pluridisciplinaire Hubert Curien, France

Funding: Financial support was provided in part by KAUST baseline research funds (to MLB), KAUST award numbers USA00002 and KSA 00011 (to SRT), and the U.S. National Science Foundation (OCE 0825148 to SRT and GBS). The funders had no role in study design, data collection and analysis, decision to publish, or preparation of the manuscript.

Competing Interests: The authors have declared that no competing interests exist.

* E-mail: cbraun@whoi.edu

Introduction

The recent Appendix II listing by CITES of the genus *Manta* has focused attention on this enigmatic group of large pelagic rays and the human threats that have led to their vulnerable status. Both manta ray species have life history characteristics that make populations particularly vulnerable to directed and by-catch fishing [1]. Adequate conservation actions for mantas require data on horizontal and vertical movements as population connectivity plays an important role in determining the spatial scale of significant human threats [2]. While researchers had traditionally lacked the ability to track large pelagic animals *in situ*, the development of electronic tag technology has provided a wealth of information on the movements of these animals [3]. These studies have revealed a remarkable array of behaviors from ocean basin migrations [4] to individual fish dive profiles as deep as 2,000 m [5]. Yet, despite these efforts we remain remarkably ignorant of the movement patterns in many pelagic fishes, perhaps none more so than the myliobatoid (devil) rays.

The reef manta ray (*Manta alfredi*) is a small, coastal mobulid (Family: Mobulidae) that exhibits typical K-selected life history traits including slow maturation rate, small, infrequent litters, and long lifespan [1]. Sighting records suggest preferential occupation of nearshore tropical waters with strong site affinity and limited movements [6], although horizontal excursions >500 km have been documented [7]. Movements of *M. alfredi* coincide with predictable manta aggregations at several known locations in tropical and subtropical waters [1] associated with mating, cleaning, food availability, and currents [7–10].

Few aspects of manta biology and ecology have been adequately described, yet recent efforts have dramatically increased our knowledge about this enigmatic group [1]. Some information on horizontal movements of reef manta rays is now available [7,8,11,12], but vertical habitat use remains largely undescribed and unstudied. An understanding of vertical movements is important for understanding reef manta ecology, including site occupation, feeding and mating behavior, habitat use, and potential interactions with humans. *Manta alfredi* was recently designated "Vulnerable to Extinction" by the IUCN and is listed on Appendix II of the Convention on International Trade in Endangered Species and Appendices I and II of the Convention on Migratory Species.

In the Red Sea, schools of *Manta* individuals are listed as significant features of two marine protected areas in Sudan [13], yet information on *Manta* ecology and behavior is lacking [14,15]. Ecological information is therefore necessary for developing adequate management and conservation strategies as well as ensuring that existing protective measures are effective.

Satellite telemetry has been used to study the ecology of a diverse group of marine vertebrates [3,16,17]. Recent advances in this technology have enabled longer term tracking over significantly larger spatial scales than previous techniques. For instance, the use of pop-up satellite archival transmitting (PSAT) tags has facilitated the investigation of diving behavior in basking sharks by archiving high-resolution depth records for up to a year [16]. Studies on the vertical movements of *M. alfredi* have, however, been limited to active acoustic tracks, which provided high-resolution depth data over short timescales (<103 hours continuously) [11,12,18].

Our study provides the first use of satellite telemetry techniques to document broad-scale vertical behavior of *Manta alfredi*. Here, we seek to characterize temporal trends, habitat use, and diving in relation to manta ecology.

Methods

Nine *M. alfredi* were tagged with PSAT tags off the coast of Al Lith, Saudi Arabia in the south-central Red Sea (Fig. 1) during spring 2011 and 2012 (5 miniPAT, 4 MK10-AF; Wildlife Computers Inc., USA) (Table 1). Tags were tethered to a stainless steel dart with small diameter cable and applied by a freediver with a sling spear into the dorsal musculature close to where the pectoral muscles attach to the body. This research was carried out under the general auspices of King Abdullah University of Science and Technology's (KAUST) arrangements for marine research with the Saudi Arabian Coast Guard and the Presidency of Meteorology and Environment. The animal use protocol was performed in accordance with the Woods Hole Oceanographic Institution's Animal Care and Use Committee (IACUC) protocol #16518 and approved by KAUST's Biosafety and Ethics Committee.

Tags archived light level, depth, and water temperature every 10 and 15 seconds for the MK10-AF (MK10) and miniPAT (mP), respectively. Data were processed onboard the tag to summarize information for transmission via the Argos satellite system. Depth and temperature data were summarized every 24 (mP) and 12 (MK10) hours into 14 (mP) or 16 (MK10) bins. High-resolution time series depth data were recorded continuously (every 450 sec - mP) or cycled with one week on and two weeks off (every 75 sec - MK10) for the duration of tag deployment. At pre-programmed deployment durations of 100 (MK10) or 180 (mP) days, tags automatically popped off the animal using a corrosive burn wire. After the tags released and floated to the surface, packets of summarized data were transmitted to Argos satellites until battery failure.

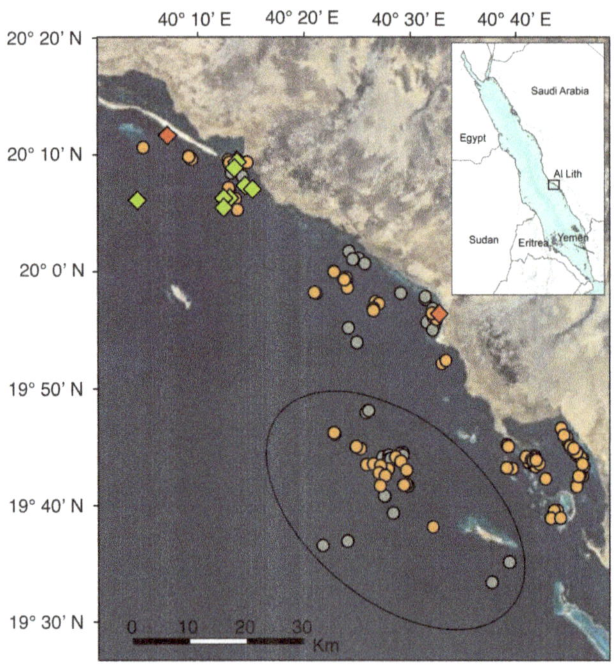

Figure 1. *Manta alfredi* tagging locations and habitat characterization. Locations for MA111 (orange) and MA112 (grey-green). Black oval indicates the locations characterized as "offshore" (>15 km from the coast). Green triangles indicate tagging locations for all individuals and red diamonds indicate pop-off locations for MA111 and MA112. Inset: Location of study site near Al Lith in the eastern Red Sea. Sources: Bing, U.S. Geological Survey, and ESRI.

Transmitted data were decoded with manufacturer software (WC-DAP 3.0, Wildlife Computers, Inc., Redmond, WA), and all subsequent analyses were conducted in the R Statistical Environment [19]. Values are expressed as mean ± SD unless stated otherwise. Dive data were stratified into day (6 am–6 pm) and night (6 pm–6 am) periods. Diel differences in depth distribution were assessed with a Two-sample Kolmogorov-Smirnov (K-S) test, and a Student's t-test was used to compare mean day and nighttime depths. Linear least-squares regression was used to assess correlations between mean nightly depth and moon phase for each individual and a pooled dataset of all individuals. Lunar illumination data were obtained from the United States Naval Observatory (http://aa.usno.navy.mil/data/docs/MoonFraction.php, accessed Apr 2013) and arcsine-transformed before statistical analysis [20].

We recovered two MK10-AF tags that used GPS technology to acquire accurate positions of tagged animals when at the surface for sufficient time to log a position fix [21]. The GPS tracks from these two individuals were combined with high-resolution diving records to examine differences in dive behavior between coastal and offshore habitats. The two habitat characterizations were delineated by distance from shore (>15 km indicated offshore), and dive data was only considered for inclusion in either category if a GPS location was available for that day. We calculated correlations between mean nightly depth and moon fraction, vertical habitat occupation, and diel diving trends using the methods previously described.

For those tags recovered after deployment, further analysis of raw archived data included the calculation and assessment of overall dive velocities, total vertical distance travelled, and rhythmic and deep diving behaviors. Deep dives were defined as those in excess of 150 m originating or ending at depths <15 m. Ascent and descent velocities ($m.s^{-1}$) were calculated from archived datasets (tags MA106, MA111, and MA112) as depth change over the 10 (MK10) or 15 (mP) second interval recorded by the PSAT tag. Total vertical distance traveled for each diel cycle was computed as the sum of depth changes per day and night period. Dive series were examined for rhythmic elements using autocorrelation functions in R.

Results

Of the 9 mantas tagged, PSAT deployments ranged from 102–188 days. Four PSATs transmitted summarized data representing 5.3–54.7% of the deployment durations; two MK10-AF tags did not report, and three tags were physically retrieved. The latter tags yielded >2.6 million archived records each comprising depth, temperature, and light level records collected every 10 (MK10) or 15 (mP) seconds over the duration of tag deployment (Table 1).

Daily vertical habitat use by the tagged mantas indicated nearly constant occupation of the upper 50 m of the water column (Figs. 2,3) with sporadic use of deeper water down to 432 m

Table 1. Satellite tagging summary for nine reef manta rays (*Manta alfredi*) from the Saudi Arabian Red Sea.

Manta ID	Disc Width (cm)	Tag Type	Tag Date	Tag Lat (°N)	Tag Long (°E)	Deployment Duration (days)	Popup Latitude (°N)	Popup Longitude (°E)	Track Distance (km)
MA102		miniPAT	2-Apr-11	20.129	40.217	188	20.026	40.416	28
MA103	200	miniPAT	20-Apr-11	20.129	40.217	133	18.999	41.145	169
MA104		miniPAT	13-Apr-11	20.129	40.217	178	20.015	40.439	30
MA105	200	miniPAT	17-Apr-11	20.129	40.217	172	20.023	40.431	22.5
MA106*	250	miniPAT	16-Apr-11	20.129	40.217	155	20.166	40.200	4
MA111*	200	MK10-AF	21-Apr-11	20.154	40.229	102	20.158	40.158	15
MA112*	250	MK10-AF	22-Apr-11	20.136	40.237	102	19.928	40.549	45
MA204	180	MK10-AF	25-Apr-12	20.1542	40.2310	DNR			
MA205	200	MK10-AF	25-Apr-12	20.1558	40.2308	DNR			

Track distance refers to straight-line distance from tagging to popup locations. Two satellite tags did not report (DNR). *indicates tag was physically recovered.

(Fig. 2c). Mantas also exhibited diel vertical behavior during which the daytime depth distribution was skewed toward the upper layers and differed significantly from the nocturnal occupation of deeper waters (K-S Test $p = 0.000$; Fig. S1). In addition, mean daytime depths for each individual (Table 2) and all individuals combined (21.7±23.4 m) were significantly shallower than nighttime depths (34.2±21.0 m; t-test $p = 0.000$). Finally, regression analyses indicated that mean nightly depth was unrelated to lunar illumination for 6 of 7 individuals (Table 2, Fig. S2) and for the aggregate nightly depth data ($R^2 = 0.000$, Fig. S2). Only one individual exhibited a significant but weak relationship between mean nighttime depth and lunar illumination (MA007, $p<0.03$, $R^2 = 0.047$).

Three recovered tags facilitated further analysis of high-resolution data sets. Vertical behaviors were dominated by slow ascents and descents (<0.25 $m.s^{-1}$; Fig. 4a,b) with maximum rates of 1.87 and 2.27 $m.s^{-1}$, respectively. Despite slow depth changes, these fish exhibited highly variable cumulative daily vertical movements that ranged from 2870–10,300 m. One manta (MA111) exhibited significantly more vertical movement during the day than the night (mean total daytime movement 2804±725 m and night 2528±589 m, $p = 0.003$) while the other two exhibited no diel differences. In addition, no rhythmic components were found using autocorrelation analyses.

We recorded 14 partial deep dive profiles from transmitted tag data and 62 complete deep dives (>150 m) from archival depth records that were distributed randomly throughout a daily cycle. Most descent rates derived from the cumulative deep dive record were between 0–0.10 $m.s^{-1}$ while the ascent rate frequency distribution was skewed toward slightly faster depth changes (Fig. 4c,d). Isolation of a single series of consecutive dives highlighted the higher ascent rates compared to descent rates (Fig. 5).

Finally, a comparison of coastal and offshore diving behavior demonstrated marked differences between habitats (Fig. 1). A total of 65 days were classified as coastal based on GPS locations (49

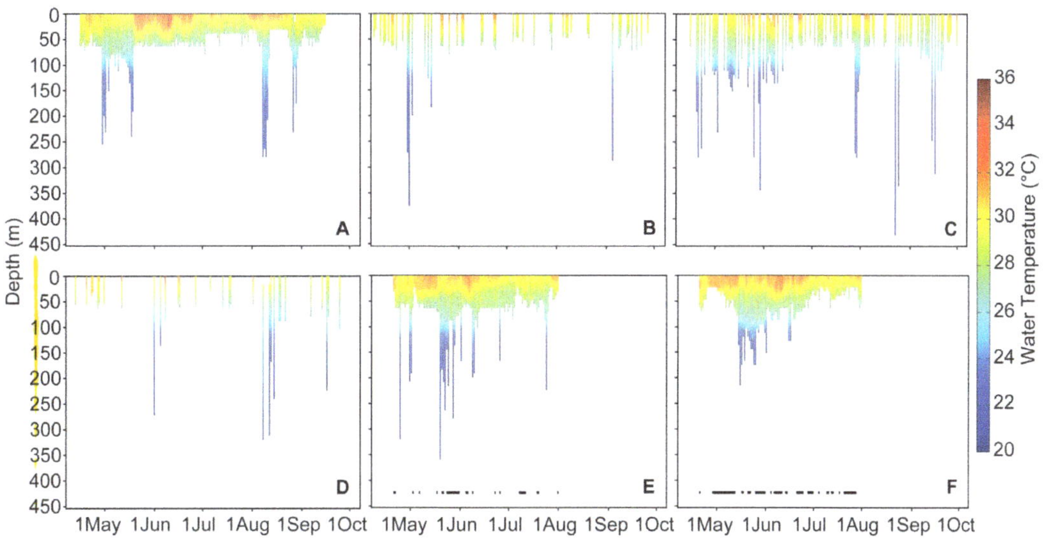

Figure 2. Daily ambient water temperature and depth. Daily ambient water temperature (indicated by color) and depth (y-axis) experienced by *Manta alfredi* for the duration of satellite tag deployment in 2011 near Al Lith, Saudi Arabia. (A) Manta identification number MA106 (B) MA102 (C) MA105 (D) MA104 (E) MA112 (F) MA111. Black points at the bottom of panels (E) and (F) indicate days for which the satellite tags resolved FastLoc GPS locations.

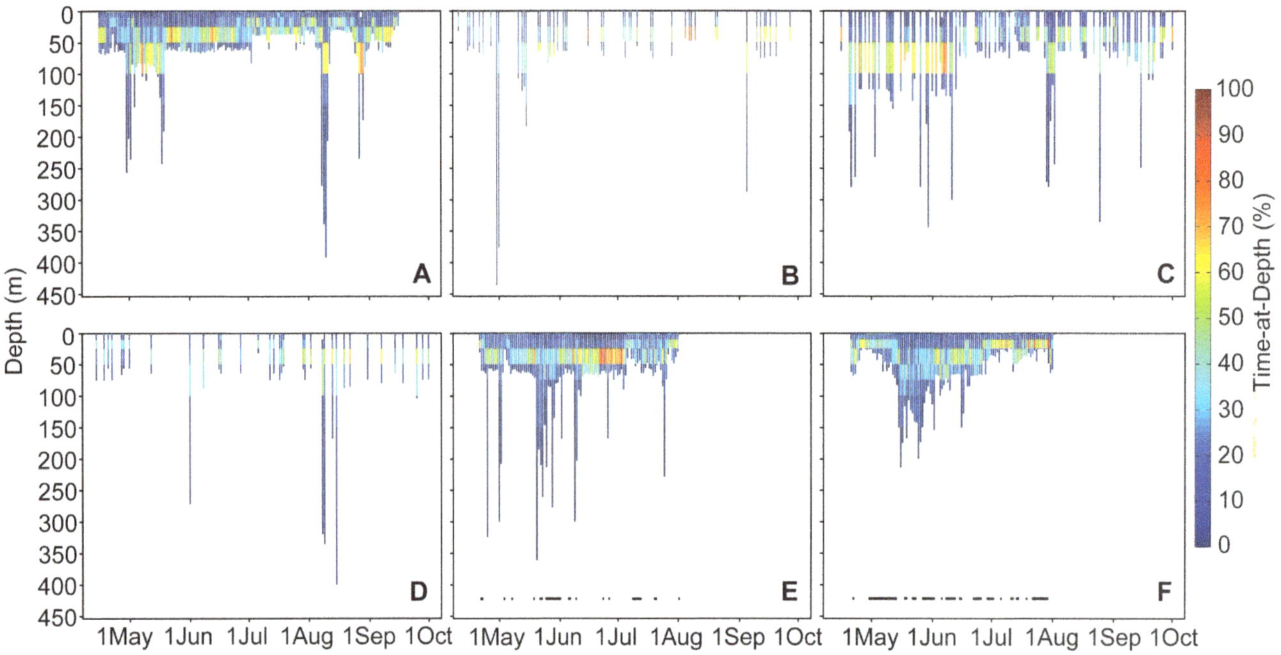

Figure 3. Percent daily time at depth. Daily time-at-depth (indicated by color across y-axis) by *Manta alfredi* for the duration of satellite tag deployment in 2011 near Al Lith, KSA. (A) Manta identification number MA106 (B) MA102 (C) MA105 (D) MA104 (E) MA112 (F) MA111. Black points at the bottom of panels (E) and (F) indicate days for which the satellite tags resolved FastLoc GPS locations.

and 16 days for MA111 and MA112, respectively) while 25 days were considered offshore (12 and 13 days, respectively). Vertical use of the water column while in coastal habitats indicated similar depth preferences as those demonstrated when mantas were offshore but was skewed shallower, presumably as an artifact of bathymetric constraints (Fig. 6). However, regression of moon fraction (arcsine-transformed, as above) with mean nightly depth indicated a strong positive relationship between moon fraction and nighttime depth at offshore locations ($R^2 = 0.50$, $p<0.001$), while a significant but weaker relationship was apparent in coastal environments ($R^2 = 0.098$, $p = 0.01$, Fig. 7). Diving behavior over the diel cycle provided further contrast among coastal and offshore sites (Fig. 8). Coastal diving was indicative of reverse diel vertical migration (rDVM) in which these individuals spent much of their time at depth overnight that was punctuated by a rise to and descent from the surface at dawn and dusk, respectively. Conversely, when mantas moved offshore they exhibited a similar rDVM but with a period of deeper occupation during midday that was not evident in the coastal environment.

Table 2. Statistical results for vertical habitat use and diving behaviors of seven PSAT-tagged reef manta rays in the Saudi Arabian Red Sea.

Manta ID	Day Depth (m)		Night Depth (m)			Max Velocity ($m.s^{-1}$)	
	Mean±SE	Max	Mean±SE	Max	Lunar (R)	Descent	Ascent
MA102	24.3±0.48	341	36.0±0.37	434	0.003	-	-
MA103	23.7±0.65	59	35.0±0.61	61	0.002	-	-
MA104	23.1±0.50	363	33.9±0.37	260	0.081	-	-
MA105	35.1±0.36	312	47.6±0.30	247	0.000	-	-
MA106	21.2±0.04	392	35.3±0.03	318	0.000	1.77	1.53
MA111	20.0±0.03	214	32.7±0.03	175	0.047^{Ψ}	2.27	1.87
MA112	23.5±0.04	362	34.7±0.03	326	0.000	1.37	1.83

All individuals exhibited a daytime depth distribution that was skewed toward upper layers compared to nocturnal depth occupation (p = 0.000). Lunar (R) values indicate the correlation coefficient for a linear regression between mean nightly depth and fraction of moon illuminated.
$^{\Psi}$indicates p<0.05.

Discussion

Reef mantas exhibited frequent use of the upper epipelagic layer (<60 m) of the Red Sea throughout the diel cycle. This is similar to vertical behavior reported by the few existing studies conducted on other mobulid species [11,22,23]. However, in contrast to other mobulids, mantas in this study demonstrated strong rDVM within the upper layer, occupying surface habitats during daytime hours and moving deeper at night. Several active and passive acoustic studies have suggested movement away from reef environments and into deeper water at night. Anderson et al. [9] speculated, based on field observations in the Maldives, that mantas regularly used a deep zooplankton layer at night. During major upwelling events, this layer appears to reach the surface where large groups of mantas were observed feeding. Similarly, manta detections on passive acoustic arrays have been shown to drastically decrease during nocturnal periods, leading the researchers to draw similar nighttime feeding conclusions [8]. In addition, Clark [11] found all actively tracked mantas made similar offshore excursions to deeper waters during the night. Until now, however, this behavior was observed in a few actively tracked animals over a relatively short duration (up to 150 hours) [11] or inferred from passive acoustics [8,12]. Our study quantified diel depth changes over several months, and we conclude that this behavior is likely common in reef mantas throughout their range.

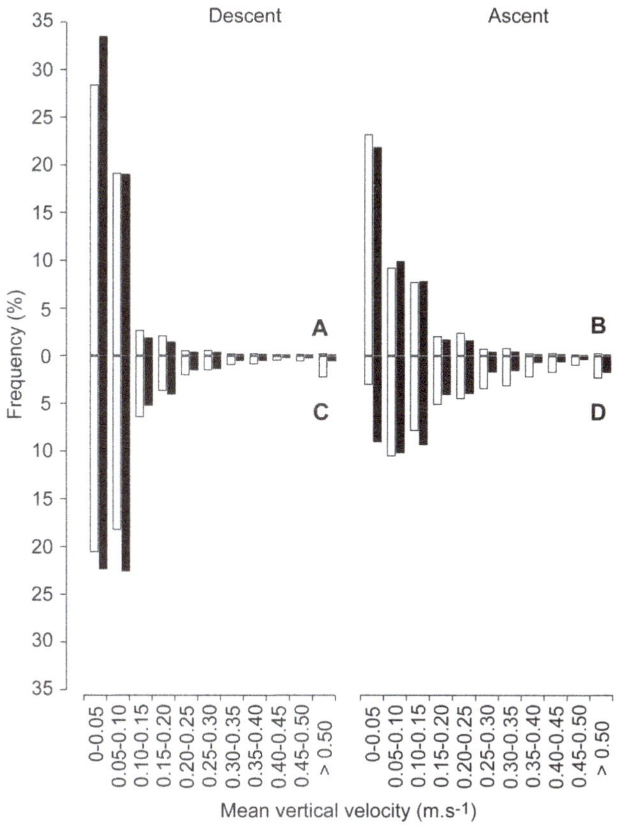

Figure 4. Vertical velocity during normal and deep diving. Average vertical velocity (m.s^{-1}) from three archival satellite tag datasets collected during normal (<150 m) (A) descent and (B) ascent and deep (>150 m) (C) descent and (D) ascent dive behavior of *Manta alfredi*. Zero velocities (no depth change) were dropped from plotting to facilitate visualization of non-zero values. White and black bars indicate day and night, respectively.

Reverse DVM behavior similar to that reported here has also been shown in whale sharks at Ningaloo Reef [24]. However, mantas in this study exhibited no particular affinity for surface waters specifically during dusk periods as reported for whale sharks. Rather, mantas spent most of their time at the surface during all daylight hours before descending around dusk. Previous studies have suggested that nocturnal occupation of deeper water may increase foraging opportunities in other marine species, including filter feeding elasmobranchs [25–28], and may be a strategy for mantas to exploit rising layers of mesopelagic zooplankton that move into the photic zone at night [29,30] or demersal plankton emerging from the reef benthos [30,31]. In the Red Sea, it is clear that significant plankton biomass moves shallower at night from deeper layers in the pelagic environment and emerges from the benthos in reef environments [30,32,33]. Both likely provide valuable food resources in an oligotrophic environment like the Red Sea. Vertical behaviors in this study suggested that mantas occupy both offshore and reef habitats in a way that facilitates exploitation of emergent reef and rising pelagic plankton. This hypothesis is supported by isotopic analysis of reef manta muscle tissue that indicated a strong demersal, reef-derived diet source [34]. Feeding on emerging reef plankton may also explain the origination of unique fatty acids found in manta tissues [35].

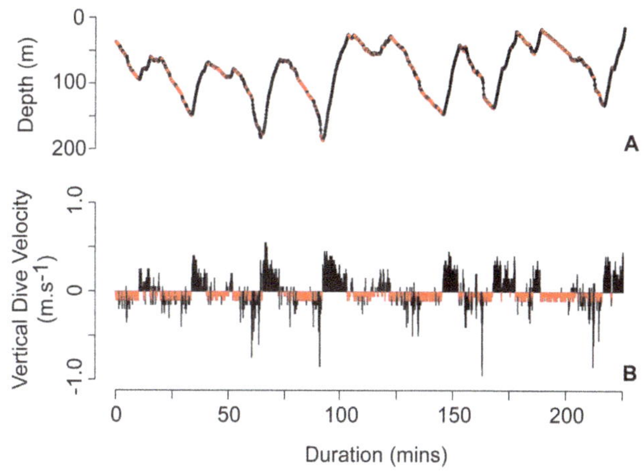

Figure 5. Gliding during deep dives. (A) Consecutive deep diving behavior exhibited by MA006 initiated at 0900 on 22 May 2011 and (B) the corresponding mean vertical velocity histogram. Negative velocities describe descents. Red segments indicate descent velocity between depth points <0.1 m.s^{-1} which we suggest correspond to gliding behavior.

We were able to characterize offshore diving behavior in two mantas based on high-resolution GPS locations. These individuals occupied a single offshore area at the same time in late May 2011. During this time, both individuals exhibited midday occupation of deeper water in contrast to the rDVM shown in coastal habitats. We can only speculate as to why these rays where deeper in the water column during midday hours. Possible explanations include visits to cleaning stations [7], reproductive behavior [36,37], or perhaps most likely the exploitation of an ephemeral food resource [9,10]. More work will be needed to distinguish among these hypotheses.

Although light likely influences the vertical behavior of many large pelagic fishes, including billfishes [38,39] and sharks [40], the

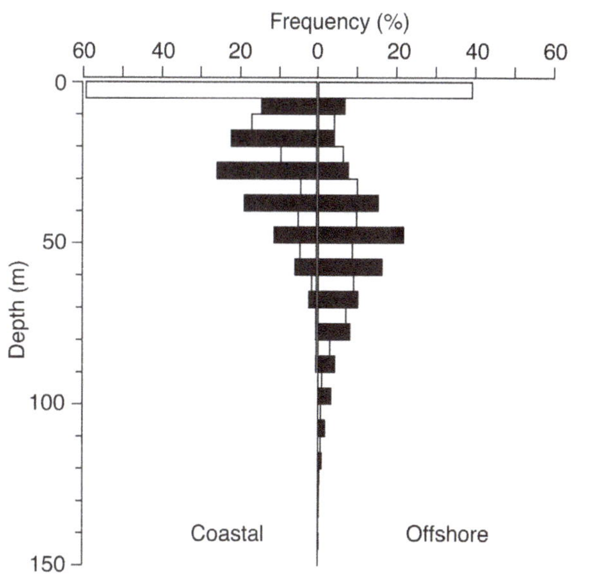

Figure 6. Coastal and offshore depth occupation. Frequency histogram comparing coastal (left) and offshore (right) depth occupation over a diel cycle. Bars are in 10 m bins. White and black bars indicate day and night, respectively.

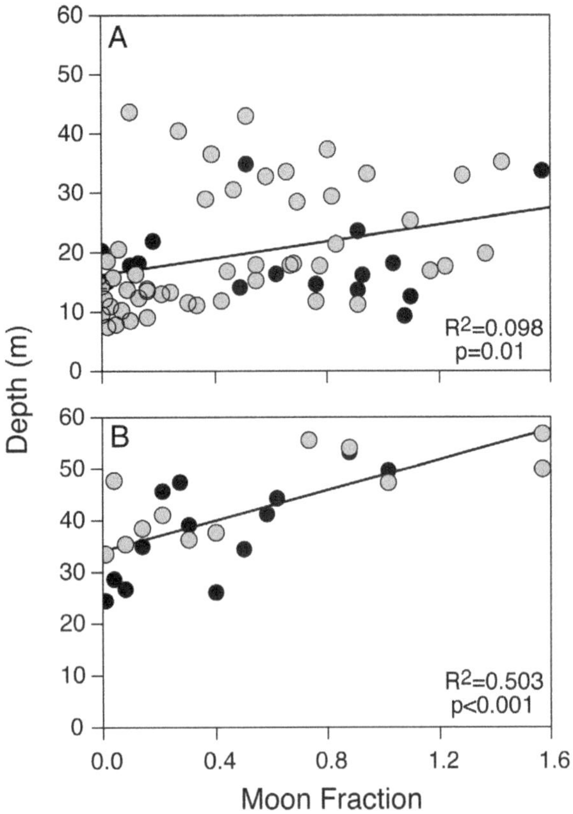

Figure 7. Coastal and offshore lunar regression. Moon fraction regression with mean daily nighttime depth for both GPS-tagged mantas (MA111 = grey, MA112 = black) comparing (A) coastal and (B) offshore nocturnal depth occupation.

direct effects of lunar illumination on mobulids is unclear. To date, the extent to which the lunar cycle influences manta behavior has been inferred from short-term passive acoustic studies [8,10,11]. Working with a more comprehensive dataset, we found that mantas overall did not alter vertical habitat use in association with the lunar cycle. However, during offshore movements, mantas in our study occupied deeper water at night as the moon phases progressed toward full. This behavior is consistent with the vertical movements of pelagic plankton as a strategy to avoid visual predation during increasing surface irradiance around the full moon [30] and indicates mantas are potentially exploiting these planktonic resources while in offshore waters. In contrast, nighttime depths of mantas exploiting coastal, reef-derived plankton as a nocturnal food source are probably unaffected by lunar illumination [30].

Mantas inhabit the full temperature range in the epi- and mesopelagic layers of the Red Sea (range 21.6–34.2°C) with dives >400 m, including a 432 m dive that extends the current depth range for *M. alfredi* by over 100 m [41]. Deep dives occurred relatively infrequently throughout all but one deployment and were characterized by asymmetrical V-shaped dive profiles with extremely short bottom times, relatively slow descents, and faster ascents. Although these dives may allow the rays to exploit the relatively rich zooplankton layers below 200 m in the Red Sea [33], this motivation seems unlikely based on the dive profiles. If the mantas were searching for food, they would presumably be successful at least occasionally, cease diving, and occupy that layer for some period of time. Yet we never observed this leveling out behavior in any of the 76 deep dives performed by 6 individuals in this study. Alternatively, these deep dives may be associated with travelling behavior in which mantas perform gliding dives to conserve energy and maximize movement efficiency. Although our tags didn't record pitch angle and tail-beat frequency data, vertical descent velocities from the three high-resolution profiles were skewed toward slower descents than ascents. The flattened descent profile of a single series of consecutive deep dives that

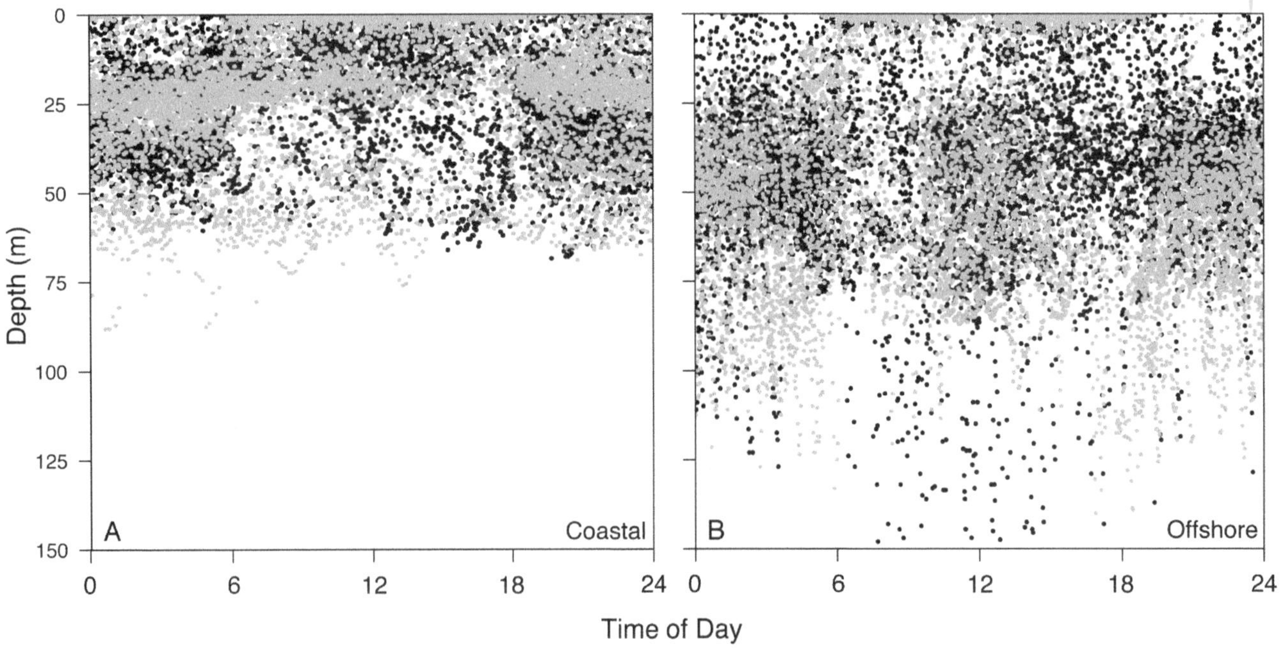

Figure 8. Coastal and offshore diel diving behavior. Combined diel diving behavior for MA111 (grey) and MA112 (black) during coastal (A) and offshore (B) habitat occupation. Note y-axis is truncated in (B) to facilitate comparison.

results from this behavior provided further visual support for the glide hypothesis and demonstrates what appeared to be a largely passive descent followed by active ascent. Interestingly, these descent rates are similar to those recorded for Japanese flounder [42]. In addition, manta ascent and descent velocities exhibited the same relationship as lateral acceleration during ascent and descent in whale sharks with very slow descent rates while consistently reaching 0.10 $m.s^{-1}$ upon ascent [43].

Gliding in fishes involves harvesting kinetic energy from potential that is derived from the fish's negative buoyancy upon descent. This is an efficient strategy for movement in an aqueous medium and may decrease drag threefold compared to active swimming [44]. Exceptional glide performance has been demonstrated in both Japanese flounder (*Paralichthys olivaceus*) [42] and whale sharks (*Rhincodon typus*) [43], and is a prevalent behavior among white sharks [45]. *R. typus* has large, flat pectoral fins that aid in passive descent while *P. olivaceus* exhibits significant dorso-ventral flattening capable of producing substantial lift. Mobulid rays are also dorso-ventrally flattened. Their wing-like pectoral fins can generate significant lift [46], and, therefore, they possess body types ideally suited for maximizing glide efficiency. Morphological traits for efficient movement combined with negative buoyancy make powerless glide diving followed by active ascent a more energetically efficient strategy for traversing horizontally than continuous swimming [42] and may provide significant energetic benefits to mantas during travel. Further insight and confirmation of manta gliding will require the deployment of accelerometers that can quantify three-dimensional movement at high temporal resolution [47].

Manta rays have remained largely enigmatic despite their iconic status with divers around the world. Our tag results extend the habitat of reef mantas well into the mesopelagic zone of the global ocean. Recent PSAT tagging studies have similarly extended the depth ranges of other elasmobranchs through the mesopelagic and even into bathypelagic depths [5,16]. Prey at these depths below the epipelagic zone may provide valuable food resources in oligotrophic environments. This diving behavior may provide a new ecological connection between the surface ocean and the deep sea.

Supporting Information

Figure S1 Diel changes in depth occupation. Distribution of percent time at depth (m) during day (white) and night (black) from aggregate depth records for nine *Manta alfredi* in the south-central Red Sea.

Figure S2 Lunar regression with mean nightly depth. Moon fraction regression with mean daily nighttime depth for individual reef mantas (A) MA106 (B) MA102 (C) MA103 (D) MA105 (E) MA104 (F) MA112 (G) MA111 and H) all mantas tagged with satellite tags in the Saudi Arabian Red Sea.

Acknowledgments

The authors thank J Cochran for useful discussion of the data; Dream Divers for logistical support; J Cochran, M Khalil, P De La Torre, T Sinclair-Taylor, M Priest, and the Reef Ecology Lab at King Abdullah University of Science and Technology (KAUST) for assistance in the field. The manuscript was improved by feedback from M Meekan, B Jones, S Kaartvedt, and Y Ropert-Coudert.

Author Contributions

Conceived and designed the experiments: CDB GBS SRT MLB. Performed the experiments: CDB GBS SRT MLB. Analyzed the data: CDB. Contributed reagents/materials/analysis tools: GBS SRT MLB. Wrote the paper: CDB.

References

1. Couturier LI, Marshall AD, Jaine FR, Kashiwagi T, Pierce SJ, et al. (2012) Biology, ecology and conservation of the Mobulidae. J Fish Biol 80: 1075–1119.
2. Cowen RK, Gawarkiewicz GG, Pineda J, Thorrold SR, Werner FE (2007) Population connectivity in marine systems: an overview. Oceanography 20: 14–21.
3. Block B, Jonsen I, Jorgensen S, Winship A, Shaffer S, et al. (2011) Tracking apex marine predator movements in a dynamic ocean. Nature 475: 86–90.
4. Bonfil R, Meyer M, Scholl MC, Johnson R, O'Brien S, et al. (2005) Transoceanic migration, spatial dynamics, and population linkages of white sharks. Science 310: 100–103.
5. Hueter RE, Tyminski JP, de la Parra R (2013) Horizontal Movements, Migration Patterns, and Population Structure of Whale Sharks in the Gulf of Mexico and Northwestern Caribbean Sea. PLoS One 8.
6. Marshall AD, Compagno LJV, Bennett MB (2009) Redescription of the genus Manta with resurrection of Manta alfredi (Krefft, 1868) (Chondrichthyes; Myliobatoidei; Mobulidae). Zootaxa: 1–28.
7. Couturier LIE, Jaine FRA, Townsend KA, Weeks SJ, Richardson AJ, et al. (2011) Distribution, site affinity and regional movements of the manta ray, Manta alfredi (Krefft, 1868), along the east coast of Australia. Marine and Freshwater Research 62: 628–637.
8. Dewar H, Mous P, Domeier M, Muljadi A, Pet J, et al. (2008) Movements and site fidelity of the giant manta ray, Manta birostris, in the Komodo Marine Park, Indonesia. Marine Biology 155: 121–133.
9. Anderson RC, Adam MS, Goes JI (2011) From monsoons to mantas: seasonal distribution of Manta alfredi in the Maldives. Fisheries Oceanography 20: 104–113.
10. Jaine FRA, Couturier LIE, Weeks SJ, Townsend KA, Bennett MB, et al. (2012) When Giants Turn Up: Sighting Trends, Environmental Influences and Habitat Use of the Manta Ray Manta alfredi at a Coral Reef. PLoS One 7: e46170.
11. Clark TB (2010) Abundance, home range, and movement patterns of manta rays (*Manta alfredi, M. birostris*) in Hawai'i: University of Hawai'i.
12. Deakos MH, Baker JD, Bejder L (2011) Characteristics of a manta ray Manta alfredi population off Maui, Hawaii and implications for management. Marine Ecology Progress Series 420: 245–260.
13. Gladstone W, Krupp F, Younis M (2003) Development and management of a network of marine protected areas in the Red Sea and Gulf of Aden region. Ocean & Coastal Management 46: 741–761.
14. Berumen M, Hoey A, Bass W, Bouwmeester J, Catania D, et al. (2013) The status of coral reef ecology research in the Red Sea. Coral Reefs: 1–12.
15. Spaet JL, Thorrold SR, Berumen ML (2012) A review of elasmobranch research in the Red Sea. Journal of Fish Biology 80: 952–965.
16. Skomal GB, Zeeman SI, Chisholm JH, Summers EL, Walsh HJ, et al. (2009) Transequatorial migrations by basking sharks in the western Atlantic Ocean. Current Biology 19: 1019–1022.
17. Womble JN, Gende SM (2013) Post-Breeding Season Migrations of a Top Predator, the Harbor Seal (Phoca vitulina richardii), from a Marine Protected Area in Alaska. PLoS One 8: e55386.
18. Papastamatiou YP, DeSalles PA, McCauley DJ (2012) Area-restricted searching by manta rays and their response to spatial scale in lagoon habitats. Marine Ecology Progress Series 456: 233–244.
19. R Core Team (2013) R: A Language and Environment for Statistical Computing.
20. Zar JH (1996) Biostatistical Analysis. Upper Saddle River, NJ, USA: Prentice-Hall. 662 p.
21. Moen R, Pastor J, Cohen Y (1997) Accuracy of GPS telemetry collar locations with differential correction. The Journal of Wildlife Management 61: 530–539.
22. Rubin RD, Kumli KR, Chilcott G (2008) Dive characteristics and movement patterns of acoustic and satellite-tagged manta rays (*Manta birostris*) in the Revillagigedos Islands of Mexico. Joint Meeting of Ichthyologists and Herpetologists. Montreal, Canada.
23. Canese S, Cardinali A, Romeo T, Giusti M, Salvati E, et al. (2011) Diving behavior of the giant devil ray in the Mediterranean Sea. Endangered Species Research 14: 171–176.
24. Gleiss AC, Wright S, Liebsch N, Wilson RP, Norman B (2013) Contrasting diel patterns in vertical movement and locomotor activity of whale sharks at Ningaloo Reef. Marine Biology 160: 2981–2992.
25. Nelson DR, McKibben JN, Strong WR, Lowe CG, Sisneros JA, et al. (1997) An acoustic tracking of a megamouth shark, Megachasma pelagios: a crepuscular vertical migrator. Environmental Biology of Fishes 49: 389–399.

26. Graham RT, Roberts CM, Smart JCR (2006) Diving behaviour of whale sharks in relation to a predictable food pulse. Journal of the Royal Society Interface 3: 109–116.
27. Shepard ELC, Ahmed MZ, Southall EJ, Witt MJ, Metcalfe JD, et al. (2006) Diel and tidal rhythms in diving behaviour of pelagic sharks identified by signal processing of archival tagging data. Marine Ecology Progress Series 328: 205–213.
28. Sims DW, Witt MJ, Richardson AJ, Southall EJ, Metcalfe JD (2006) Encounter success of free-ranging marine predator movements across a dynamic prey landscape. Proceedings of the Royal Society B: Biological Sciences 273: 1195–1201.
29. Gliwicz MZ (1986) Predation and the evolution of vertical migration in zooplankton. Nature 320: 746–748.
30. Yahel R, Yahel G, Berman T, Jaffe JS, Genin A (2005) Diel pattern with abrupt crepuscular changes of zooplankton over a coral reef. Limnology and Oceanography: 930–944.
31. Osada K (2010) Relationship of zooplankton emergence, manta ray abundance and SCUBA diver usage Kona Hawaii: University of Hawai'i at Hilo.
32. Yahel R, Yahel G, Genin A (2005) Near-bottom depletion of zooplankton over coral reefs: I: diurnal dynamics and size distribution. Coral Reefs 24: 75–85.
33. Klevjer TA, Torres DJ, Kaartvedt S (2012) Distribution and diel vertical movements of mesopelagic scattering layers in the Red Sea. Marine Biology 159: 1833–1841.
34. Couturier LI, Rohner CA, Richardson AJ, Marshall AD, Jaine FR, et al. (2013) Stable Isotope and Signature Fatty Acid Analyses Suggest Reef Manta Rays Feed on Demersal Zooplankton. PLoS One 8.
35. Couturier L, Rohner C, Richardson A, Pierce S, Marshall A, et al. (2013) Unusually High Levels of n-6 Polyunsaturated Fatty Acids in Whale Sharks and Reef Manta Rays. Lipids 48: 1029–1034.
36. Marshall A, Bennett M (2010) Reproductive ecology of the reef manta ray Manta alfredi in southern Mozambique. Journal of Fish Biology 77: 169–190.
37. Deakos MH (2012) The reproductive ecology of resident manta rays (Manta alfredi) off Maui, Hawaii, with an emphasis on body size. Environmental Biology of Fishes 94: 443–456.
38. Carey FG, Robison BH (1981) Daily patterns in the activities of swordfish, Xiphias gladius, observed by acoustic telemetry. Fishery Bulletin 79: 277–292.
39. Abascal FJ, Mejuto J, Quintans M, Ramos-Cartelle A (2010) Horizontal and vertical movements of swordfish in the Southeast Pacific. ICES Journal of Marine Science 67: 466–474.
40. Musyl MK, Brill RW, Curran DS, Fragoso NM, McNaughton LM, et al. (2011) Postrelease survival, vertical and horizontal movements, and thermal habitats of five species of pelagic sharks in the central Pacific Ocean. Fishery Bulletin 109: 341–368.
41. Marshall A, Kashiwagi T, Bennett MB, Deakos M, Stevens G, et al. (2011) *Manta alfredi*. IUCN Red List of Threatened Species. Version 2012.2 ed.
42. Takagi T, Kawabe R, Yoshino H, Naito Y (2010) Functional morphology of the flounder allows stable and efficient gliding: an integrated analysis of swimming behaviour. Aquatic Biology 9: 149–153.
43. Gleiss AC, Norman B, Wilson RP (2011) Moved by that sinking feeling: variable diving geometry underlies movement strategies in whale sharks. Functional Ecology 25: 595–607.
44. Weihs D (1973) Mechanically efficient swimming techniques for fish with negative buoyancy. New Haven, CT: Yale University Press. 16 p.
45. Del Raye G, Jorgensen SJ, Krumhansl K, Ezcurra JM, Block BA (2013) Travelling light: white sharks (Carcharodon carcharias) rely on body lipid stores to power ocean-basin scale migration. Proceedings of the Royal Society B: Biological Sciences 280.
46. Fontanella JE, Fish FE, Barchi EI, Campbell-Malone R, Nichols RH, et al. (2013) Two-and three-dimensional geometries of batoids in relation to locomotor mode. Journal of Experimental Marine Biology and Ecology 446: 273–281.
47. Wilson RP, Shepard E, Liebsch N (2008) Prying into the intimate details of animal lives: use of a daily diary on animals. Endangered Species Research 4: 123–137.

Indigenous Knowledge and Science Unite to Reveal Spatial and Temporal Dimensions of Distributional Shift in Wildlife of Conservation Concern

Christina N. Service[1,2,3,4], Megan S. Adams[1,2,4], Kyle A. Artelle[2,4,5], Paul Paquet[1,2], Laura V. Grant[1,3], Chris T. Darimont[1,2,3,4]*

1 Department of Geography, University of Victoria, Victoria, BC, Canada, 2 Raincoast Conservation Foundation, Denny Island, BC, Canada, 3 Spirit Bear Research Foundation, Klemtu, BC, Canada, 4 Hakai Beach Institute, Heriot Bay, BC, Canada, 5 Earth to Ocean Research Group, Department of Biological Sciences, Simon Fraser University, Burnaby, BC, Canada

Abstract

Range shifts among wildlife can occur rapidly and impose cascading ecological, economic, and cultural consequences. However, occurrence data used to define distributional limits derived from scientific approaches are often outdated for wide ranging and elusive species, especially in remote environments. Accordingly, our aim was to amalgamate indigenous and western scientific evidence of grizzly bear (*Ursus arctos horribilis*) records and detail a potential range shift on the central coast of British Columbia, Canada. In addition, we test the hypothesis that data from each method yield similar results, as well as illustrate the complementary nature of this coupled approach. Combining information from traditional and local ecological knowledge (TEK/LEK) interviews with remote camera, genetic, and hunting data revealed that grizzly bears are now present on 10 islands outside their current management boundary. LEK interview data suggested this expansion has accelerated over the last 10 years. Both approaches provided complementary details and primarily affirmed one another: all islands with scientific evidence for occupation had consistent TEK/LEK evidence. Moreover, our complementary methods approach enabled a more spatially and temporally detailed account than either method would have afforded alone. In many cases, knowledge already held by local indigenous people could provide timely and inexpensive data about changing ecological processes. However, verifying the accuracy of scientific and experiential knowledge by pairing sources at the same spatial scale allows for increased confidence and detail. A similarly coupled approach may be useful across taxa in many regions.

Editor: Marco Festa-Bianchet, Université de Sherbrooke, Canada

Funding: The authors are grateful for support from the Habitat Conservation Trust Foundation, Hakai Network, Tides Canada, Tula Foundation, and Wilburforce Foundation. Additionally, the authors are thankful for the in-kind support provided by SEI Industries. CNS, KAA and MSA were supported by the Natural Sciences and Engineering Research Council of Canada [CNS and KAA by NSERC CGSA (Alexander Graham Bell Supplements) and MSA by an NSERC IPS]. CTD acknowledges a NSERC Discovery Grant (435683). Websites: http://www.hctf.ca/; http://hakai.org/; http://tidescanada.org/; http://tula.org/; http://www.wilburforce.org/; http://www.nserc-crsng.gc.ca/; http://www.sei-ind.com/. This support does not alter the authors adherence to PLOS ONE policies on sharing data and materials. The funders had no role in study design, data collection and analysis, decision to publish, or preparation of the manuscript.

Competing Interests: The authors have declared that no competing interests exist.

* Email: darimont@uvic.ca

Introduction

Distributions of organisms are shaped and re-shaped over geological and ecological timescales. Broadly structured by a suite of natural processes and their interactions, changes to species distributions (hereafter 'range shifts'), can be driven by abiotic factors (*e.g.*, CO_2 enrichment, nitrogen deposition, climate; [1]), biotic processes (*e.g.*, competition and facilitation [2]), and dispersal capability [3,4]. In recent history, distributional shifts have often been rapid and associated with human-caused drivers. Causes include climate change, habitat modification, over-exploitation, persecution, introduction of exotic species, and re-introduction of native species (*e.g.*, [5–10]). The increased pace of range shifts caused by humans can impose ecological effects on other species, communities, and ecosystems by exposing recipients to novel predation pressure, competition, and diseases [11]. For example, the rapid, human-aided range expansion of the brown tree snake (*Boiga irregularis*) to Guam decimated native bird populations, which in turn reduced the reproductive success of vertebrate-pollinated native plant species [12].

Given such potential impacts, current and accurate knowledge of species distributions comprises a fundamental and important dimension in conservation management. For example, many policy-relevant processes, such as protected areas design, mapping of critical habitat, and land-use planning require distributional data [13,14]. Moreover, current distribution information can also inform proactive conservation intervention in the face of climate change and other stressors (*e.g.*, [10,15]). More generally, detecting shifts in species' distributions can signal underlying ecological changes within an ecosystem, providing managers with early insight that changes might be occurring in other species and communities.

Identifying contemporary ecological change requires knowledge of the past. A complementary methods approach that combines traditional and local ecological knowledge with conventional scientific methods can provide data that not only offer detailed occurrence data across large areas but also over long time periods [16]. Traditional ecological knowledge (TEK) of indigenous people is transmitted through generations and revolves around a cumulative body of knowledge, practice, and belief surrounding the relationships of living and nonliving beings with their environment and one another [17]. Local ecological knowledge (LEK), often but not exclusively associated with indigenous people, also provides information about ecosystem change, but is gained from observations over lifetimes and not via inter-generational transmission [18]. However, in practice the distinction between TEK and LEK is often imprecise as they may share many similarities [18,19]. In contrast, wildlife science uses a variety of empirical techniques that span differing temporal and spatial resolutions and can provide detailed and quantitative information on populations and individuals [18]. Such data, however, are often very expensive to acquire and temporally and spatially limited. Accordingly TEK/LEK data – potentially spanning decades or longer – can be summarized and analyzed to yield information on elusive species across large areas that are otherwise too expensive or difficult to monitor with conventional scientific tools (*e.g.*, [20,21]). Employing TEK/LEK and wildlife science approaches together might yield more comprehensive and detailed information about changes over time and space than either method alone [22,23]. Importantly, incorporating TEK/LEK into ecological research also can facilitate the engagement of communities [24,25]. Social research components of LEK and TEK necessarily include local people and affirms the importance of their contributions [[24]]; in turn, a collaborative methods approach can be a critical first step in establishing more collaborative management [17,19,25].

Grizzly bear (*Ursus arctos horribilis*) distribution on the remote and now sparsely populated central coast of British Columbia (BC), Canada, provides an ideal system to examine the temporal and spatial components of potential range shifts using a complementary methods approach. The provincial government's current management boundary, delineated as the western (*i.e.*, seaward) extent of the so-called Grizzly Bear Population Units (GBPUs), is the spatial scale at which grizzly bears are managed for hunting, habitat protection, and human-wildlife conflict [26]. Whereas formalized species accounts do not include details of coastal grizzly distribution at an appropriate spatial scale for this study [27], the current management boundary suggests grizzlies are functionally absent from all but five of the dozens of islands in the vast archipelago ([26,28–32]; T. Hamilton, BC Ministry of Environment, pers. comm; Figure 1). In contrast, local people in the area, including several First Nations communities that still rely heavily on subsistence activities and travel across their expansive Territories, now commonly observe grizzlies on islands. If present, subsistence hunters and fishers are likely to sight and remember large-bodied, diurnal and iconic wildlife, like grizzly bears.

Our overarching aim was to combine TEK, LEK, and western scientific methods (human-caused mortality records, non-invasive genetic sampling, and remote camera data) to record grizzlies on islands and detail a potential range shift. Understanding the potential colonization and occupancy of grizzly bears onto islands in the region has significant conservation implications because these animals possess tremendous ecological, cultural, and economic importance in this area and indeed, where they still exist throughout their global range (*e.g.*, [33–36]). More broadly applicable to other areas and taxa, we also test the hypothesis that data from each method yield similar spatial and temporal patterns. Finally, we explore how these approaches reveal complementary spatial and temporal dimensions of data and emerge with broadly applicable conclusions relevant to many systems.

Methods

Study area

The islands and nearby mainland of the central coast of BC (8800 km^2) occur within a nearly road-less and now sparsely populated region extending from its southern boundary of Calvert Island (51.58° N, 127.81° W) north to Princess Royal Island (53.21° N, 128.05° W) (Figures 1a & 1b). A complex matrix of landmasses, the central coast as a whole is composed of mainland valleys divided by extensive fjords and various sized islands (< 1 km^2 to >2200 km^2) separated by tidal waters [37]. Eighteen major islands (*i.e.*, greater than 45 km^2) were included in the study area. The closest major island to the mainland (Yeo) is separated by 230 m of tidal water. The Coastal Western Hemlock biogeoclimatic zone dominates low elevations of the region [38]. Potential foods for bears are well distributed across the coastal mainland and islands and include spring (sedges and forbs), summer (berries), and fall (spawning salmonids) resources [39,40]. Since colonization by Europeans, most people in the area now live in the communities of Bella Bella (Heiltsuk Nation, population ~2200) and Klemtu (Kitasoo/Xai'xais Nation, population ~400). Our study area comprised four islands (of five) now recognized by the current management limit and 14 islands beyond this management range (Figures 1 & 2; Table 1).

Data types

Interview data. Ethics Statement: The interview component of this study was approved by the Human Research Ethics Board at the University of Victoria (Victoria, BC, Canada - Protocol # 12-385), Heiltsuk Integrated Resource Management Department, and Kitasoo/Xai'xais Integrated Resource Authority. All participants provided written informed consent. Field sampling was approved by the Heiltsuk Integrated Resource Management Department, Kitasoo/Xai'xais Integrated Resource Authority, and BC Parks.

We conducted 22 LEK and seven TEK interviews using a "snowball sampling" method [41]. For TEK interviewees, the Kitasoo/Xai'xais Stewardship Department and the Heiltsuk Integrated Resource Management Department recommended initial participants, who hold traditional oral and observational history knowledge, and who in turn suggested additional experts to interview (University of Victoria Human Research Ethics Approval # 12-385) [22,42]. The LEK participant pool included mostly indigenous (19 of 22) candidates, including subsistence and commercial fishers, hunters, eco-tour operators, salmon counters, bear viewing guides, and biologists.

Using adapted general guidelines of TEK/LEK data collection, we guided participants through target questions while also recording additional comments [22]. To assess historical grizzly distributions, comprising occupancy over the past century and beyond, we asked TEK participants about their experience and cultural transmission of knowledge about bears. We asked them to depict on their Traditional Territory map where, based on their Nation's oral histories, they would historically expect to see grizzly bears (hereafter 'Pre 1992'). Given inter-generational transmission of knowledge, we estimate that this TEK 'temporal window' provides information about island occupancy that spans centuries or more.

We used a mapping approach to complement interview data. Specifically, LEK participants indicated on a map the islands on which they had observed a grizzly bear. For each observation,

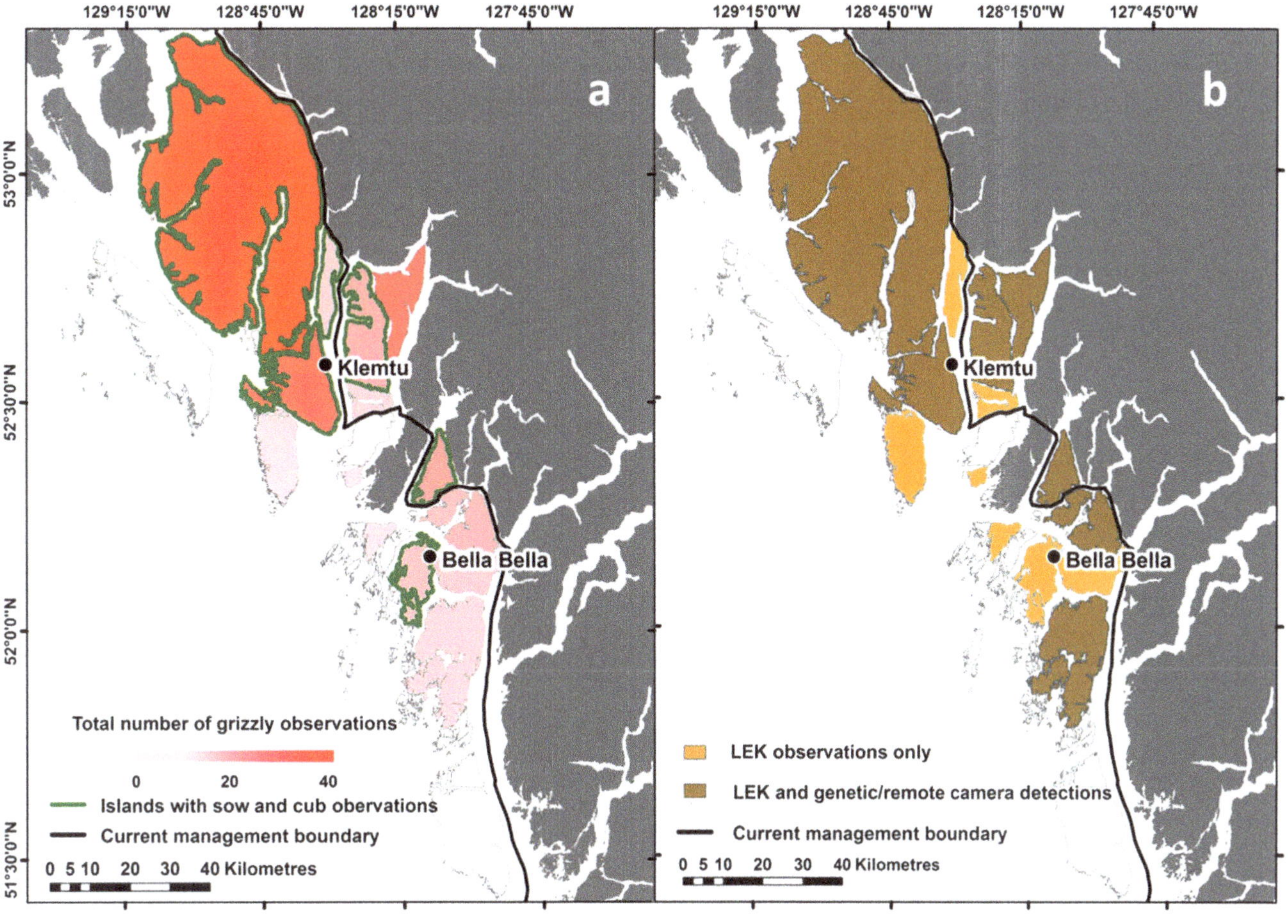

Figure 1. Spatial pattern of evidence of island occupancy and detection type. a) Weight of evidence of grizzly bear (*Ursus arctos horribilis*) occupancy at the island scale within Heiltsuk and Kitasoo/Xai'xais Territories in coastal British Columbia, Canada. Each data type (local ecological knowledge observation, mortality record, genetic ID and remote camera observation) was weighted equally to provide an indication of occupancy rather than bear density. Dark grey areas were not included in our study. The 'current management boundary' is the westward (i.e., seaward extent) of the Province of British Columbia's Grizzly Bear Population Units, the spatial scale at which grizzly bears are managed in the province. Spatial pattern in data types used to detect grizzly bears (*Ursus arctos horribilis*) in sampled areas within Heiltsuk and Kitasoo/Xai'xais Territories. Eighteen islands were included in the study.

participants provided supporting information including the island name, year, and season. We also asked LEK participants to draw on a map the areas they would define as coastal grizzly bear distribution during two timespans (1992–2002, 2003–2012). In addition, an estimate of survey effort for each LEK participant was quantified spatially – by indicating the extent of their area of expertise on a map – and temporally – as the estimated number of years and days per year the participant was in the field and potentially able to observe bears. All TEK/LEK data were verified with interview participants through follow up workshops. Once verified, interview data in transcript form were returned to both the participants and the Nations' Resource Stewardship offices in digital and print versions.

Genetic, remote camera, and mortality data. We used genetic data, remote camera images, and mortality records to identify location, date, and – in some cases sex, individual identity, and age class – of grizzly bears detected in the same area from 2009–2012. We sampled genetic data from non-invasive hair-snagging stations baited with a non-reward bait [43,44]. These stations (n = 33) were part of longitudinal carnivore monitoring programs across 10 islands in Kitasoo/Xai'xais (2012) and Heiltsuk (2009–2012) Territories (Table 1 & 2). In addition, 18 remote trail cameras (n = 1268 trap nights) were deployed on four islands in 2012 (Table 1 & 2). Finally, we queried island locations within the BC Ministry of Environment's kill records from their Compulsory Inspection Database [39,40], yielding dates, locations, and sexes of known human-killed grizzly bears. All island mortality records (n = 3) were hunted individuals rather than animal control (i.e., human-wildlife conflict) kills (Table 1 & 2).

Analyses

Occupancy. We summed the number of grizzly bear observations from all data sources on each island to assess the weight of evidence for contemporary occupancy. We accounted for observations of (genetically) undocumented bears from sightings and camera images equally with known, specific individuals from mortality and genetic data. Accordingly, it is possible that multiple observations were from the same individual bear. However, we did not expect bias across candidate islands in the spatial pattern where such multiple counting might have occurred. We calculated detection-per-unit-effort (DPUE) for all data types except mortality data (Table 1 and Table 2). For remote camera and genetic data, we quantified DPUE for each island by dividing the number of observations by the cumulative number of

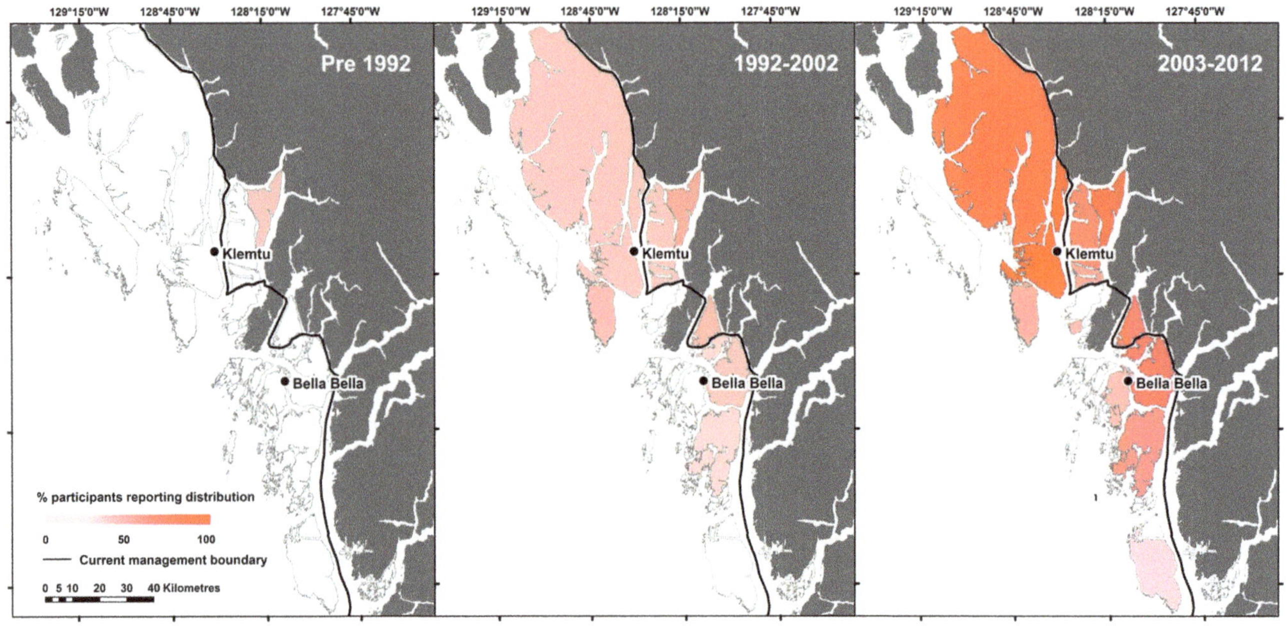

Figure 2. Time series of island grizzly bear occupation over three time periods. Local residents' perception of grizzly bear (*Ursus arctos horribilis*) distribution throughout Heiltsuk and Kitasoo/Xai'xais Territories in coastal British Columbia, Canada during the Pre 1992 (traditional ecological knowledge data), 1992–2002 (local ecological knowledge data) and 2003–2012 (local ecological knowledge data) time periods. Data are reported at the island scale (n = 18 islands) as the percentage of participants who indicated the island was within their area of observation and supported grizzly bears. Dark grey areas were not included in our study and the 'current management boundary' is the westward (i.e., seaward extent) of the Province of British Columbia's Grizzly Bear Population Units, the spatial scale at which grizzly bears are managed in the province.

days each camera (n = 18) or hair snag (n = 33) was employed (Table 2). We standardized LEK observations by dividing the total number of observations for each island by the total number of survey days each interviewee estimated s/he spent within their identified geographic area of expertise. Both survey effort and total number of observations were summed for all participants at the island scale. As level of experience could not be reliably estimated, we assumed the same observational abilities of all participants. To assess which islands contain reproducing females, we used both remote camera and LEK observations to identify all detections of sows with cubs.

Temporal Trends in Occupancy. We assessed temporal occupancy across three intervals: Pre 1992 (n = 7 interviewees), 1992–2002 (n = 27 interviewees), and 2003–2012 (n = 27 interviewees). We estimated 'Pre 1992' occupancy from TEK data, and 1992–2002 and 2003–2012 periods from LEK data. Because LEK observations are likely increasingly comprehensive with increasing proximity to communities of residence [18], for each island we only used interview data from the closest community. We indexed the total number of observations across all islands by dividing them by the total number of survey days in each year. Survey effort was calculated annually using Equation 1:

$$\text{Survey Effort} = \frac{\sum(\text{Survey days per participant})}{\#\text{ of participants} \times 365} \qquad (1)$$

where # of participants is the number of LEK interviewees, 365 denotes the total possible number of survey days a year, and survey days per person indicates the total number of survey days for each participant. The number of island LEK observations per year was then divided by annual survey effort across all islands. This metric allows survey effort to be calculated per annum across all participants for all islands.

Results

We identified 149 grizzly bear observations across 15 major islands, including 10 islands outside the current grizzly bear management boundary (Figure 1a, Table 1). LEK yielded the largest number of data points relative to other sources, with 110 observations across 15 major islands (Table 1 & 2). Twelve islands hosted more than one observation (mean = 9.93, range = 1 to 39) (Table 1; Figure 1a). Across all LEK participants, 86% (19 of 22) reported at least one observation of an island grizzly (Table 1). Data on known individuals (genetic data and mortality records) indicated the presence of at least nine unique individuals (three female, five male and one of unknown gender) across 10 locations on seven islands. Remote cameras yielded a total of 28 observations of grizzlies on four islands (Table 1), two of which are outside of the current management boundary. Sow and cubs were detected in 23 LEK observations and seven remote camera observations across six islands (Table 1).

TEK/LEK data sources revealed a similar pattern of island occupation as did evidence from genetics, cameras, and hunting records; all newly occupied islands identified by scientific evidence were affirmed by complementary TEK/LEK observations (Figure 1b; Table 1). Of the seven islands solely associated with TEK/LEK evidence, all but one (Denny Island) lacked scientific evidence (Figure 1b).

Indigenous knowledge additionally yielded historical information suggesting that bear colonization of islands occurred recently, and that the process has increased in pace. Only one of seven TEK interviewees identified grizzly bears on an island before 1992. Accounts of island occurrence, within and outside the currently recognized distribution, were increasingly common and spatially widespread in 1992–2002 and 2003–2012 periods (encompassing 12 and 15 islands, respectively; Figure 2). The total island

Table 1. Detections of island grizzly bears (*Ursus arctos horribilis*) by data type (presented as raw observations and as detection-per-unit-effort (DPUE) values) within Heiltsuk and Kitasoo/Xai'xais Territories in coastal British Columbia, Canada (n = 18 islands).

Island	Total Remote Camera Observations	Remote Camera Detections Per Trap Night[a]	Total Hair Snag Observations	Hair Snag Year	Hair Snag Detections Per Trap Night[b,d]	LEK Observations	Total Sow/Cub Observations	Mortality Observation
Aristazabal	-	-	-	-	-	0	0	0
Calvert	-	-	-	-	-	0	0	0
Campbell	-	-	0	-	-	6	0	0
Chatfield	-	-	2	2009–2012	0.03	6	0	0
Cunningham	-	-	1	2009–2012	0.01	7	3	0
Denny	-	-	0	2009–2012	0	5	0	0
Dufferin	-	-	-	-	-	1	0	0
Hecate	-	-	-	-	-	0	0	0
Hunter	-	-	-	-	-	1	0	1
Yeo**	-	-	1	2009–2012	0.04	12	4	0
Lady Douglas	-	-	-	-	-	1	0	0
Pooley**	7	0.02	0	2012	0.01	15	0	0
Price	0	0	0	2012	0	1	0	0
Princess Royal	19	0.03	3	2012	0.02	16	17	1
Roderick**	1	0.01	0	2012	0.01	12	3	1
Sarah	-	-	-	-	-	9	3	0
Susan**	-	-	-	-	-	2	0	0
Swindle	1	0.03	1	2012	0.03	22	15	0

[a]For islands with remote cameras.
[b]For islands with hair snags.
[c]Indicates number of observations, not number of individual grizzly bears, across study period.
[d]Summed across seasons and years.
** Indicates islands within the current management boundary'.

Table 2. Sampling effort of island grizzly bears (*Ursus arctos horribilis*) by data type within Heiltsuk and Kitasoo/Xai'xais Territories in coastal British Columbia, Canada (n = 18 islands).

Island	Number of Remote Cameras	Total Remote Camera Nights[a]	Number of Hair Snag Sites	Hair Snag Year	Total Hair Snag Nights	Number of LEK Interviewees Who Reported on Island[b]	Average Number of Report Days Per Year[c]
Aristazabal	0	-	0	-	-	13	120
Calvert	0	-	0	-	-	9	120
Campbell	0	-	0	-	-	14	142
Chatfield	0	-	2	2009–2012	159	14	142
Cunningham	0	-	4	2009–2012	304	14	142
Denny	0	-	2	2009–2012	178	13	148
Dufferin	0	-	0	-	-	13	148
Hecate	0	-	0	-	-	9	120
Hunter	0	-	0	-	-	12	146
Yeo**	0	-	3	2009–2012	221	15	145
Lady Douglas	0	-	0	-	-	14	137
Pooley**	3	357	3	2012	416	10	136
Price	1	23	1	2012	23	13	140
Princess Royal	9	661	9	2012	1217	8	140
Roderick**	4	192	3	2012	398	11	139
Sarah	0	-	0	-	-	8	128
Susan**	0	-	0	-	-	14	145
Swindle	1	135	1	2012	121	11	140

[a]For islands with remote cameras.
[b]Across entire study period.
[c]Average number of 'report days' (i.e., days interviewees reported on for each island) per year from 1992–2012. Data derived from all participants who included. the island in their area of expertise.
** Indicates islands within the "current management boundary".

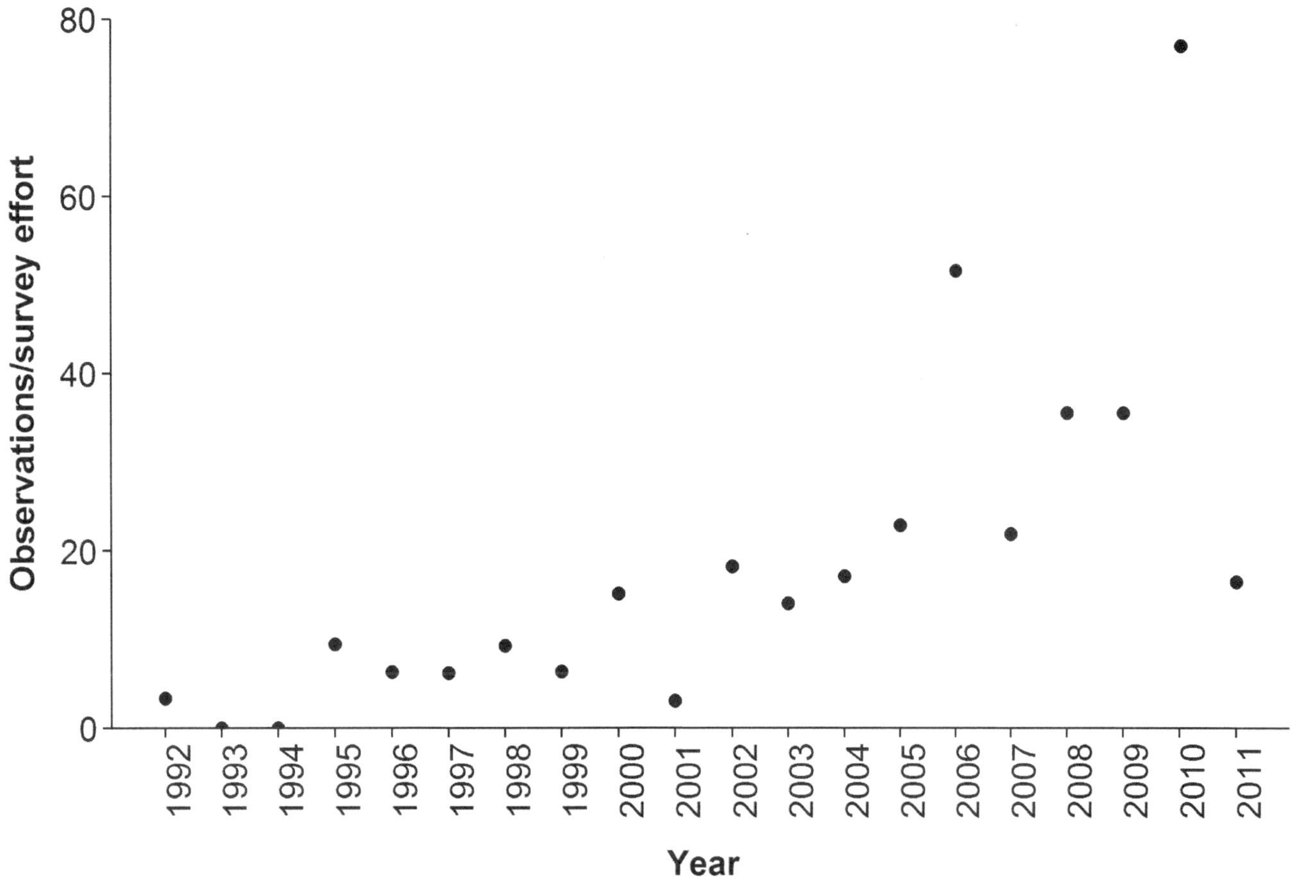

Figure 3. Trend of island grizzly occupation over time as derived from LEK data. Yearly number of island grizzly bear (*Ursus arctos horribilis*) observations per unit survey effort (a summation of all participants' observations across all islands; local ecological knowledge data only) between 1992–2011 in the Traditional Territories of the Heiltsuk and Kitasoo/Xai'xais First Nations in British Columbia, Canada. These observations encompass islands within and outside the current management boundary (n = 18 islands).

observations per year, indexed by survey effort, increased between 1992–2011 (Figure 3).

Discussion

Our approach coupling indigenous knowledge and western science offered multiple lines of evidence that grizzly bears of both sexes have recently colonized 10 islands outside of the currently accepted distribution. Moreover, camera and interview data provided strong evidence for reproductive individuals and their offspring on four islands. Collectively, this suggests a distributional process beyond any exploratory extra-range movements of mobile males. Such an abrupt distributional shift of an apex predator may invoke cascading ecological consequences (*e.g.*, [45]). More broadly, the drivers of such change may also act on other less iconic and recognizable species and processes at differing spatial and temporal scales [46,47].

Though trends in these data are apparent, our approach also presents some inherent limitations. Specifically, although highly useful, TEK/LEK data are not systematic in coverage [23]. Accordingly, use of islands by interviewees likely varies in their mode of use (hunting, fishing, ecotourism, etc.) and as a result the habitat types people frequent may differ. As such, some interviewees may be more likely to encounter grizzly bears than others. Additionally, it is possible that black bears may have been mistaken for grizzly bears by some interviewees. Furthermore, TEK/LEK may not always be appropriately responsive to detect re-colonization and abandonment events that could accompany dynamic distributional processes. Snowball sampling methods may have also excluded some knowledgeable participants who were outside of the social networks of interviewees [48]. Finally, we note that recent observations might be more likely to be remembered and reported. More likely, however, we postulate that older observations of grizzlies outside their known range would have been interpreted as surprising and, accordingly, equally or even more memorable. Despite these limitations, the inclusion of TEK/LEK data provides important process values through the engagement of local people and as a result this approach may facilitate collaborative rather than antagonistic conservation efforts (*e.g.*, [49–51]).

Conventional scientific tools, such as remote cameras and genetic identification through hair snagging, also present limitations. Although more systematic in their deployment, the temporal resolution of these data types are limited, both on an annual basis (*i.e.*, spring sampling) and over longer timeframes (most ecological monitoring programs span years rather than decades or centuries) [20]. Additionally, these tools are spatially static and can be impacted by poor placement. Indeed, in the nearly road-less landscape we study, most snag station are set up close to shorelines where they can be accessed by boat. Importantly, though all data sources have potential weaknesses, uniting these independent sources provides increased temporal and spatial detail of the range shift we describe.

Interviewee comments and relevant literature allow us to offer working hypotheses about potential drivers of this shift. Specifically, modifications to the abundance and distribution of food resources as well as changes to intra- and inter- specific competition might be relevant. Many interviewees cited the reduction of salmon (*Oncorhynchus* spp.), which has declined throughout coastal BC, especially since 2000 [52]. As a critical resource that influences ursid body condition, reproductive output, and population dynamics [36,53], salmon abundance could influence bear distribution; individuals within other carnivore species have expanded or abandoned their range following declines in prey and subsequent increased intraspecific competition (*e.g.*, [54–56]). Increasing berry abundance and accessibility, the result of recent logging on islands, might cause individuals on exploratory forays to islands to remain. Grizzly bears in other areas select recently logged habitat to exploit diverse food resources offered in early regeneration stages of disturbed habitat [57]. Alternatively, black bears (*Ursus americanus*) are thought to limit grizzlies via exploitative competition on coastal islands where food resources are more dispersed and more difficult to defend. Acting alone or synergistically with human-caused mortality from trophy hunting of grizzlies, such competition between species has been proposed as a mechanism for excluding grizzly bears from islands [31] or reducing their densities elsewhere [58]. Recent reductions of grizzly bear trophy hunting on the mainland (BC Ministry of Environment, unpublished data), and/or changes in competition from black bears, might have reduced the demographic constraints on grizzly bears, thereby allowing dispersal to nearby islands. Whatever the cause(s) in our system, range shifts observed in closely monitored species may signal changes to underlying ecological conditions and provide resource managers the opportunity to consider and plan for wider implications.

The ecological, cultural, and economic consequences of this distribution shift in coastal grizzly populations might be considerable. As a larger species, grizzly bears outcompete co-occurring black bears (*U. americanus*) for salmon through interference competition [59] and also kill juvenile and adult black bears [60]. In addition, 'spirit bears' (*Ursus americanus kermodei*), a rare and geographically restricted white morph of black bears that occur with relatively high frequency on several islands in the area [61,62], are not only revered culturally but are also economically prized as the foundation of wildlife eco-tourism. Consequently, any mortality or increased avoidance behaviour due to new grizzly bear presence might impact cultural and economic values. Finally, forest management plans for grizzlies – a 'species of conservation concern' in BC [63] – carries legally-binding measures to protect high quality habitat wherever grizzly bears occur [30]. Such multifaceted implications would also be likely for range shifts in other high-profile fauna. Indeed, reintroductions of wolves (*Canis lupus*) to the Yellowstone Ecosystem provide a flagship example of the tremendous ecological, cultural, and economic ramifications associated with a change in distribution [64,65]. Similarly, expansions to sea otter (*Enhydra lutris*) ranges, a system well-suited to TEK/LEK study, carry comparably broad implications for people and ecosystems [66].

Beyond its regional relevance, this study illustrates the synergistic benefits of combining science with TEK/LEK over concurrent and complementary spatial and temporal scales. In our study, both data types predominately affirmed one another; islands without scientific evidence for occupation also lacked concurrent TEK/LEK observations. The co-affirmation of data sources in locations where they overlapped added confidence in patterns observed by LEK where they did not overlap [16]. Genetic and camera data provided precise information on individuals captured, including individual identities, gender, age class, location, and time of visit. These data were captured in all weather and at all hours each day. Costs of these scientific methods, however, necessarily limit their spatial and temporal coverage. Moreover, they can only be employed to capture data in real time. In contrast, although yielding less detailed information on individual bears, TEK/LEK data afforded broader information across a larger area and longer timespan. Finally, despite the inherent value of TEK/LEK, it remains largely ignored as a source of ecological data [19].This case study provides one of the few examples we detected in the literature that illustrates the value in uniting TEK/LEK with scientific methods to provide meaningful input into wildlife management [18,24,25,67].

With careful consideration a TEK/LEK method alone could perform well in other wildlife systems. Such an approach, however, requires careful assessment. Specifically, LEK information might be most useful for questions of distribution but may lack the detail required for other population parameters, such as absolute abundance, without calibration from scientific sources [18,68]. The case for a TEK/LEK approach alone might be particularly compelling with conspicuous and culturally important focal species in data-deficient regions for which species distributions impose serious management implications and funds for science are scarce. Moreover, as many indigenous governments and societies across the globe play increasingly prominent roles in resource management again [69–71], methods that emphasize an integration of, or focus on, existing local knowledge might emerge as default approaches. Such a transition might help overcome the dual common barriers of lack of conservation action due to inadequate data and data deficiency resulting from scarcity of funds. In this way existing indigenous knowledge can be proactively incorporated into management. Such consideration may support a transition to implement management action more rapidly, the timescale in which conservation action is often required.

Acknowledgments

We thank interview participants, the Kitasoo/Xai'xais Integrated Resources Authority and the Heiltsuk Integrated Resource Management Department, and are grateful to Jennifer Carpenter, Tony Hamilton, William Housty, Wayne McCrory, Douglas Neasloss, and Laurie Whitehead for comments on this manuscript. We also acknowledge and thank our colleagues in the field: Harvey Brown Sr, Vernon Brown, Heather Bryan, Collin Reid, Ian Reid, Brandon Robinson, Krista Robinson, Leonard Robinson, Medrick Robinson, Shawn Robinson, Howard Humchitt, and Whitney Starr.

Author Contributions

Conceived and designed the experiments: CNS CTD PP. Performed the experiments: CNS MSA LVG. Analyzed the data: CNS MSA KAA LVG. Wrote the paper: CNS CTD PP KAA MSA LVG. GIS analysis support: KAA. Interview data collection: CNS LVG MSA. Field collection of remote camera and genetic data: CTD CNS LVG MSA KAA.

References

1. Tylianakis JM, Didham RK, Bascompte J, Wardle DA. (2008) Global change and species interactions in terrestrial ecosystems. Ecol Lett 11: 1351–1363.
2. Boulangeat I, Gravel D, Thuiller W (2012) Accounting for dispersal and biotic interactions to disentangle the drivers of species distributions and their abundances. Ecol Lett 15: 584–593.
3. MacArthur RH, Wilson EO (1967) The theory of island biogeography: monographs in population biology. Princeton, NJ, USA: Princeton University Press.
4. Pulliam HR (2000) On the relationship between niche and distribution. Ecol Lett 3: 349–361.
5. Robinson JG, Bennett EL (2000) Hunting for sustainability in tropical forests. New York, NY, USA: Columbia University Press.
6. Warren M, Chikin Y, Danilov I, King C, Morris N, et al. (2001) Rapid responses of British butterflies to opposing forces of climate and habitat change. Nature 414: 65–69.
7. Ripple WJ, Beschta RL (2004) Wolves and the ecology of fear: Can predation risk structure ecosystems? Bioscience 54: 755–766.
8. Whitfield D, Fielding AH, Mcleod DRA, Haworth PF (2004) The effects of persecution on age of breeding and territory occupation in golden eagles in Scotland. Biol Conserv 118: 249–259.
9. Chen I-C, Hill JK, Ohlemüller R, Roy DB, Thomas CD (2011) Rapid range shifts of species associated with high levels of climate warming. Science 333: 1024–1026.
10. Bennie J, Hogdgson JA, Lawson,CR, Holloway, CTR, Roy, DB,Brereton,T., Thomas, CD, and Wilson RJ (2013) Range expansion through fragmented landscapes under a variable climate. Ecol Lett 16: 921–929.
11. Vermeij GJ (1991) When biotas meet: Understanding biotic interchange. Science 253: 1099–1104.
12. Mortensen HS, Dupont YL, Olesen JM (2008) A snake in paradise: Disturbance of plant reproduction following extirpation of bird flower-visitors on Guam. Biol Conserv 141: 2146–2154.
13. Araújo MB, Williams PH (2000) Selecting areas for species persistence using occurrence data. Biol Conserv 96: 331–345.
14. Laliberte AS, Ripple WJ (2004) Range contractions of North American carnivores and ungulates. Bioscience 54: 123–138.
15. Guisan A, Thuiller W (2005) Predicting species distribution: offering more than simple habitat models. Ecol Lett 8: 993–1009.
16. Huntington HP, Suydam RS, Rosenberg DH (2004) Traditional knowledge and satellite tracking as complementary approaches to ecological understanding. Environ Conserv 31: 177–180.
17. Berkes F, Colding J, Folke C (2000) Rediscovery of traditional ecological knowledge as adaptive management. Ecol Appl 10: 1251–1262.
18. Gilchrist G, Mallory M, Merkel F (2005) Can local ecological knowledge contribute to wildlife management? Case studies of migratory birds. Ecol Soc 10.
19. Brook RK, McLachlan SM (2008) Trends and prospects for local knowledge in ecological and conservation research and monitoring. Biodivers Conserv 17: 3501–3512.
20. Anadón JD, Giménez A, Ballestar R, Pérez I (2009) Evaluation of local ecological knowledge as a method for collecting extensive data on animal abundance. Conserv Biol 23: 617–625.
21. Carter BTG, Nielsen EA (2011) Exploring ecological changes in Cook Inlet beluga whale habitat though traditional and local ecological knowledge of contributing factors for population decline. Mar Policy 35: 299–308.
22. Huntington HP (2000) Using traditional ecological knowledge in science: Methods and applications. Ecography (Cop) 10: 1270–1274.
23. Moller H, Berkes F, Lyver POB, Kislalioglu M (2004) Combining science and traditional ecological knowledge: Monitoring populations for co-management. Ecol Soc 9.
24. Brook RK, Mclachlan SM (2005) On using expert-based science to "test " Local Ecological Knowledge. Hum Ecol 10: 2–4.
25. Berkes F (2004) Rethinking community-based conservation. Conserv Biol 18: 621–630.
26. Ministry of Environment (2012) British Columbia Grizzly Bear Population Estimates for 2012. Available: http://www.env.gov.bc.ca/fw/wildlife/big_game_info.html#GrizzlyBear.
27. Committee on the Status of Endangered Wildlife in Canada (2012) COSEWIC assessment and status report on the Grizzly Bear, western population, in Canada. Ottawa, Canada.
28. McCrory WP, Bergdah J, Paquet PC, Cross B (2001) A conservation area design for protection of the white black bear (Ursus americanus kermodei) on the central coast of British Columbia: 25.
29. Schwartz CC, Miller SD, Haroldson M (2003) Grizzly bear. In: Feldhamer G, Thompson B, Chapman J, editors. Wild Mammals of North America. Baltimore, MD, USA: The Johns Hopkins University Press. pp. 556–586.
30. Ministry of Forests, Lands and NRO (2013) Ministerial Order: Central and North Coast Order Schedule 2 Grizzly Bear Habitat April 2013 Amendments.
31. Mattson DJ, Herrero S, Merrill T (2005) Are black bears a factor in the restoration of North American grizzly bear populations? Are black bears a factor in the restoration of North American grizzly bear populations? Ursus 16: 11–30.
32. Servheen C (1990) The Status and Conservation of the Bears of the World. International Conference of Bear Research and Management. Victoria, Canada. pp. 11–20.
33. Rockwell D (1991) Giving voice to bear: North American indian rituals, myths, and images of the bear. Lanham, Maryland, USA: Rowman & Littlefield.
34. Nevin OT, Gilbert BK (2005) Perceived risk, displacement and refuging in brown bears: positive impacts of ecotourism? Biol Conserv 121: 611–622.
35. Helfield JM, Naiman RJ (2006) Keystone interactions: Salmon and bear in riparian forests of Alaska. Ecosystems 9: 167–180.
36. Levi T, Darimont CT, Macduffee M, Mangel M, Paquet P, et al. (2012) Using grizzly bears to assess harvest-ecosystem tradeoffs in salmon fisheries. PLoS Biol 10: e1001303.
37. Darimont, C.T & Paquet PC (2002) The gray wolves, Canis lupus, of British Columbia's central and north coast: distribution and conservation assessment. Can F Nat 116: 416–422.
38. Krajina VJ (1965) Biogeoclimatic Zones and Classification of British Columbia. Ecology of Western North America. Ecol 1: 1: 1–17.
39. Pojar J, McKinnon A (1994) Plants of the Pacific Northwest Coast: Washington, Oregon, British Columbia, and Alaska. Auburn, WA, USA: Lone Pine Publishing.
40. Darimont CT, Bryan HM, Carlson SM, Hocking MD, MacDuffee M, et al. (2010) Salmon for terrestrial protected areas. Conserv Lett 3: 379–389.
41. Davis A, Wagner JR (2003) Who knows? On the importance of identifying " experts" when researching local ecological knowledge. Hum Ecol 31: 463–489.
42. Noy C (2008) Sampling knowledge: The hermeneutics of snowball sampling in qualitative research. Int J Soc Res Methodol 11: 327–344.
43. Press A, Woods JG, Paetkau D, Lewis D, Mclellan BN, et al. (1999) Genetic tagging of black bears. Wildl Soceity Bull 27: 616–627.
44. Bryan HM, Darimont CT, Paquet PC, Wynne-Edwards KE, Smits JE (2013) Stress and reproductive hormones in grizzly bears reflect nutritional benefits and social consequences of a salmon niche. PLoS One: In press
45. Beschta RL, Ripple WJ (2009) Large predators and trophic cascades in terrestrial ecosystems of the western United States. Biol Conserv 142: 2401–2414.
46. Carroll C, Noss RF, Paquet PC (2001) Carnivores as focal species for conservation planning in the Rocky Mountain region. Ecol Appl 11: 961–980.
47. Niemi GJ, McDonald ME (2004) Application of Ecological Indicators. Annu Rev Ecol Evol Syst 35: 89–111.
48. Biernacki P, Waldorf D, Methods S (1981) Snowball sampling: Problems and techniques of chain referral sampling. Sociol Methods Res 10. doi:10.1177/004912418101000205
49. Adams M, Carpenter J, Housty J, Neasloss D, Paquet P, et al. (2014) Towards increased engagement between academic and indigenous community partners in ecological research. Ecol Soc: In press.
50. Dowie M (2009) Conservation refugees: the hundred-year conflict between global conservation and native peoples. Cambridge, Massachusetts, USA: MIT Press.
51. Grove R (1995) Green imperialism: colonial expansion, tropical island edens and the origins of environmentalism. Cambridge, England: Cambridge University Press.
52. Price MHH, Darimont CT, Temple NF, MacDuffee SM (2008) Ghost runs: management and status assessment of Pacific salmon (Oncorhynchus spp.) returning to British Columbia's central and north coasts. Can J Fish Aquat Sci 65: 2712–2718.
53. Hilderbrand GV, Schwartz CC, Robbins CT, Jacoby ME, Hanley TA, et al.(1999) The importance of meat, particularly salmon, to body size, population productivity, and conservation of North American brown bears. Can J Zool 7: 132–138.
54. Ward RMP, Krebs CJ (1985) Behavioural responses of lynx to declining snowshoe hare abundance. Can J Zool 63: 2817–2824.
55. Knick ST (1990) Ecology of bobcats relative to exploitation and a prey decline in Southeastern Idaho. Wildl Monogr 108: 3–42.
56. Schmidt K (2008) Behavioural and spatial adaptation of the Eurasian lynx to a decline in prey availability. Acta Theriol (Warsz) 53: 1–16.
57. Nielsen SE, Boyce MS, Stenhouse GB (2004) Grizzly bears and forestry. For Ecol Manage 199: 51–65.
58. Mowat G, Heard DC, Schwarz CJ (2013) Predicting grizzly bear density in western North America. PLoS One 8: e82757.
59. Belant JL, Kielland K, Follmann EH, Adams LG (2006) Interspecific resource partitioning in sympatric ursids. Ecol Appl 16: 2333–2343.
60. Mattson DJ, Knight RR, Blanchard BM (1992) Cannilbalism and predation on black bears by grizzly bears in the Yellowstone ecosystem, 1975–1990. J Mammal 73: 422–425.
61. Ritland K, Newton C, Marshall HD (2001) Inheritance and population structure of the white-phased "Kermode" black bear. Curr Biol 11: 1468–1472.
62. Klinka DR, Reimchen TE (2009) Adaptive coat colour polymorphism in the Kermode bear of coastal British Columbia. Biol J Linn Soc 98: 479–488.
63. Austin MA, Heard DC, Hamilton AN (2004) Grizzly bear (Ursus arctos) harvest management in British Columbia. Victoria, BC, Canada: Ministry of Water, Land and Air Protection. Available: www.env.gov.bc.ca/wld/documents/gb_harvest_mgmt.pdf.

64. Morell V (2008) Wolves at the door of a more dangerous world. Science (80) 319: 890–892.
65. Wilson MA (1997) The wolf in Yellowstone: Science, symbol, or politics? Deconstructing the conflict between environmentalism and wise use. Soc Nat Resour 10: 453–468.
66. Riedman ML, Estes JA (1988) A Review of the History, Distribution and Foraging Ecology of Sea Otters. In: VanBlaricom G, Estes J, editors. The Community Ecology of Sea Otters. Berlin, Germany: Springer. pp. 4–21.
67. Abraão MB, Nelson BW, Baniwa JC, Yu DW, Shepard Jr GH (2008) Ethnobotanical ground-truthing: indigenous knowledge, floristic inventories and satellite imagery in the upper Rio Negro, Brazil. J Biogeogr 35: 2237–2248.
68. Anadón JD, Giménez A, Ballestar R (2010) Linking local ecological knowledge and habitat modelling to predict absolute species abundance on large scales. Biodivers Conserv 19: 1443–1454.
69. Folke C, Carpenter S, Elmqvist T, Gunderson L, Holling CS, et al. (2002) Resilience and sustainable development: building adaptive capacity in a world of transformations. Ambio 31: 437–440.
70. Houde N (2007) The six faces of traditional ecological knowledge: challenges and opportunities for Canadian co-management arrangements. Ecol Soc 12: 34.
71. Wyatt S (2008) First Nations, forest lands, and "aboriginal forestry" in Canada: from exclusion to co management and beyond. Can J For Res 38: 171–180.

Mapping Reef Fish and the Seascape: Using Acoustics and Spatial Modeling to Guide Coastal Management

Bryan Costa[1,2]*, J. Christopher Taylor[3], Laura Kracker[1], Tim Battista[1], Simon Pittman[1,4]

1 Biogeography Branch, National Centers for Coastal Ocean Science, National Oceanic and Atmospheric Administration, Silver Spring, Maryland, United States of America, **2** Consolidated Safety Services, Fairfax, Virginia, United States of America, **3** Center for Coastal Fisheries and Habitat Research, National Centers for Coastal Ocean Science, National Oceanic and Atmospheric Administration, Beaufort, North Carolina, United States of America, **4** Marine Institute, University of Plymouth, Plymouth, United Kingdom

Abstract

Reef fish distributions are patchy in time and space with some coral reef habitats supporting higher densities (i.e., aggregations) of fish than others. Identifying and quantifying fish aggregations (particularly during spawning events) are often top priorities for coastal managers. However, the rapid mapping of these aggregations using conventional survey methods (e.g., non-technical SCUBA diving and remotely operated cameras) are limited by depth, visibility and time. Acoustic sensors (i.e., splitbeam and multibeam echosounders) are not constrained by these same limitations, and were used to concurrently map and quantify the location, density and size of reef fish along with seafloor structure in two, separate locations in the U.S. Virgin Islands. Reef fish aggregations were documented along the shelf edge, an ecologically important ecotone in the region. Fish were grouped into three classes according to body size, and relationships with the benthic seascape were modeled in one area using Boosted Regression Trees. These models were validated in a second area to test their predictive performance in locations where fish have not been mapped. Models predicting the density of large fish (≥29 cm) performed well (i.e., AUC = 0.77). Water depth and standard deviation of depth were the most influential predictors at two spatial scales (100 and 300 m). Models of small (≤11 cm) and medium (12–28 cm) fish performed poorly (i.e., AUC = 0.49 to 0.68) due to the high prevalence (45–79%) of smaller fish in both locations, and the unequal prevalence of smaller fish in the training and validation areas. Integrating acoustic sensors with spatial modeling offers a new and reliable approach to rapidly identify fish aggregations and to predict the density large fish in un-surveyed locations. This integrative approach will help coastal managers to prioritize sites, and focus their limited resources on areas that may be of higher conservation value.

Editor: Judi Hewitt, University of Waikato (National Institute of Water and Atmospheric Research), New Zealand

Funding: This research was funded by the National Oceanic and Atmospheric Administration's Coral Reef Conservation Program (http://coralreef.noaa.gov/) and Office of Coast Survey (http://www.nauticalcharts.noaa.gov/). The funders had no role in study design, data analysis, decision to publish, or preparation of the manuscript.

Competing Interests: The authors have declared that no competing interests exist.

* E-mail: bryan.costa@noaa.gov

Introduction

The rapid emergence of place-based management strategies, such as marine protected areas (MPAs), has increased the demand for reliable information describing the distribution of fish across large portions (i.e., 10 s to 100 s km^2) of the ocean [1,2,3]. Patchiness of fish populations in time and space, combined with resource constraints on management, often requires that coastal managers identify spatial priorities. A common strategy to identify these priorities is to select locations of high conservation value based on biological characteristics and relative vulnerability [2,4,5,6]. Fish aggregations (i.e., locations where a suite of environmental conditions interact to support high densities of fish) are typically given high priority in MPA network design and marine spatial planning [7,8,9]. However, locating and characterizing fish aggregations can be challenging over broad geographic areas (i.e., 10 s to 100 s km^2), especially when they occur in waters too deep for surveys using conventional SCUBA diving; where turbidity impairs visual surveys; or when aggregations are transient and only detectable at night. Additional challenges arise when animal distribution patterns need to be expanded from fine-scale visual surveys (covering $<100\ m^2$) to broader spatial scales (covering 10 s to 100 s km^2) that are operationally relevant to coastal and marine management [10,11,12]. Scaling up patterns from fine-scale surveys is challenging because there is no single scale at which ecological patterns should be studied since organisms show variability at a range of spatial, temporal, and organizational scales [13].

Although underwater acoustic technology is not new, rarely have coral reef ecosystem studies simultaneously mapped and quantified the locations and size of fish along with the three-dimensional structure of the surrounding seafloor. These data provide an opportunity to model fish-seascape relationships at multiple spatial scales that are appropriate for studies of highly mobile organisms. In both tropical and temperate waters, seafloor structure (derived from bathymetry) has been established as a useful predictor of fish distributions [9,14,15]. Water depth and seafloor topography (e.g., rugosity, slope-of-the-slope, slope, curvature), sometimes combined with relative across-shelf position, have repeatedly emerged as excellent predictors for fish

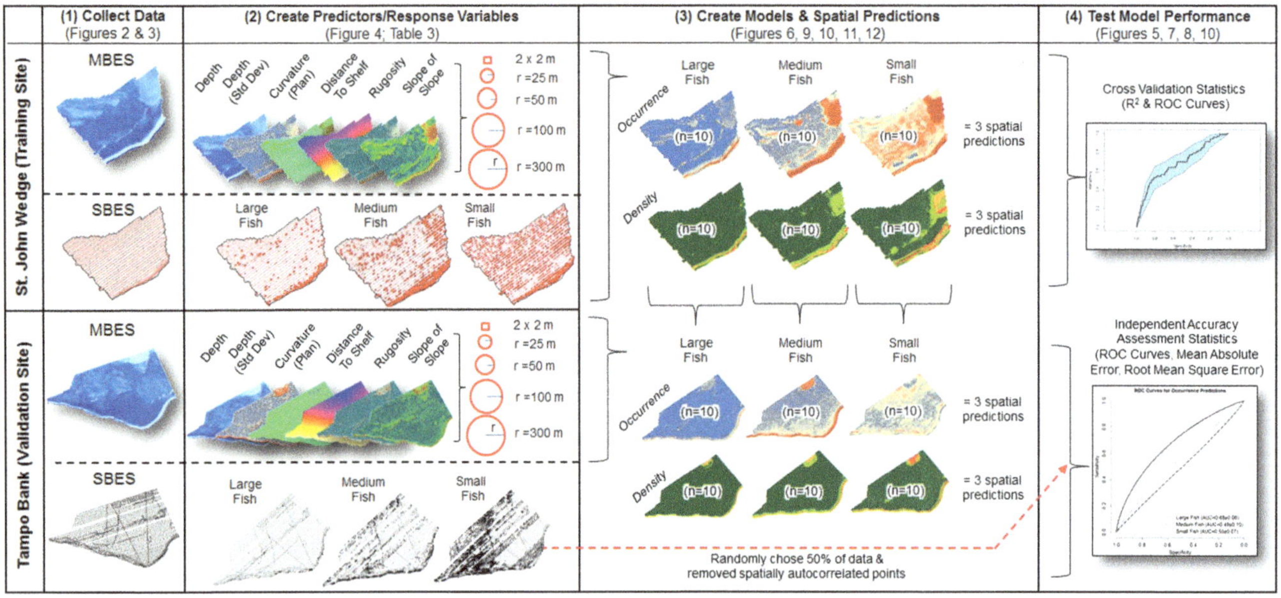

Figure 1. Diagram of modeling process. Steps used to train and validate the Boosted Regression Tree models developed from the SBES and MBES datasets.

[9,14,15,16]. These data can be analyzed to produce descriptive maps of fish distributions for discrete size-classes, providing a unique opportunity to bridge the informational gap between ecology and management [17].

The importance of structural complexity for maintaining the integrity and function of coral reef ecosystems is well established at fine spatial scales (i.e., centimeters to meters) [18,19,20]. The importance of this complexity at broader spatial scales is less well studied and understood. To address this research gap, we quantified seafloor structure (and distance to the shelf edge) at multiple spatial scales to explore how fish in different sizes classes are distributed across the seascape [20,21,22]. We acquired spatially and temporally coincident data with a multibeam echosounder (MBES) to map seafloor terrain and a splitbeam echosounder (SBES) to map fish in the water column in two areas in the U.S. Virgin Islands (USVI). These two areas were chosen because they contained spawning aggregation sites for commercially important snapper and grouper species. We then used Boosted Regression Trees (BRT), a machine learning algorithm, and Geographic Information Systems (GIS) software, to model and map the spatial relationship between seafloor structure and the density of fishes in three size classes (Fig. 1). The key questions addressed were:

1. Which seafloor structure and distance surfaces were the best predictors of fish occurrence and density?
2. Does fish-habitat data derived from acoustic sensors provide sufficient information to develop useful spatial predictions of fish distributions across the seascape?

Methods

Study Areas

This research was conducted in two areas (i.e., St. John Wedge and Tampo Bank) south of St. John in the USVI (Fig. 2). St. John Wedge is 22 km^2 and Tampo Bank is 62 km^2. Both areas were mapped using multibeam and splitbeam echosounders. No specific permissions were required to survey these locations because they are not actively managed by the territorial or federal government, and this study did not involve endangered or protected species. These areas were chosen because they were in close proximity to known spawning sites (i.e., Grammanik Bank) for commercially important fish species (e.g., *Ocyurus chrysurus* and *Epinephelus guttatus*) [9,6]. Tampo Bank is also suspected to be spawning site for Mutton Snapper (*Lutjanus analis*) [23]. BRT models were trained in St. John Wedge, and validated in Tampo Bank. Depths in these two areas ranged from 22 to 100 m (Table 1), although over 70% of each site was shallower than 55 m. Fish density measurements ranged from 0 to 33 fish per 100 m^2.

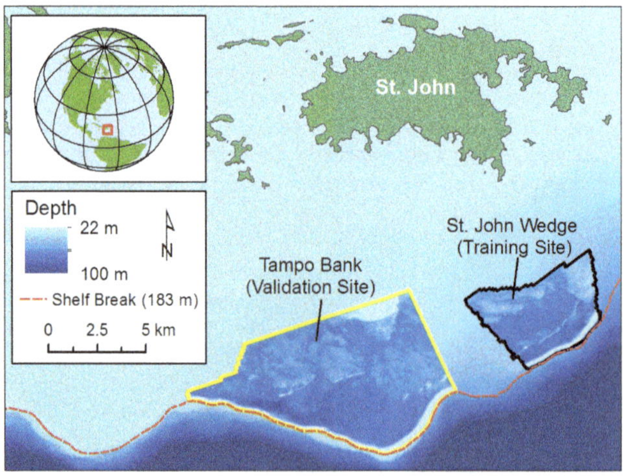

Figure 2. Map of study areas. Location of the study sites in the U.S. Virgin Islands. The spatial predictions were developed in one site (St. John Wedge) and validated in the other site (Tampo Bank).

Table 1. Depths in the training and validation sites.

Depths	% Area	
(m)	St. John Wedge	Tampo Bank
22<35	3.7	1.8
35<45	16.7	3.8
45<55	73.0	64.8
55<65	5.7	29.2
65<75	0.4	0.4
75<85	0.3	0.0
85<100	0.3	0.0

Depths found in St. John Wedge and Tampo Bank ranged from 22 to 100 m. However, greater than 70% of both areas were less than 55 meters deep.

Mapping Fish and the Seafloor

Mapping fish using a splitbeam echosounder. SBES data describing fish sizes, densities and distributions were collected in St. John Wedge from March 29 – April 16, 2011, and in Tampo Bank from March 18 – April 6, 2010 using a Simrad EK-60 (120 kHz) splitbeam echosounder. This scientific echosounder was calibrated using a tungsten carbide sphere, allowing for accurate measurements of fish size. Splitbeam echosounders detect fish in the water column by rapidly transmitting pulses of sound (pings) that reflect off objects of differing densities than the surrounding water. An internal air filled sac (called the swim bladder) is the primary contributor to a fish's acoustic reflection [24]. Larger fish reflect more sound because they have larger swim bladders. Survey lines were acquired parallel to depth contours, and spaced to provide complete MBES coverage of the seafloor. Only a small percentage of each survey line was sampled by the SBES because it had a narrower swath (about 7°) than the MBES (about 120°).

SBES data were processed to detect individual fish using Echoview software version 5.3. The water-seafloor interface was delineated to separate this acoustic signal from fish detections. Other acoustic interference (e.g., air bubbles) and faint echoes, likely representing plankton and other non-fish targets, were masked or eliminated from the data. Vessel speed (ca 7 kts) and rapid ping rate (3–8 pulses s^{-1}) resulted in multiple, sequential echoes from each individual fish. These sequential echoes were grouped using a target tracking algorithm [25] and retained as individual fish targets. Each fish was assigned a central geographic position, a depth below the water surface and average, calibrated target strength measured in decibels (dB). Very few controlled observations are available to determine specific relationships between target strength and total length for coral reef fish [26]. Here, we used a species-independent, generalized formula to convert target strength into fish length [9,27]. Data along the survey path were binned into 100 m^2 intervals to normalize for the variation in beam width caused by changing depths. Fish densities were calculated for all fish exceeding −50 dB or a length of about 6 cm. The final dataset was exported as an ArcGIS point shapefile (referenced to North American Datum 1983 Universal Transverse Mercator 20 North) with each point representing the centroid of a 100 m^2 bin.

The species of individual fish cannot be identified from a single SBES frequency. Instead, fish targets were sorted into three size classes with the goal of separating them into ecological groups. These groups were initially chosen based on size estimates for species and groups from visual censuses for fish communities in the region [9,28]. Large fish (≥29 cm) are comprised of many important fishery species (e.g., *Serranidae* and *Lutjanidae*) and other large predatory species (Table 2) [28]. Medium fish (12–28 cm) include large reef residents and juvenile or small adults of fishery species (Table 3) [28]. Small fish (≤11 cm) represent small reef resident species, small planktivores and possibly juveniles of fishery species (Table 4) [28]. Maps of fish density were created for each size class in the training and validation sites (Fig. 3).

Mapping the seafloor using a multibeam echosounder. Bathymetry (i.e., depth) was collected concurrently with the SBES data in St. John Wedge and Tampo Bank using a hull-mounted Reson SeaBat 7125 SV1 MBES system [29,30]. MBES systems measure seafloor depth by transmitting multiple pulses of sound several times a second and then recording the time and angle of the acoustic returns. These two pieces of information are used to create highly resolved images of seafloor depth and topography. Each study area was mapped using the 400 kHz frequency, producing 2×2 m depth surface. Depth surfaces were corrected for sensor offsets, latency, roll, pitch, yaw, static draft, influence of tides and the changing speed of

Table 2. Large fish (≥29 cm) commonly found <55 m deep around St. John.

Number	Species Scientific Name	Species Common Name	Inhabited Depths (m)	Preferred Habitat
1	*Caranx crysos*	Blue runner	<100	water-column/seafloor (hardbottom)
2	*Lutjanus griseus*	Gray snapper	0–180	seafloor (hardbottom)
3	*Epinephelus guttatus*	Red hind	2–100	seafloor (hardbottom)
4	*Ocyurus chrysurus*	Yellowtail snapper	10–180	water-column/seafloor (hardbottom)
5	*Lutjanus analis*	Mutton snapper	40–95	water-column/seafloor (hardbottom)
6	*Lutjanus apodus*	Schoolmaster	2–63	water-column/seafloor (hardbottom)
7	*Pomacanthus paru*	French angelfish	3–100	water-column/seafloor (hardbottom)
8	*Cephalopholis fulva*	Coney	2–150	seafloor (hardbottom)
9	*Melichthys niger*	Black durgeon	0–75	water-column/seafloor (hardbottom)
10	*Bodianus rufus*	Spanish hogfish	3–70	seafloor (hardbottom)

Species of large fish commonly found in depths <55 m around St. John. The 55 m cutoff was used because >70% of both the training and validation sites were shallower than this depth. The most commonly observed species were identified from surveys conducted from 2001 to 2011 around St. John [28]. These fish species may represent the species of fish detected in the SBES data.

Table 3. Medium-sized fish (12–28 cm) commonly found <55 m deep around St. John.

Number	Species Scientific Name	Species Common Name	Inhabited Depths (m)	Preferred Habitat
1	*Clepticus parrae*	Creole wrasse	8–100	water-column/seafloor (hardbottom)
2	*Haemulon flavolineatum*	French grunt	0–60	seafloor (hardbottom)
3	*Cephalopholis fulva*	Coney	2–150	seafloor (hardbottom)
4	*Halichoeres garnoti*	Yellowhead wrasse	4–60	seafloor (hardbottom)
5	*Ocyurus chrysurus*	Yellowtail snapper	10–70	water-column/seafloor (hardbottom)
6	*Decapterus macarellus*	Mackerel scad	40–200	water-column
7	*Pseudupeneus maculatus*	Spotted goatfish	0–90	seafloor (softbottom)
8	*Epinephelus guttatus*	Red hind	2–100	seafloor (hardbottom)
9	*Lutjanus apodus*	Schoolmaster	2–63	water-column/seafloor (hardbottom)
10	*Myripristis jacobus*	Blackbar soldierfish	2–100	water-column/seafloor (hardbottom)

Species of medium-sized fish commonly found in depths <55 m around St. John. The 55 m cutoff was used because >70% of both the training and validation sites were shallower than this depth. The most commonly observed species were identified from surveys conducted from 2001 to 2011 around St. John [28]. These fish species may represent the species of fish detected in the SBES data.

sound in the water column. Both surfaces met International Hydrographic Organization Order 1 standards [31], and had a maximum horizontal uncertainty of ±10.0 m and vertical uncertainty of ±1.39 m. All data were referenced to North American Datum 1983 Universal Transverse Mercator 20 North projection and Mean Lower Low Water vertical coordinate system.

Surfaces describing the three dimensional structure of the seafloor were derived from these depth surfaces using ArcGIS's Spatial Analyst Toolbox and DEM Surface Toolbox [32] (Fig. 4). The surfaces, including standard deviation of depth, plan (or cross-sectional) curvature, rugosity and slope of slope (Table 5), were selected based on their demonstrated utility for predicting coral reef fish abundances and distributions [14,18]. The Spatial Analyst Toolbox was also used to calculate the geographic distance of the center of each grid cell to the shelf edge (i.e., the 183 m isobath). These six surfaces (i.e., depth, standard deviation of depth, plan curvature, rugosity, slope of slope, and distance to the shelf edge) were computed at four additional spatial scales (i.e., mean values within a radius of 25, 50, 100 and 300 m) to examine the influence of scale on fish distributions. These spatial scales were chosen based on previous research, which showed strong fish-seascape relationships at similar spatial scales [14,18,33,34]. A total of 24 predictors (i.e., 6 surfaces x 4 spatial scales) were included in the modeling process and used to develop spatial predictions in the training and validation sites.

Predicting Fish Distributions and Densities

Boosted regression trees. Boosted regression trees (BRT) is a machine learning technique used effectively in ecology to model the complex, non-linear relationships between organisms and their environment. BRTs model these complex relationships by developing many (sometimes hundreds to thousands) simple models based on random subsets of the data [35,36]. These simple models are then combined linearly to produce one final aggregate (i.e., ensemble) model [37]. The fitted values in this ensemble model are more stable than values from an individual model, improving its overall predictive performance [37,38]. The

Table 4. Small fish (≤11 cm) commonly found <55 m deep around St. John.

Number	Species Scientific Name	Species Common Name	Inhabited Depths (m)	Preferred Habitat
1	*Stegastes partitus*	Bicolor damselfish	0–100	seafloor (hardbottom)
2	*Chromis cyanea*	Blue chromis	10–60	water column/seafloor (hardbottom)
3	*Halichoeres garnoti*	Yellowhead wrasse	4–80	seafloor (hardbottom)
4	*Serranus tortugarum*	Chalk bass	8–90	water column/seafloor (hardbottom)
5	*Clepticus parrae*	Creole wrasse	8–100	water column/seafloor (hardbottom)
6	*Chromis multilineata*	Brown chromis	0–60	water column/seafloor (hardbottom)
7	*Sparisoma atomarium*	Greenblotch parrotfish	20–55	seafloor (hardbottom/softbottom)
8	*Cryptotomus roseus*	Bluelip parrotfish	0–60	seafloor (softbottom)
9	*Ocyurus chrysurus*	Yellowtail snapper	10–180	water column/seafloor (hardbottom)
10	*Gramma loreto*	Fairy basslet	1–60	seafloor (hardbottom)

Species of small fish commonly found in depths <55 m around St. John. The 55 m cutoff was used because >70% of both the training and validation sites were shallower than this depth. The most commonly observed species were identified from surveys conducted from 2001 to 2011 around St. John [28]. These fish species may represent the species of fish detected in the SBES data.

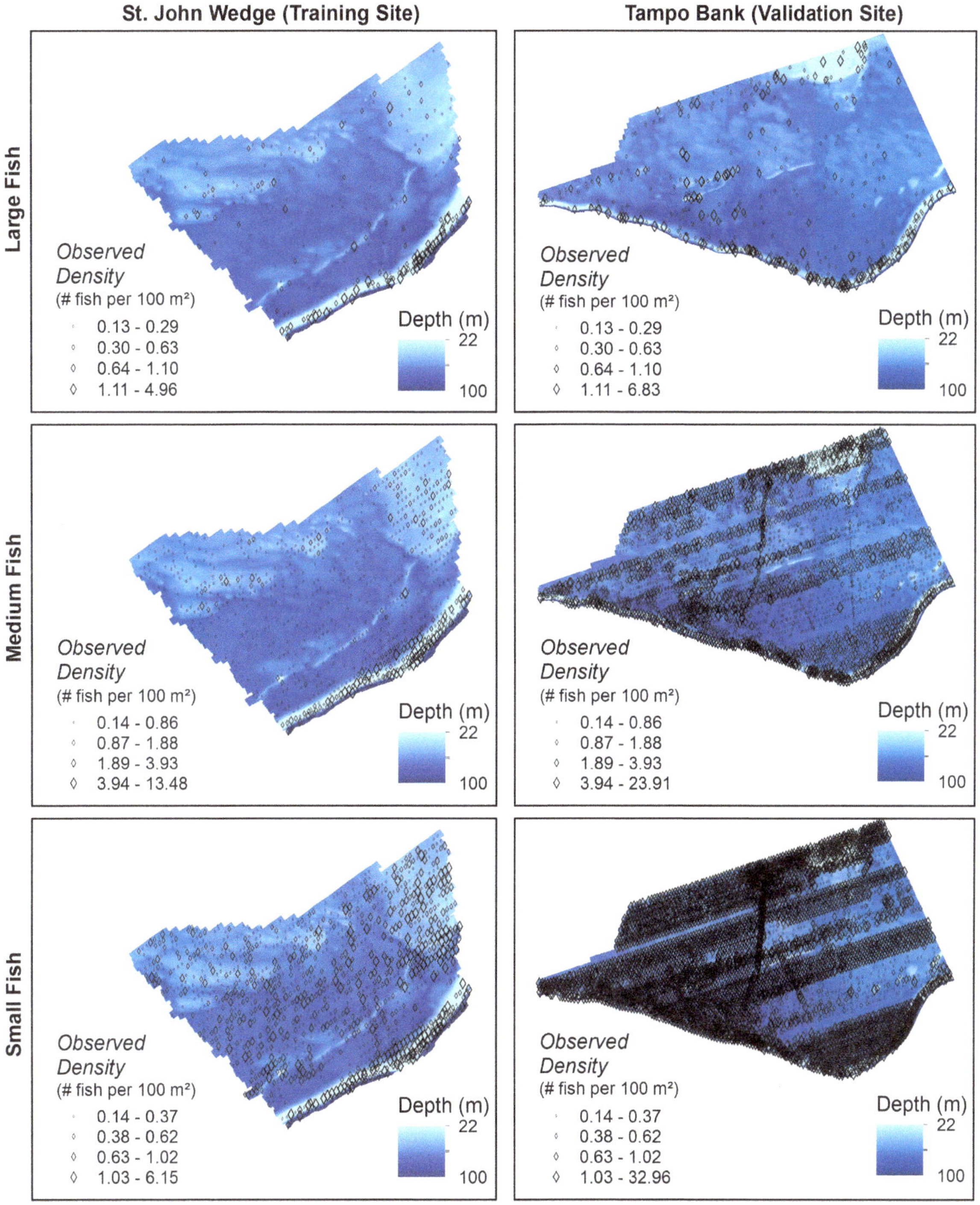

Figure 3. Maps of SBES data. Presence and density of large, medium and small fish in the training site (left) and validation site (right).

BRT approach to spatial modeling was used in this study because it can deal with data that is not normally distributed, is robust to missing data values, can handle interactions among predictors and compared favorably (both in terms of predictive performance and accuracy) to other modeling techniques [14,37,38].

Model development. For this study, 60 BRT models were generated from MBES and SBES data. Ten of these models predicted large fish occurrence, 10 predicted medium fish occurrence, 10 predicted small fish occurrence, 10 predicted large fish density, 10 predicted medium fish density, and 10 predicted small fish density. Multiple models for occurrence and density were created to avoid fitting one model too closely to the data, and to better understand and quantify the stability of BRT's variable selection and predictive performance [37]. BRT models were developed using the "gbm.step" function in the "dismo" package version 0.7 [39] implemented in R software version 2.15. Each BRT model was trained using a different random 50% of the St.

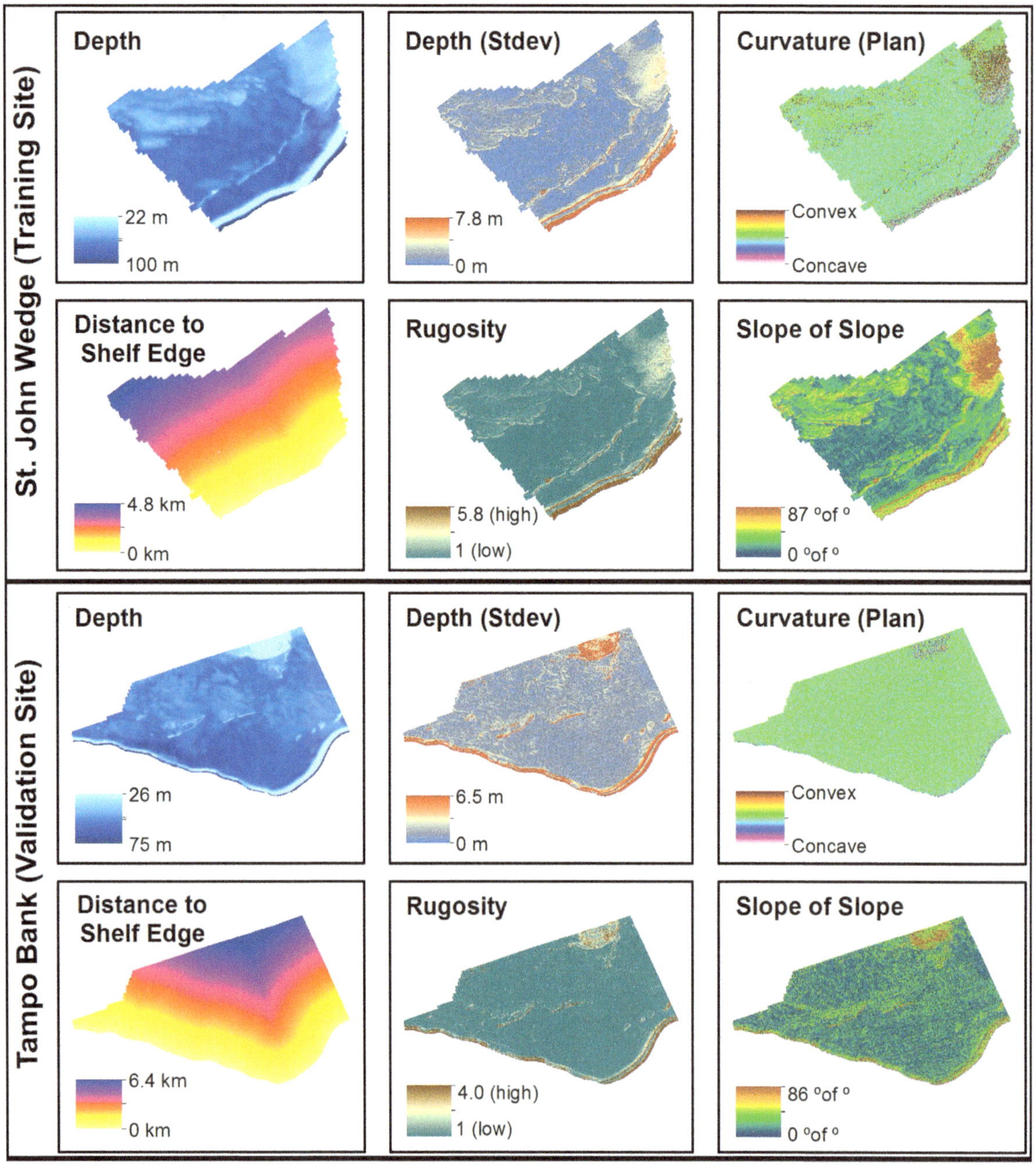

Figure 4. Maps of predictor data. Six environmental variables used as predictors in both the training and validation sites. These six variables were also included in the modeling process at four additional spatial scales. A total of 24 predictors were used to create each boosted regression tree model and spatial prediction.

John Wedge SBES data (n = 1,641 points representing 100 m^2 bins). The remaining 50% were used for cross validation (CV).

In each of the St. John Wedge and Tampo Bank areas, six predictive surfaces (i.e., three surfaces predicting fish occurrence by size class and three predicting fish density by size class) were produced by averaging each group of 10 BRT models. Spatial predictions were developed using the "raster" package version 2.0 [40] implemented in R version 2.15. Spatial predictions for fish occurrence denote the probability that a large, medium or small fish is present in a 2×2 m area. Spatial predictions for fish density denote the number of large, medium and small fish predicted to be in a 2×2 m area. These spatial predictions were independently validated using the Tampo Bank SBES data to simulate and evaluate how well they would perform in areas that had not been surveyed with a SBES.

Model performance. When evaluating a model's performance, both its discrimination capacity and the reliability should be assessed. Discrimination capacity refers to the ability of the model to differentiate between classes (e.g., presences and absences), while reliability describes the agreement between the predicted and observed values at specific locations [41,42]. The discrimination capacity of the BRT models for large, medium and small fish occurrence and density was assessed using receiver operating characteristic (ROC) curves, and their reliability was

Table 5. Descriptions of predictors.

Predictor Dataset	Unit	Description	Tool Used
1. Depth	Meters	Water depth	-
2. Depth (Standard Deviation)	Meters	Dispersion of water depth values about the mean (in a 3×3 cell neighborhood)	Focal statistic function in ArcGIS's Spatial Analyst
3. Curvature (Plan or Cross-Sectional)	Concave (−) & Convex (+)	Curvature of the surface perpendicular to the slope direction (in a 3×3 cell neighborhood)	Curvature function in ArcGIS's Spatial Analyst
4. Distance to Shelf Edge	Kilometers	Distance of the centroid of each pixel to the 183 m (100 fathom) isobath	Euclidean distance function in ArcGIS's Spatial Analyst
5. Rugosity	Ratio value	Ratio of surface area to planar area (in a 3×3 cell neighborhood)	Surface Area and Ratio function in DEM Surface Toolbox
6. Slope of the Slope	Degrees of degrees	Maximum rate of maximum slope change (in a 3×3 cell neighborhood)	Slope function in ArcGIS's Spatial Analyst

Environmental variables used to predict large, medium and small fish occurrence and density. Each variable was also included in the modeling process at four additional spatial scales (i.e., using circles with radii of 25, 50, 100 and 300 m).

evaluated using mean absolute error (MAE) and root mean square error (RMSE). MAE and RMSE both measure the average magnitude of the predictive errors (independent of their direction). However, MAE weights each error equally in the average, while RMSE weights large errors much more heavily. Both metrics are reported here so that the impact on different management applications can be explored.

The other model performance metric, called ROC curves, measure a model's performance by comparing its sensitivity (i.e., true positive prediction rate) to its specificity (i.e., false positive prediction rate) over the continuous range of predicted values. The diagonal y = x line in a ROC curves denotes how a randomly generated model would perform. ROC curves above this line perform better than a random model, while ROC curves below this line have useful information but are applying it incorrectly [43]. The area under the curve (AUC) statistic was also calculated, which describes the overall predictive performance of a model compared to a random guess. It is equal to the probability that a model will rank a randomly chosen presence higher than a randomly chosen absence. AUC values ranging from 0.5 to 0.6 suggest the model is no better at discriminating classes than random chance; values from 0.6 to 0.7 denote "poor" model performance; values ranging from 0.7 to 0.8 denote "acceptable" model performance; values from 0.8 to 0.9 denote "excellent" model performance, and values greater than 0.9 denote "outstanding" model performance [44].

ROC curves have several advantages over traditional accuracy assessment techniques, including confusion matrices. One notable advantage is that ROC curves are independent of binary thresholds (i.e., break points where animals are defined as present or absent) that are often chosen subjectively [45,46]. ROC curves do not require that a predictive threshold be selected because they describe a model's performance over the complete range of predicted values. The other important advantage of ROC curves is that they are unaffected by changes in animal prevalence (i.e., unequal amounts of presences and absences) [45,46] because they are based on ratios (and not summaries) of true presences to false presences. This independence is particularly important when developing models for rare animals (i.e., that have low prevalence, like large fish) because it is possible to get high overall model accuracy by predicting such animals are absent everywhere [45].

Evaluating model performance. ROC curves and correlation coefficients were developed in R using 10-fold cross validation data in the St. John Wedge area. In the Tampo Bank area, ROC curves were developed (along with MAE and RMSE) using an independent SBES dataset. This independent dataset (n = 5,269) was used solely for assessing the performance of the final spatial predictions. A subset of this validation dataset (n = 2,634) was chosen randomly to avoid biasing the evaluation process in R. Spatially autocorrelated points were then removed from this data subset because positive autocorrelation violates the assumption of independence and biases statistical tests by effectively overestimating the true sample size [47,48].

These autocorrelated points were identified in the large, medium and small fish datasets by detrending them using local polynomial regression, developing three empirical semi-variograms from the residuals, and fitting spherical models to the variograms using the "stats" and "geoR" package version 1.7 in R [49,50]. The ranges for the large, medium and small fish variogram models were 280 m, 503 m and 272 m, respectively.

Table 6. Frequency of Large Fish Records by Density Class.

Fish Density Class	Fish Density Threshold (# fish/100 m^2)	Percent of Total Records (Training)	Percent of Total Records (Validation)
Absent to Low	≤0.29	93.1%	91.9%
Low	0.30 – 0.63	4.0%	5.0%
Medium	0.64 –1.10	1.2%	1.6%
High	≥1.11	1.8%	1.5%

Frequency of large fish records by density class in both the training and validation areas. These classes were determined using Jenks Natural Breaks.

Table 7. Frequency of Medium Fish Records by Density Class.

Fish Density Class	Fish Density Threshold (# fish/100 m^2)	Percent of Total Records (Training)	Percent of Total Records (Validation)
Absent to Low	≤0.86	89.5%	64.0%
Low	0.87 – 1.88	6.3%	18.8%
Medium	1.89 – 3.93	2.9%	12.4%
High	≥3.94	1.3%	4.8%

Frequency medium fish records by density class in both the training and validation areas. These classes were determined using Jenks Natural Breaks.

Points closer together than these distances were assumed to be spatially autocorrelated, and were removed from the validation process using Matlab. This step removed 2201, 2479 and 2219 points from the large, medium and small fish validation datasets. The remaining 432, 154 and 415 spatially independent points were used to create ROC curves and calculate MAE and RMSE for the large, medium and small fish occurrence and density predictions.

ROC curves for the fish density predictions were created differently than those for the fish occurrence predictions because ROC curves are not designed to handle validation data that is continuous (i.e., densities). To address this issue, the large, medium and small fish density data was divided into four classes (i.e., absent to low, low, medium and high) using Jenks natural breaks in ArcGIS (Tables 6, 7, 8). This method was chosen because it is well suited for grouping data with large variances [51]. Six ROC curves were then created for each of the large, medium and small fish density predictions by comparing the four density classes in a pair-wise fashion (i.e., absent vs. low, absent vs. medium, absent vs. high, low vs. medium, low vs. high and medium vs. high). AUC was calculated for each curve as well as for the entire multiclass prediction using the method defined by Hand and Till, 2001 [52].

Spatial distribution of model errors. While ROC curves, MAE and RMSE describe the discrimination capacity and reliability of models, they do not describe the spatial distribution of model errors [53,54]. Analyzing the spatial location and arrangement of errors can be important because they may offer clues about missing ecological or biological variables and their spatial structure [54]. A model with spatially clustered errors (versus randomly distributed) may indicate that there are unaccounted for spatially structured variables. To better understand this spatial structure, the large, medium and small fish validation datasets were subtracted from their corresponding occurrence and density predictions. Cluster and outlier analysis in ArcGIS's Spatial Analyst Toolbox was then used to describe the spatial distribution and clustering of the residual model errors. This tool identifies statistically significant spatial clusters of high values, low values and outliers using inverse distance weighting and the Anselin Local Moran's I statistic.

Contribution of predictor variables. Two different metrics were used to quantify how much each predictor contributed to the BRT models. The first metric (i.e., 'the relative influence of each predictor variable') is based on the number of times that a predictor is selected for splitting. This sum is weighted by how much the model is improved by each split, averaged across all the trees and scaled so that the sum equals 100 [37]. The higher the scaled number, the more influence a predictor has on the model and vice versa. The top three predictors from this analysis were examined in the discussion for each spatial prediction. The second metric (i.e., 'partial dependence plots') examines how fish occurrences and densities change over the continuous range of values for a predictor (after accounting for the average effects of all other predictors in the model). These plots can be used to identify thresholds or peaks in the presence and density of large, medium and small fish for each predictor [37].

Results

Fish Occurrence Models

Large fish. Large fish were observed in 15% of the SBES records in the St. John Wedge area and 19% of the SBES records in the Tampo Bank area. In Tampo Bank, the AUC value for the large fish occurrence prediction (0.68±0.06) indicated 'poor' model performance (Fig. 5). The average difference between the predicted and observed probability of occurrence values was MAE = 30.0% and RMSE = 36.0% (Fig. 6).The majority (79.4%) of model errors were positive and ≤MAE, indicating that the BRT model more commonly over-predicted (vs. under-predicted) the probability of occurrence for large fish (Fig. 7). Negative errors (i.e., where the model under-predicted the probability of occurrence) were much less common (15.3%), but were always larger than the MAE. Large, positive errors comprised the remaining 5.3% of the model errors, and were located mainly along the shelf edge and over hard bottom in the northeast

Table 8. Frequency of Small Fish Records by Density Class.

Fish Density Class	Fish Density Threshold (# fish/ 100 m^2)	Percent of Total Records (Training)	Percent of Total Records (Validation)
Absent to Low	≤0.37	62.7%	35.4%
Low	0.38 – 0.62	18.0%	5.6%
Medium	0.63 – 1.02	10.8%	7.1%
High	≥1.03	8.5%	51.9%

Frequency of small fish records by density class in both the training and validation areas. These classes were determined using Jenks Natural Breaks.

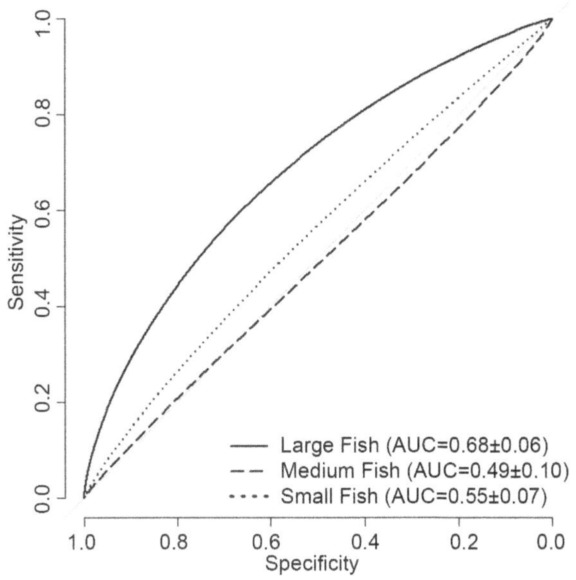

Figure 5. ROC curve for fish occurrence predictions. These ROC curves were developed using the independent dataset in the Tampo Bank area. Area under the curve (AUC) values (at the 95% confidence interval) are listed in the lower right hand corner.

quadrant of Tampo Bank. Large negative errors were located throughout Tampo Bank, but were clustered primarily along the shelf edge and over a linear reef in the southwest quadrant of Tampo Bank. A description the partial dependence plots and influence of each predictor is not provided because the occurrence model for large fish performed poorly.

Medium and small fish. Medium and small fish were observed in 45% and 66% of the SBES records in the St. John Wedge area and 74% and 79% of the SBES records in the Tampo Bank area, respectively. In Tampo Bank, the AUC values for the medium fish (0.49±0.10) and small fish (0.55±0.07) occurrence predictions indicated that they performed no better than a random model (Fig. 5). This weaker model performance is also reflected in the larger MAE and RMSE values (Fig. 6), which were 3% to 15% higher for medium and small fish than for large fish (MAE = 40.0% and RMSE = 45.0% for medium fish, and MAE = 33.0% and RMSE = 41.0% for small fish). The majority of errors were negative and <MAE for medium and small fish (72.7% and 78.5%, respectively) (Fig. 7). All of the positive model errors were >MAE and clustered mainly on the insular shelf for both predictions. A description the partial dependence plots and influence of each predictor is not provided because the occurrence models for medium and small fish did not perform better than would be expected by random chance.

Fish Density Models

Large fish. High densities of large fish were rare (<1.8% of the SBES records) in both the St. John Wedge and Tampo Bank areas (Table 6). The multi-class AUC value (0.77) for the large fish density prediction indicated 'good' overall model performance (Fig. 8), outperforming the large fish occurrence model. Pairwise comparisons between density classes indicated that the BRT model was able to reliably distinguish the absent to low and the low density classes from the medium (AUC = 0.73±0.20; AUC = 0.70±0.24, respectively) and high (AUC = 0.87±0.05; AUC = 0.73±0.16, respectively) density classes. The model also reliably differentiated the medium from the high density class (AUC = 0.73±0.19), but not the absent to low from the low class (AUC = 0.53±0.16). The average difference between the predicted and observed large fish density values was MAE = 0.16 and RMSE = 0.26 fish per 100 m^2 (Fig. 9).The majority (78.0%) of model errors were positive and <MAE, indicating that the BRT model more commonly over-predicted large fish densities (Fig. 10). Negative errors were much less common (12.7%), and were about equally above and below the MAE. Large, positive errors comprised the remaining 9.3% of the model errors, and were located mainly along the shelf edge and over hard bottom in the northeast quadrant of Tampo Bank. Large negative errors were located about equally on the insular shelf (n = 12) and along the insular shelf edge (n = 10) in Tampo Bank, although more clustering occurred along the shelf edge.

Medium fish. High densities were also uncommon for medium fish (i.e., <4.8%) in both project areas, although medium fish were more often found in low and medium densities (>17%) than large fish (Table 7). The medium fish density prediction performed poorly (multi-class AUC = 0.68; Fig. 8). Pairwise comparisons among the absent to low, the low and the medium densities classes also indicated that the medium fish density BRT model performed poorly or no better than a random model (AUC = 0.55 to 0.62). However, the BRT model was able to reliably distinguish the high density class from the absent to low, the low and the medium density classes (AUC = 0.77±0.15; AUC = 0.80±0.13; AUC = 0.73±0.17, respectively). The average difference between the predicted and observed medium fish density values was MAE = 1.0 and RMSE = 1.65 fish per 100 m^2 (Fig. 9). Model errors were nearly equally split between being negative (43.5%) and positive (56.5%), indicating that the BRT model did not systematically under or over-predict medium fish densities (Fig. 10). The majority of errors (71.4%) were <MAE. Of the 28.6% of errors >MAE, more were negative (21.4%) than positive (7.1%). The large positive errors were located mainly along the shelf edge and in the northeast quadrant of Tampo Bank. The large negative errors were located in all four quadrants of the Tampo Bank area, but they were clustered along the shelf edge.

Small fish. Small fish were commonly found at medium and high densities (>19%) in both areas (Table 8). The BRT model was no better at predicting small fish densities than a random model (multi-class AUC = 0.53; Fig. 8), and could not reliably distinguish between any of the density classes (AUC<0.55). This weaker model performance is also reflected in the larger MAE and RMSE values (2.1 and 3.6 fish per 100 m^2, respectively; Fig. 9). The majority (67.6%) of model errors were negative and clustered mainly in the southwest and northwest quadrants of Tampo Bank (Fig. 10). Approximately 31.6% of these negative errors were > MAE. Positive errors were all <MAE, and located mainly in the northeast and southeast quadrants. Given that the density models for small and medium fish both performed poorly, a description the partial dependence plots and influence of each predictor is not provided for either model.

High to medium densities of large fish were predicted along the shelf edge in both the St. John Wedge and Tampo Bank areas (Fig. 9). Low densities were predicted shoreward at the Tampo Bank area along hard bottom features with moderate amounts of structural complexity. Absent to low densities of large fish were predicted shoreward of the shelf edge over areas with low amounts of structure. Depth (at two different spatial scales) and standard deviation of depth were the top three most important environmental variables influencing the density and distribution of large fish (Fig. 11). These three predictors each explained between 7.8

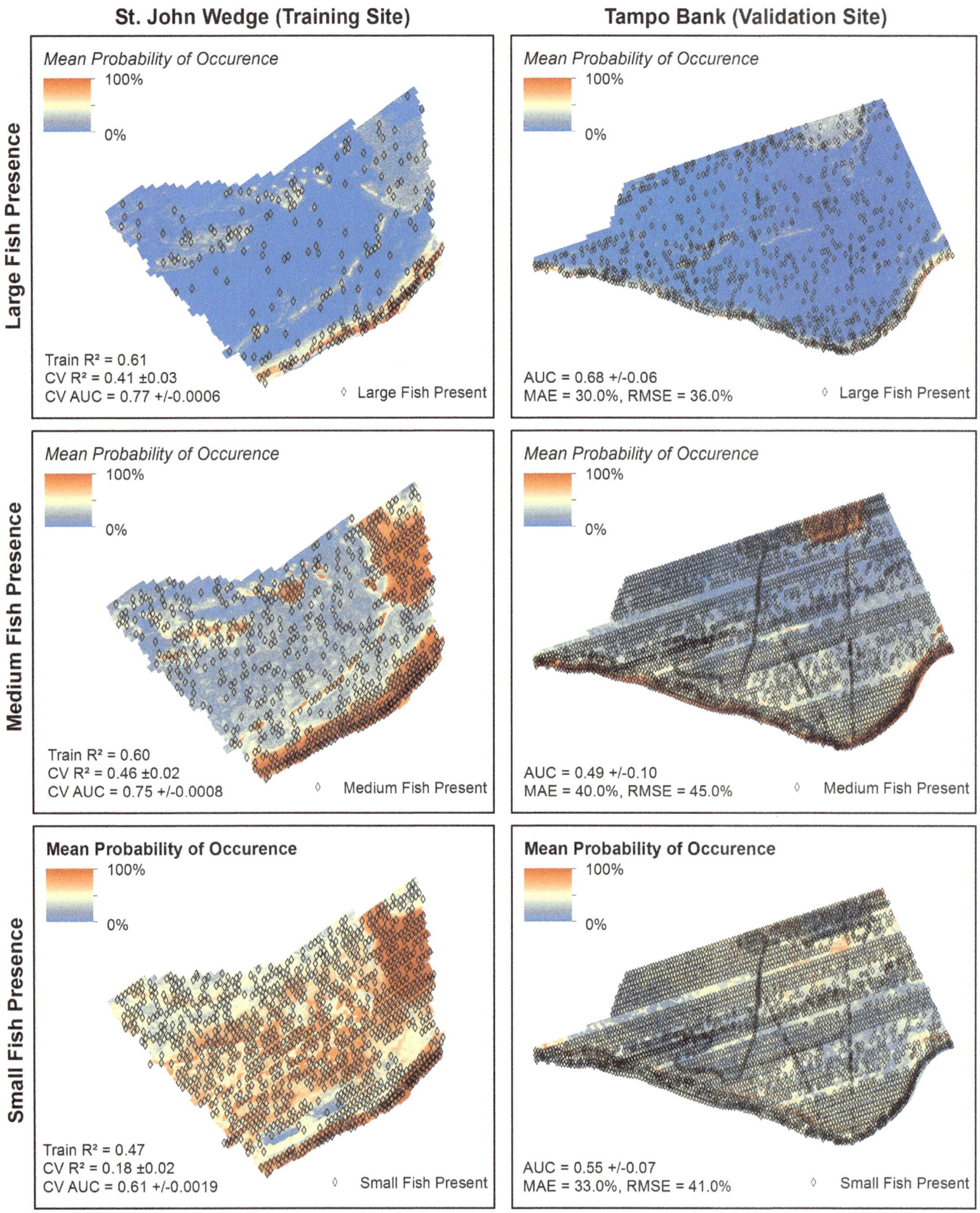

Figure 6. Fish occurrence spatial predictions. Three spatial predictions denoting the probability of occurrence for large, medium and small fish were created for both the training and validation sites. The spatial outputs from 10 different BRT models were averaged to create the spatial predictions seen here. Metrics describing the performance and accuracy of these models and spatial predictions are located in the lower left corner of each map.

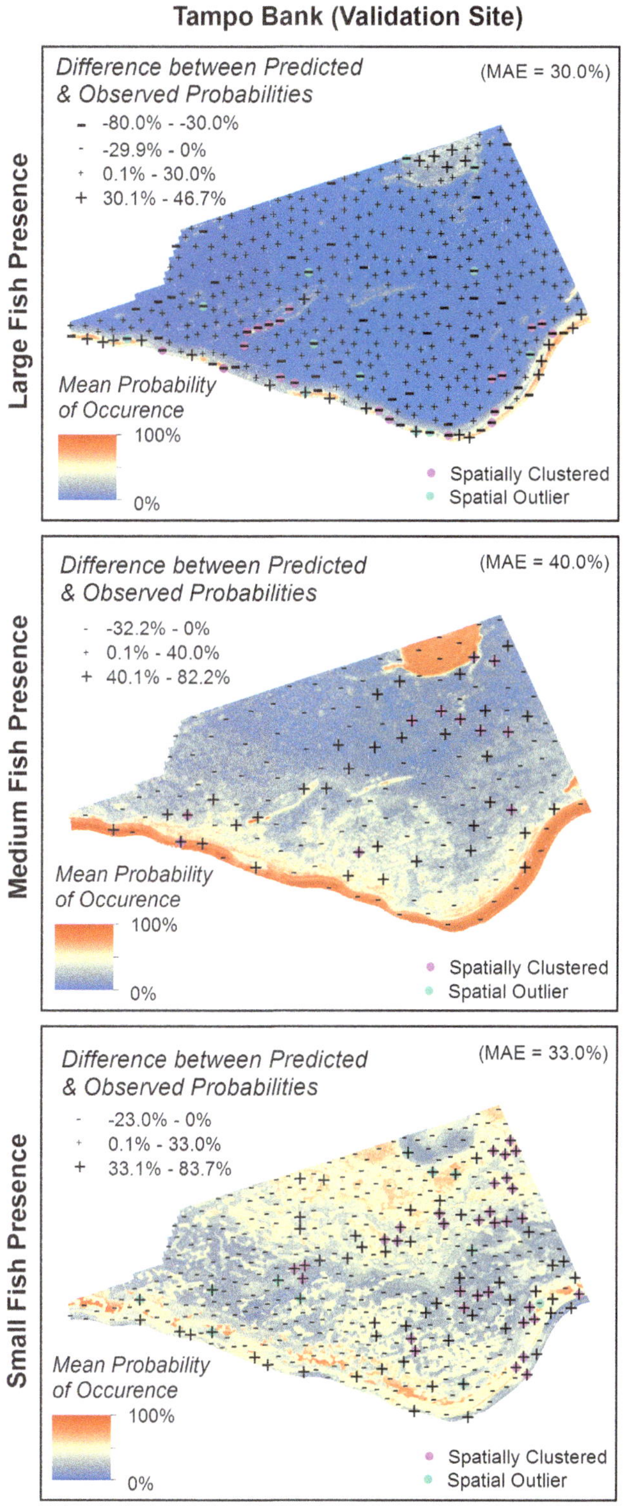

Figure 7. Errors for fish occurrence spatial predictions. The magnitude and spatial structure of the errors for the fish occurrence predictions were calculated by subtracting the predicted from observed probability of occurrence values. Negative values indicate that the model under-predicted, and positive values indicate the model over-predicted the probability of occurrence for large, medium and small fish. The error data was divided into classes based on the MAE. Spatial autocorrelation of the residuals were analyzed using Anselin Local Moran's I statistic. Analyzing the spatial location and arrangement of errors can be important because they may offer clues about missing ecological or biological variables.

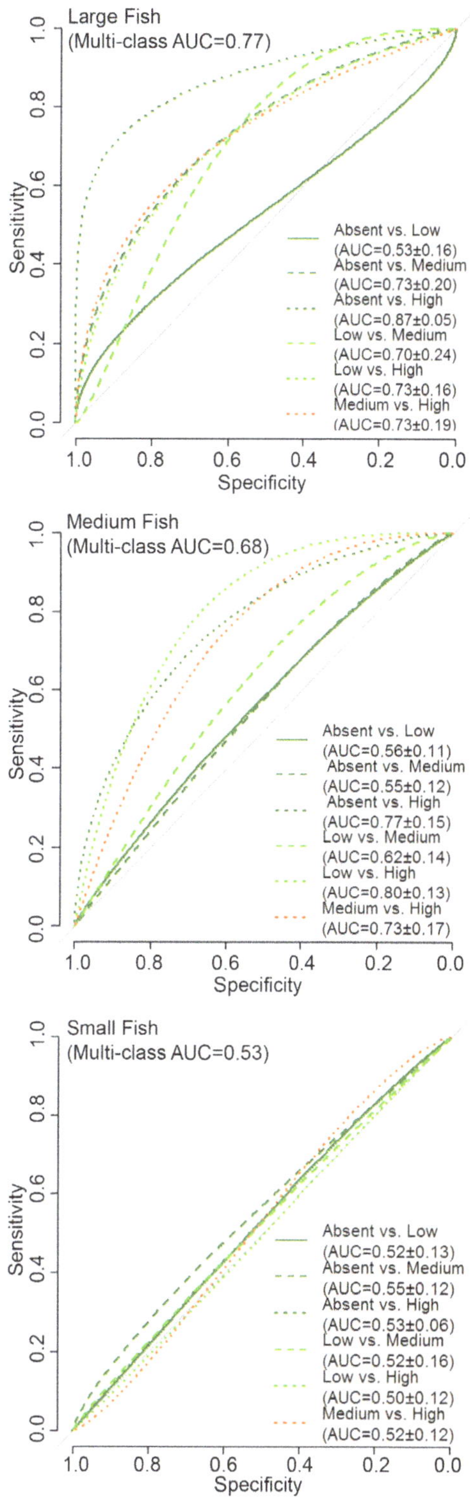

Figure 8. ROC curves for fish density predictions. These ROC curves were developed using the independent dataset in the Tampo Bank area. The multiclass AUC values are listed at the top of the figure for the large, medium and small fish predictions, and the AUC values (at the 95% confidence interval) are listed in the lower right hand corner for each pairwise comparison.

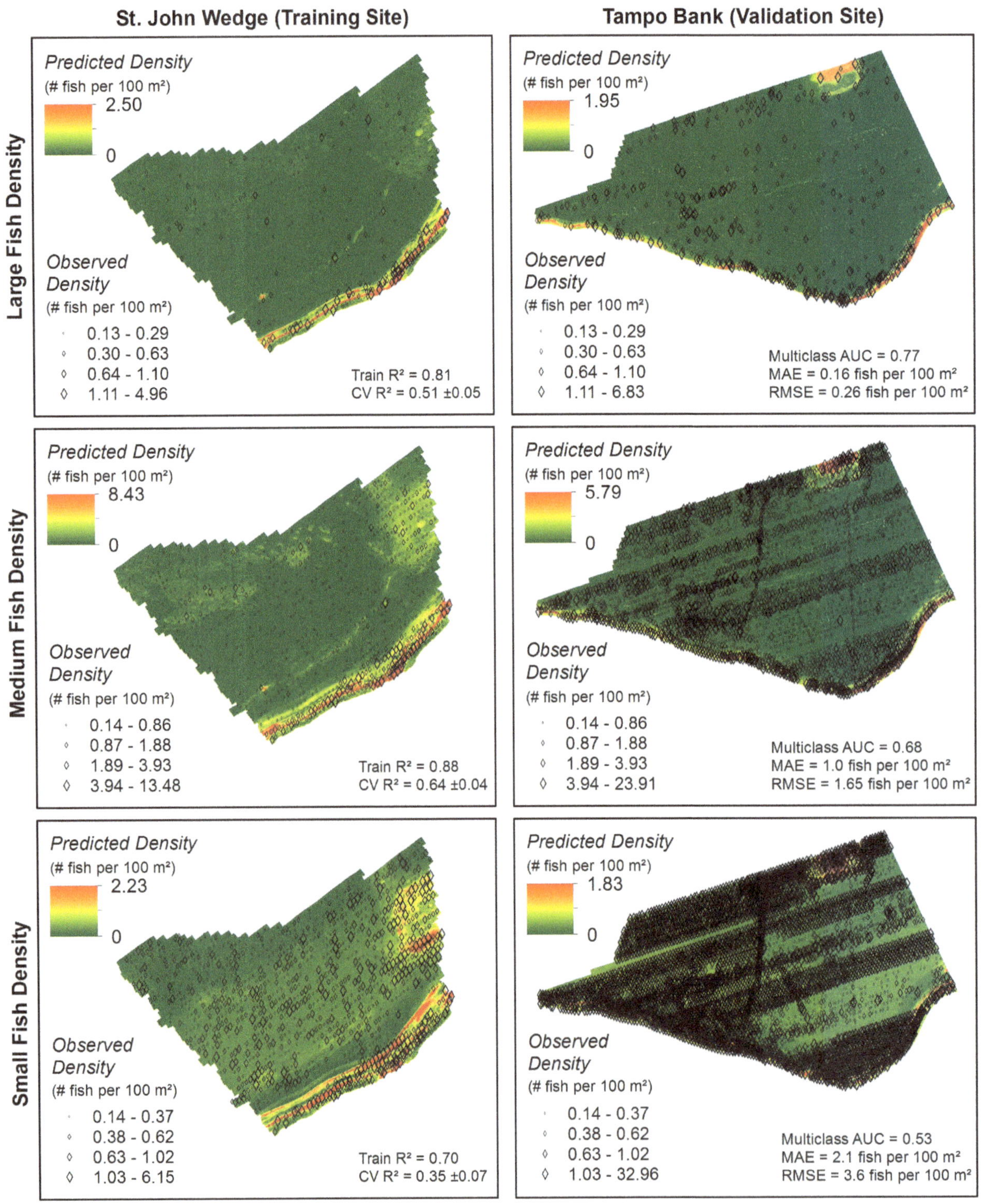

Figure 9. Fish density spatial predictions. Three spatial predictions denoting the density of large, medium and small fish were created for both the training and validation sites. The spatial outputs from 10 different BRT models were averaged to create the spatial predictions seen here. Metrics describing the performance and accuracy of these models and spatial predictions are located in the right left corner of each map.

and 14.6% of the variance in the large fish density data. All of these predictors had spatial scales ≥100 m, suggesting that the BRT density models were more heavily influenced by these variables at broad spatial scales. Partial dependence plots for these three predictors showed clear peaks and breakpoints in the response data (Fig. 12). When all other variables were held at their average values, large fish were more likely to occur at high densities where the seafloor was shallower (<38 m) and more complex (i.e., areas where the depth deviated by >0.29 m) (Figs. 12 a, b and c).

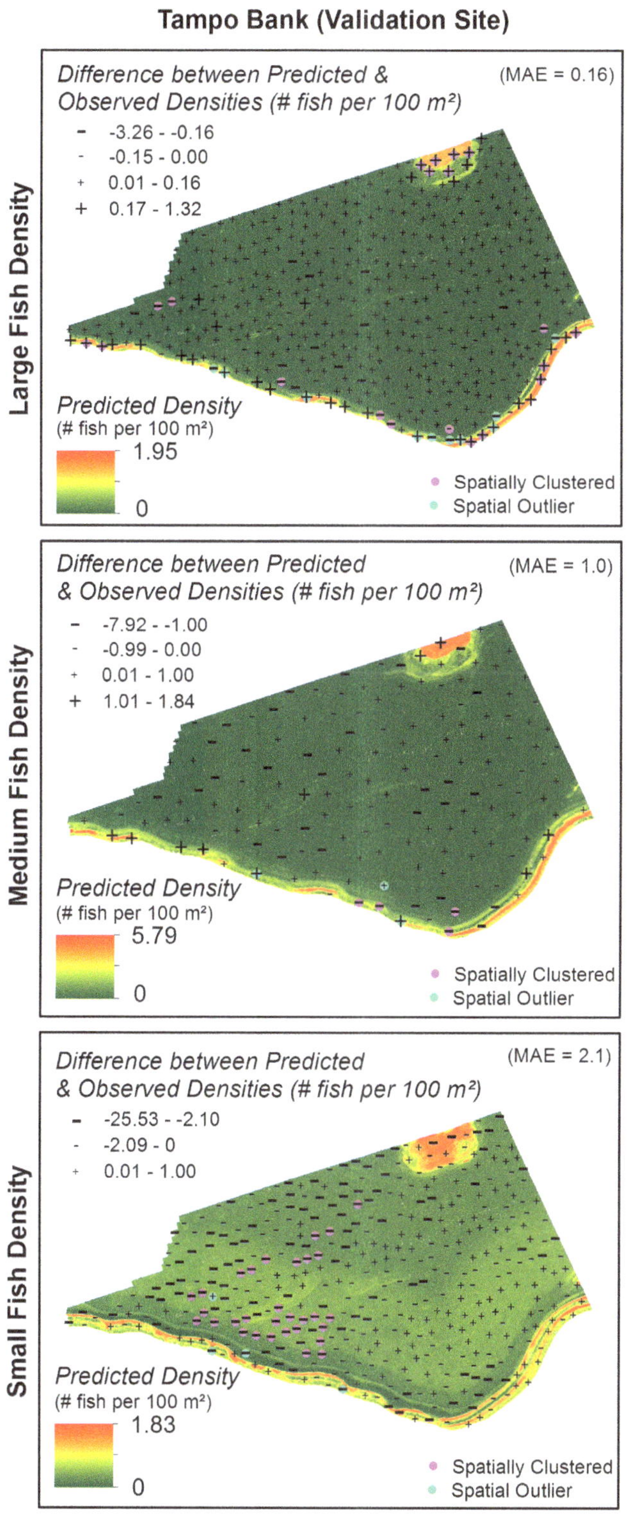

Figure 10. Errors for fish density spatial predictions. The magnitude and spatial structure of the errors for the fish density predictions were calculated by subtracting the predicted from the observed fish density values. Negative values indicate that the model under-predicted, and positive values indicate the model over-predicted the density of large, medium and small fish. The error data was divided into classes based on the MAE. Spatial autocorrelation of the residuals were analyzed using Anselin Local Moran's I statistic. Analyzing the spatial location and arrangement of errors can be important because they may offer clues about missing ecological or biological variables and their spatial structure.

Discussion & Conclusions

The novel integration of acoustic sensors offers a new approach to rapidly acquire spatial data across broad extents to identify both fish aggregations and areas of low fish occurrence. When combined with predictive modeling, it also offers a reliable method for predicting the density of large fish in areas where fish distributions have not been mapped. While fish acoustic systems are currently unable to identify fish species, they can collect many thousand observations (>5000 data points in each location) in a short amount of time (i.e., days to weeks versus months to years for SCUBA diver based surveys). These large acoustic datasets can support analysis—across a range of fish size classes and spatial scales—linking fish densities with seafloor structure and proximity to the insular shelf edge. They can also provide insights into geographic areas that are important for reef and non-reef associated species, helping coastal managers focus their efforts and limited resources on locations that may be of higher conservation value.

Model Performance

Fish size classes. The model results indicate that we can reliably predict the density of large fish (in areas up to 100 m deep). Confusion between the absent to low and the low density classes was most likely due to the threshold chosen, and could be removed by merging the two classes. These results are comparable to those previously developed for shallow areas (<30 m) in a similar tropical coral reef ecosystem using fish distribution data from SCUBA diver surveys [14,18,33,55,56]. However, we were unable to produce an 'acceptable' model predicting the probability of occurrence for large fish. The large fish occurrence model may have performed better if we were able to divide the presence and absence data by species or trophic groups (instead of size class), as in previous studies.

Predictions for medium fish occurrence and for small fish occurrence and density were no more accurate than would be expected by random chance alone. Models for medium fish density performed somewhat better (particularly for the absent to low, the low and the medium versus high density classes), but their overall performance was still below the acceptable range. We attribute the poor performance of these models to the unequal prevalence of small to medium fish in the validation area versus the training area. No attempt was made to choose areas with similar prevalences because the main goal of this modeling effort was to predict fish distributions in un-surveyed (i.e., where animals' distributions and prevalences are unknown). However, this result highlights the need for caution when applying predictive models to new locations, and the need for an independent assessment of their accuracy before using them to make management decisions. In this case, it is difficult to know why smaller fish were more prevalent in the validation area. However, one possibility is the distinct shape of the shelf edge in the Tampo Bank area, which protrudes further out into deeper waters than the St. John Wedge area (Fig. 2). The importance of promontories (i.e., bends in the shelf edge, where the steep terrain protrudes into deeper waters) and shelf edge habitats are discussed in more detail in section 4.4.

In addition to unequal prevalences, we also attribute the poor performance of the medium and small fish models to the more even distribution and higher prevalence of smaller fish (45–79%) versus large fish (15–19%) overall. Smaller fish were most likely more prevalent because they experience less fishing pressure than larger fish, and they were more evenly distributed because they exploit a wider range of habitats than larger fish. The latter half of

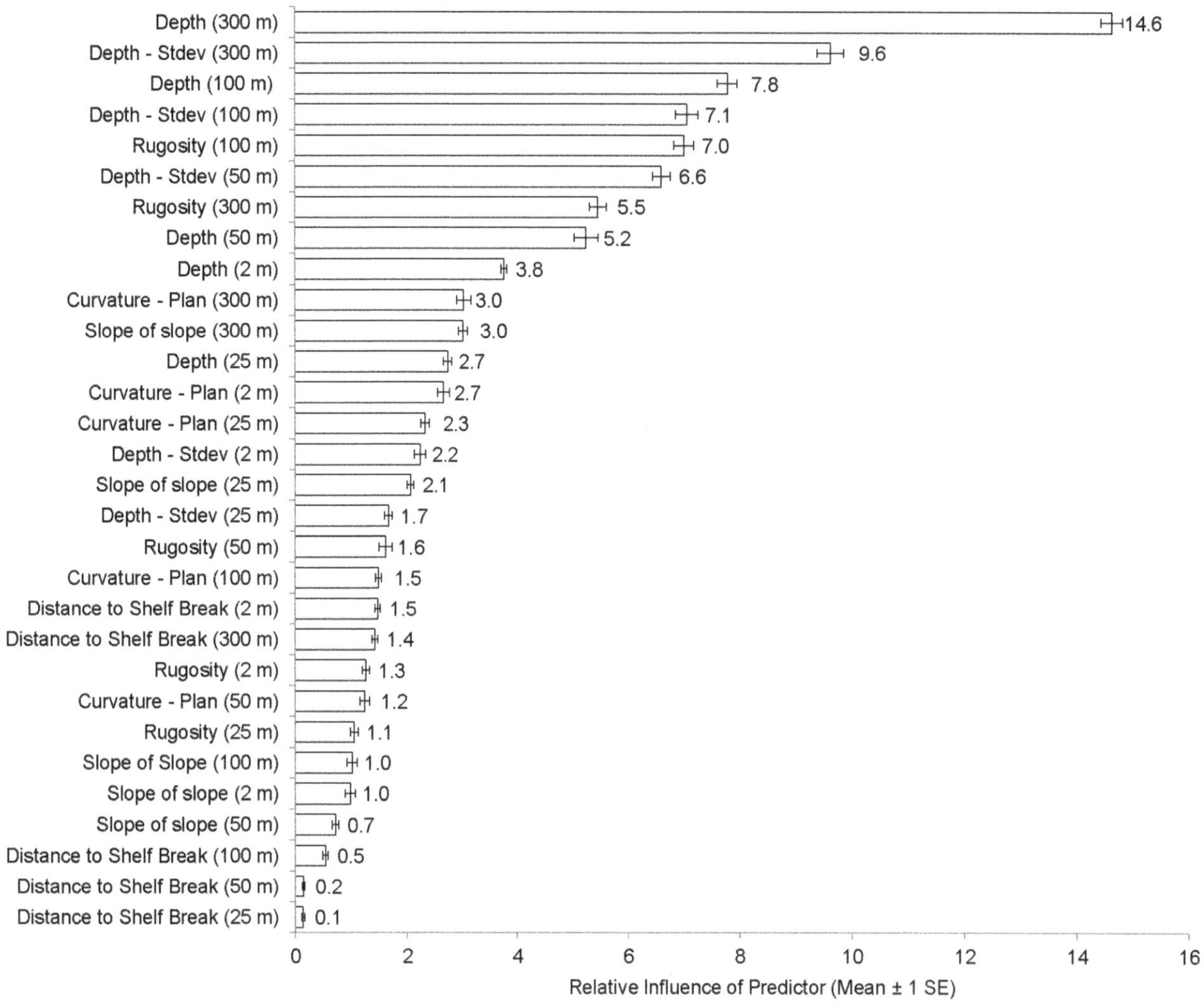

Figure 11. Influence of predictors on large fish density models. This figure shows the relative influence of each environmental variable on the large fish density spatial prediction. Cross validation data were used to calculate these values and standard errors, which were averaged 10 BRT model replicates.

this explanation is supported by the fact that the most commonly seen small and medium fish groups included species with more varied habitat preferences and feeding habits than those found in the large fish class. Notably in Table 2, the most commonly seen large fish species were associated with only two types of habitats, whereas the most commonly seen medium and small fish species (Tables 3 and 4) were associated with three and four types of habitats, respectively.

Seafloor complexity predictors. The most influential factors for predicting large fish density were depth and variation

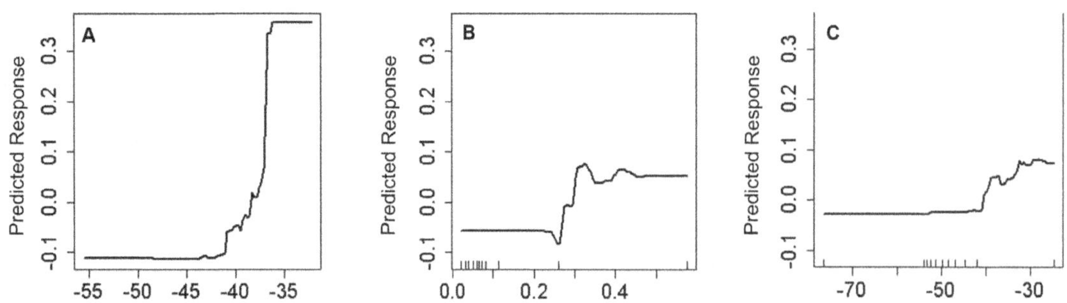

Figure 12. Partial dependence plots for large fish density models. Response curves for the three environmental variables that had the greatest influence on the prediction of large fish density. Collectively, these variables explain approximately 32% of the variance in the response data. All other variables were held at their average value.

in depth. Combined, these predictors explained 32% of the variance in the large fish density data. These influential predictors are similar to those identified by other reef fish modeling studies. Notably, depth explained over 10% of the variance in the occurrence of several fish species, and the abundance and biomass of piscivores in southwestern Puerto Rico [14,18]. Slope of slope was also a common predictor for these individual species and community metrics [18], but it was not an influential factor in our models (<5% relative influence). Its influence may have been masked by the variation and overlap in habitat preferences among fish species grouped by size class. It is likely that both pelagic and demersal fish were included in our estimates of large fish occurrence and density, confounding the link between seafloor structure and large fish densities.

Spatial scale. The top model predictors for large fish density were important at relatively broad spatial scales (100 and 300 m). These relationships are similar to other modeling exercises, which also found linkages between larger fish and habitats at similarly broad spatial scales (i.e., 100, 200 m and 500 m) [14,18,34,55]. However, some of these same studies also found that smaller fish responded to habitats at much finer (<25 m) spatial scales whereas we did not [18,34]. One explanation for this difference may have to do with the timing of the fish surveys. Several of these studies used data collected during the day [18,34,55], whereas here, we used data collected between dusk and dawn. The time of day may be an important factor because many species make nocturnal migrations, feeding in habitats adjacent to structured reefs and hard bottom habitats [57,58,59]. Tagging studies have found that several reef fish species move 300 m or more diurnally [59] during these migrations.

Importance of the shelf-edge reefs. The relative influence of proximity to the shelf edge was relatively low compared to other factors (e.g., depth) for predicting large fish density. This reduced influence is most likely because depth and distance to shelf edge are highly correlated (i.e., the seafloor becomes deeper further from shore), and are most likely interchangeable as predictors. It remains that high fish occurrence and densities were observed and predicted at the shelf edge reefs in both St. John Wedge and Tampo Bank. High fish densities has also been noted at shelf-edge reefs in the Great Barrier Reef [60,61] and other reef systems in the western Atlantic [62,63,64,65]. The shelf-edge reefs are considered an important ecotone, where shelf waters containing land-based sources of nutrients converge with clear, oligotrophic ocean currents. Juveniles of many species, which use near-shore reefs and vegetated habitats, migrate to shelf edge reefs when they become adults presumably to rest, forage, and reproduce [66].

Our measure of distance from the shelf edge may also be a surrogate for other environmental or geophysical parameters that we did not measure. This explanation is supported by the fact that the biggest model errors (>MAE) for the large fish density prediction were spatially clustered along the shelf edge. This spatial clustering suggests that other physically and biologically important variables (e.g., nutrients, currents, thermoclines, prey abundances, fishing pressure) correlated with the shelf edge were missing from this modeling process. These variables were intentionally excluded from this study to investigate whether seafloor complexity would explain much of the variance in fish distributions. However, in future modeling iterations, additional oceanographic variables should be included at the very least, since nutrient supply and photic depth, combined with relatively stable, warm ocean waters appear to support high abundance of both oceanic and shelf species across a broad range of trophic guilds in this area [66]. Our observations of fish of all size classes show that densities are not the same along the entire shelf edge. Though we did not include a predictor to formally assess this pattern, higher densities were apparent along promontories in both regions. The promontories and submerged capes are notable geomorphologic features on the insular shelf of the U.S. Caribbean, and are common features where spawning aggregations for reef fish occur [6,66]. These features may possess other qualities that support high densities of large fish during non-spawning periods. While we anticipated that shelf edge reefs and promontories would be important habitats, further research is needed to better understand the ecological processes behind these preferences.

Management Implications and Future Developments

Splitbeam echosounders can rapidly survey fish distributions over large areas (10 s to 100 s km^2) and at relatively fine spatial resolutions (<100 m^2). This capability may make predictive models unnecessary in some cases. However in other cases, seafloor structure has been mapped in many areas without accompanying data describing fish distributions or densities. Model predictions could be used to provide first-order maps of large fish densities in these areas. These first order maps could potentially help managers focus their energies on areas that may be critical for large fish and that require additional study, as well as save resources by identifying broad areas that may not require visual surveys (e.g., over 93% of the sites presented here). These models could also be used to forecast how habitat use patterns for larger fish may change under different reef disturbance and flattening scenarios.

This study further emphasized reef complexity as an important geophysical feature in coral ecosystems, particularly at the ecotone of the insular shelf-edge. The shelf-edge habitats are also popular fishing grounds for pelagic and reef-associated species. Our findings at St. John Wedge and Tampo Bank have identified areas of high fish density that may benefit from long-term conservation and management actions to sustain fish populations. Visual surveys can also be conducted in these areas to better understand the environmental conditions attracting higher densities of fish, as well as to obtain better information on species comprising these assemblages.

This study also suggests that fish-seascape relationships and spatial predictions derived from fish acoustic surveys are similar to those derived from visual observations, although more research is needed directly comparing the two [14]. We are particularly encouraged by the performance of our model predicting large fish density. Coastal and fisheries managers are often most interested in the distribution of large, commercially valuable and vulnerable reef fish to identify essential fish habitat, including spawning aggregation sites. Identifying where large fish are most abundant will help coastal managers to prioritize sites and focus their efforts and limited resources on specific areas that may be of the highest conservation value. This type of targeted resource allocation will be particularly important as budgets are continually stretched and reef habitats become increasingly vulnerable, affecting the health and sustainability of reef fish populations. We propose wider use of these acoustic remote sensing tools, coupled with continued improvements in predictive modeling, to map and monitor fish aggregations in sensitive ecosystems.

Acknowledgments

We thank the many people that helped to collect and process the data used in this study, including Eric Ebert, Mike Stecher and the officers and crew of the NOAA ship *Nancy Foster*. We also thank three internal reviewers and two anonymous reviewers for their helpful comments.

Author Contributions

Conceived and designed the experiments: BC JCT. Performed the experiments: BC JCT. Analyzed the data: BC JCT LK TB SP. Contributed reagents/materials/analysis tools: BC JCT LK TB SP. Wrote the paper: BC JCT LK TB SP.

References

1. Sale PF, Cowen RK, Danilowicz BS, Jones GP, Kritzer JP, et al. (2005) Critical science gaps impede use of no-take fishery reserves. Trends in Ecology & Evolution 20: 74–80.
2. Crowder L, Norse E (2008) Essential ecological insights for marine ecosystem-based management and marine spatial planning. Marine Policy 32: 772–778.
3. Foley MM, Halpern BS, Micheli F, Armsby H, Caldwell MR, et al. (2010) Guiding ecological principles for marine spatial planning. Marine Policy 34: 955–966.
4. Myers N, Mittermeier RA, Mittermeier CG, Da Fonseca GA, Kent J (2000) Biodiversity hotspots for conservation priorities. Nature 403: 853–858.
5. Lourie SA, Vincent ACJ (2004) Using Biogeography to help set priorities in marine conservation. Conservation Biology 18: 1004–1020.
6. Kobara S, Heyman WD, Pittman SJ, Nemeth RS (2013) Biogeography of transient reef-fish spawning aggregations in the Caribbean: a Synthesis for future research and management. Oceanography and Marine Biology 51: 281–326.
7. Roberts CM, Branch G, Bustamante RH, Castilla JC, Dugan J, et al. (2003) Application of ecological criteria in selecting marine reserves and developing reserve networks. Ecological Applications 13: 215–228.
8. Geselbracht L, Torres R, Cumming GS, Dorfman D, Beck M, et al. (2008) Identification of a spatially efficient portfolio of priority conservation sites in marine and estuarine areas of Florida. Aquatic Conservation: Marine and Freshwater Ecosystems 19: 408–420.
9. Kracker LM, Taylor JC, Ebert EF, Battista TA, Menza C (2011) Integration of fisheries acoustics surveys and bathymetric mapping to characterize mid-water seafloor habitats of US Virgin Islands and Puerto Rico (Technical Memorandum NOS NCCOS 130). Silver Spring: National Oceanic and Atmospheric Administration. 44 p.
10. Guisan A, Thuiller W (2005) Predicting species distribution: offering more than simple habitat models. Ecology Letters 8: 993–1009.
11. Valavanis VD, Pierce GJ, Zuur AF, Palialexis A, Saveliev A, et al. (2008) Modelling of essential fish habitat based on remote sensing, spatial analysis and GIS. Hydrobiologia 612: 5–20.
12. Elith J, Leatherwick JR (2009) Species Distribution Models: Ecological Explanation and Prediction Across Space and Time. Ann Rev Evol Syst 40: 677–97.
13. Levin SA (1992) The problem of pattern and scale in ecology. Ecology 73: 1943–67.
14. Pittman SJ, Brown KA (2011) Multi-Scale Approach for Predicting Fish Species Distributions across Coral Reef Seascapes. PLoS One 6: 1–12.
15. Collin A, Archambault P, Long B (2011) Predicting Species Diversity of Benthic Communities within Turbid Nearshore Using Full-Waveform Bathymetric LiDAR and Machine Learners. PLoS ONE 6: 1–16.
16. Lucieer V, Hill NA, Barrett NS, Nichol S (2013) Do marine substrates 'look' and 'sound' the same? Supervised classification of multibeam acoustic data using autonomous underwater vehicle images. Estuarine, Coastal and Shelf Science 117: 94–106.
17. Heyman WD, Wright DJ (2011) Marine geomorphology in the design of marine reserve networks. The Professional Geographer 63: 429–442.
18. Pittman SJ, Costa BM, Battista TA (2009) Using LiDAR Bathymetry and Boosted Regression Trees to Predict the Diversity and Abundance of Fish and Corals. Journal of Coastal Research 25: 27–38.
19. Sale PF (1998) Appropriate spatial scales for studies of reef-fish ecology. Australian Journal of Ecology 23: 202–208.
20. Nash KL, Graham NAJ, Wilson SK, Bellwood DR (2012) Cross-scale habitat structure drives fish body size distributions on coral Reefs. Ecosystems 16: 478–490.
21. Kendall MS, Christensen JD, Hillis-Starr Z (2003) Multi-scale data used to analyze the spatial distribution of French grunts, Haemulon flavolineatum, relative to hard and soft bottom in a benthic landscape. Environmental Biology of Fish 66: 19–26.
22. Pittman SJ, McAlpine CA, Pittman KM (2004) Linking fish and prawns to their environment: a hierarchical landscape approach. Marine Ecology Progress Series 283: 233–254.
23. Nemeth R, Hall D (2013) Identifying and defining management areas for queen triggerfish and mutton snapper spawning aggregation sites in the US Virgin Islands and assessing effectiveness of area closures on mutton snapper populations. NOAA Marfin FY3 Full Proposal: NOAA-NMFS-SE-2013-2003417. 21 p.
24. Simmonds EJ, MacLennan D (2005) Fisheries acoustics: Theory and practice, 2nd Ed. Oxford: Blackwell Scientific Publications. 437 p.
25. Echoview (2013) Fish tracking algorithm: Echoview's Alpha-Beta tracking algorithm. Available: http://support.echoview.com/WebHelp/Reference/Algorithms/Fish_tracking_module/Fish_tracking_algorithms.htm. Accessed 5 September 2013.
26. Johnston SV, Rivera JA, Rosario A, Timko MA, Nealson PA, et al. (2006) Hydroacoustic evaluation of spawning red hind (Epinephelus guttatus) aggregations along the coast of Puerto Rico in 2002 and 2003. NOAA Professional Papers Series 5: 10–17.
27. Love RH (1977) Target strength of an individual fish at any aspect. Journal of the Acoustical Society of America 62: 1397–1403.
28. FishBase (2013) FishBase: A Global Information System on Fishes. Available: http://fishbase.org/search.php. Accessed 21 November, 2013.
29. Battista TA, Stecher M (2010) MBES data acquisition and processing report (Project M-I907-NF-10). Silver Spring: National Oceanic and Atmospheric Administration. 46 p.
30. Battista TA, Stecher M (2011) MBES data acquisition and processing report (Project M-I907-NF-11 USVI). Silver Spring: National Oceanic and Atmospheric Administration. 55 p.
31. IHO (International Hydrographic Organization) (2008) IHO standards for hydrographic surveys: Special publication N° 44, 5th Ed. Available: http://www.iho.int/iho_pubs/standard/S-44_5E.pdf. Accessed 12 December 2013.
32. Jenness J (2010) DEM Surface Tools. Available: http://www.jennessent.com/arcgis/surface_area.htm. Accessed 12 July, 2013.
33. Grober-Dunsmore R, Frazer TK, Lindberg WJ, Beets J (2007) Reef fish and habitat relationships in a Caribbean seascape: the importance of reef context. Coral Reefs 26: 201–216.
34. Grober-Dunsmore R, Frazer TK, Beets JP, Lindberg WJ, Zwick P, et al. (2008) Influence of landscape structure on reef fish assemblages. Landscape Ecology 23: 37–53.
35. Chapire RE, Freund Y (2012) Boosting: Foundations and Algorithms. Cambridge: The Massachusetts Institute of Technology Press. 496 p.
36. Breiman L, Friedman JH, Olshend RA, Stone CJ (1984) Classification and regression trees. Belmont: Wadsworth International Group. 324 p.
37. Elith J, Leathwick JR, Hastie T (2008) A working guide to boosted regression trees. Journal of Animal Ecology 77: 802–81.
38. Elith J, Graham CH, Anderson RP, Dudik M, Ferrier S, et al. (2006) Novel methods improve prediction of species' distributions from occurrence data. Ecography 29: 129–151.
39. Hijmans RJ, Phillips S, Leathwick J, Elith J (2012) Species distribution modeling package (dismo) v0.7-0.17. Available: http://cran.r-project.org/web/packages/dismo/index.html. Accessed 12 July, 2013.
40. Hijmans RJ, van Etten J (2012) Geographic data analysis and modeling package (raster) v2.0-31. Available: http://cran.r-project.org/web/packages/raster/index.html. Accessed 12 July, 2013.
41. Pearce J, Ferrier S (2000) Evaluating the predictive performance of habitat models developed using logistic regression. Ecological Modelling 133: 225–245.
42. Liu C, White M, Newell G (2009) Measuring the accuracy of species distribution models: a review. 18th World IMACS/MODSIM Congress, 13-17 July 2009, Cairns, Australia. 4241–4247.
43. Flach P, Wu S (2003) Repairing Concavities in ROC Curves. In: Proc. 2003 UK Workshop on Computational Intelligence. University of Bristol, pp. 38–44.
44. Hosmer DW, Lemeshow S (2000) Applied logistic regression, 2nd ed. New York: John Wiley & Sons. 375 p.
45. Fielding and Bell (1997) A review of methods for the assessment of prediction errors in conservation presence/absence models. Environmental Conservation 24: 38–49.
46. Fawcett T (2006) An introduction to ROC analysis. Pattern Recognition Letters 27: 861–874.
47. Legendre P (1993) Spatial Autocorrelation: Trouble or New Paradigm?. Ecology 74: 1659–1673.
48. Aubry P, Debouzie D (2000) Geostatistical estimation of variance for the spatial mean in two-dimensional systematic sampling. Ecology 81: 543–553.
49. R Core Team (2012) The R Stats Package (stats) version 2.15.2. Available: http://www.r-project.org/. Accessed 12 July, 2013.
50. Ribeiro PJ Jr., Diggle PJ (2012) Analysis of geostatistical data (geoR) version 1.7-4. Available: http://cran.r-project.org/web/packages/geoR/index.html. Accessed 12 July, 2013.
51. Jenks GF (1967) The data model concept in statistical mapping. International Yearbook of Cartography 7: 186–190.
52. Hand DJ, Till RJ (2001) A Simple Generalization of the Area Under the ROC Curve for Multiple Class Classification Problems. Machine Learning 45: 171–186.
53. Pontius RG Jr., Schneider LC (2001) Land-cover change model validation by an ROC method for the Ipswich watershed, Massachusetts, USA. Agriculture, Ecosystems and Environment 85: 239–248.
54. Lobo JM, Jiménez-Valverde A, Real R (2007) AUC: a misleading measure of performance of predictive distribution models. Global Ecology and Biogeography 17: 145–151.
55. Pittman SJ, Christensen JD, Caldow C, Menza C, Monaco ME (2007) Predictive mapping of fish species richness across shallow-water seascapes in the Caribbean. Ecological Modeling 204: 9–21.

56. Knudby A, LeDrew E, Brenning A (2010) Predictive mapping of reef fish species richness, diversity and biomass in Zanzibar using IKONOS imagery and machine-learning techniques. Remote Sensing of Environment 114: 1230–1241.
57. Ogden JC, Ehrlich PR (1977) The behavior of heterotypic resting schools of juvenile grunts (Pomadasyidae). Marine Biology 42:273–280.
58. Meyer JL, Schultz ET (1985) Migrating haemulid fish as a source of nutrients and organic matter on coral reefs. Limnology and Oceanography 30:146–156.
59. Hitt S, Pittman SJ, Nemeth RS (2011) Diel movements of fish linked to benthic seascape structure in a Caribbean coral reef ecosystem. Marine Ecology Progress Series 427: 275–291.
60. Cappo M, De'ath G, Spear P (2007) Inter-reef vertebrate communities of the Great Barrier Reef Marine Park determined by baited remote underwater video stations. Marine Ecology Progress Series 350:209–221.
61. Mellin C, Bradshaw CJA, Meekan MG, Caley MJ (2010) Environmental and spatial predictors of species richness and abundance in coral reef fish. Global Ecology and Biogeography 19:212–222.
62. Parker RO, Mays RW (1998) Southeastern U.S. Deep-water Reef Fish Assemblages, Habitat Characteristics, Catches, and Life History Summaries (Technical Memorandum NMFS 138). Seattle: National Oceanic and Atmospheric Administration. 41 p.
63. Sedberry GR, Cooksey CL, Crowe SF, Hyland J, Jutte PC, et al. (2004) Characterization of Deep Reef Habitat off the Southeastern U.S. with Particular Emphasis on Discovery, Exploration and Description of Reef Fish Spawning Sites. Charleston: South Carolina Department of Natural Resources. 76 p.
64. Olavo G, Costa PAS, Martins AS, Ferreira BP (2011) Shelf-edge reefs as priority areas for conservation of reef fish diversity in the tropical Atlantic. Aquatic Conservation: Marine and Freshwater Ecosystems 21: 199–209.
65. Koenig C, Coleman FC, Grimes CB, Fitzhugh GR, Scanlon KM, et al. (2000) Protection of fish spawning habitat for the conservation of warm temperate reef fish fisheries of shelf-edge reefs of Florida. Bulletin of Marine Science 66: 593–616.
66. Heyman WD, Kjerfve B (2008) Multi-species reef fish spawning aggregations at Gladden Spit, Belize. Bulletin of Marine Science 83: 531–551.

Pushing the Pace of Tree Species Migration

Eli D. Lazarus[1]*, Brian J. McGill[2]

1 Environmental Dynamics Lab, Earth Surface Processes Group, School of Earth and Ocean Sciences, Cardiff University, Cardiff, Wales, United Kingdom, **2** School of Biology and Ecology, University of Maine, Orono, Maine, United States of America

Abstract

Plants and animals have responded to past climate changes by migrating with habitable environments, sometimes shifting the boundaries of their geographic ranges by tens of kilometers per year or more. Species migrating in response to present climate conditions, however, must contend with landscapes fragmented by anthropogenic disturbance. We consider this problem in the context of wind-dispersed tree species. Mechanisms of long-distance seed dispersal make these species capable of rapid migration rates. Models of species-front migration suggest that even tree species with the capacity for long-distance dispersal will be unable to keep pace with future spatial changes in temperature gradients, exclusive of habitat fragmentation effects. Here we present a numerical model that captures the salient dynamics of migration by long-distance dispersal for a generic tree species. We then use the model to explore the possible effects of assisted colonization within a fragmented landscape under a simulated tree-planting scheme. Our results suggest that an assisted-colonization program could accelerate species-front migration rates enough to match the speed of climate change, but such a program would involve an environmental-sustainability intervention at a massive scale.

Editor: Gil Bohrer, The Ohio State University, United States of America

Funding: This research was supported by the U.S. National Science Foundation award EPS-0904155. The funders had no role in study design, data collection and analysis, decision to publish, or preparation of the manuscript.

Competing Interests: The authors have declared that no competing interests exist.

* Email: LazarusED@cf.ac.uk

Introduction

Plants and animals have responded to past climatic changes by shifting the spatial boundaries of their geographic ranges. Such migratory responses to climate change have proved remarkably successful for many species. For example, in North America, no species of small mammal [1] and only one species of tree [2] went extinct in the dozen-plus glacial cycles over the Quaternary. With the approximately 7°C warming that occurred 20–10 kya (punctuated by rapid rates of temperature change over shorter intervals within that period), plant and animal ranges in North America shifted up to the order of 1000 km [3–5]. Pollen records show evidence that temperate tree species migrated at rapid rates (100–1000 m yr^{-1}) following the last Ice Age [6]. Warming of approximately 3.5°C is expected over the next 100 years [7]. As a first approximation, linearly extrapolating the ratio of warming (7°C) to range adjustment in the early Holocene (1000 km) suggests that ranges could shift an average of 500 km in the coming century (5 km yr^{-1}) [8]. Reconciling paleo-rates with observed spreading patterns of plant taxa has motivated a wealth of ecology research into mechanisms for rapid migration, such as long-distance dispersal [9,10]. However, paleo-range shifts occurred in the absence of anthropogenically dominated landscapes. How human uses of otherwise suitable habitat will affect climatically driven range migration across landscapes is an open, fundamental question in environmental sustainability. Here we explore this question in terms of temperate-forest tree species.

Field measurements of typical seed dispersal distances would suggest that tree fronts migrate across a landscape by a process of local diffusion, at rates significantly slower than the velocities reflected in pollen data [9]. But seeds are occasionally carried long distances from their source by wind or by animals [11]. If those seeds mature into trees that in turn dispense seeds, the plant species may migrate at rates that far exceed diffusive propagation [6,9]. Existing models of tree migration by long-distance dispersal produce migration rates between approximately 100–200 m yr^{-1} [9,10,12]. A global analysis of temperature-change rates across geographic gradients and biomes finds that temperate broad-leaf and mixed forests, which includes the North American taxa that spread by wind-blown dispersal [10], will need to shift at a mean velocity of 350 m yr^{-1} [13]. These required migration rates appear to exceed the fastest modeled rates, but may fall within the ranges empirically observed in the last deglaciation. Migration rates barely sufficient to track with climate, combined with the well-documented effect that landscape fragmentation further impedes migration [14–19], points to the apparently unequivocal conclusion that climatic change will outpace the migration of wind-dispersed tree species through human-dominated landscapes. One of the few known cases in which climate-driven species migration was impeded comes from Europe, where east-west mountain ranges and the Mediterranean Sea prevented trees and plants from advancing far enough south during Pleistocene glaciation, resulting in a high proportion of extinction [20]. By extension, understanding how human fragmented landscapes interfere with migration rates might mean the difference between minimal extinction rates and massive extinction rates in next few hundred years.

Assisted migration, assisted colonization, and species translocation are already common conservation practices applied to variety of plant and animal species [21,22]. Unwitting or unintended human introduction of alien, invasive species to new environs via long-distance transport pathways is arguably a version of the same idea. Assuming disparity between rates of climate change and temperate-forest migration is an inherent condition of the future sets up a new question: how much human intervention might windblown-dispersal taxa need to keep pace with climate change? Although assisted migration remains controversial [23], here we neither advocate for nor reject it. We simply ask: if humans were to deliberately augment the migration rates of temperate-forest tree species, what would the scale of that intervention need to be relative to the natural dynamics driving the migration front?

Here we present results from a probabilistic spatial model that simulates the passage of a generic, wind-dispersed tree species through a two-state, cellular landscape in which cells are either suitable or unsuitable for trees (Fig. 1 and fig. S1; additional model explanation, parameters, and source code are provided in File S1; data shown in figures are provided in File S2.). Carefully abstracted models that capture the salient dynamics of well-defined systems are useful tools for insight into how natural systems may respond to changes in forcing, and by design are more exploratory than explicitly predictive [24,25]. Cellular architecture has been used in several idealized spatial models of migration patterns through heterogeneous habitats and fragmented landscapes [16,26–28]. We use our model to test how augmenting the number of successful seedlings accelerates the rate of tree front migration. Our results suggest that relatively small changes in the annual number of successful seedlings could have a comparatively large effect on biome migration rates.

Model Design

Spatial domain

Our model operates on a gridded domain with a width-to-length ratio of 1:200 km and a periodic (i.e. wrap-around) boundary in the width dimension (fig. S1). The periodic boundary condition makes the domain infinitely wide; the extended length dimension allows a fixed aerial view (200 km) large enough to frame potentially rapid incoming migration fronts (> km yr $^{-1}$). Each grid cell is 10×10 m, and can accommodate a single tree. Although these cell dimensions are coarse for small, young trees, they are consistent with crown widths of large trees measured via remote sensing in a North American temperate mixed forest [29]. We run the model for 1000 years so that initial conditions dissipate and migration dynamics may play out in full (e.g. until available area is fully populated, or all trees die out).

Seed dispersal

Accounting for infrequent long-distance dispersal events requires a statistical function with a "heavy tail," so that the maximum dispersal distance may be very large even if the average dispersal distance is short. Other models have used various heavy-tailed probability distribution functions to drive dispersal distance [9,10,12,16,30–32]. Seed dispersal distance in our model is generated by a *t*-distribution with a degrees-of-freedom variable that changes the "heaviness" of the distribution's tail: the minimum degrees of freedom (DOF = 1) allows the longest dispersal distances; by comparison, a high degree of freedom (e.g. DOF = 100) approximates a Gaussian distribution, with a dispersal pattern limited to local diffusion (Fig. 2). We use the *t*-distribution because of the straightforward means by which the distribution can be shifted from Gaussian to heavy tailed for comparative scenarios. In the results presented here, the initial model domain (Fig. 1) is populated with trees according to a *t*-distribution (with DOF = 1) centered on the far left edge.

Tree characteristics

Each tree in our model is assigned biological characteristics for tree fecundity (seeds tree $^{-1}$ yr $^{-1}$, as a function of tree age), seed dispersal distance, and seedling survival according to probability-distribution functions constrained by empirically derived parameters for North American wind-dispersed trees (see Table S1) [10]. Maturation age (yr) and seed-crop interval (yr) are sampled from normal probability distributions. Longevity (yr) is sampled from a gamma distribution, which allows a few long-lived trees to persist among a community of trees whose average age is younger. Each model tree broadcasts some number of "successful" seeds (seeds that will germinate into seedlings) every few years as a function of its maturation age, seed-crop interval, and fecundity. We treat tree fecundity as an increasing parabolic function of age; the number of viable seedlings per tree is then a percentage of that fecundity, determined by a probability of seedling survival for a given tree's seed crop in a given model year. Each seed is assigned a randomly selected compass direction and a dispersal distance that corresponds to a grid cell in the domain, which, if unoccupied, becomes a new tree with parametric values resampled from the probability-distribution functions for biological characteristics.

Landscape disturbance

We create disturbance patterns by first generating a domain-scale layer of random values between 0–1 that are either highly spatially autocorrelated (e.g. high values occur near other high values, and likewise for low values) or are spatially uncorrelated (spatially random), with the degree of correlation controlled by the Hurst exponent [17,19,33]. We use Hurst exponents of $H = 0.99$ and $H = 0$ for our correlated and uncorrelated landscapes, respectively. We control the percentage of landscape disturbance by isolating cells in the grid that exceed a given threshold value between 0–1; cells whose values exceed the threshold are rendered "disturbed" and assigned a value $= 1$; all other cells (undisturbed) are set $= 0$. For comparability across scenarios, the underlying Hurst-model landscape (for a given value of H) is kept constant. The threshold changes, and therefore the overall pattern of disturbance, but retaining the same underlying Hurst model means that a patch disturbed at a high threshold (yielding a low percentage of disturbance) will persist at lower thresholds. Furthermore, the disturbance pattern is held constant through time within a model run. Trees may occupy any undisturbed cells in the domain; seedlings distributed to disturbed cells are removed.

Assisted colonization

Assisted colonization occurs in the model when new trees are "planted" at the end of a model year, after all existing, active trees in the domain have broadcast their seeds (fig. S1). Grid cells in which new trees are planted are randomly selected from among all undisturbed cells not already occupied by an existing tree. We do not account for ways in which proximal areas of disturbance might impinge upon tree health, or disturbance "edge effects" [34,35]: planted trees are governed by the same parameters as trees growing "naturally" in the model.

The number of trees planted is based on the average number of new trees that grow in the model when (a) there is no landscape disturbance and (b) dispersal distances are generated by a heavy-tailed *t*-distribution function (DOF = 1). In the results shown here, that average number is 2000 trees yr $^{-1}$. These conditions constitute the model's the ecological baseline. Assistance is thus a function of

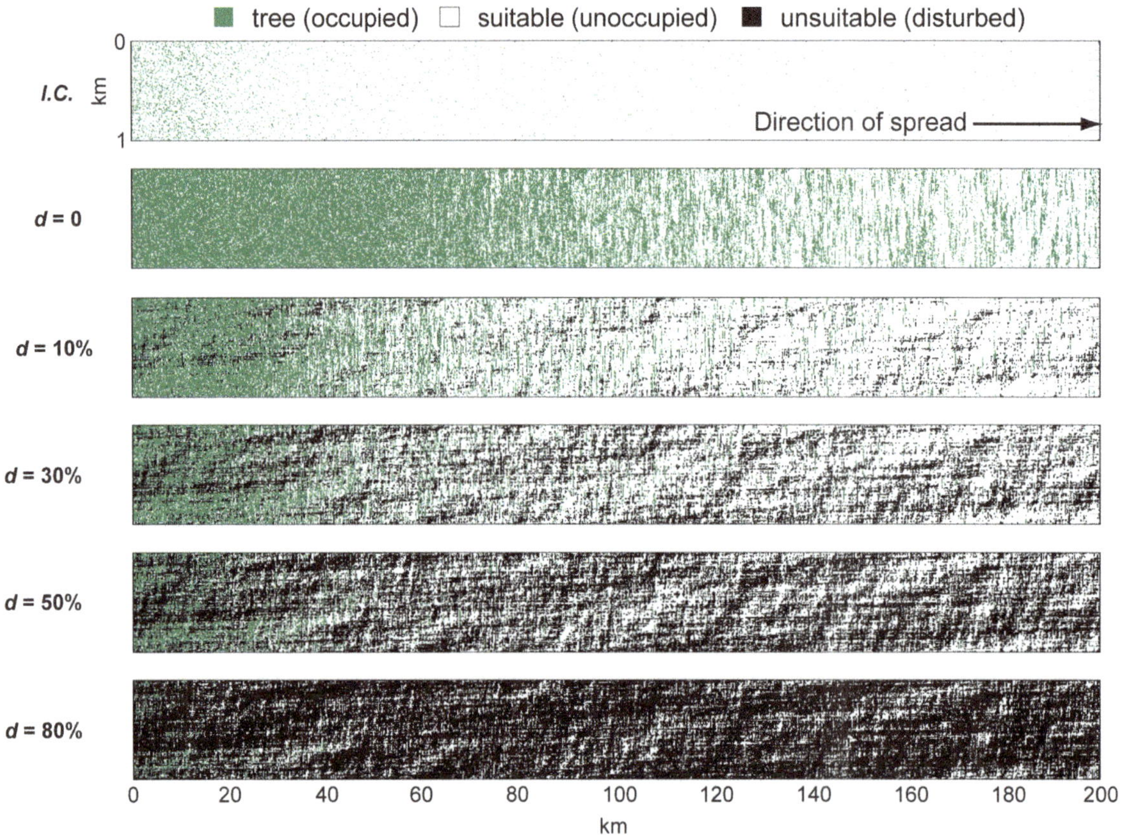

Figure 1. Representative model domains. Top panel shows the model's initial condition (*I.C.*) for tree positions in an undisturbed landscape. Subsequent panels show the resulting domains after 1000 years given different extents of disturbance (*d*), where disturbance is spatially autocorrelated (Hurst exponent $H = 0.99$). Apparent lineations in the ordinate dimension are a distortion in the aspect ratio, not an artifact of the model.

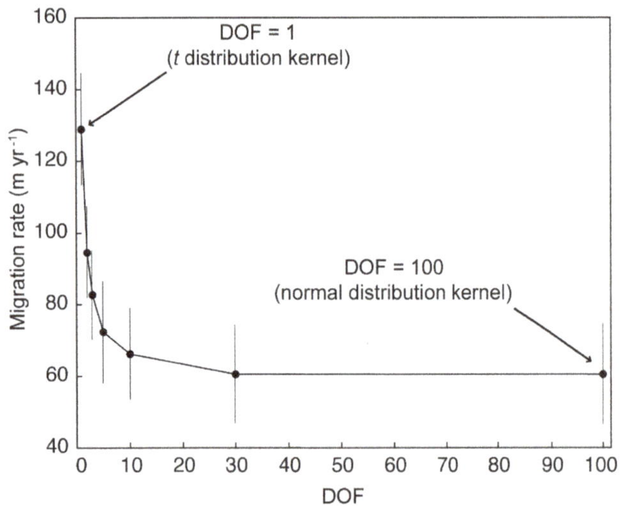

Figure 2. Mean migration rates with increasing degrees of freedom (DOF) in a *t*-distribution kernel for long-distance dispersal. Bars indicate envelope of one standard deviation. When DOF = 1, the dispersal kernel is "heavy tailed." When DOF is >>1 (e.g. DOF = 100), the dispersal kernel has the comparatively limited range of a Gaussian distribution, and resulting mean migration rates represent the rate of spreading by local diffusion.

increasing the fixed number of new trees per year by a percentage *K*.

Planting in the model is not explicitly organized in the direction of migration: available sites at the distal edge of the domain are not prioritized over other available sites. However, because the distal edge of the initial model domain is comparatively empty, distal planting sites are more numerous and therefore more likely to be selected from among the sites available, especially when the rate of assistance is high.

Results

When the entire model domain is hospitable to seedling growth, heavy-tailed dispersal distances drive front migration rates of approximately 130 m yr^{-1} that are nearly double those driven by diffusive dispersal (Fig. 2). Front migration rates in this model, measured as the change in mean tree position across the domain, are comparable to rates reported for other, analogous long-distance dispersal models [9,12], including the study from which our ecological parameters derive [10]. This agreement in rates is not a foregone conclusion. Although the independent biological variables in the model are constrained by parametric distributions that are similar to (but not the same as) those used in other studies, migration rate is an implicit variable that arises from the collective state of the full model domain over time. The agreement we find provides a good validation of our modeling approach.

Fragmenting the model domain with landscape disturbance shows that migration rate and disturbance are negatively

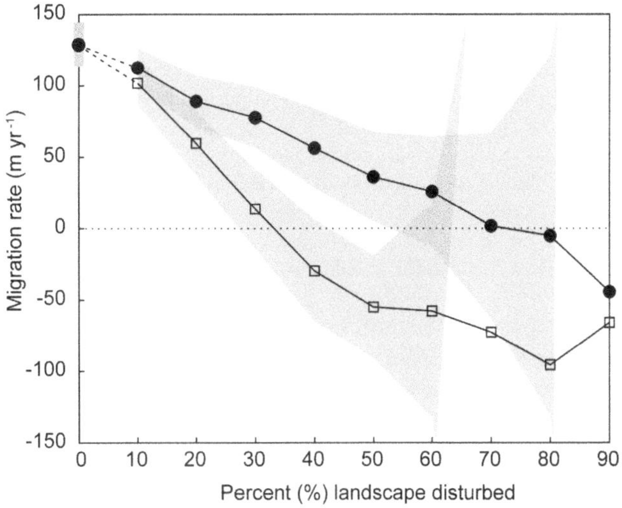

Figure 3. Mean migration rates (via long-distance dispersal, given *t*-distribution kernel with DOF = 1) through increasing proportions of landscape disturbance when disturbance is spatially autocorrelated (black dots; *H* = 0.99) and spatially random (open squares; *H* = 0). Gray regions denote envelope of one standard deviation. Negative values indicate conditions under which the tree front receded from the domain (died back) rather than advanced across it.

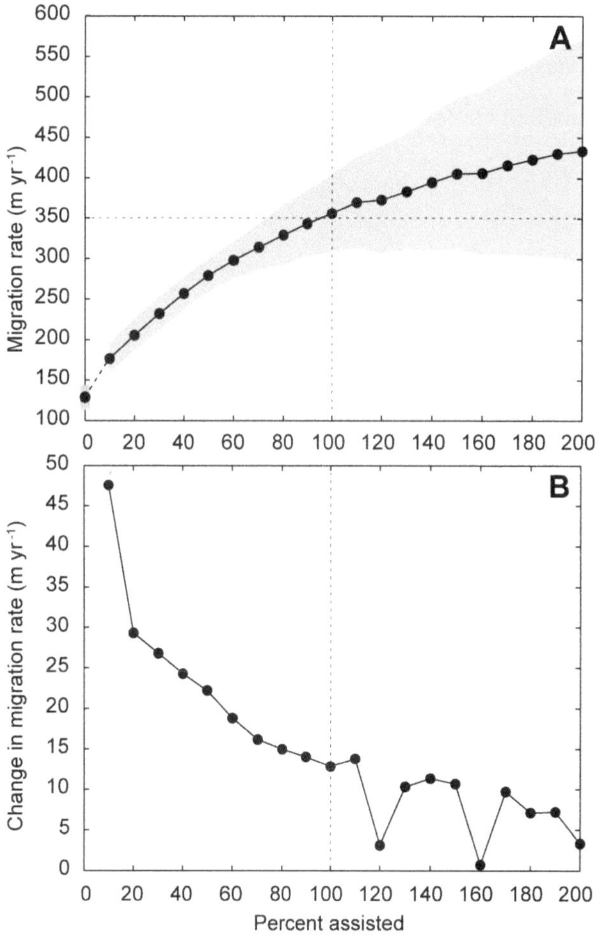

Figure 4. Mean migration rates with assistance. (A) Mean migration rate through an undisturbed landscape with increasing levels of assisted colonization. Gray regions denote envelope of one standard deviation. When the number of naturally successful seedlings is doubled (*K* = 100%), migration rates become comparable to the measured mean velocity of range shift for temperate deciduous forests [13]. (B) First derivative of migration rates plotted in (A), showing that increasing assistance has a diminishing return on migration-rate acceleration, especially for *K*>100%.

correlated (Fig. 3), as other disturbance models have also shown [16,17,19]. Furthermore, the presence or absence of pattern in the disturbance regime affects the migration rate. When disturbance across the model landscape is spatially random ($H = 0$), the migration front ceases to advance once the total area disturbed reaches 40%. This propagation result grants some comparison to percolation models for square lattices. The percolation threshold for a square lattice is approximately 59% [36], meaning a fully connected route through the lattice exists as long as disturbance to the lattice is less than 41%. Long-distance dispersal in model does not necessarily succeed where random walks cannot: even a species capable of dispersal between non-adjacent sites does not advance through a randomly fragmented domain ($H = 0$) in which disturbance exceeds the percolation threshold. When disturbed areas are highly autocorrelated in the model ($H = 0.99$), leaving contiguous reaches of potential habitat (corridors) relatively intact, migration rates decrease more slowly, tolerating up to 70% total disturbance (Fig. 3). Most importantly, none of these migration rates—including the maximum rates in our idealized, undisturbed scenario (Fig. 3)—come close to matching the 350 m yr^{-1} "velocity of climate change" modeled for temperate broad-leaf and mixed forests [13].

That apparent lag motivates our simplified version of assisted colonization to accelerate migration rates. The key assumption we make is that all of the trees deliberately introduced to the domain can grow to be viable adults, subject to the probabilistic limits of tree longevity and fecundity imposed in the model. To consider the effect of assisted colonization on rates of tree front migration through an undisturbed landscape (Fig. 4), we systematically increase the number of planted trees by $K = 10$–200% in increments of 10% (e.g. when $K = 10\%$, 200 trees are planted annually). Planting begins in model year = 200 and continues until year = 1000. For these experiments, we calculate mean migration rate over the first hundred years of assisted colonization (model years 201–300); after 100 years of planting, the migration front effectively halts as available spaces in the model domain fill up (fig. S2).

These simulations suggest that doubling the number of annually successful trees ($K = 100\%$) in an undisturbed landscape would result in migration rates equal to the necessary mean velocity of range shift of 350 m yr^{-1}. We also find that even the weakly organized strategy of assisted colonization in this model has a pronounced, accelerating effect on migration rates in fragmented landscapes where disturbance is highly autocorrelated ($H = 0.99$) (Fig. 5). This result is not necessarily surprising. Unlike stochastic seed dispersal as a means of colonizing undisturbed areas, planting ensures that trees occupy available space. In a highly disturbed landscape, planting makes colonization a comparatively targeted process, which only makes migration rates faster. However, a caveat associated with determining the migration rate within a disturbed domain is that the calculation is based on fewer trees overall than are present in an undisturbed domain. A high migration rate can appear to be maintained through large extents

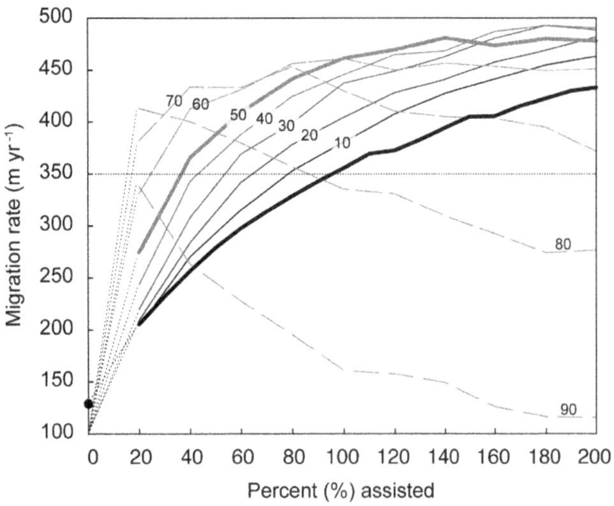

Figure 5. Assisted migration rates through landscapes with increasing levels of disturbance (lines labeled 10–90%), where disturbance is spatially autocorrelated ($H = 0.99$). Bold line shows assisted migration rates through an undisturbed landscape. Even small amounts of assisted colonization have a strong effect on migration rates as landscape disturbance increases.

of disturbance with less intervening assistance (Fig. 5), but the total population of trees is smaller.

Discussion

Model uncertainty

Migration rates in the model depend principally on the presence or absence of a heavy tail in the distribution of dispersal distances (Fig. 2), and on the empirical constraints on the probability-distribution functions [10]. We ran our model using other heavy-tailed probability-distribution functions for dispersal distances (e.g. power-law functions) constrained by the same statistical ranges and found that, although the resulting landscapes differed in detail, the model produced comparable migration rates with similar migration-front dynamics.

Given that our interest is in the migration rate of a generic pattern at the domain scale, our model is more phenomenological than mechanistic and shares the same parametric sensitivities as others of its kind [32]. The spatial distribution of surviving seedlings, not just the distribution of dispersal distances, is an important component of migration, but here we do not account for potential differences in seedling success (e.g., germination, competition, survivorship, seed production) with distance from the parent tree [32,37].

Furthermore, although climate change motivates our interest in modeling tree-species migration through fragmented landscapes, we do not attempt to model climate-change effects on ecological traits [38] or on habitat traits such as soil conditions [39]. To keep our model focused on dispersal, we treat climate change here as the exogenous driver of the minimally required rate of migration needed to avoid extinction. Assuming that the trees in this model migrate into and across a landscape already ecologically suitable for them to inhabit emphasizes the relative effects of disturbance and assistance on migration dynamics. We thus treat climate change as an implicit migration-rate threshold: even in this "fastest-case" scenario, for a given pair of disturbance and assistance conditions, are migration rates faster or slower than the projected rate they must achieve to survive?

Of course, the inherent uncertainty in our model derives far less from tree ecology than from patterns of landscape disturbance and the application of assisted migration, the abstracted human activities we represent. As Higgins et al. [32] remark in their review of plant-migration forecasting, "inherent uncertainty cannot be reduced by improving model or parameter uncertainty." By design, our model is a suitable tool for reasoning by analysis rather than by analogy [40]: a means of exploring inherent uncertainty through the encapsulation of relevant dynamics rather than in a scaled simulacrum of the real world.

Landscape corridors and assisted migration

Protecting contiguous reaches of undisturbed habitat by concentrating new disturbance near existing disturbed areas is the premise behind ecological-corridor conservation practices [23,41,42]. Some researchers have discussed the extent to which the mechanism of long-distance dispersal could enable a tree species, in the absence of an uninterrupted habitat corridor, to migrate along a series of proximal habitat "islands" [16,17]. Our results suggest that even when available habitat is strongly autocorrelated, once half the landscape domain has been disturbed, long-distance dispersal advances the front no faster than diffusive passage through an undisturbed landscape (Figs. 2 and 3). Moreover, beyond approximately 70% disturbance, the migration front fails to advance even with long-distance dispersal and in the presence of autocorrelated habitat islands (Fig. 3).

Indeed, if migration fronts effectively halt with disturbances exceeding 70% (Fig. 3), then migration by assisted colonization might be the only means by which a front of trees could move through a landscape so heavily disturbed. Moreover, such planting and could accelerate migration rates within decades (fig. S2), commensurate with time scales relevant to climate-change projections and mitigation actions [7]. We find that migration rates in our model increase with increased assistance, but augmenting the number of annually successful trees by more than 60–80% has a diminishing effect on acceleration (Fig. 4). Within a fixed domain (e.g. a spatial domain that fills as the migration front advances, rather than tracking with the front) an economy of scale therefore may govern the rate-accelerating effects of assisted colonization. Assuming tree planting has an associated fixed cost (i.e. each tree requires time and labor to be planted; trees for planting need to be raised in nurseries, etc.), at least in the context of a coordinated, top-down climate-adaptation program [43,44], this diminishing advantage in the relative effect of planting more trees could manifest in both ecological and practical terms.

Representing a tree-planting strategy as a stochastic, opportunistic process is at least as reasonable as our stylized treatment of landscape fragmentation, and likewise is as reasonable as assuming that in a real tree-planting campaign every site available ecologically would be available in practice. Targeting planting efforts at the far edge of a species' geographic range is one way to accelerate migration, but trees planted beyond the advancing front are not the only ones capable of affecting the migration rate. A consequence of long-distance dispersal is that even a tree planted behind the migrating front is capable of broadcasting a large number of viable seeds toward the distal edges of the geographic range. Perhaps the primary benefit of assisting tree-species colonization and migration, as opposed to managing animal species that roam, is the ability to selectively and specifically establish sensitive taxa in undisturbed environs irrespective of fragmentation intensity.

Implications

The lag between tree migration rates and the velocities of their preferred environments raises interesting implications for ecological dynamics at ecotonal boundaries. For example, temperate-broadleaf and mixed-forest biome environments in North America will need to move to higher latitudes at an average rate of 350 m yr^{-1} to track with climate, but boreal forest and taiga environments, also moving northward, are shifting at a faster average of 430 m yr^{-1} [13]. Ecotones between biomes demarcate zones where individuals are living at the fringe of their environmental tolerances and are arguably the most sensitive to relative differences in surface-temperature velocities [45]. Habitat fragmentation in a shifting ecotonal zone is therefore especially complicated and problematic [45,46]. Conceivably, habitat fragmentation across an ecotone could result in one principal biome departing before the next can establish, making the liminal landscape a kind of anthropogenic biome with its own ecological signature [46].

An intentional planting program therefore represents a compromise: a direct intervention that enables an erstwhile natural biome to persist under anthropogenic pressure. Our model results suggest that this would have to occur on a massive scale to be effective for a continental ecotone. Targeted plantings at specific locations assumes a top-down approach to climate-change adaptation, but such adaptation is at least as likely to occur—and is perhaps more realistic—through a bottom-up approach in which people act locally (i.e., plant trees in their neighborhood) [43,44]. An environmental conservation campaign resembling the simplified one in our model—a tree-planting campaign without explicit spatial targets but enacted at a large enough scale to potentially accelerate the migration rates of a forest biome—is not without precedent. The Green Belt Movement, founded by the late Nobel Laureate Prof. Wangari Maathai, is credited with having planted more than 51 million trees in Kenya since 1977 through grass-roots community organization and mobilization [47]. The Billion Tree Campaign, a United Nations Environment Programme initiative that Maathai also championed, has registered over 12 billion trees planted worldwide since 2006 [48]. Given that landscape fragmentation and land-use changes will only further hinder slow-moving tree fronts as they follow fast-moving climate change, adding various common tree types to the list of migrating species that large-scale and local conservation organizations actively assist is both a necessary action and an environmental sustainability problem with, it would seem, a viable solution.

Supporting Information

Figure S1 Schematics of our tree migration model: (A) representative initial condition without disturbance (arrows indicate how the periodic boundary operates on seed dispersal); final landscapes (B) without disturbance and (C) with disturbance; (D) sequence in a model year in which (D2) new trees are added by natural dispersal and (D3) by assisted colonization, resulting in (D4) the final landscape as shown. Actual model outputs are shown in Fig. 1.

Figure S2 Plot of mean front position versus time for increasing levels of assisted migration in the absence of landscape disturbance. Black line shows the natural, background migration rate (~130 m yr^{-1}), given a *t*-distribution dispersal kernel with DOF = 1. In this baseline scenario, approximately 2000 new trees grow in the domain per year. Blue lines show the effects of assisted migration regimes in which an additional K% trees per year are "planted", for $K = 10–200\%$. Bold blue lines denote 50% increments; $K = 200\%$ is shown in red.

File S1 Document file containing model source code and explanatory text.

File S2 Spreadsheet file containing data shown in the published figures.

Acknowledgments

We thank A. Payson and D. Hall for helpful discussions, the Sustainability Solutions Initiative at the University of Maine, and the anonymous reviewers who contributed their comments.

Author Contributions

Conceived and designed the experiments: EDL BJM. Performed the experiments: EDL. Analyzed the data: EDL BJM. Contributed reagents/materials/analysis tools: EDL BJM. Contributed to the writing of the manuscript: EDL BJM.

References

1. McGill BJ, Hadly EA, Maurer BA (2005) Community inertia of quaternary small mammal assemblages in North America. P Natl Acad Sci USA 102: 16701–16706.
2. Jackson ST, Weng CY (1999) Late quaternary extinction of a tree species in eastern North America. P Natl Acad Sci USA 96: 13847–13852.
3. Davis MB, Shaw RG (2001) Range shifts and adaptive responses to Quaternary climate change. Science 292: 673–679.
4. Lyons SK (2003) A quantitative assessment of the range shifts of Pleistocene mammals. J Mammal 84: 385–402.
5. Williams JW, Shuman BN, Webb T, Bartlein PJ, Leduc PL (2004) Late-quaternary vegetation dynamics in North America: scaling from taxa to biomes. Ecol Monogr 74: 309–334.
6. Huntley B, Birks HJB (1983) An atlas of past and present pollen maps for Europe 0–13000 years ago. Cambridge: Cambridge University Press. 667 p.
7. IPCC (2007) Contribution of Working Groups I, II and III to the Fourth Assessment Report of the Intergovernmental Panel on Climate Change, Core Writing Team. Fourth Assessment Report of the Intergovernmental Panel on Climate Change. Geneva, Switzerland: IPCC.
8. Parmesan C, Yohe G (2003) A globally coherent fingerprint of climate change impacts across natural systems. Nature 421: 37–42.
9. Clark JS, Fastie C, Hurtt G, Jackson ST, Johnson C, et al. (1998) Reid's paradox of rapid plant migration: dispersal theory and interpretation of paleoecological records. Bioscience 48: 13–24.
10. Nathan R, Horvitz N, He Y, Kuparinen A, Schurr FM, et al. (2011) Spread of North American wind-dispersed trees in future environments. Ecol Lett 14: 211–219.
11. Nathan R (2006) Long-distance dispersal of plants. Science 313: 786–788.
12. Higgins SI, Richardson DM (1999) Predicting plant migration rates in a changing world: the role of long-distance dispersal. Am Nat 153: 464–475.
13. Loarie SR, Duffy PB, Hamilton H, Asner GP, Field CB, et al. (2009) The velocity of climate change. Nature 462: 1052–1055.
14. Schwartz MW (1993) Modeling effects of habitati fragmentation on the ability of trees to respond to climatic warming. Biodivers Conserv 2: 51–61.
15. Hill MF, Caswell H (1999) Habitat fragmentation and extinction thresholds on fractal landscapes. Ecol Lett 2: 121–127.
16. Collingham YC, Huntley B (2000) Impacts of habitat fragmentation and patch size upon migration rates. Ecol Appl 10: 131–144.
17. Pearson RG, Dawson TP (2005) Long-distance plant dispersal and habitat fragmentation: identifying conservation targets for spatial landscape planning under climate change. Biol Conserv 123: 389–401.
18. Pearson RG (2006) Climate change and the migration capacity of species. Trends Ecol Evol 21: 111–113.
19. McInerny G, Travis JMJ, Dytham C (2007) Range shifting on a fragmented landscape. Ecol Inform 2: 1–8.
20. Latham RE, Ricklefs RE (1993) Continental comparisons of temperate-zone tree species diversity. In: Ricklefs RE, Schluter D (eds), Species diversity in ecological

commmunities: historical and geographical perspectives. Chicago: University of Chicago Press. 294–314 p.
21. Hoegh-Guldberg O, Hughes L, McIntyre S, Lindenmayer DB, Parmesan C, et al. (2008) Assisted colonization and rapid climate change. Science 321: 345–346.
22. Vitt P, Havens K, Kramer AT, Sollenberger D, Yates E (2010) Assisted migration of plants: changes in latitudes, changes in attitudes. Biol Conserv 143: 18–27.
23. Simberloff D, Farr JA, Cox J, Mehlman DW (1992) Movement corridors: conservation bargains or poor investments? Conserv Biol 6: 493–504.
24. Murray AB (2007) Reducing model complexity for explanation and prediction. Geomorphology 90: 178–191.
25. Guisan A, Thuiller W (2005) Predicting species distribution: offering more than simple habitat models. Ecol Lett 8: 993–1009.
26. Travis JMJ (2003) Climate change and habitat destruction: a deadly anthropogenic cocktail. P Roy Soc B-Biol Sci 270: 467–473.
27. Carey PD (1996) DISPERSE: A cellular automaton for predicting the distribution of species in a changed climate. Global Ecol Biogeogr 5: 217–226.
28. Iverson LR, Prasad A, Schwartz MW (1999) Modeling potential future individual tree-species distributions in the eastern United States under a climate change scenario: a case study with Pinus virginiana. Ecol Model 115: 77–93.
29. Popescu SC, Wynne RH, Nelson RF (2003) Measuring individual tree crown diameter with lidar and assessing its influence on estimating forest volume and biomass. Can J Remote Sens 29: 564–577.
30. Johst K, Brandl R, Eber S (2002) Metapopulation persistence in dynamic landscapes: the role of dispersal distance. Oikos 98: 263–270.
31. Iverson LR, Schwartz MW, Prasad AM (2004) How fast and far might tree species migrate in the eastern United States due to climate change? Global Ecol Biogeogr 13: 209–219.
32. Higgins SI, Clark JS, Nathan R, Hovestadt T, Schurr F, et al. (2003) Forecasting plant migration rates: managing uncertainty for risk assessment. J Ecol 91: 341–347.
33. Hastings HM, Sugihara G (1993) Fractals: a user's guide for the natural sciences. New York: Oxford University Press 235 p.
34. Alston KP, Richardson DM (2006) The roles of habitat features, disturbance, and distance from putative source populations in structuring alien plant invasions at the urban/wildland interface on the Cape Peninsula, South Africa. Biol Conserv 132: 183–198.
35. Saunders DA, Hobbs RJ, Margules CR (1991) Biological consequences of ecosystem fragmentation: a review. Conserv Biol 5: 18–32.
36. Ziff RM, Sapoval B (1986) The efficient determination of the percolation threshold by a frontier-generating walk in a gradient. J Phys A Math Gen 19: L1169–1172.
37. Clark DA, Clark DB (1984) Spacing dynamics of a tropical rainforest tree: evaluation of the Janzen-Connell model. Am Nat 124: 769–788.
38. Morin X, Viner D, Chuine I (2008) Tree species range shifts at a continental scale: new predictive insights from a process-based model. J Ecol 96: 784–794.
39. Pastor J, Post WM (1988) Response of northern forests to CO_2-induced climate change. Nature 334: 55–58.
40. Paola C, Straub K, Mohrig D, Reinhardt L (2009) The "unreasonable effectiveness" of stratigraphic and geomorphic experiments. Earth-Science Reviews 97: 1–43.
41. Earn DJD, Levin SA, Rohani P (2000) Coherence and conservation. Science 290: 1360–1364.
42. Damschen EI, Haddad NM, Orrock JL, Tewksbury JJ, Levey DJ (2006) Corridors increase plant species richness at large scales. Science 313: 1284–1286.
43. Rayner S (2010) How to eat an elephant: a bottom-up approach to climate policy. Clim Policy 10: 615–621.
44. Wilbanks TJ, Kates RW (1999) Global change in local places: How scale matters. Climatic Change 43: 601–628.
45. Neilson RP (1993) Transient ecotone response to climatic change: some conceptual and modeling approaches. Ecol Appl 3: 385–395.
46. Lloyd KM, McQueen AAM, Lee BJ, Wilson RCB, Walker S, et al. (2000) Evidence on ecotone concepts from switch, environmental and anthropogenic ecotones. J Veg Sci 11: 903–910.
47. The Greenbelt Movement. Available: http://www.greenbeltmovement.org. Accessed February 2013.
48. United Nations Environment Programme (UNEP) Billion Tree Campaign. Available: http://www.plant-for-the-planet-billiontreecampaign.org/. Accessed February 2013.

PERMISSIONS

 Every chapter published in this book has been scrutinized by our experts. Their significance has been extensively debated. The topics covered herein carry significant findings which will fuel the growth of the discipline. They may even be implemented as practical applications or may be referred to as a beginning point for another development.

The contributors of this book come from diverse backgrounds, making this book a truly international effort. This book will bring forth new frontiers with its revolutionizing research information and detailed analysis of the nascent developments around the world.

We would like to thank all the contributing authors for lending their expertise to make the book truly unique. They have played a crucial role in the development of this book. Without their invaluable contributions this book wouldn't have been possible. They have made vital efforts to compile up to date information on the varied aspects of this subject to make this book a valuable addition to the collection of many professionals and students.

This book was conceptualized with the vision of imparting up-to-date information and advanced data in this field. To ensure the same, a matchless editorial board was set up. Every individual on the board went through rigorous rounds of assessment to prove their worth. After which they invested a large part of their time researching and compiling the most relevant data for our readers.

The editorial board has been involved in producing this book since its inception. They have spent rigorous hours researching and exploring the diverse topics which have resulted in the successful publishing of this book. They have passed on their knowledge of decades through this book. To expedite this challenging task, the publisher supported the team at every step. A small team of assistant editors was also appointed to further simplify the editing procedure and attain best results for the readers.

Apart from the editorial board, the designing team has also invested a significant amount of their time in understanding the subject and creating the most relevant covers. They scrutinized every image to scout for the most suitable representation of the subject and create an appropriate cover for the book.

The publishing team has been an ardent support to the editorial, designing and production team. Their endless efforts to recruit the best for this project, has resulted in the accomplishment of this book. They are a veteran in the field of academics and their pool of knowledge is as vast as their experience in printing. Their expertise and guidance has proved useful at every step. Their uncompromising quality standards have made this book an exceptional effort. Their encouragement from time to time has been an inspiration for everyone.

The publisher and the editorial board hope that this book will prove to be a valuable piece of knowledge for researchers, students, practitioners and scholars across the globe.

LIST OF CONTRIBUTORS

Qinfeng Guo
USDA FS, Eastern Forest Environmental Threat Assessment Center, Asheville, North Carolina, United States of America

Julian D. Olden
School of Aquatic and Fishery Sciences, University of Washington, Seattle, Washington, United States of America

Jianliang Zhang, Fangzheng Liu and Guofa Cui
College of Nature Conservation, Beijing Forestry University, Beijing, China

Sascha A. Ismail and Jaboury Ghazoul and Chris J. Kettle
ETH Zürich, Institute of Terrestrial Ecosystems, Ecosystem Management, Zürich, Switzerland

Gudasalamani Ravikanth
Ashoka Trust for Research in Ecology and the Environment, Royal Enclave, Srirampura, Jakkur Post, Bangalore, India

Cheppudira G. Kushalappa
College of Forestry, University of Agricultural Sciences (Bangalore), Ponnampet, Kodagu district, Karnataka, India

Ramanan Uma Shaanker
Ashoka Trust for Research in Ecology and the Environment, Royal Enclave, Srirampura, Jakkur Post, Bangalore, India
Department of Crop Physiology and School of Ecology and Conservation, University of Agricultural Sciences, GKVK Campus, Bangalore, India

David G. Angeler, Emma Gö the, Richard K. Johnson
Department of Aquatic Sciences and Assessment, Swedish University of Agricultural Sciences, Uppsala, Sweden

Su Yun Kang, James McGree and Kerrie Mengersen
Mathematical Sciences School, Queensland University of Technology, Brisbane, Queensland, Australia,
Cooperative Research Centre for Spatial Information, Melbourne, Victoria, Australia

William B. Monahan and Nicholas A. Fisichelli
National Park Service, Natural Resource Stewardship & Science, Fort Collins, Colorado, United States of America

Francesco Maria Sabatini, Sabina Burrascano and Carlo Blasi
Department of Environmental Biology, Sapienza, University of Rome, Rome, Italy

Hanna Tuomisto
Department of Biology, University of Turku, Turku, Finland

Jérôme Rousselet, Charles-Edouard Imbert, Anissa Dekri, Jacques Garcia, Francis Goussard, Bruno Vincent, Olivier Denux, Christelle Robinet and Alain Roques
Unité de Recherche 633 Zoologie Forestiére, Institut National Recherche Agronomique, Centre d'Orléans, France

Jean-Pierre Rossi and Franck Dorkeld
Unité Mixte de Recherche 1062 Centre de Biologie pour la Gestion des Populations, Institut National Recherche Agronomique, Centre de Montpellier, France

Mark E. Monaco, Matthew S. Kendall, Matthew Poti, Randall D. Clark and Chris Caldow
U.S. National Oceanic and Atmospheric Administration, Center for Coastal Monitoring and Assessment, Silver Spring, Maryland, United States of America

Richard S. Nemeth
Center for Marine and Environmental Studies, University of the Virgin Islands, St. Thomas, Virgin Islands, U.S. Virgin Islands

Bryan Legare
Center for Marine and Environmental Studies, University of the Virgin Islands, St. Thomas, Virgin Islands, U.S. Virgin Islands
Texas Parks and Wildlife, Dickinson Marine Lab, Dickinson, Texas, United States of America

Simon J. Pittman
U.S. National Oceanic and Atmospheric Administration, Center for Coastal Monitoring and Assessment, Silver Spring, Maryland, United States of America
Center for Marine and Environmental Studies, University of the Virgin Islands, St. Thomas, Virgin Islands, U.S. Virgin Islands
Centre for Marine and Coastal Policy Research, The Marine Institute, Marine Building, Plymouth University, Plymouth, Devon, United Kingdom

Alan M. Friedlander
Department of Biology, Fisheries Ecology Research Laboratory, University of Hawai'i at Mānoa, Honolulu, Hawai'i, United States of America

Lisa M. Wedding
U.S. National Oceanic and Atmospheric Administration, Center for Coastal Monitoring and Assessment, Silver Spring, Maryland, United States of America
Center for Ocean Solutions, Woods Institute for the Environment, Stanford University, Monterey, California, United States of America

Mayuko Nakamaru and Kanan Miura
Department of Value and Decision Science, Tokyo Institute of Technology, Tokyo, Japan

Takenori Takada
Graduate School of Environmental Earth Science, Hokkaido University, Sapporo, Japan

Akiko Ohtsuki and Sayaki U. Suzuki
Department of Value and Decision Science, Tokyo Institute of Technology, Tokyo, Japan
The Graduate University for Advanced Studies, Kanagawa, Japan

Kazuki Tsuji
Faculty of Agriculture, University of the Ryukyus, Okinawa, Japan

Arjun Srivathsa
Post-Graduate Programme in Wildlife Biology and Conservation, Wildlife Conservation Society, India Program, National Centre for Biological Sciences (Tata Institute of Fundamental Research), Bangalore, India
Centre for Wildlife Studies, Bangalore, India

Devcharan Jathanna
Centre for Wildlife Studies, Bangalore, India

Krithi K. Karanth
Centre for Wildlife Studies, Bangalore, India
Wildlife Conservation Society, Bronx, New York, United States of America

K. Ullas Karanth
Centre for Wildlife Studies, Bangalore, India
Wildlife Conservation Society, Bronx, New York, United States of America
Nicholas School of Environment, Duke University, Durham, North Carolina, United States of America

N. Samba Kumar
Centre for Wildlife Studies, Bangalore, India
Wildlife Conservation Society, India Program, Bangalore, India

Mauricio M. Petrucio
Graduate Program in Ecology of Department of Ecology and Zoology, Institute of Biological Sciences, Federal University of Santa Catarina, Florianó polis, Santa Catarina, Brazil

Renan S. Rezende
Graduate Program in Ecology of Department of Ecology and Zoology, Institute of Biological Sciences, Federal University of Santa Catarina, Florianó polis, Santa Catarina, Brazil
Department of Ecology, Institute of Biology, University of Brasília, Brasília, Federal District, Brazil

José F. Gonc¸alves and Jr.
Department of Ecology, Institute of Biology, University of Brasília, Brasília, Federal District, Brazil

Sonia Kéfi
Institut des Sciences de l'Evolution, CNRS, Universite´ de Montpellier II, Montpellier, France

Vishwesha Guttal
Centre for Ecological Sciences, Indian Institute of Science, Bangalore, India

William A. Brock
Department of Economics, University of Wisconsin, Madison, Wisconsin, United States of America
Department of Economics, University of Missouri, Columbia, Missouri, United States of America

Stephen R. Carpenter
Department of Economics, University of Missouri, Columbia, Missouri, United States of America
Center for Limnology, University of Wisconsin, Madison, Wisconsin, United States of America

Aaron M. Ellison
Harvard Forest, Harvard University, Petersham, Massachusetts, United States of America

Valerie N. Livina
National Physical Laboratory, Hampton Road, Teddington, United Kingdom

David A. Seekell
Department of Environmental Sciences, University of Virginia, Charlottesville, Virginia, United States of America

Marten Scheffer and Egbert H. van Nes
Department of Aquatic Ecology and Water Quality Management, Wageningen University, Wageningen, The Netherlands

Vasilis Dakos
Integrative Ecology Group, Estacion Biologica de Donana, Sevilla, Spain

Dexiecuo Ai, Chengjin Chu and Gang Wang
State Key Laboratory of Grassland and Agro-Ecosystems, School of Life Sciences, Lanzhou University, Lanzhou, China

Dominique Gravel
Département de Biologie, Chimie et Géographie, Université du Québec á Rimouski, Rimouski, Canada

Gretchen J. A. Hansen, M. Jake Vander Zanden, Marit Izzo, Matthew S. Kornis, Peter B. McIntyre, Monica Papeş, Sapna Sharma and Erika Nilsson
Center for Limnology, University of Wisconsin-Madison, Madison, Wisconsin, United States of America

Michael J. Blum
Department of Ecology & Evolutionary Biology, Tulane University, New Orleans, Louisiana, United States of America

Murray K. Clayton
Departments of Plant Pathology and Statistics, University of Wisconsin-Madison, Madison, Wisconsin, United States of America

Ernie F. Hain
Department of Biology, North Carolina State University, Raleigh, North Carolina, United States of America

Jennifer Hauxwell
Science Services, Wisconsin Department of Natural Resources, Madison, Wisconsin, United States of America

Alison Mikulyuk
Center for Limnology, University of Wisconsin-Madison, Madison, Wisconsin, United States of America
Science Services, Wisconsin Department of Natural Resources, Madison, Wisconsin, United States of America

Julian D. Olden
School of Aquatic and Fishery Sciences, University of Washington, Seattle, Washington, United States of America

Matthew P. Hammond and Jurek Kolasa
Department of Biology, McMaster University, Hamilton, Ontario, Canada

Jacob L. Johansen
ARC Centre of Excellence for Coral Reef Studies, and School of Marine and Tropical Biology, James Cook University, Townsville, Queensland, Australia

Dewei Yang, Tao Luo, Tao Lin, Quanyi Qiu and Yunjian Luo
Key Lab of Urban Environment and Health, Institute of Urban Environment, Chinese Academy of Sciences, Xiamen, China
Xiamen Key Lab of Urban Metabolism, Xiamen, China

Jillian L. S. Ooi
School of Plant Biology, Faculty of Natural and Agricultural Sciences, The University of Western Australia, Perth, Western Australia, Australia
School of Earth and Environment, Faculty of Natural and Agricultural Sciences, The University of Western Australia, Perth, Western Australia, Australia
The UWA Oceans Institute, The University of Western Australia, Perth, Western Australia, Australia
Department of Geography, Faculty of Arts and Social Sciences, Universiti Malaya, Kuala Lumpur, Malaysia

Kimberly P. Van Niel
School of Earth and Environment, Faculty of Natural and Agricultural Sciences, The University of Western Australia, Perth, Western Australia, Australia
The UWA Oceans Institute, The University of Western Australia, Perth, Western Australia, Australia

Gary A. Kendrick
School of Plant Biology, Faculty of Natural and Agricultural Sciences, The University of Western Australia, Perth, Western Australia, Australia
The UWA Oceans Institute, The University of Western Australia, Perth, Western Australia, Australia

Karen W. Holmes
School of Earth and Environment, Faculty of Natural and Agricultural Sciences, The University of Western Australia, Perth, Western Australia, Australia
Centre for Ecohydrology, School of Environmental Engineering, The University of Western Australia, Perth, Western Australia, Australia

Michael L. Berumen
Red Sea Research Center, King Abdullah University of Science and Technology Thuwal, Kingdom of Saudi Arabia

Camrin D. Braun
Red Sea Research Center, King Abdullah University of Science and Technology Thuwal, Kingdom of Saudi Arabia

Biology Department, Woods Hole Oceanographic Institution, Woods Hole, Massachusetts, United States of America

Gregory B. Skomal
Massachusetts Division of Marine Fisheries, New Bedford, Massachusetts, United States of America

Simon R. Thorrold
Biology Department, Woods Hole Oceanographic Institution, Woods Hole, Massachusetts, United States of America

Christina N. Service and Chris T. Darimont
Department of Geography, University of Victoria, Victoria, BC, Canada
Raincoast Conservation Foundation, Denny Island, BC, Canada
Spirit Bear Research Foundation, Klemtu, BC, Canada
Hakai Beach Institute, Heriot Bay, BC, Canada

Megan S. Adams
Department of Geography, University of Victoria, Victoria, BC, Canada
Raincoast Conservation Foundation, Denny Island, BC, Canada
Hakai Beach Institute, Heriot Bay, BC, Canada

Kyle A. Artelle
Raincoast Conservation Foundation, Denny Island, BC, Canada
Hakai Beach Institute, Heriot Bay, BC, Canada
Earth to Ocean Research Group, Department of Biological Sciences, Simon Fraser
University, Burnaby, BC, Canada

Paul Paquet
Department of Geography, University of Victoria, Victoria, BC, Canada
Raincoast Conservation Foundation, Denny Island, BC, Canada

Laura V. Grant
Department of Geography, University of Victoria, Victoria, BC, Canada
Spirit Bear Research Foundation, Klemtu, BC, Canada

Laura Kracker and Tim Battista
Biogeography Branch, National Centers for Coastal Ocean Science, National Oceanic and Atmospheric Administration, Silver Spring, Maryland, United States of America

Bryan Costa
Biogeography Branch, National Centers for Coastal Ocean Science, National Oceanic and Atmospheric Administration, Silver Spring, Maryland, United States of America
Consolidated Safety Services, Fairfax, Virginia, United States of America

J. Christopher Taylor
Center for Coastal Fisheries and Habitat Research, National Centers for Coastal Ocean Science, National Oceanic and Atmospheric Administration, Beaufort, North Carolina, United States of America

Simon Pittman
Biogeography Branch, National Centers for Coastal Ocean Science, National Oceanic and Atmospheric Administration, Silver Spring, Maryland, United States of America
Marine Institute, University of Plymouth, Plymouth, United Kingdom

Eli D. Lazarus
Environmental Dynamics Lab, Earth Surface Processes Group, School of Earth and Ocean Sciences, Cardiff University, Cardiff, Wales, United Kingdom

Brian J. McGill
School of Biology and Ecology, University of Maine, Orono, Maine, United States of America

Index